A VISION FOR DYNAMICS IN THE 21ST CENTURY

A large international conference celebrated the 50-year career of Anatole Katok and the body of research across smooth dynamics and ergodic theory that he touched. In this book many leading experts provide an account of the latest developments at the research frontier and together set an agenda for future work, including an explicit problem list. This includes elliptic, parabolic, and hyperbolic smooth dynamics, ergodic theory, smooth ergodic theory, and actions of higher-rank groups. The chapters are written in a readable style and give a broad view of each topic; they blend the most current results with the developments leading up to them and give a perspective on future work. This book is ideal for graduate students, instructors, and researchers across all research areas in dynamical systems and related subjects.

DANIJELA DAMJANOVIĆ is an Associate Professor at the Royal Institute of Technology (KTH), Sweden. She received her PhD in mathematics from Pennsylvania State University, USA in 2004. Her research is in the area of dynamics of smooth group actions. She serves on the editorial board of *Journal of Modern Dynamics*.

BORIS HASSELBLATT is a Professor at Tufts University, USA. He received his PhD from California Institute of Technology in 1989. He serves on the editorial board of *Journal of Modern Dynamics* and is a co-author of several books in the field of smooth dynamical systems, notably *Introduction to the Modern Theory of Dynamical Systems*. He is also (co-)editor of several large volumes, *Handbook of Dynamical Systems*, 1A and 1B, and *The Collected Works of Anatole Katok*, Volumes 1 and 2. He is Fellow and the Secretary of the American Mathematical Society.

ANDREY GOGOLEV is a Professor of Mathematics at The Ohio State University, USA. He received his PhD from Pennsylvania State University in 2009. His primary research interest is smooth dynamics with hyperbolic behavior. He serves on the editorial board of *Journal of Modern Dynamics*.

YAKOV PESIN is Distinguished Professor of Mathematics at Pennsylvania State University, USA. He is director of the Anatole Katok Center for Dynamical Systems and Geometry, a member of the European Academy (Academia Europea) and the American Academy of Arts and Sciences, a Fellow of the American Mathematical Society, and an editor of the journal *Ergodic Theory and Dynamical Systems*. He received his PhD in mathematics from Gorky State University in 1979. He has (co)authored six books covering smooth dynamics, ergodic theory, dimension theory, and fractal geometry, and a dozen surveys on these topics.

A VISION FOR DYNAMICS IN THE 21ST CENTURY

The Legacy of Anatole Katok

Edited by

DANIJELA DAMJANOVIĆ
KTH Royal Institute of Technology

BORIS HASSELBLATT
Tufts University

ANDREY GOGOLEV
The Ohio State University

YAKOV PESIN
Pennsylvania State University

CAMBRIDGE
UNIVERSITY PRESS

CAMBRIDGE
UNIVERSITY PRESS

Shaftesbury Road, Cambridge CB2 8EA, United Kingdom

One Liberty Plaza, 20th Floor, New York, NY 10006, USA

477 Williamstown Road, Port Melbourne, VIC 3207, Australia

314–321, 3rd Floor, Plot 3, Splendor Forum, Jasola District Centre, New Delhi – 110025, India

103 Penang Road, #05–06/07, Visioncrest Commercial, Singapore 238467

Cambridge University Press is part of Cambridge University Press & Assessment, a department of the University of Cambridge.

We share the University's mission to contribute to society through the pursuit of education, learning and research at the highest international levels of excellence.

www.cambridge.org
Information on this title: www.cambridge.org/9781009278904

DOI: 10.1017/9781009278898

First published 2024

A catalogue record for this publication is available from the British Library.

A Cataloging-in-Publication data record for this book is available from the Library of Congress

ISBN 978-1-009-27890-4 Hardback

Contents

Editor's Note

The book cover shows the Palace of Będlewo, a conference center of the Polish Academy of Sciences where a memorial conference for the great dynamicist Anatole Katok took place in August 2019.* This conference inspired this book, and many of the authors participated. During this conference we celebrated 50 years of Katok's leadership by doing what he had always fostered: creating a connected vision of the field of dynamical systems and ergodic theory and shaping an agenda for future research. These are the aims of this book, and its chapters give direction to their respective subjects, as well as a broad view and a contemporary account. A separate problem list provides specific suggestions for investigation.

* The conference was supported by the Polish Academy of Sciences, its Banach Center, and by NSF grant DMS-1900171.

Contributors

Adam Abrams
Wroclaw University of Science and Technology

Lluís Alsedà
Autonomous University of Barcelona

Jairo Bochi
The Pennsylvania State University

Barney Bramham
Ruhr University Bochum

Vaughn Climenhaga
University of Houston

Danijela Damjanović
Royal Institute of Technology (KTH) Stockholm

Andrey Gogolev
The Ohio State University

Federico Rodriguez Hertz
The Pennsylvania State University

Oliver Jenkinson
Queen Mary University of London

Boris Kalinin
The Pennsylvania State University

Adam Kanigowski
University of Maryland

Anatole Katok
Late of the Pennsylvania State University

Svetlana Katok
The Pennsylvania State University

Philipp Kunde
University of Hamburg

Mariusz Lemańczyk
Nicolaus Copernicus University

Michał Misiurewicz
Indiana University-Purdue University Indianapolis

Rodrigo A. Pérez
Indiana University-Purdue University Indianapolis

Yakov Pesin
The Pennsylvania State University

Mark Pollicott
University of Warwick

Victoria Sadovskaya
The Pennsylvania State University

Samuel Senti
Federal University of Rio de Janeiro

Farruh Shahidi
Siemens Digital Industries Software

Ilie Ugarcovici
DePaul University

Daren Wei
National University of Singapore

Zhiyuan Zhang
Sorbonne Paris North University

1

Open Problems from 2020 Vision for Dynamics Conference in Będlewo

ADAM ABRAMS, JAIRO BOCHI, AND DANIJELA DAMJANOVIĆ

1.1 Introduction

This chapter contains a collection of open problems presented during the "2020 Vision for Dynamics" conference held in Będlewo in the fall of 2019. We have collected open problems presented in the two scheduled problem sessions: those discussed during the conference talks, and those contributed outside these events. We are grateful to all contributors for taking part in the preparation of this chapter. We thank Yakov Pesin for writing Section 1.2.15. We also thank Corinna Ulcigrai for sharing with us her notes from one of the problem sessions.

1.2 Open Problems

1.2.1 Pierre Berger

Here are two kinds of interesting measures in the phase space for (non-uniformly) hyperbolic dynamics:

- Measures of maximal entropy;
- Sinai–Ruelle–Bowen or, more generally, Gibbs measures.

The following interesting properties suffice to motivate their studies. In many contexts, the measure of maximal entropy is equidistributed on the hyperbolic periodic points, while the Gibbs measure has its unstable dimension, which coincides with one of its supports.

Are there similar measures *in the parameter space*?

For holomorphic families of dynamics of polynomial maps of $\mathbb{C}$, the bifurcation measure μ^{biff} is the canonical counterpart of the maximal entropy measure: it is equidistributed on the set of parameters for which all the critical points are strictly preperiodic, among other equidistribution properties; see [19, 44, 20] for a survey. Such a measure can be consistently defined for any properly embedded family of polynomials. Interestingly, at the μ^{biff}-a.e. parameter the dynamics satisfy the Collet–Eckmann condition [25, 47].

Problem 1. *When is it possible to define a counterpart of* μ^{biff} *for the real family of dynamics of the interval?*

Let us also recall the following:

Problem 2. *Find a counterpart of the bifurcation measure for surface mappings (with similar equidistribution properties).*

By a theorem of Makarov [37], the bifurcation measure for the quadratic family has Hausdorff dimension 1 and its support is the Mandelbrot set. Nevertheless, by Shishikura's theorem [46], the Hausdorff dimension of the Mandelbrot set is 2.

Problem 3. *Define a natural measure on the parameter space of the quadratic family supported by the Mandelbrot set and with full Hausdorff dimension* $(= 2)$.

A solution of this problem would lead to a parametric counterpart of the Gibbs measure. This might be helpful for studying the Lebesgue measure of the Mandelbrot set (which is conjecturally 0 following Shishikura [46]).

1.2.2 Jairo Bochi

Problem 4. *Is there a* C^∞ *conservative (i.e., volume-preserving) Anosov diffeomorphism of* $\mathbb{T}^3$ *with a 2-dimensional unstable manifold whose top Lyapunov exponent (with respect to volume) is strictly bigger than the top Lyapunov exponent of the linear Anosov map in the same homotopy class?*

By a result of Brin, Burago, and Ivanov [7], the unstable bundle cannot admit a dominated splitting.

Added in proof: The question was recently answered affirmatively by Carrasco and Saghin [10].

1.2.3 Danijela Damjanović

The questions here aim at more precise understanding of what happens with symmetry groups under perturbations. Typically they should not be preserved, but can one describe precisely what happens with symmetry groups after perturbation?

For a smooth diffeomorphism $f \in \mathrm{Diff}^\infty(\mathrm{M})$ of a compact smooth manifold M, let $Z^\infty(f) = \{g \in \mathrm{Diff}^\infty(\mathrm{M})\colon g \circ f = f \circ g\}$ denote the smooth centralizer of f in $\mathrm{Diff}^\infty(\mathrm{M})$. Generically in C^1 topology, it was conjectured by Smale and proved by Bonatti, Crovoisier, and Wilkinson [6] that $Z^\infty(f)$ contains only powers of f. However, for algebraic transformations, the (algebraic) centralizer can be large. One class of such examples is ergodic automorphisms of nilmanifolds. Such automorphisms are necessarily partially hyperbolic since ergodicity of a nilmanifold automorphism is given by ergodicity of the induced map of the abelianization, and ergodic toral automorphisms are always partially hyperbolic. Let f_0 be such an ergodic (partially hyperbolic) automorphism of a nilmanifold. Centralizers of such maps within the space of automorphisms can often be very well understood by using algebraic methods, especially under extra conditions of irreducibility, and predominantly these centralizers have rank greater than 2. In particular, these automorphisms are very atypical in the sense of Smale.

Problem 5 (Local centralizer classification). *Let f_0 be an ergodic (partially hyperbolic) automorphism of a nilmanifold. Let f be a C^1-small smooth perturbation of f_0. Can we give a finite list of possibilities for $Z^\infty(f)$ for every f in a small C^1 neighborhood of f_0 and show that every possibility on this list actually happens?*

If the centralizer $Z^\infty(f)$ for a perturbation f is (up to a finite index subgroup) isomorphic to $Z_{\mathrm{aff}}(f_0)$, and $Z_{\mathrm{aff}}(f_0)$ is a group containing at least $\mathbb{Z}^2$, then we can hope in some cases that $Z^\infty(f)$ being large can give some more information on f. To make the question more precise, we may even assume that the action $Z^\infty(f)$ is homotopic to the linear part of $Z_{\mathrm{aff}}(f_0)$.

Problem 6 (Local centralizer rigidity). *Assume that $Z^\infty(f)$ is homotopic to the linear part of $Z_{\mathrm{aff}}(f_0)$. For which f_0 does this imply that f is C^∞ conjugate to f_0?*

For some initial steps in these directions we refer the reader to [17] and references therein, and [28]. One of the results proved in [17] is the following dichotomy:

Theorem 1 ([17]). *Let f_0 be the time-1 map of the geodesic flow on a closed, negatively curved, locally symmetric manifold. The centralizer $Z_\infty(f_0)$ is $\mathbb{R}$. Let f be a C^1, volume-preserving perturbation of f_0. Then $Z_\infty(f)$ is either virtually $\mathbb{Z}$ or it contains a smooth flow and is virtually $\mathbb{R}$.*

Related to this, we ask the following:

Problem 7. *Is it true that a sufficiently small perturbation of any volume-preserving Anosov flow either has a virtually trivial centralizer, or embeds in a smooth flow?*

1.2.4 Manfred Denker

Problem 8. *In cases where a normal conditional local limit theorem holds, prove convergence of the local time of an integer-valued function to the local time of Brownian motion (at least if the dynamics is Gibbs–Markov).*

Background information can be extracted from [18], which may be seen as a first attempt to describe local times in dynamical systems.

Problem 9. *Can one formulate a suitable theory for the local time for real-valued functions and in the case of flows?*

A particular case in this problem is that of geodesic flows.

1.2.5 Bassam Fayad

The question of existence of a smooth, area-preserving diffeomorphism of the disc that is mixing with zero (metric or topological) entropy was raised by Fayad and Katok in [23].

Problem 10. *Prove that a conservative (area-preserving) transitive diffeomorphism of the disk with zero topological entropy is not mixing.*

Problem 11. *Does there exist a conservative diffeomorphism of the disk which has zero topological entropy but positive metric entropy?*

Problem 12. *On interval maps, there exists a dichotomy: an interval map on* $[0, 1]$ *that fixes the boundary has growth that is either exponential or below* n^2*. Is this true for zero-entropy maps on surfaces?*

The next question is about affine parabolic abelian actions.

Problem 13. *Take two parabolic commuting affine maps of the torus:* $f = A + \alpha$ *and* $g = B + \beta$*. Their linear parts* A *and* B *are given by commuting unipotent matrices in* $SL(n, \mathbb{Z})$*, and* α *and* β *are in* $\mathbb{R}^n$*. Consider the* $\mathbb{Z}^2$ *action generated by* f *and* g *on the torus* $\mathbb{T}^n$*. When is such an action locally rigid? More precisely, do the local rigidity results on commuting perturbations of rigid translations on the torus (which follow Moser's result [38] on the circle, such as [15], [49]) extend to some of these parabolic actions?*

Progress has been made toward this problem under extra conditions on the linear part of the action, namely when one of the action generators is a step 2 parabolic map. Then a form of local rigidity is proved by Damjanović, Fayad, and Saprykina in [16].

1.2.6 Giovanni Forni

Problem 14 (Katok). *Prove that there exists a G_δ dense set of polygons with weakly mixing billiard flow. Prove that there exists a mixing polygon.*

Kerckhoff, Masur, and Smillie [32] proved ergodicity for a G_δ dense set of polygons. To prove mixing, one suggestion is to look at an arc of a circle around a point, consider its orbit, and see whether the push-forward of circles equidistributes on the surface. This has a fairly simple proof for the torus, but for other rational polygons the argument does not work: when the surface has singularities, the arcs at time T are expected to have length $1/T$.

The result by Chaika and Hubert [12] shows that for almost all translation surfaces there is a sequence of times for which the circle equidistributes, but the sequence depends on the point, so the result does not imply a (relative) weak mixing result for the geodesic flow on the translation surface.

Recently, the first part of the problem (on weak mixing) has been solved in the preprint [11].

Problem 15. *Prove (polynomial) bounds on the decay of correlations for time-changes of higher step nilflows.*

Mixing with polynomial decay of correlations for smooth observables is known only in a few cases. For results about rate of mixing, see Fayad [21] for Kochergin flows, Forni–Ulcigrai [24] for time changes of horocycle flows, Forni–Kanigowski [23] for Heisenberg nilflows, and Ravotti for surface locally Hamiltonian flows with non-degenerate saddles (this case has logarithmic decay).

To get precise decay of correlations one needs an estimate for the shear, especially one needs lower bounds for ergodic integrals of the "roof" function, namely the time-change function. Such estimates are hard to establish for higher-step nilflows.

1.2.7 Anton Gorodetski

The following problem was popularized by logician Matt Foreman. The general question is

Can one prove that the set of dynamical systems is impossible to classify?

More precisely:

Problem 16. *Consider $\mathit{Diff}^\infty(M)$ with $\dim M \geq 2$. Then*

$$\mathit{Diff}^\infty(M) \times \mathit{Diff}^\infty(M)$$

is a complete metric space. Let $E \subseteq Diff^\infty(M) \times Diff^\infty(M)$ be the set of pairs of diffeomorphisms that are topologically conjugate. Can you prove that E is not Borel?

If the answer is yes, then this would imply that $Diff^\infty(M)$ is not classifiable, in the sense that there is no "countable" way to determine whether two diffeomorphisms are conjugate; see [22] for details.

The next problem concerns the "geometry" of hyperbolicity in a space of dynamical systems. The perception of the "size" of hyperbolicity has changed over the decades; see the figure.

Let $f: M \to M$ be a smooth dynamical system (e.g., an irrational rotation of the circle). Let $A_t: M \to SL(2, \mathbb{R})$ be a typical 1-parameter family of cocycles (the blue curve in the following figure). Let

$$\Sigma = \{t\colon\ (f, A_t) \text{ is not uniformly hyperbolic}\}$$

be the intersection of this curve with the set of (parameters of) non-hyperbolic cocycles.

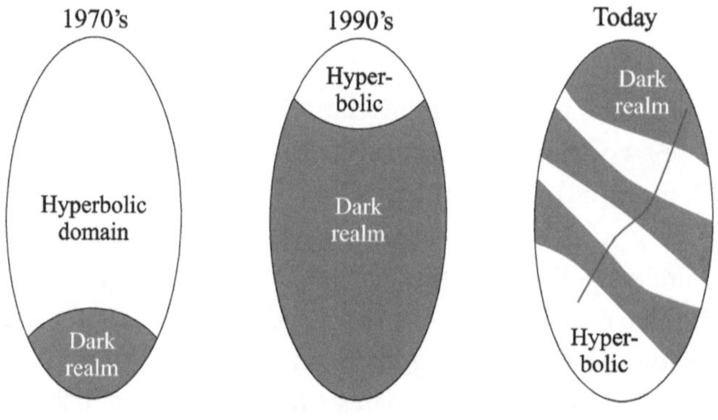

For any specific dynamical system one can ask the following question:

Problem 17. *What is the structure of the set Σ for a typical 1-parameter family of cocycles over a given dynamical system? Does it have to be a finite number of intervals? A Cantor set? A Cantorval? In particular, in the case of $SL(2, \mathbb{R})$ cocycles over an irrational rotation of the circle, is it true that typically the set Σ is either empty or a Cantor set of positive measure?*

The problem is motivated by spectral theory of ergodic discrete Schrödinger operators, where the family of Schrödinger cocycles is parameterized by energy, and due to Johnson's theorem [31] the set Σ corresponds to the spectrum of an operator. In the case of irrational rotation of the circle, the problem can be considered as a dynamical analog of the Ten Martini Problem in spectral theory [3]. Also in that context, the case of C^0 cocycles over a strictly ergodic dynamical system (e.g., odometer) was studied in [1].

1.2.8 Colin Guillarmou

Let g_0 be a fixed metric of negative curvature on M. Denote by $\mathcal{C}$ the set of free homotopy classes. For $c \in \mathcal{C}$, define *marked length spectrum* $L_g\colon \mathcal{C} \to \mathbb{R}^+$ that maps c to the length of the shortest geodesic γ_c^g in the class c.

Define $\mathcal{L}_g\colon \mathcal{C} \to \mathbb{R}^+$ by

$$\mathcal{L}_g(c) := L_g(c)/L_{g_0}(c),$$

where $c \in \mathcal{C}$.

Consider the map

$$\Phi\colon g \in C^\infty(M; S_+^2 T^*M) \mapsto \mathcal{L}_g \in L^\infty(\mathcal{C}).$$

Problem 18. *What can be said about the range of* Φ*? Is the range a submanifold in some appropriate sense?*

1.2.9 Federico Rodriguez Hertz

Let X be a compact manifold (or a finite CW-complex, or any space that has a nice finite-dimensional homology theory). Let $T\colon X \to X$ be a homeomorphism. We say a matrix $A \in GL(n, \mathbb{R})$ is *quasi-unipotent* if all eigenvalues have modulus one.

Problem 19. *If* T *is minimal, is* $T^*\colon H^k(X) \to H^k(X)$ *quasi-unipotent for all* k*?*

Using results of Mañé , it is known that if T^* is hyperbolic, then T is not minimal. But if T^* is partially hyperbolic only, this is unknown.

Note that the minimal positive entropy example of Herman does not help here, because it is homotopic to the identity.

1.2.10 Steven Hurder

A Cantor action $(\mathfrak{X}, G, \Phi)$ is called *stable* if the chain $\{\widehat{K}_\ell\colon \ell \geq 1\}$ is bounded, that is, if there exists ℓ_0 so that $\widehat{K}_\ell = \widehat{K}_{\ell+1}$ for $\ell \geq \ell_0$. The action is called *wild* if the chain $\{\widehat{K}_\ell\colon \ell \geq 1\}$ is unbounded.

A monodromy action $(\mathfrak{X}, G, \Phi)$ is *locally quasi-analytic (LQA)* if there exists $\epsilon > 0$ so that, if U adapted and $\mathrm{diam}_{\mathfrak{X}}(U) < \epsilon$, then for all clopen $V \subset U$ we have

$$\Phi(g)\big|_V = Id \quad \Longrightarrow \quad \Phi(g)\big|_U = Id \qquad \text{for all } g \in G_U,$$

that is, the action of $\mathcal{H}_U^\Phi$ on U is *topologically free.*

Problem 20. *How to classify Cantor actions of a finitely generated nilpotent group G? Can one use invariants of the associated cross-product C^*-algebra? Is a classification possible in terms of the representations of G?*

Problem 21. *If an action is wild, when is the action non-LQA?*

Problem 22. *For which number fields and polynomials f is the action of the absolute Galois group* $\mathrm{Gal}_{\mathrm{arith}}(f)$*, on the boundary of the tree of iterated solutions, non-LQA?*

Problem 23. *If G is a higher-rank lattice and the action is effective, must it be stable?*

Problem 24. *If G is a higher-rank lattice and the action is wild, must it be flat wild?*

For more details on Cantor actions, we refer the reader to [29], [30].

1.2.11 Yulij Ilyashenko

Consider diffeomorphisms of a closed manifold. An attractor is "thick" if both the attractor and its complement are of positive measure.

Problem 25 (Genericity of existence of thick attractors – simplified version). *Is there an open set in Diff(M) such that every f in this set has a thick attractor?*

There is a theorem from the 1940s saying that in the space of diffeomorphisms of a manifold with boundary there is an open set of diffeomorphisms that preserve the boundary and these diffeomorphisms have thick attractors. So, genericity of existence of thick attractors is established for manifolds with boundary.

There are at least three kinds of "attractors":

1. The *maximal attractor* (or traditional attractor) for a map $f : X \to X$ is
$$\bigcap_{n=0}^{\infty} f^n(X).$$
This idea of an attractor is not reasonable for some maps. For example, a map of circle with one semi-stable (parabolic) fixed point could be considered to have that point as an attractor, but not with this definition.
2. If X is a metric space, the *Milnor attractor* of $f : X \to X$ is the minimal closed set with the property that
$$d(f^n x, A) \to 0 \text{ for a.e. } x.$$
The point of the definition is that A is the biggest global attractor.
3. There are also *statistical attractors*.

Problem 26. *Is the non-coincidence of different definitions of attractors generic? To what extent is the situation of having one kind of attractor be strictly smaller than another generic?*

Recently, Ivan Shilin proved in [45] that non-coincidence of maximal and Milnor attractor is topologically generic in some domain.

Problem 27. *Is the coexistence of an infinite number of attracting periodic orbits "truly prevalent"?*

Newhouse showed coexistence of an infinite number of attracting periodic orbits in some open set of diffeomorphisms. Pierre Berger found open sets of dynamics where coexistence of an infinite number of attracting periodic orbits is typical in the sense of Arnold. But this notion of typicality has some drawbacks: if you apply it to $\mathbb{R}^n$, then there are metrically typical subsets of $\mathbb{R}^n$ with zero measure. The strongest notion of typicality is called prevalence and was introduced by Hunt, Sauer, and Yorke [43].

1.2.12 Raphaël Krikorian

Problem 28. *Take a twist real-analytic (C^ω) symplectic diffeomorphism of an annulus with $h_{\text{top}}(f) = 0$. Does it follow that (up to a finite quotient) f is the time-1 map of a real-analytic flow (as in the figure)?*

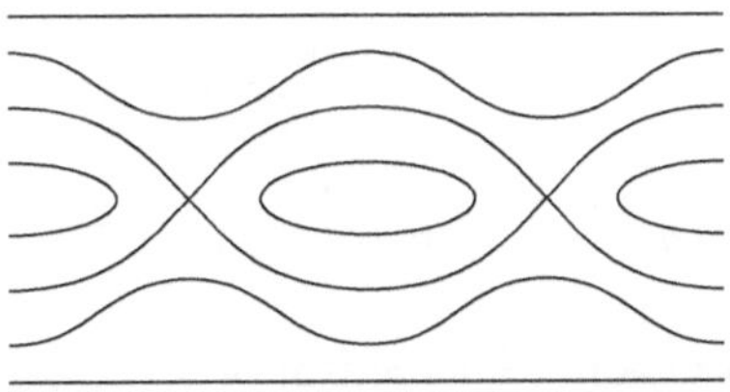

The following question is related to the problems stated in Section 17. Consider quasi-periodic cocyles on $\mathbb{T}^d \times SL(2,\mathbb{R})$ given by $(x,y) \mapsto (x+\alpha, A(x)y)$. A cocycle is *non-uniformly hyperbolic (NUH)* if it has nonzero Lyapunov exponents. A cocycle is *almost reducible* if it cannot be conjugated to a constant but you can come arbitrarily close:

$$(\cdot, B_n) \circ (\alpha, A) \circ (\cdot, B_n)^{-1} \to (\alpha, \text{const}).$$

A cocycle is *stable* if you cannot approximate it by NUH cocycles.

Problem 29. *Does stable imply almost reducible?*

Next problem is asking about possible generalization of the Herman–Yoccoz theorem to tori of any dimension.

Recall the Herman–Yoccoz theorem on $\mathbb{T}$. Let $f\colon \mathbb{T} \to \mathbb{T}$ be orientation-preserving and smooth diffeomorphism of the circle. We can define rotation number $\rho(f)$. By Denjoy's theorem, since f is at least C^2, we know that, if $\rho(f)$ is irrational, then f is topologically conjugate to rotation by $\rho(f)$. The Herman–Yoccoz theorem says that if f is smooth and $\rho(f)$ is Diophantine, then the conjugacy h is smooth.

The following question concerns generalization of the Herman–Yoccoz theorem to $f\colon \mathbb{T}^d \to \mathbb{T}^d$ with $d \geq 2$.

Problem 30. *Suppose $f \in \mathit{Diff}_0^\infty(\mathbb{T}^d)$ is homotopic to the identity and is conjugate to a translation T_α by vector α via a homeomorphism h. If α is Diophantine, is h necessarily smooth?*

A local result is that this is true if $|f - T_\alpha| < \epsilon(\alpha)$. The semi-local version would be to have ϵ not depending on α, but this is not known.

1.2.13 François Ledrappier

Let M be a Banach manifold, $f\colon M \to M$ be a diffeomorphism, and let μ be an invariant probability measure with compact support. Assume that $\int \log^+ \|D_x f\|\, d\mu(x)$ is finite. Then, by the subadditive ergodic theorem,

$$\frac{1}{n} \log \|D_x f^n\| \to \text{some } \lambda \qquad \mu\text{-almost everywhere.}$$

Problem 31 (Katok, 1987). *Does $\lambda = 0$ imply $h_\mu(f) = 0$?*

When $f\colon M \to M$ is a C^1-diffeomorphism of a compact Riemannian manifold and μ is an f-invariant probability measure on M, the Ruelle inequality gives an upper bound on entropy $h_\mu(f)$ in terms of Lyapunov exponents, which in particular implies that the answer to the preceding question is yes in this case.

When M is finite-dimensional but non-compact, F. Riquelme in [42] gave an example where f is a diffeomorphism of M and μ is a finite f-invariant measure such that $\lambda = 0$, but $h_\mu(f) > 0$; this example satisfies the integrability condition, but the support of μ is non-compact.

P. Thieullen in his PhD thesis [48] showed an estimate on entropy which implies that the answer to this question is yes in the case of Banach manifolds, under extra compactness assumptions (when the quasi-compactness exponent is negative).

1.2.14 Mariusz Lemańczyk

Problem 32 (Katok, 2004 or earlier). *Is it true that the von Neumann flows have finite multiplicity?*

A von Neumann flow is a special flow over a rotation in which the roof function has finitely many discontinuities and the sum of jumps is not equal to zero. These von Neumann flows are now known to have a variation of Ratner's property.

Let $\{T_t\}_{t\in\mathbb{R}}$ be a flow with Ratner's property. The *essential centralizer* of such a flow is defined as $C(T_t)/\{T_t\}_{t\in\mathbb{R}}$, where $C(T_t)$ is the total centralizer in the space of all automorphisms of the given measure space.

Problem 33. *What is the essential centralizer of a flow with Ratner's property? Is it always finite? (Lemańczyk and Kanigowski [33] proved it is at most countable.)*

It is expected that the essential centralizer of flows with Ratner's property is not always finite.

Remark: Unipotent flows in $SL(2)$ have Ratner's property, but in $SL(3)$ they do not, essentially because their homogeneous centralizer is a group larger than the flow itself. They do have the generalized shearing property, namely that the fastest relative motion is along the same direction as the centralizer of the flow.

1.2.15 Yakov Pesin

I state and discuss here two long-standing open problems in smooth ergodic theory related to genericity of diffeomorphisms with nonzero Lyapunov exponents. They first appeared in [39] and were also mentioned in [4, 14, 40].

Let M be a compact, smooth Riemannian manifold without boundary and f a volume-preserving diffeomorphism of M. Recall that given a point $x \in M$ and a tangent vector $v \in T_xM$, the *Lyapunov exponent* $\chi_f(x, v)$ at x in the direction of v is given by

$$\chi_f(x, v) := \lim_{n\to\infty} \frac{1}{n} \log \|df_x^n v\|, \ x \in M, \ v \in T_xM.$$

Denote by μ the Riemannian volume on M and by $\mathrm{Diff}^r(M, \mu)$ the space of all C^r volume-preserving diffeomorphisms on M. We say that a diffeomorphism $g \in \mathrm{Diff}^r(M, \mu)$ has nonzero Lyapunov exponents on a set $A = A(g) \subset M$ of positive (respectively full) volume if

1. $\mu(A) > 0$ (respectively, $\mu(A) = 1$);
2. for μ-almost every $x \in A$ (with respect to volume) and every $v \in TM$ one has $\chi_g(x, v) \neq 0$;
3. there are $v_1, v_2 \in T_xM$ such that $\chi_g(x, v_1) < 0$ and $\chi_g(x, v_2) > 0$.

Problem 34. *Let f be a C^2 volume-preserving diffeomorphism of M. Assume that f has nonzero Lyapunov exponents on a set of full volume. Then there exists a neighborhood $\mathcal{U} \subset \mathit{Diff}^2(M, \mu)$ of f and a G_δ subset $\mathcal{B} \subset \mathcal{U}$ such that every $g \in \mathcal{B}$ has nonzero Lyapunov exponents on a set $A = A(g) \subset M$ of positive volume.*

Problem 35. *Let f be a C^2 volume-preserving diffeomorphism of M (possibly with some zero Lyapunov exponents on a set of positive or even full volume). Then arbitrarily close to f in $\mathit{Diff}^2(M, \mu)$ there is a diffeomorphism g, which has nonzero Lyapunov exponents on a set $A = A(g) \subset M$ of positive volume.*

Remark 1. The requirement that the map f is of class C^2 can be relaxed by assuming that f is of class $C^{1+\alpha}$, namely it is of class C^1 and the differential df_x is Hölder continuous in x. It is, however, crucial that f is not simply of class C^1 due to the following two results:

1. *On any compact surface there is a dense G_δ subset $\mathcal{U} \subset \mathit{Diff}^1(M, \mu)$ such that any $f \in \mathcal{U}$ is either (1) Anosov (and is ergodic) or (2) has zero Lyapunov exponents on a set of full volume*; see [5].
2. *On any compact manifold there is a dense G_δ subset $\mathcal{U} \subset \mathit{Diff}^1(M, \mu)$ such that any $f \in \mathcal{U}$ is either (1) non-uniformly hyperbolic (and is ergodic; in particular, f has nonzero Lyapunov exponents on a set of full volume) or (2) has zero Lyapunov exponent*; see [27] for the case $\dim M = 3$ and [2] for the general case.

Remark 2. In Problem 1, while the map f is assumed to have nonzero Lyapunov exponents on a set of full volume, one should expect its small perturbations to have nonzero Lyapunov exponents only on a set of positive (not necessarily full) volume. This is due to a phenomenon known as *essential coexistence*; see [14]. More precisely, a map f is said to exhibit an essential coexistence of regular and chaotic behavior if

1. M can be split into two invariant disjoint Borel subsets A and B of positive volume – the chaotic and regular regions for f;
2. for almost every $x \in A$ and every $v \in T_xM$ the Lyapunov exponent $\chi_f(x, v) \neq 0$;
3. for every $x \in B$ and every $v \in T_xM$ the Lyapunov exponent $\chi_f(x, v) = 0$;
4. $f|A$ is ergodic.

The entropy formula gives that $h_{\text{top}}(f|A) > 0$ and the Margulis–Ruelle inequality implies that $h_{\text{top}}(f|B) = 0$.

Essential coexistence is said to be of type I if the set A is dense in M and of type II otherwise. In the former case the regular and chaotic regions for f cannot be topologically separated. The essential coexistence phenomenon is an obstruction to having nonzero Lyapunov exponents on a set of full volume.

The concept of essential coexistence of type I was inspired by the work of Cheng–Sun [13], Herman [26], Xia [50], and Yoccoz [51], who have shown that on any manifold M and for any sufficiently large r one has what can be regarded as a discrete version of the classical KAM theory phenomenon in the volume-preserving category – there are open sets of volume-preserving C^r diffeomorphisms of M all of which possess positive volume sets of codimension-1 invariant tori; on each such torus the diffeomorphism is C^1 conjugate to a Diophantine translation; all Lyapunov exponents are zero at every point in any invariant torus. The set of invariant tori is nowhere dense and has positive volume.

Examples of essential coexistence of type II can be found in works of Przytycki [41] and of Liverani [34], where they constructed surface diffeomorphisms with nonzero Lyapunov exponents such that some of their arbitrary small perturbations have elliptic islands.

1.2.16 E. Arthur Robinson, Jr.

The flow corresponding to the substitution $\{a \to abbb,\ b \to a\}$, suspended by its left Perron–Frobenius eigenvector, has purely singular diffraction spectrum. We can think of this as a tiling (or set of tilings) of the real line by intervals whose lengths are 1 and the Perron–Frobenius eigenvalue.

Problem 36. *Does this flow actually have purely singular dynamical spectrum? (Is the measure of maximal spectral type singular to Lebesgue?) What about the spectrum of the discrete substitution* $\{a \to abbb,\ b \to a\}$ *itself?*

Note that the measure of maximal spectral type is continuous (except for the single atom corresponding to constant functions) because the flow is weakly mixing. However, it cannot be purely absolutely continuous (because no substitution with finite local complexity can be strongly mixing).

Showing that the dynamical spectrum equals the diffraction spectrum would amount to showing that a function supported on a set near the tile endpoints of the tiles has maximal spectral type.

Remark: This problem has now essentially been solved by Bufetov and Solomyak; see [8]. See also [9], which deals with the case of discrete substitutions.

Problem 37. *Is the Conway–Radin pinwheel tiling strongly mixing or not? It is easily shown to be weakly mixing. Is it topologically mixing?*

Problem 38. *Is there a general approach to prove a primitive Hilbert cube substitution is uniquely ergodic?*

Problem 39. *Is there some kind of Markov partition for a geodesic flow on a compact surface of constant negative curvature that will make the horocycle flow into an infinite local complexity tiling substitution? Can this be done for Anosov flows in general?*

Note that in the case of finite local complexity tiling substitutions, there is always a Markov partition hiding in the background, often for an Anosov diffeomorphism, but in general for a "Smale space" in the terminology of Putnam. For example, Penrose tilings come from looking at a Markov partition for a hyperbolic automorphism of $\mathbb{T}^4$ restricted to its 2-dimensional stable manifold. Conversely, every Smale space gives rise to a tiling substitution (at least if we allow for disconnected tiles).

We know that in case $d > 2$ Markov partitions for hyperbolic toral automorphisms "usually" have fractal boundaries. (The Penrose case mentioned earlier avoids this because it is essentially a product of two 2-dimensional automorphisms.)

Problem 40. *When does a hyperbolic toral automorphism have a Markov partition with connected partition elements? When does the resulting tiling have finite local complexity?*

1.2.17 Klaus Schmidt

The questions raised here are related to homoclinic points for abelian actions. We begin by looking at a few examples.

Example 1. The matrix $\left(\begin{smallmatrix} 0 & 1 \\ 1 & 1 \end{smallmatrix}\right)$ induces a map f on $\mathbb{T}^2$. The eigenlines of the matrix induce two f-invariant foliations with dense leaves on the torus. Each point x where these leaves intersect gives $\lim_{|n|\to\infty} f^n x = 0$ but will deviate significantly from 0 for many values of n.

Now suppose we have another hyperbolic toral automorphism g that commutes with f. Then f and g generate a $\mathbb{Z}^2$ action α on the torus. Given a $\underline{n} = (n_1, n_2) \in \mathbb{Z}^2$, consider $\alpha^{\underline{n}} = f^{n_1} g^{n_2} \in Aut(\mathbb{T}^2)$. We say that x is a *homoclinic* point if $\lim_{\|\underline{n}\| \to \infty} \alpha^{\underline{n}} x = 0$.

If a $\mathbb{Z}^2$ action by automorphisms of the torus is higher-rank, there are no nonzero homoclinic points. Also, higher-rank actions have zero action-entropy.

Example 2. Ledrappier's example: the shift-invariant subspace $X \subset (\mathbb{Z}/2\mathbb{Z})^{(\mathbb{Z}^2)}$ given by

$$X = \left\{ (x_{k,\ell})\colon\ x_{k,\ell} + x_{k+1,\ell} + x_{k,\ell+1} \equiv 0 \bmod 2 \text{ for all } k, \ell \right\}.$$

There are no homoclinic points for the multi-dimensional shift σ on X (the $\underline{n}$-th coordinate of $\sigma^{\underline{m}}(x)$ is defined as $x_{\underline{m}+\underline{n}}$).

Example 3. If we replace the alphabet $\mathbb{Z}/2\mathbb{Z}$ with the circle $\mathbb{T} = \mathbb{R}/\mathbb{Z}$, but keep the same rule, we get a large space for X. (If we fix one row we can uniquely fill in one half of the plane.) This action has actually positive entropy; see [35] for the entropy calculation. Also, this action has interesting homoclinic points. Moreover, it also has so-called "summable homoclinic points" (SHP), meaning here that the conference to 0 happens at a reasonable speed (not exponential decay but in ℓ^1). See [36] for existence of SHP.

Example 4. We can also consider the following generalizations of Ledrappier's example. Let

$$X_f = \ker\big(f(\sigma)\big),$$

where, for example, $f(\sigma) = \mathrm{Id} + \sigma^{(1,0)} + \sigma^{(0,1)}$ for $f = 1 + z_1 + z_2$. For the following Laurent polynomials, we have

$$\begin{aligned} 1 + z_1 + z_2 &\text{ has SHP,} \\ 3 - z_1 - z_1^{-1} - z_2 - z_2^{-1} &\text{ does not have homoclinic points, no SHP,} \\ 4 - z_1 - z_1^{-1} - z_2 - z_2^{-1} &\text{ does have homoclinic points, has SHP,} \\ 5 - z_1 - z_1^{-1} - z_2 - z_2^{-1} &\text{ does have homoclinic points.} \end{aligned}$$

The existence of SHP is the measure of degree of non-expansiveness of the shift on X_f. The shift is expansive if and only if f is a Laurent polynomial seen as an element of $\ell^1(\mathbb{Z}^2)$ is invertible.

Expansiveness is equivalent to

$$U(f) = \big\{ \underline{c} \in \mathbb{T}^2 : f(\underline{c}) = 0 \big\}$$

being non-empty. The existence of SHP is equivalent to $\dim U(f) \leq d - 2$.

Why should we be interested in the existence of SHP? Firstly, there are examples where the existence of SHP is the only way of proving certain dynamical properties, for example, a very strong version of specification. Also, the existence of SHP implies positive entropy.

Problem 41. *How can we prove properties such as the ones just mentioned,* without *using summable homoclinic points?*

References

[1] A. Avila, J. Bochi, D. Damanik, Opening gaps in the spectrum of strictly ergodic Schrödinger operators, J. Eur. Math. Soc. 14(2012), 61–106.

[2] A. Avila, S. Crovisier, A. Wilkinson, Diffeomorphisms with positive metric entropy, Publ. Math. IHES, 124(2016), 589–602.

[3] A. Avila, S. Jitomirskaya, The Ten Martini Problem, Ann. Math. (2) 170(2009), 303–342.

[4] L. Barreira, Y. Pesin, *Smooth Ergodic Theory and Nonuniformly Hyperbolic Dynamics*, Handbook of Dynamical Systems, 1B, edited by B. Hasselblatt and A. Katok, Elsevier, 2005.

[5] J. Bochi, Genericity of zero Lyapunov exponents, Ergod. Theory Dyn. Syst., 22(2002), 1667–1696.

[6] C. Bonatti, S. Crovisier, A. Wilkinson, The C^1 generic diffeomorphism has trivial centralizer. Publ. Math. Inst. Hautes Études Sci. 109(2009), 185–244.

[7] M. Brin, D. Burago, S. Ivanov, Dynamical coherence of partially hyperbolic diffeomorphisms of the 3-torus, J. Mod. Dyn., 3(2009), 1–11.

[8] A. Bufetov, B. Solomyak, A spectral cocycle for substitution systems and translation flows, J. Anal. Math. 141 (2020), no. 1, 165–205.

[9] A. Bufetov, B. Solomyak, On substitution automorphisms with pure singular spectrum Bufetov, Alexander I.; Solomyak, Boris Math. Z. 301 (2022), no. 2, 1315–1331.

[10] P. D. Carrasco, R. Saghin. Extended flexibility of Lyapunov exponents for Anosov diffeomorphisms. Trans. Amer. Math. Soc. 375(2022), 3411–3449.

[11] J. Chaika, G. Forni, Weakly Mixing polygonal billiards, preprint arXiv:2003 .00890.

[12] J. Chaika, P. Hubert, Circle averages and disjointness in typical translation surfaces on every Teichmüller disc, Bull. Lond. Math. Soc. 49(2017), 755–769.

[13] C.-Q. Cheng, Y.-S. Sun, Existence of invariant tori in three dimensional measure-preserving mappings, Celestial Mech. Dynam. Astronom., 47(1989/90), 275–292.

[14] V. Climenhaga, Y. Pesin, Open problems in the non-uniform hyperbolicity theory, Discrete. Contin., 27(2010) 589–607 (special issue on "Trends and Developments in DE/Dynamics").

[15] D. Damjanović, B. Fayad, On local rigidity of partially hyperbolic affine $\mathbb{Z}^k$ actions, J. reine angew. Math., 751(2019), 1–26, preprint arXiv:2303.04367.

[16] D. Damjanović, B. Fayad, M. Saprykina, KAM-rigidity for parabolic affine abelian actions, preprint arXiv:2303.04367.

[17] D. Damjanović, A. Wilkinson, D. Xu, Pathology and asymmetry: centralizer rigidity for partially hyperbolic diffeomorphisms, Duke Math. J., 170(2021), 3815–3890.

[18] M. Denker, Z. Zheng. On local times for stationary processes with conditional local limit theorems, Stoch. Proc. Appl. 128(2018), 2448–2262.

[19] A. Douady, J. Hubbard, Étude dynamique des polynômes complexes, Publications Mathématiques d'Orsay, Université de Paris-Sud, Orsay, 1985.

[20] R. Dujardin, Bifurcation currents and equidistribution in parameter space, *Frontiers in Complex Dynamics*, Princeton Mathematical Series 51, 515–566.

[21] B. Fayad, Polynomial decay of correlations for a class of smooth flows on the two torus, Bull. Soc. Math. 129(2001), 487–503.

[22] M. Foreman, What is a Borel reduction? Notices Amer. Math. Soc. 65(2018), 1263–1268.

[23] G. Forni, A. Kanigowski, Time-changes of Heisenberg nilflows, Astérisque (2020), no. 416, 253–299.

[24] G. Forni, C. Ulcigrai, Time changes of horocycle flows, J. Mod. Dyn. 6 (2012), 251–273.

[25] J. Graczyk, G. Swiatek, Harmonic measure and expansion on the boundary of the connectedness locus. Invent. Math. 142(2000), 605–629.

[26] M. Herman, Une méthode pour minorer les exposants de Lyapounov et quelques exemples montrant le caractère local d'un théorème d'Arnold et de Moser sur le tore de dimension 2, Comment. Math. Helv., 58(1983), 453–502.
[27] Federico R. Hertz, Jana R. Hertz, A. Tahzibi, Raul Ures, A criterion for ergodicity of non-uniformly hyperbolic diffeomorphisms, *Electronic Research Announcements of the American Mathematical Society*, 14(2007), 74–81.
[28] F. Rodriguez Hertz, Z. Wang, Global rigidity of higher rank abelian Anosov algebraic actions, *Invent. Math.* 198(2014), 165–209.
[29] S. Hurder, O. Lukina, Nilpotent Cantor actions, Proc. Am. Math. Soc. 150(2022), 289–304.
[30] S. Hurder, O. Lukina, Limit group invariants for non-free Cantor actions, Ergod. Theory Dyn. Syst., 41(2021), 1751–1794.
[31] R. Johnson, Exponential dichotomy, rotation number, and linear differential operators with bounded coefficients, J. Diff. Eq., 61(1986), 54–78.
[32] S. Kerckhoff, H. Masur, J. Smillie, Ergodicity of billiard flows and quadratic differentials, Ann. Math. (2) 124(1986), 293–311.
[33] M. Lemańczyk, A. Kanigowski, Flows with Ratner's property have discrete essential centralizer, Stud. Math. 237(2017), 185–194.
[34] C. Liverani, Birth of an elliptic island in a chaotic sea, Math. Phys. Electron. J., 10(2004), n. pag.
[35] D. Lind, K. Schmidt, T. Ward, Mahler measure and entropy for commuting automorphisms of compact groups, Invent. Math. 101(1990), 593–629.
[36] D. Lind, K. Schmidt, E. Verbitski, Homoclinic points, atoral polynomials, and periodic points of algebraic $\mathbb{Z}^d$-actions, Ergod. Theory Dyn. Syst. 33(2013), 1060–1081.
[37] N. G. Makarov, Metric properties of harmonic measures. Proceedings of the International Congress of Mathematicians, Vol. 1, 2 (Berkeley, CA, 1986), 766–776, American Mathematical Society, Providence, RI, 1987.
[38] J. Moser, On commuting circle mappings and simultaneous Diophantine approximations, Math. Z., 205(1990), 105–121.
[39] Y. Pesin, Characteristic Lyapunov exponents and smooth ergodic theory, Russ. Math. Surveys, 32(1977), 55–114.
[40] Y. Pesin, Existence and genericity problems for dynamical systems with nonzero Lyapunov exponents, Regul. Chaotic Dyn., 12(2007), 476–489.
[41] F. Przytycki, Examples of conservative diffeomorphisms of the two-dimensional torus with coexistence of elliptic and stochastic behavior, Ergod. Th. Dynam. Syst., 2(1982), 439–463.
[42] F. Riquelme, Counterexamples to Ruelle's inequality in the noncompact case, Annales de l'nstitut Fourier, 67(2017), 23–41.
[43] B. R. Hunt, T. Sauer, J. A. Yorke, Prevalence: a translation invariant "almost every" on infinite dimensional spaces, Bull. Amer. Math. Soc. (N.S.) 27(1992), 217–238.
[44] N. Sibony, Iteration of polynomials. Lecture notes (unpublished), UCLA, 1984.
[45] I. Shilin, Locally topologically generic diffeomorphisms with Lyapunov unstable Milnor attractors, Mosc. Math. J., 17(2017), 511–553.
[46] M. Shishikura, The Hausdorff dimension of the boundary of the Mandelbrot set and Julia sets, Ann. of Math. (2) 147(1998), 225–267.
[47] S. Smirnov, Symbolic dynamics and the Collet–Eckmann condition, Internat. Math. Res. Notices 2000, no. 7, 333–351.
[48] P. Thieullen, Fibres dynamiques asymptotiquement compacts exposants de Lyapunov. Entropie. Dimension, Ann. Inst. Henri Poincaré, 4(1987), 49–97.

[49] A. Wilkinson, J. Xue, Rigidity of some abelian-by-cyclic solvable group actions on $\mathbb{T}^N$, Commun. Math. Phys., 376(2020), 1223–1259.
[50] Z. Xia, Existence of invariant tori in volume-preserving diffeomorphisms, Ergod. Theory Dyn. Syst. 12 (1992) 621–631.
[51] J.-C. Yoccoz, Travaux de Herman sur les tores invariants, Séminaire Bourbaki, Vol. 1991/92, Astérisque, 206 Exp. 754 Talk no. 4 (1992) 311–344.

2

Rigidity of Topological Entropy of Boundary Maps Associated to Fuchsian Groups

ADAM ABRAMS, SVETLANA KATOK*, AND ILIE UGARCOVICI

Dedicated to the memory of Anatole Katok

Given a closed, orientable surface of constant negative curvature and genus $g \geq 2$, we study a family of generalized Bowen–Series boundary maps and prove the following rigidity result: in this family the topological entropy is constant and depends only on the genus of the surface. We give an explicit formula for this entropy and show that the value of the topological entropy also stays constant in the Teichmüller space of the surface. The proofs use conjugation to maps of constant slope.

2.1 Introduction

The notion of topological entropy was introduced by Adler, Konheim, and McAndrew in [5]. Their definition used covers and applied to compact Hausdorff spaces; Dinaburg [11] and Bowen [8] gave definitions involving distance functions and separated sets, which are often more suitable for calculations. While these formulations of topological entropy were originally intended for continuous maps acting on compact spaces, Bowen's definition can actually be applied to piecewise continuous, piecewise monotone maps on an interval, as explained in [19]. The theory naturally extends to maps of the circle, where piecewise monotonicity is understood to mean local monotonicity or, equivalently, having a piecewise monotone lift to $\mathbb{R}$.

In [21], following his seminal work [20] on Markov maps, Parry showed that a piecewise monotone, (strongly) transitive interval map with positive topological entropy is conjugate to a constant slope map. In [17], Milnor and Thurston used kneading theory to prove a semiconjugacy result for continuous, piecewise monotone, but not necessarily transitive, interval maps. In [7], following [6], Alsedà and Misiurewicz give a simpler proof that also generalizes to piecewise continuous, piecewise monotone interval maps.

* S. K. was partially supported by the NSF grant DMS-16-02409.

In this chapter we apply the results of [21, 7] to a multiparameter family of piecewise continuous, piecewise monotone maps of the circle, the so-called "boundary maps" for surfaces of constant negative curvature, as in [15]. Some particular maps in this family – including those considered by Bowen and Series [9] and further studied by Adler and Flatto [4] – are Markov, and the topological entropy can be calculated as the logarithm of the maximal eigenvalue of a transition matrix [22, Theorem 7.13] in these cases. However, not all maps in our family admit a Markov partition, and yet we prove the following rigidity result: in this family, the topological entropy is constant and depends only on the genus of the surface. Therefore, the topological entropy in these non-Markov cases is the same logarithmic expression. We also show that topological entropy stays constant in the Teichmüller space of the surface.

Let Γ be a finitely generated cocompact Fuchsian group of the first kind acting freely on the unit disc $\mathbb{D} = \{ z \in \mathbb{C} : |z| < 1 \}$ endowed with hyperbolic metric $\frac{2|dz|}{1-|z|^2}$ such that $S = \Gamma\backslash\mathbb{D}$ is a surface of genus $g \geq 2$.

A classical (Ford) fundamental domain for Γ is a $4g$-sided regular polygon centered at the origin. In [4], Adler and Flatto used another fundamental domain – an $(8g-4)$-sided polygon $\mathcal{F}$ – that was much more convenient for their purposes. Its sides are geodesic segments which satisfy the *extension condition*: the geodesic extensions of these segments never intersect the interior of the tiling sets $\gamma\mathcal{F}$, $\gamma \in \Gamma$.

We denote the endpoints of the oriented infinite geodesic that extends side k to the circle at infinity $\partial\mathbb{D}$ by P_k and Q_{k+1}, where $1 \leq k \leq 8g-4$ is considered mod $8g-4$ throughout this chapter (see Figure 2.1). The counterclockwise order of endpoints on $\partial\mathbb{D}$ is the following:

$$P_1, Q_1, P_2, Q_2, \ldots, Q_{8g-4}.$$

The identification of the sides of $\mathcal{F}$ is given by the side pairing rule:

$$\sigma(k) := \begin{cases} 4g - k \bmod (8g-4) & \text{if } k \text{ is odd,} \\ 2 - k \bmod (8g-4) & \text{if } k \text{ is even.} \end{cases}$$

The generators T_k of Γ associated to this fundamental domain are Möbius transformations satisfying the following properties: denoting $\rho(k) = \sigma(k) + 1$ and with V_k as the vertex of $\mathcal{F}$ where sides $k-1$ and k meet,

$$T_k(V_k) = V_{\rho(k)}, \qquad T_{\sigma(k)}T_k = \mathrm{Id}, \qquad T_{\rho^3(k)}T_{\rho^2(k)}T_{\rho(k)}T_k = \mathrm{Id}.$$

Remark. As functions on $\overline{\mathbb{D}} \subset \mathbb{C}$, the generators T_k are Möbius transformations, but restricted to the boundary $\mathbb{S}$ they are real functions of the arguments (but *not* fractional linear transformations of the arguments). To simplify notation we will use "T_k" in both situations: $T_k(z)$ with $z \in \partial\mathbb{D}$ for complex (multiplicative) notation

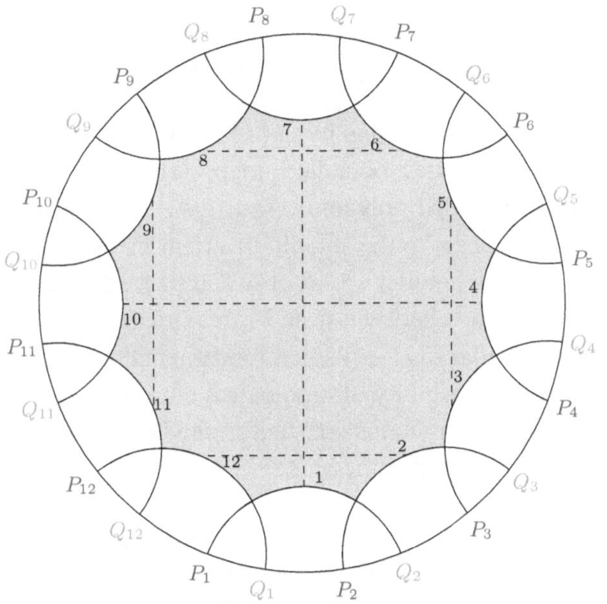

Figure 2.1 Fundamental polygon $\mathcal{F}$ for genus $g = 2$.

and $T_k(x) := \arg(T_k(e^{ix}))$ with $x \in \mathbb{S} = \mathbb{R}/2\pi\mathbb{Z}$ for real (additive) notation. See Figure 2.5(a) for a plot of $y = T_k(x)$ with $x, y \in [-\pi, \pi]$.

Notice that in general the polygon $\mathcal{F}$ need not be regular. In fact, one of the definitions of Teichmüller space, used in [3], is the space of all marked canonical hyperbolic $(8g-4)$-gons in the unit disk $\mathbb{D}$ (up to an isometry of $\mathbb{D}$) such that side k and side $\sigma(k)$ have equal length and the internal angles at vertices V_k and $V_{\sigma(k)+1}$ sum to π. (The topology on the space of polygons is as follows: $\mathcal{P}_n \to \mathcal{P}$ if and only if the lengths of all sides converge and the measures of all angles converge.)

If $\mathcal{F}$ is regular, it is the Ford fundamental domain, that is, the geodesic from P_k to Q_{k+1} (which we denote as just P_kQ_{k+1}) is the isometric circle for T_k, and $T_k(P_kQ_{k+1}) = Q_{\sigma(k)+1}P_{\sigma(k)}$ is the isometric circle for $T_{\sigma(k)}$ so that the inside of the former isometric circle is mapped to the outside of the latter, and all internal angles of $\mathcal{F}$ are equal to $\pi/2$. See [2] for more details and Section 2.2 for additional properties of the generators T_k.

The object of our study is the family of generalized Bowen–Series boundary maps studied in [15, 2, 1, 3] and defined by the formula

$$f_{\bar{A}}(x) = T_k(x) \quad \text{if } x \in [A_k, A_{k+1}), \tag{2.1}$$

where

$$\bar{A} = \{A_1, A_2, ..., A_{8g-4}\} \quad \text{and} \quad A_k \in [P_k, Q_k].$$

When all $A_k = P_k$, we denote the map by $f_{\bar{P}}$ (this map is what Adler and Flatto [4] refer to as "the Bowen–Series boundary map," although Bowen and Series' construction [9] used $4g$-sided polygons). In [3] we analyzed how the measure-theoretic entropy with respect to the smooth invariant measure $\mu_{\bar{A}}$ of maps in this family changes in the Teichmüller space of S and proved a flexibility result: the entropy $h_{\mu_{\bar{A}}}$ takes all values between 0 and a maximum that is achieved on the surface that admits a regular $(8g-4)$-sided fundamental polygon. In contrast, the main result of this paper is rigidity of topological entropy: its value depends only on the genus of the surface, remains constant in the Teichmüller space $\mathcal{T}(S)$, and does not depend on the (multi)parameter $\bar{A}$.

Theorem 1 (Main Theorem). *Let Γ be a cocompact torsion-free Fuchsian group such that $S = \Gamma\backslash\mathbb{D}$ is a surface of genus $g \geq 2$. For any $\bar{A} = \{A_1, ..., A_{8g-4}\}$ with $A_k \in [P_k, Q_k]$, the map $f_{\bar{A}} : \mathbb{S} \to \mathbb{S}$ has topological entropy $h_{\text{top}}(f_{\bar{A}}) = \log(4g-3+\sqrt{(4g-3)^2-1})$.*[1]

Remark. Most previous results on boundary maps $f_{\bar{A}} : \mathbb{S} \to \mathbb{S}$ require the parameters $\bar{A}$ to be in a smaller class: [4] uses only $\bar{A} = \bar{P}$ and $\bar{A} = \bar{Q}$, [1] focuses on extremal parameters and their duals, and [2, 3] require that the parameters have the short cycle property. In this chapter Theorem 1 applies to *all* parameters $\bar{A}$ with $A_k \in [P_k, Q_k]$. Although our result shows that all maps $f_{\bar{A}}$ have the same topological entropy for a given genus g, they are not necessarily topologically conjugate, since, according to [15], the combinatorial structure of the orbits associated to the discontinuity points A_k could differ.

The chapter is organized as follows. In Sections 2.2–2.4 we restrict ourselves to the case when Γ admits a regular $(8g-4)$-sided fundamental polygon. In Section 2.2 we give the formulas for generators T_k as functions on $\overline{\mathbb{D}} \subset \mathbb{C}$ (Proposition 2) and prove two additional symmetric properties of generators as functions on $\partial\mathbb{D}$. In Section 2.3 we compute the maximal eigenvalue of the transition matrices for all "extremal" parameters and hence the topological entropy for these Markov cases. In Section 2.4 we prove some symmetric properties of the map $\psi_{\bar{P}}$ conjugating $f_{\bar{P}}$ to a constant slope map. We conclude that $\psi_{\bar{P}}$ actually conjugates all $f_{\bar{A}}$ to constant slope maps, and in Section 2.5 we use this to prove Theorem 1, first for Γ admitting regular $(8g-4)$-sided fundamental polygons and then in the fully general case. A technical result stated and used in Section 2.4 is proved in Appendix A.

[1] The quantity $\log(4g-3+\sqrt{(4g-3)^2-1})$ can also be expressed as $\operatorname{arccosh}(4g-3)$, but logarithm expressions are more common for entropies in general and especially for shifts, so we use the longer expression.

2.2 Additional Properties of Generators

Proposition 2. *If the* $(8g-4)$*-sided fundamental polygon* $\mathcal{F}$ *is regular, then the generators* T_k *of the group* Γ *are given as functions on* $\overline{\mathbb{D}} \subset \mathbb{C}$ *by*

$$T_k(z) = (-1)^{k+1} \frac{e^{\mathrm{i}(1-k)\alpha} z + \mathrm{i}\sqrt{\cos\alpha}}{(-\mathrm{i}\sqrt{\cos\alpha})z + e^{\mathrm{i}(k-1)\alpha}}, \quad \textit{where } \alpha := \frac{2\pi}{8g-4}. \tag{2.2}$$

Proof. We derive a formula for $T_k(z)$ based on some geometric considerations also presented in [14, Section 4.3] and [15, Appendix].

Let $T_k(z) = (az+\overline{c})/(cz+\overline{a})$, where $|a|^2 - |c|^2 = 1$. The isometric circle $P_k Q_{k+1}$ of T_k, also denoted $I(T_k)$, is given by the equation $|cz + \overline{a}| = 1$ and has center O_k located at $-\overline{a}/c$ with $\arg(-\overline{a}/c) = -\frac{\pi}{2} + (k-1)\alpha$ and radius $R = 1/|c|$.

Let $d = |a|/|c|$ be the distance from the origin O to the center O_k of $I(T_k)$. The following formula for R was obtained in [15, Appendix]:

$$R = \frac{\sqrt{2}\sin(\alpha/2)}{\sqrt{\cos\alpha}} = \frac{\sqrt{1-\cos\alpha}}{\sqrt{\cos\alpha}}.$$

This implies that

$$|c| = \frac{1}{R} = \frac{\sqrt{\cos\alpha}}{\sqrt{1-\cos\alpha}} \quad \text{and} \quad |a| = d\,|c| = \frac{1}{\sqrt{1-\cos\alpha}}.$$

The isometric circle $I(T_k)$ is mapped by T_k to the isometric circle of $T_k^{-1} = T_{\sigma(k)}$ with center located at a/c. We analyze two cases:

- If k is odd and $k < 4g$, then $T_k^{-1} = T_{4g-k}$. The oriented angle $\angle O_k O O_{4g-k} = (4g-2k)\alpha = \pi + (2-2k)\alpha$, so $a/c = e^{\mathrm{i}(\pi+(2-2k)\alpha)}(-\overline{a}/c)$, which implies that $\arg(a) = \pi + (1-k)\alpha$. From $\arg(-\overline{a}/c) = -\frac{\pi}{2} + (k-1)\alpha$, we get $\arg(c) = \pi/2$. Thus,

$$a = \frac{-e^{\mathrm{i}(1-k)\alpha}}{\sqrt{1-\cos\alpha}} \quad \text{and} \quad c = \frac{\mathrm{i}\sqrt{\cos(\alpha)}}{\sqrt{1-\cos\alpha}},$$

  and, after simplifying the common term $-\sqrt{1-\cos\alpha}$, we get relation (2.2).
- If k is odd and $k > 4g$, then $\sigma(k) = 4g - k \pmod{8g-4} = 12g - k - 4$, so $T_k^{-1} = T_{12g-k-4}$. The oriented angle $\angle O_k O O_{12g-k-4} = (12g-2k-4)\alpha = 3\pi + (2-2k)\alpha$, so $a/c = e^{\mathrm{i}(3\pi+(2-2k)\alpha)}(-\overline{a}/c)$, which implies that $\arg(a) = 2\pi + (1-k)\alpha$. From $\arg(-\overline{a}/c) = -\frac{\pi}{2} + (k-1)\alpha$, we get $\arg(c) = -\pi/2$. Thus

$$a = \frac{e^{\mathrm{i}(1-k)\alpha}}{\sqrt{1-\cos\alpha}} \quad \text{and} \quad c = -\frac{\mathrm{i}\sqrt{\cos(\alpha)}}{\sqrt{1-\cos\alpha}},$$

  and, after simplifying the common term $\sqrt{1-\cos\alpha}$, we get relation (2.2).

The case when k is even can be treated similarly. □

Proposition 3. *For all* $x \in \mathbb{S}$, $T_k(x+\alpha) = T_{k-1}(x) + (4g-3)\alpha$.

Proof. Let $\beta = (4g-3)\alpha$. Then in complex (multiplicative) notation, the claim is

$$T_k(e^{\mathrm{i}\alpha}z) = e^{\mathrm{i}\beta}T_{k-1}(z).$$

Note that $e^{\mathrm{i}\beta} = e^{\mathrm{i}\left(\frac{4g-3}{8g-4}\right)2\pi} = -e^{-\mathrm{i}\alpha}$. Then, using (2.2),

$$\begin{aligned}
T_k(e^{\mathrm{i}\alpha}z) &= (-1)^{k+1}\frac{(e^{\mathrm{i}(1-k)\alpha})(e^{\mathrm{i}\alpha}z) + \mathrm{i}\sqrt{\cos\alpha}}{(-\mathrm{i}\sqrt{\cos\alpha})(e^{\mathrm{i}\alpha}z) + e^{\mathrm{i}(k-1)\alpha}} \cdot \frac{-e^{\mathrm{i}\beta}}{e^{-\mathrm{i}\alpha}} \\
&= (-1)^{k+1}(-e^{\mathrm{i}\beta})\frac{(e^{\mathrm{i}(1-k)\alpha}e^{\mathrm{i}\alpha})z + \mathrm{i}\sqrt{\cos\alpha}}{(-\mathrm{i}\sqrt{\cos\alpha}) + (e^{\mathrm{i}(k-1)\alpha}e^{-\mathrm{i}\alpha})} \\
&= (-1)^{k}(e^{\mathrm{i}\beta})\frac{(e^{\mathrm{i}(-k)\alpha})z + \mathrm{i}\sqrt{\cos\alpha}}{(-\mathrm{i}\sqrt{\cos\alpha})z + e^{\mathrm{i}(k-2)\alpha}} \\
&= e^{\mathrm{i}\beta} \cdot T_{k-1}(z).
\end{aligned}$$

□

A function $r(x)$ is said to be *centrally symmetric around* c if $r(c+x) + r(c-x)$ is constant for all $x \in \mathbb{S}$ (this constant will be $2r(c)$). This property is equivalent to saying that the graph of a lift of r to $\mathbb{R}$ restricted to any rectangle $[c-\delta, c+\delta] \times [r(c-\delta), r(c+\delta)]$ is symmetric under rotation by π around the center of that rectangle. If the circle is modeled as $\partial\mathbb{D} \subset \mathbb{C}$, then the analogous property is that $r(cz) \cdot r(c/z)$ is constant for all $z \in \partial\mathbb{D}$.

Denote by C_k the midpoint of the segment $[P_k, Q_{k+1}]$. The next proposition asserts that the graph of T_k is centrally symmetric around C_k.

Proposition 4. *For all* $x \in \mathbb{S}$, $T_k(C_k+x) + T_k(C_k-x) = -2C_k$.

Proof. In complex (multiplicative) notation, the claim is that $T_k(C_k z) \cdot T_k(C_k/z) = 1/C_k^2$ for all $z \in \partial\mathbb{D}$. Since $C_k \in \partial\mathbb{D}$ is the midpoint of the counterclockwise arc of the circle from P_k to Q_{k+1}, it satisfies $C_k^2 = P_k \cdot Q_{k+1}$ as complex numbers. The isometric circle of $T_k(z)$ connects P_k to Q_{k+1} and consists of those $z \in \overline{\mathbb{D}}$ for which $|T_k'(z)| = 1$, so $P_k, Q_{k+1} \in \partial\mathbb{D}$ are the complex numbers z, $|z| = 1$, satisfying

$$\left|\frac{-\mathrm{i}\sqrt{\cos\alpha}}{\sqrt{1-\cos\alpha}}z + \frac{e^{\mathrm{i}(k-1)\alpha}}{\sqrt{1-\cos\alpha}}\right| = 1.$$

The solutions to this equation are $-e^{\mathrm{i}(k-1)\alpha}w$ and $e^{\mathrm{i}(k-1)\alpha}\overline{w}$, where $w = \sqrt{1-\cos\alpha} + \mathrm{i}\sqrt{\cos\alpha}$. The product of the two solutions is

$$\begin{aligned}
P_k \cdot Q_{k+1} &= e^{\mathrm{i}(k-1)\alpha}w \cdot e^{\mathrm{i}(k-1)\alpha}(-\overline{w}) = (e^{\mathrm{i}(k-1)\alpha})^2(-|w|^2) \\
&= e^{-\mathrm{i}\pi}e^{\mathrm{i}(2k-2)\alpha} = e^{\mathrm{i}((2k-2)\alpha-(4g-2)\alpha)} = (e^{\mathrm{i}\alpha})^{2k-4g}.
\end{aligned}$$

Since C_k is the midpoint of the smaller of the two arcs comprising $\partial\mathbb{D}\backslash\{P_k, Q_{k+1}\}$, we have that

$$C_k = (e^{\mathrm{i}\alpha})^{k-2g}. \tag{2.3}$$

To prove Theorem 4, we use the alternative form

$$C_k = e^{\mathrm{i}(k-2g)\alpha} = e^{-\mathrm{i}(2g-1)\alpha}e^{\mathrm{i}(k-1)\alpha} = e^{-\mathrm{i}(\pi/2)}e^{\mathrm{i}(k-1)\alpha} = -\mathrm{i}e^{\mathrm{i}(k-1)\alpha}$$

to compute

$$\begin{aligned} T_k(C_k z) &= (-1)^{k+1}\frac{e^{\mathrm{i}(1-k)\alpha}(-\mathrm{i}e^{\mathrm{i}(k-1)\alpha}z) + \mathrm{i}\sqrt{\cos\alpha}}{(-\mathrm{i}\sqrt{\cos\alpha})(-\mathrm{i}e^{\mathrm{i}(k-1)\alpha}z) + e^{\mathrm{i}(k-1)\alpha}} \\ &= \frac{(-1)^{k+1}\cdot \mathrm{i}}{e^{\mathrm{i}(k-1)\alpha}}\cdot\frac{-z+\sqrt{\cos\alpha}}{(-\sqrt{\cos\alpha})z+1} = \frac{(-1)^k}{C_k}\cdot\frac{z-\sqrt{\cos\alpha}}{(-\sqrt{\cos\alpha})z+1} \end{aligned}$$

and then

$$T_k(C_k z)\cdot T_k(C_k/z) = \left(\frac{(-1)^k}{C_k}\frac{z-\sqrt{\cos\alpha}}{(-\sqrt{\cos\alpha})z+1}\right)\left(\frac{(-1)^k}{C_k}\frac{z^{-1}-\sqrt{\cos\alpha}}{(-\sqrt{\cos\alpha})z^{-1}+1}\right) = \frac{1}{C_k^2}$$

as claimed. □

Corollary 5. *For all* $x \in \mathbb{S}$, $T_k(-x) = -T_{4g-k}(x)$.

Proof. Applying Theorem 3 repeatedly gives

$$T_k(x) = T_{k+n}(x + n\alpha) - n\beta$$

for any $n \in \mathbb{Z}$, where $\beta = (4g-3)\alpha$. Using $n = 2g - k$, we have

$$\begin{aligned} T_k(-x) &= T_{2g}\big(-x + (2g-k)\alpha\big) - (2g-k)\beta \\ &= -T_{2g}\big(x - (2g-k)\alpha\big) - (2g-k)\beta \quad \text{by Proposition 4} \\ &= -T_{2g-(k-2g)}(x) + (2g-k)\beta - (2g-k)\beta = -T_{4g-k}(x). \end{aligned}$$ □

2.3 Markov Matrices for Extremal Parameters

Definition 6. A parameter choice $\bar{A} = \{A_1, ..., A_{8g-4}\}$ with $A_k \in [P_k, Q_k]$ is called *extremal* if, for each k, either $A_k = P_k$ or $A_k = Q_k$.

Extremal parameters were first introduced in [1], in which several results of [15, 16, 2] for parameters with "short cycles" were extended to extremal parameters. Note that the classical cases $\bar{A} = \bar{P}$ and $\bar{A} = \bar{Q}$ are examples of extremal parameter choices.

Since, for all $k = 1, ..., 8g - 4$, $f_{\bar{A}}(P_k)$ and $f_{\bar{A}}(Q_k)$ belong to the set $\bar{P} \cup \bar{Q}$ (see [15, Proposition 2.2], originally [4, Theorem 3.4]), the partition of $\mathbb{S}$ into intervals $I_1, ..., I_{16g-8}$ given by

$$I_{2k-1} := [P_k, Q_k], \qquad I_{2k} := [Q_k, P_{k+1}], \qquad k = 1, ..., 8g - 4,$$

is a Markov partition for $f_{\bar{A}}$ for every extremal $\bar{A}$. Each extremal $\bar{A}$ has a transition matrix $M_{\bar{A}} = (m_{i,j})$ with

$$m_{i,j} := \begin{cases} 1 & \text{if } f_{\bar{A}}(I_i) \supset I_j \\ 0 & \text{otherwise,} \end{cases}$$

and an infinite sequence $(\omega_0, \omega_1, ...)$ or finite sequence $(\omega_0, \omega_1, ..., \omega_n)$ over the alphabet $\{1, ..., 16g - 8\}$ is called $\bar{A}$*-admissible* if all $m_{\omega_i, \omega_{i+1}} = 1$ for all $i \geq 0$ (and $i < n$ for finite).

For each extremal parameter $\bar{A}$, we can define the shift space[2]

$$X_{\bar{A}} = \overline{\left\{ \omega \in \{1, ..., 16g - 8\}^{\mathbb{N}} : \omega \text{ is admissible} \right\}}$$

on which we have the left-shift $\sigma_{\bar{A}} : X_{\bar{A}} \to X_{\bar{A}}$ and the essentially bijective *coding map* $\phi_{\bar{A}} : X_{\bar{A}} \to \mathbb{S}$ given by

$$\phi_{\bar{A}}(\omega) = \bigcap_{i=0}^{\infty} f_{\bar{A}}^{-i}(I_{\omega_i}) \tag{2.4}$$

so that the following diagram is commutative:

$$\begin{array}{ccc} X_{\bar{A}} & \xrightarrow{\sigma_{\bar{A}}} & X_{\bar{A}} \\ {\scriptstyle \phi_{\bar{A}}}\downarrow & & \downarrow{\scriptstyle \phi_{\bar{A}}} \\ \mathbb{S} & \xrightarrow{f_{\bar{A}}} & \mathbb{S} \end{array}.$$

In the case where $f_{\bar{A}}$ is Markov, the system $(X_{\bar{A}}, \sigma_{\bar{A}})$ is a topological Markov chain.

The following formulas use [2, Proposition 3.1 and Lemma 3.2]. For odd indices $2k - 1$, depending on whether $A_k = P_k$ or $A_k = Q_k$ we have, respectively, either

$$f_{\bar{A}}(I_{2k-1}) = T_k(I_{2k-1}) = [Q_{\sigma(k)+1}, Q_{\sigma(k)+2}] = I_{2\sigma(k)+2} \cup I_{2\sigma(k)+3} \tag{2.5}$$

or

$$\begin{aligned} f_{\bar{A}}(I_{2k-1}) = T_{k-1}(I_{2k-1}) = [P_{\sigma(k-1)-1}, P_{\sigma(k-1)}] &= [P_{\sigma(k)+4g-2}, P_{\sigma(k)+4g-1}] \\ &= I_{2\sigma(k)+8g-5} \cup I_{2\sigma(k)+8g-4}. \end{aligned} \tag{2.6}$$

In either case, $f_{\bar{A}}(I_{2k-1})$ is the union of two consecutive Markov partition elements. For even indices $2k$, we know that for any extremal $\bar{A}$ the image

2 In [2] the notation $X_{\bar{A}}$ is used for a space of sofic sequences in $8g - 4$ symbols. Here we use it for a Markov shift on $16g - 8$ symbols.

$$\begin{pmatrix}
0&0&0&0&0&0&0&0&0&0&0&0&0&0&0&1&1&0&0&0&0&0&0&0\\
1&1&1&1&1&1&1&1&1&1&0&0&0&0&0&0&0&1&1&1&1&1&1&1\\
0&1&1&0&0&0&0&0&0&0&0&0&0&0&0&0&0&0&0&0&0&0&0&0\\
0&0&0&1&1&1&1&1&1&1&1&1&1&1&1&1&1&1&1&1&0&0&0&0\\
0&0&0&0&0&0&0&0&0&0&0&1&1&0&0&0&0&0&0&0&0&0&0&0\\
1&1&1&1&1&1&0&0&0&0&0&0&0&1&1&1&1&1&1&1&1&1&1&1\\
0&0&0&0&0&0&0&0&0&0&0&0&0&0&0&0&0&0&0&0&0&1&1&0\\
1&1&1&1&1&1&1&1&1&1&1&1&1&1&1&1&0&0&0&0&0&0&0&1\\
0&0&0&0&0&0&0&1&1&0&0&0&0&0&0&0&0&0&0&0&0&0&0&0\\
1&1&0&0&0&0&0&0&0&1&1&1&1&1&1&1&1&1&1&1&1&1&1&1\\
0&0&0&0&0&0&0&0&0&0&0&0&0&0&0&0&0&1&1&0&0&0&0&0\\
1&1&1&1&1&1&1&1&1&1&1&1&0&0&0&0&0&0&0&1&1&1&1&1\\
0&0&0&1&1&0&0&0&0&0&0&0&0&0&0&0&0&0&0&0&0&0&0&0\\
0&0&0&0&0&1&1&1&1&1&1&1&1&1&1&1&1&1&1&1&1&1&0&0\\
0&0&0&0&0&0&0&0&0&0&0&0&0&1&1&0&0&0&0&0&0&0&0&0\\
1&1&1&1&1&1&1&1&0&0&0&0&0&0&0&1&1&1&1&1&1&1&1&1\\
1&0&0&0&0&0&0&0&0&0&0&0&0&0&0&0&0&0&0&0&0&0&0&1\\
0&1&1&1&1&1&1&1&1&1&1&1&1&1&1&1&1&1&0&0&0&0&0&0\\
0&0&0&0&0&0&0&0&0&1&1&0&0&0&0&0&0&0&0&0&0&0&0&0\\
1&1&1&1&0&0&0&0&0&0&0&1&1&1&1&1&1&1&1&1&1&1&1&1\\
0&0&0&0&0&0&0&0&0&0&0&0&0&0&0&0&0&0&0&1&1&0&0&0\\
1&1&1&1&1&1&1&1&1&1&1&1&1&1&0&0&0&0&0&0&0&1&1&1\\
0&0&0&0&0&1&1&0&0&0&0&0&0&0&0&0&0&0&0&0&0&0&0&0\\
0&0&0&0&0&0&0&1&1&1&1&1&1&1&1&1&1&1&1&1&1&1&1&1
\end{pmatrix}$$

(a)

$$\begin{pmatrix}
1&1&0&0&0&0&0&0&0&0&0&0&0&0&0&0&0&0&0&0&0&0&0&0\\
1&1&1&1&1&1&1&1&1&1&0&0&0&0&0&0&0&1&1&1&1&1&1&1\\
0&0&0&0&0&0&0&0&0&0&1&1&0&0&0&0&0&0&0&0&0&0&0&0\\
0&0&0&1&1&1&1&1&1&1&1&1&1&1&1&1&1&1&1&1&0&0&0&0\\
0&0&0&0&0&0&0&0&0&0&0&0&0&0&0&0&0&0&0&0&1&1&0&0\\
1&1&1&1&1&1&0&0&0&0&0&0&0&1&1&1&1&1&1&1&1&1&1&1\\
0&0&0&0&0&0&1&1&0&0&0&0&0&0&0&0&0&0&0&0&0&0&0&0\\
1&1&1&1&1&1&1&1&1&1&1&1&1&1&1&1&0&0&0&0&0&0&0&1\\
0&0&0&0&0&0&0&0&0&0&0&0&0&0&0&0&1&1&0&0&0&0&0&0\\
1&1&0&0&0&0&0&0&0&1&1&1&1&1&1&1&1&1&1&1&1&1&1&1\\
0&0&1&1&0&0&0&0&0&0&0&0&0&0&0&0&0&0&0&0&0&0&0&0\\
1&1&1&1&1&1&1&1&1&1&1&1&0&0&0&0&0&0&0&1&1&1&1&1\\
0&0&0&0&0&0&0&0&0&0&0&0&1&1&0&0&0&0&0&0&0&0&0&0\\
0&0&0&0&0&1&1&1&1&1&1&1&1&1&1&1&1&1&1&1&1&1&0&0\\
0&0&0&0&0&0&0&0&0&0&0&0&0&0&0&0&0&0&0&0&0&0&1&1\\
1&1&1&1&1&1&1&1&0&0&0&0&0&0&0&1&1&1&1&1&1&1&1&1\\
0&0&0&0&0&0&0&0&1&1&0&0&0&0&0&0&0&0&0&0&0&0&0&0\\
0&1&1&1&1&1&1&1&1&1&1&1&1&1&1&1&1&1&0&0&0&0&0&0\\
0&0&0&0&0&0&0&0&0&0&0&0&0&0&0&0&0&0&1&1&0&0&0&0\\
1&1&1&1&0&0&0&0&0&0&0&1&1&1&1&1&1&1&1&1&1&1&1&1\\
0&0&0&0&1&1&0&0&0&0&0&0&0&0&0&0&0&0&0&0&0&0&0&0\\
1&1&1&1&1&1&1&1&1&1&1&1&1&1&0&0&0&0&0&0&0&1&1&1\\
0&0&0&0&0&0&0&0&0&0&0&0&0&0&1&1&0&0&0&0&0&0&0&0\\
0&0&0&0&0&0&0&1&1&1&1&1&1&1&1&1&1&1&1&1&1&1&1&1
\end{pmatrix}$$

(b)

Figure 2.2 Transition matrices $M_{\bar P}$ (a) and $M_{\bar Q}$ (b) for $g = 2$.

$$\begin{aligned} f_{\bar A}(I_{2k}) &= T_k(I_{2k}) = [Q_{\sigma(k)+2}, P_{\sigma(k)-1}] \\ &= I_{2\sigma(k)+4} \cup I_{2\sigma(k)+5} \cup \cdots \cup I_{2\sigma(k)-4} = \mathbb{S} \setminus \bigcup_{\ell=2\sigma(k)-3}^{2\sigma(k)+3} I_\ell \end{aligned} \tag{2.7}$$

is the union of $(16g-8)-7 = 16g-15$ consecutive intervals on the circle. Recall that the indices are mod $16g-8$; for example, with $g = 2$ and $k = 1$ we get

$$f_{\bar A}(I_2) = I_{18} \cup I_{19} \cup \cdots \cup I_{10} = I_{18} \cup I_{19} \cup \cdots \cup I_{24} \cup I_1 \cup \cdots \cup I_{10}.$$

The matrices $M_{\bar P}$ and $M_{\bar Q}$ for genus 2 are shown in Figure 2.2.

Proposition 7. *For any extremal $\bar A$, the maximal eigenvalue of $M_{\bar A}$ is*

$$\lambda = 4g - 3 + \sqrt{(4g-3)^2 - 1}.$$

Proof. Gelfand's Formula [13] states that $\lim_{n\to\infty} \left\| M_{\bar A}^n \right\|^{1/n}$ equals the maximal eigenvalue of $M_{\bar A}$, where $\|\cdot\|$ is any matrix norm. The "entrywise norm" $\left|M_{\bar A}^n\right|$ given by the sum of (absolute values of) all entries in $M_{\bar A}^n$ counts the total number of admissible sequences of length $n+1$, that is,

$$\left|M_{\bar A}^n\right| = \#\{\, (\omega_0, \omega_1, ..., \omega_n) \colon m_{\omega_i,\omega_{i+1}} = 1 \text{ for } i = 0, ..., n-1 \,\}.$$

To compress notation, we will write

$$\begin{aligned} N_n &:= \#\{\, (\omega_0, \omega_1, ..., \omega_n) \colon \text{admissible} \,\} = \left|M_{\bar A}^n\right|, \\ N_n^{\text{even}} &:= \#\{\, (\omega_0, \omega_1, ..., \omega_n) \colon \text{admissible and } \omega_n \text{ is even} \,\}, \\ N_n^{\text{odd}} &:= \#\{\, (\omega_0, \omega_1, ..., \omega_n) \colon \text{admissible and } \omega_n \text{ is odd} \,\}. \end{aligned}$$

Thus

$$N_{n+1} = 2N_n^{\text{odd}} + (16g-15)N_n^{\text{even}}. \tag{2.8}$$

Since the indices of intervals that make up $f_{\bar{A}}(I_i)$ are consecutive, $f_{\bar{A}}(I_{2k-1})$ is the union of one even-index and one odd-index Markov interval, and $f_{\bar{A}}(I_{2k})$ is the union of $8g-8$ intervals with odd indices and $8g-7$ intervals with even indices. In terms of counting sequences,

$$\begin{aligned} N_n^{\text{odd}} &= N_{n-1}^{\text{odd}} + (8g-8)N_{n-1}^{\text{even}}, \\ N_n^{\text{even}} &= N_{n-1}^{\text{odd}} + (8g-7)N_{n-1}^{\text{even}}. \end{aligned} \tag{2.9}$$

Using (2.9) and the fact that $N_{n-1}^{\text{odd}} + N_{n-1}^{\text{even}} = N_{n-1}$, we will convert (2.8) into a recurrence relation for N_n.

$$\begin{aligned} N_{n+1} &= 2N_n^{\text{odd}} + (16g-15)N_n^{\text{even}} \\ &= \big((8g-6)-(8g-8)\big)N_n^{\text{odd}} + \big((8g-6)+(8g-9)\big)N_n^{\text{even}} \\ &= (8g-6)N_n - (8g-8)N_n^{\text{odd}} + (8g-9)N_n^{\text{even}} \\ &= (8g-6)N_n - (8g-8)\big(N_{n-1}^{\text{odd}} + (8g-8)N_{n-1}^{\text{even}}\big) \\ &\quad + (8g-9)\big(N_{n-1}^{\text{odd}} + (8g-7)N_{n-1}^{\text{even}}\big) \qquad \text{by (2.9)} \\ &= (8g-6)N_n + (-1)N_{n-1}^{\text{odd}} + (-1)N_{n-1}^{\text{even}} \\ &= (8g-6)N_n - N_{n-1}. \end{aligned}$$

Any nonzero sequence $(N_0, N_1, N_2, \ldots)$ satisfying the linear recurrence relation

$$N_{n+1} = KN_n - N_{n-1}$$

has an explicit expression of the form

$$N_n = c_1 \cdot \left(K/2 + \sqrt{(K/2)^2 - 1}\right)^n + c_2 \cdot \left(K/2 + \sqrt{(K/2)^2 - 1}\right)^{-n}$$

for some constants c_1 and c_2, and therefore $\lim_{n\to\infty}(N_n)^{1/n} = K/2 + \sqrt{(K/2)^2 - 1}$. For $N_n = \left|M_{\bar{A}}^n\right|$ we have exactly this relation with $K = 8g-6$; therefore, the maximal eigenvalue of $M_{\bar{A}}$ is $\lim_{n\to\infty}\left|M_{\bar{A}}^n\right|^{1/n} = 4g-3+\sqrt{(4g-3)^2-1}$. □

Corollary 8. *For any extremal* $\bar{A}$, $h_{top}(f_{\bar{A}}) = \log\lambda$.

Proposition 9. *For any extremal* $\bar{A}$, *the right eigenvector* $v = (v_1, \ldots, v_{16g-8})$, *corresponding to the maximal eigenvalue* λ, *normalized so that* $\sum v_i = 1$, *is given by*

$$v_i = \begin{cases} \dfrac{1}{\lambda(8g-4)} & \text{if } i \text{ is odd} \\[2ex] \dfrac{\lambda - 1}{\lambda(8g-4)} & \text{if } i \text{ is even.} \end{cases}$$

Proof. From the proof of Proposition 7, for each odd i the set $f_{\bar{A}}(I_i)$ is the union of two consecutive Markov partition elements, one with an even index and one with an odd index, and thus

$$\sum_{k=1}^{8g-4} m_{i,2k-1} = 1 \quad \text{and} \quad \sum_{k=1}^{8g-4} m_{i,2k} = 1 \qquad \text{for odd } i. \tag{2.10}$$

Similarly, if i is even, then for any extremal $\bar{A}$ we know $f_{\bar{A}}(I_i)$ is the union of $8g-8$ odd indices and $8g-7$ even indices, so

$$\sum_{k=1}^{8g-4} m_{i,2k-1} = 8g-8 \quad \text{and} \quad \sum_{k=1}^{8g-4} m_{i,2k} = 8g-7 \qquad \text{for even } i. \tag{2.11}$$

We will show that the vector v' given by

$$v'_i = \begin{cases} \lambda - 8g + 7 & \text{if } i \text{ is odd} \\ 8g - 8 & \text{if } i \text{ is even} \end{cases}$$

satisfies $M_{\bar{A}} v' = \lambda v'$ by direct calculation. First, note that $\lambda = 4g-3+\sqrt{(4g-3)^2-1}$ is one root of the quadratic equation

$$\lambda(\lambda - 8g + 7) = \lambda - 1. \tag{2.12}$$

Then we have

$$\begin{aligned}
(M_{\bar{A}} v')_i &= \sum_{j=1}^{16g-8} m_{i,j} v'_j = \sum_{k=1}^{8g-4} m_{i,2k-1} v'_{2k-1} + \sum_{k=1}^{8g-4} m_{i,2k} v'_{2k} \\
&= (\lambda - 8g + 7)\left(\sum_{k=1}^{8g-4} m_{i,2k-1}\right) + (8g-8)\left(\sum_{k=1}^{8g-4} m_{i,2k}\right) \\
&= \begin{cases} (\lambda-8g+7)(1) + (8g-8)(1) & \text{if } i \text{ is odd} \\ (\lambda-8g+7)(8g-8) + (8g-8)(8g-7) & \text{if } i \text{ is even} \end{cases} \quad \text{by (2.10), (2.11)} \\
&= \begin{cases} \lambda - 1 & \text{if } i \text{ is odd} \\ \lambda(8g-8) & \text{if } i \text{ is even} \end{cases} \\
&= \begin{cases} \lambda(\lambda - 8g + 7) & \text{if } i \text{ is odd} \\ \lambda(8g-8) & \text{if } i \text{ is even} \end{cases} \quad \text{by (2.12)} \\
&= \lambda v'_i.
\end{aligned}$$

The normalized eigenvector v is then obtained by dividing v' by

$$\sum_{i=1}^{16g-8} v'_i = (8g-4)(\lambda-8g+7) + (8g-4)(8g-8) = (8g-4)(\lambda-1).$$

From (2.12), we have that $\lambda(8g-4) = (\lambda+1)^2$ and so the coordinates of v are

$$v_i = \frac{\lambda - 8g + 7}{(8g-4)(\lambda-1)} = \frac{(\lambda-1)/\lambda}{(8g-4)(\lambda-1)} = \frac{1}{\lambda(8g-4)}$$

for odd i and

$$v_i = \frac{(8g-7)-1}{(8g-4)(\lambda-1)} = \frac{(\lambda - \frac{\lambda-1}{\lambda}) - 1}{(8g-4)(\lambda-1)} = \frac{1-\frac{1}{\lambda}}{8g-4} = \frac{\lambda-1}{\lambda(8g-4)}$$

for even i. □

2.4 Conjugacy to Constant Slope Map

We begin by stating a theorem combining several results of [21, 7], stated here for circle maps instead of interval maps (as in [18]):

Theorem 10. *Given a piecewise monotone, piecewise continuous, topologically transitive map* $f: \mathbb{S} \to \mathbb{S}$ *of positive topological entropy* $h > 0$*, there exists a unique (up to rotation of* $\mathbb{S}$*) increasing homeomorphism* $\psi: \mathbb{S} \to \mathbb{S}$ *conjugating* f *to a piecewise continuous map with constant slope* e^h.

Existence follows from [7, Corollary B], and uniqueness follows from [7, Lemma 8.1, Theorem 8.2, Corollary 1], where the continuity assumption is replaced by piecewise continuity (as in [12, 10]).

The map $f_{\bar{P}}: \mathbb{S} \to \mathbb{S}$ is piecewise monotone, piecewise continuous, topologically transitive (see [9, Lemma 2.5]), and with positive topological entropy (see Corollary 8), so by Theorem 10 there exists an increasing homeomorphism $\psi_{\bar{P}}: \mathbb{S} \to \mathbb{S}$ conjugating it to a map

$$\ell_{\bar{P}} := \psi_{\bar{P}} \circ f_{\bar{P}} \circ \psi_{\bar{P}}^{-1}$$

with constant slope; see Figures 2.3 and 2.4. The map $\psi_{\bar{P}}$ is unique up to rotation of $\mathbb{S}$, and the slope of $\ell_{\bar{P}}$ is exactly $\lambda = e^{h_{\mathrm{top}}(f_{\bar{P}})}$. Although the existence of a conjugacy to a constant slope map holds for $f_{\bar{P}}$ associated to irregular fundamental polygons as well as regular, we will assume that $\mathcal{F}$ is regular for the remainder of this section.

The map $f_{\bar{Q}}: \mathbb{S} \to \mathbb{S}$, just like $f_{\bar{P}}$, is piecewise monotone, piecewise continuous, topologically transitive, and with positive topological entropy, so by Theorem 10 there exists an increasing homeomorphism $\psi_{\bar{Q}}: \mathbb{S} \to \mathbb{S}$ conjugating it to a map $\ell_{\bar{Q}}$ of constant slope, unique up to rotation of $\mathbb{S}$. By Corollary 8, both $\ell_{\bar{P}}$ and $\ell_{\bar{Q}}$ have the same slope.

Because $f_{\bar{P}}$ and $f_{\bar{Q}}$ are Markov maps, the conjugacies $\psi_{\bar{P}}$ and $\psi_{\bar{Q}}$ follow the classical construction due to Parry [20, 21] and used in the proof of [7, Lemma 5.1]. For

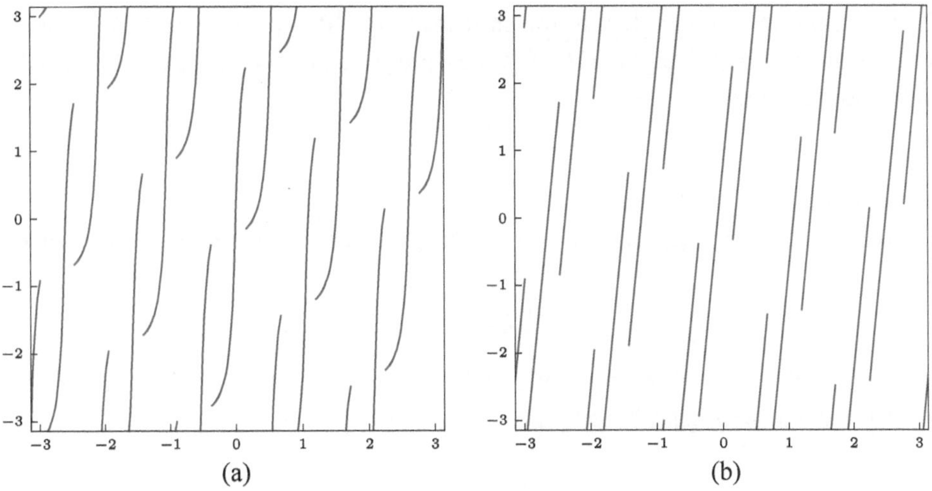

Figure 2.3 Plots of $f_{\bar{P}}(x)$ (a) and $\ell_{\bar{P}}(x)$ (b) for $g = 2$.

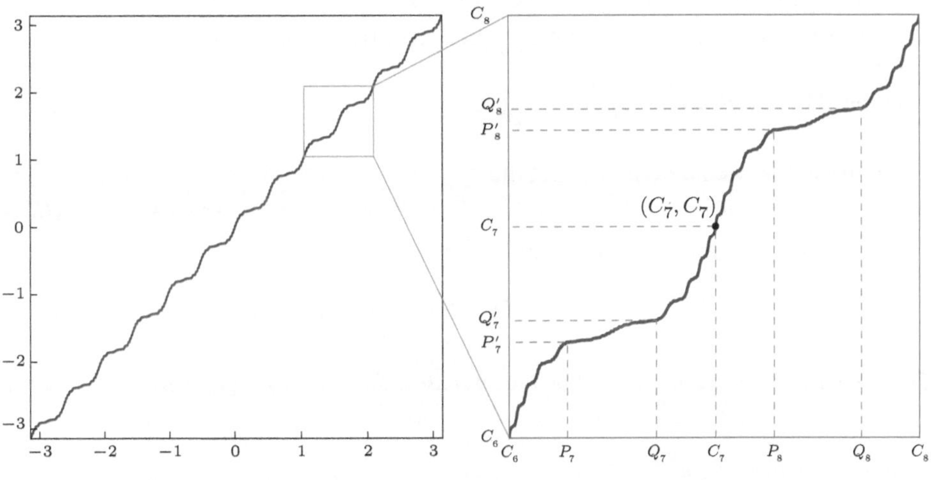

Figure 2.4 Plot of $\psi_{\bar{P}}(x)$ for $g = 2$.

each extremal parameter $\bar{A}$, we define the probability measure $\rho_{\bar{A}}$ on $X_{\bar{A}}$ as follows: let λ, v be the maximal eigenpair for the transition matrix $M_{\bar{A}}$; for an $\bar{A}$-admissible finite sequence $(\omega_0, ..., \omega_n)$, we denote the symbolic cylinder

$$C_{\bar{A}}^{(\omega_0,\ldots,\omega_n)} := \left\{ \omega' \in X_{\bar{A}} : \omega'_i = \omega_i \; \forall\, 0 \leq i \leq n \right\}$$

and define the measure $\rho_{\bar{A}}$ of this cylinder as

$$\rho_{\bar{A}}\left(C_{\bar{A}}^{(\omega_0,\ldots,\omega_n)}\right) = \frac{v_{\omega_n}}{\lambda^n}.$$

The measure $\rho_{\bar{A}}$ is equivalent to the shift-invariant "Parry measure" (the measure of maximal entropy; see [20, 21]). The measure $\rho_{\bar{A}}$ is not shift-invariant but has the "expanding property"

$$\rho_{\bar{A}}(\sigma_{\bar{A}}(C)) = \lambda\rho_A(C)$$

for all cylinders C on $(X_{\bar{A}}, \sigma_{\bar{A}})$.

Using the measure $\rho_{\bar{A}}$, one constructs the push-forward measure $\rho'_{\bar{A}}$ on $\mathbb{S}$ given by

$$\rho'_{\bar{A}}(E) = \rho_{\bar{A}}\left(\phi_{\bar{A}}^{-1}(E)\right) \qquad \text{for a Borel set } E,$$

where $\phi_{\bar{A}} : X_{\bar{A}} \to \mathbb{S}$ is the symbolic coding map (2.4). With the convention that $\psi_{\bar{A}}(0) = 0$, the conjugacy map $\psi_{\bar{A}} : \mathbb{S} \to \mathbb{S}$ is given by

$$\psi_{\bar{A}}(x) := 2\pi \cdot \begin{cases} \rho'_{\bar{A}}([0,x]) & \text{if } x \geq 0 \\ -\rho'_{\bar{A}}([x,0]) & \text{if } x < 0 \end{cases} \tag{2.13}$$

(the 2π appears because of our convention that the circle is $\mathbb{S} = [-\pi, \pi]$).

It turns out that the maps $\psi_{\bar{P}}$ and $\psi_{\bar{Q}}$ thus constructed coincide:

Theorem 11. *For all $x \in \mathbb{S}$, $\psi_{\bar{P}}(x) = \psi_{\bar{Q}}(x)$.*

To prove this, we need to connect the cylinder intervals of the two circle maps $f_{\bar{P}}$ and $f_{\bar{Q}}$. Given an $\bar{A}$-admissible sequence $\omega = (\omega_0, \omega_1, ..., \omega_n)$ with $\omega_i \in \{1, ..., 16g-8\}$, we define the corresponding *$\bar{A}$-cylinder interval*

$$I_{\bar{A}}^{(\omega_0, \omega_1, ..., \omega_n)} := I_{\omega_0} \cap f_{\bar{A}}^{-1}(I_{\omega_1}) \cap \cdots \cap f_{\bar{A}}^{-n}(I_{\omega_n}). \tag{2.14}$$

Theorem 12. *Let $\omega = (\omega_0, ..., \omega_n)$ be $\bar{P}$-admissible. There exists a $\bar{Q}$-admissible sequence $(\eta_0, ..., \eta_n, \eta_{n+1})$ such that $\eta_0 = \omega_0$, η_{n+1} is odd, and*

(i) if ω_n is odd, then

$$I_{\bar{P}}^{\omega} = I_{\bar{Q}}^{(\eta_0, ..., \eta_n, \eta_{n+1})} \cup I_{\bar{Q}}^{(\eta_0, ..., \eta_n, \eta_{n+1}+1)}; \tag{2.15}$$

(ii) if ω_n is even, then either

$$I_{\bar{P}}^{\omega} = I_{\bar{Q}}^{(\eta_0, ..., \eta_n)} = \bigcup_{i=0}^{16g-14} I_{\bar{Q}}^{(\eta_0, ..., \eta_n, 2\sigma(\eta_n/2)+4+i)} \tag{2.16}$$

or

$$I_{\bar{P}}^{\omega} = I_{\bar{Q}}^{(\eta_0, ..., \eta_n, \eta_{n+1})} \cup I_{\bar{Q}}^{(\eta_0, ..., \eta_n, \eta_{n+1}+1)} \cup \bigcup_{i=0}^{16g-16} I_{\bar{Q}}^{(\eta_0, ..., \eta_n+1, 2\sigma(\eta_n+1/2)+6+i)}. \tag{2.17}$$

The proof of Theorem 12, as well as the distinction between the two forms (2.16) and (2.17), is rather technical and is left for Appendix A .

Proof of Theorem 11. Recall from Theorem 9 that, for both $\bar{A} = \bar{P}$ and $\bar{A} = \bar{Q}$, the right-eigenvector v of $M_{\bar{A}}$ corresponding to eigenvalue λ is

$$v = c \cdot (1,\ \lambda-1,\ 1,\ \lambda-1,\ \ldots,\ 1,\ \lambda-1),$$

where $c = 1/(\lambda(8g-4))$ corresponds to $\rho'_{\bar{A}}(\mathbb{S}) = 1$. We prove $\psi_{\bar{P}} = \psi_{\bar{Q}}$ by showing that $\rho'_{\bar{P}}(I^{\omega}_{\bar{P}}) = \rho'_{\bar{Q}}(I^{\omega}_{\bar{P}})$ for all finite $\bar{P}$-admissible sequences ω. Note that, because $\phi^{-1}_{\bar{A}}$ maps a cylinder interval to a symbolic cylinder, we have

$$\rho'_{\bar{P}}(I_{\bar{P}}^{(\omega_0,\ldots,\omega_n)}) = \frac{v_{\omega_n}}{\lambda^n} \qquad \text{and} \qquad \rho'_{\bar{Q}}(I_{\bar{Q}}^{(\eta_0,\ldots,\eta_n)}) = \frac{v_{\eta_n}}{\lambda^n},$$

where $(\omega_0, \ldots, \omega_n)$ is $\bar{P}$-admissible and $(\eta_0, \ldots, \eta_n)$ is $\bar{Q}$-admissible.

Let $\omega = (\omega_0, \ldots, \omega_n)$ be $\bar{P}$-admissible, and suppose ω_n is odd. Then $v_{\omega_n} = c$, and so

$$\rho'_{\bar{P}}(I^{\omega}_{\bar{P}}) = \frac{v_{\omega_n}}{\lambda^n} = \frac{c}{\lambda^n}.$$

By Theorem 12,

$$I^{\omega}_{\bar{P}} = I^{\eta}_{\bar{Q}} \cup I^{\eta'}_{\bar{Q}}$$

for some $\eta = (\eta_0, \ldots, \eta_{n+1})$ and $\eta' = (\eta_0, \ldots, \eta_n, \eta_{n+1}+1)$. Since η_{n+1} and η'_{n+1} have different parities, we know

$$v_{\eta_{n+1}} + v_{\eta'_{n+1}} = c + (\lambda-1)c = \lambda c$$

and can compute

$$\begin{aligned}\rho'_{\bar{Q}}(I^{\omega}_{\bar{P}}) &= \rho'_{\bar{Q}}(I^{\eta}_{\bar{Q}} \cup I^{\eta'}_{\bar{Q}}) = \frac{v_{\eta_{n+1}}}{\lambda^{n+1}} + \frac{v_{\eta'_{n+1}}}{\lambda^{n+1}} = \frac{1}{\lambda^{n+1}}(v_{\eta_{n+1}} + v_{\eta'_{n+1}}) \\ &= \frac{1}{\lambda^{n+1}}(\lambda c) = \frac{c}{\lambda^n} = \rho'_{\bar{P}}(I^{\omega}_{\bar{P}}).\end{aligned}$$

If instead ω_n is even, then Theorem 12 gives $I^{\omega}_{\bar{P}} = I^{\eta^{(1)}}_{\bar{Q}} \cup \cdots \cup I^{\eta^{(16g-15)}}_{\bar{Q}}$ with exactly $8g-7$ of the final symbols $\eta^{(i)}_{n+1}$ being even (so $8g-8$ are odd). Therefore,

$$\begin{aligned}\rho'_{\bar{Q}}(I^{\omega}_{\bar{P}}) &= \rho'_{\bar{Q}}(I^{\eta^{(1)}}_{\bar{Q}} \cup \cdots \cup I^{\eta^{(16g-15)}}_{\bar{Q}}) = \frac{1}{\lambda^{n+1}}(v_{\eta^{(1)}_{n+1}} + \cdots + v_{\eta^{(16g-15)}_{n+1}}) \\ &= \frac{1}{\lambda^{n+1}}\big((8g-7)(\lambda-1)c + (8g-8)c\big) \\ &= \frac{c}{\lambda^{n+1}}\lambda(\lambda-1) \qquad \text{by (2.12)} \\ &= \frac{(\lambda-1)c}{\lambda^n} = \frac{v_{\omega_n}}{\lambda^n} = \rho'_{\bar{P}}(I^{\omega}_{\bar{P}}),\end{aligned}$$

where $v_{\omega_n} = (\lambda-1)c$ because ω_n is even.

In both cases we have $\rho'_{\bar{Q}}(I^{\omega}_{\bar{P}}) = \rho'_{\bar{P}}(I^{\omega}_{\bar{P}})$, and since $\{I^{\omega}_{\bar{P}} : \omega \text{ is } \bar{P}\text{-admissible}\}$ generates all Borel sets in $\mathbb{S}$, the two measures $\rho'_{\bar{P}}$ and $\rho'_{\bar{Q}}$ on $\mathbb{S}$ are identical. From (2.13), this implies that $\psi_{\bar{P}} = \psi_{\bar{Q}}$. □

For the remainder of Section 2.4, we deal almost exclusively with $\psi_{\bar{P}}$, although we will briefly invoke Theorem 11. We now show that $\psi_{\bar{P}}$ has translational (Proposition 13) and central (Proposition 14) symmetry. Both of these properties can be seen in Figure 2.4.

Proposition 13. *For all* $x \in \mathbb{S}$, $\psi_{\bar{P}}(x+\alpha) = \psi_{\bar{P}}(x) + \alpha$.

Proof. Define $\phi\colon \mathbb{S} \to \mathbb{S}$ recursively by

$$\phi(x) := \begin{cases} \psi_{\bar{P}}(x) & \text{if } x \in [P_1, P_2) \\ \phi(x-\alpha)+\alpha & \text{otherwise.} \end{cases}$$

Thus $\phi(x+\alpha) = \phi(x) + \alpha$ for all x by design, and since ϕ is increasing and continuous, we also have $\phi^{-1}(x+\alpha) = \phi^{-1}(x) + \alpha$ for all x.

Denote $x' = \phi(x)$. By induction on $1 \le k \le 8g-4$, we will prove that there exists b_k such that $\phi \circ T_k \circ \phi^{-1}(x') = \lambda x' + b_k$ for $x' \in [P'_k, P'_{k+1}]$. By construction the claim is true for $k=1$ since $\phi\big|_{[P_1,P_2]} = \psi_{\bar{P}}\big|_{[P_1,P_2]}$. Now assume it is true for k. Then by Theorem 3, writing $\beta = (4g-3)\alpha$, we have

$$\begin{aligned} \phi \circ T_{k+1} \circ \phi^{-1}(x') &= \phi(T_{k+1}(x)) = \phi(T_k(x-\alpha)+\beta) = \phi(T_k(x-\alpha)) + \beta \\ &= \phi(T_k(\phi^{-1}(x'-\alpha))) + \beta = \lambda(x'-\alpha) + b_k + \beta \\ &= \lambda x' + (b_k + \beta - \lambda\alpha) \end{aligned}$$

and so the claim holds for $k+1$ with $b_{k+1} = b_k + \beta - \lambda\alpha$.

Thus our map ϕ, which satisfies $\phi(x+\alpha) = \phi(x) + \alpha$ for all x by construction, conjugates $f_{\bar{P}}$ to a constant slope map on $\mathbb{S}$. By the uniqueness of $\psi_{\bar{P}}$ (Theorem 10), $\phi = \psi_{\bar{P}}$. □

Notice that $\psi_{\bar{P}}(x+n\alpha) = \psi_{\bar{P}}(x)+n\alpha$ for any integer n, as well as $\psi_{\bar{P}}^{-1}(x+n\alpha) = \psi_{\bar{P}}^{-1}(x) + n\alpha$. Since 2π is an integer multiple of α, $\psi_{\bar{P}}$ is well defined on $\mathbb{S}$, and we can choose the point where it is equal to 0 at our convenience. We will assume that $\psi_{\bar{P}}$ fixes the point $C_{2g} = 0$. Then Proposition 13 implies that $\psi_{\bar{P}}(C_k) = C_k$ for all k.

Proposition 14. *For all* $x \in \mathbb{S}$, $\psi_{\bar{P}}(C_k + x) + \psi_{\bar{P}}(C_k - x) = 2C_k$.

Proof. First, we prove that

$$\psi_{\bar{Q}}(x) = -\psi_{\bar{P}}(-x). \tag{2.18}$$

By Corollary 5 we have $T_k(-x) = -T_{4g-k}(x)$. Since $f_{\bar{P}}$ acts by the generator T_k on $x \in [P_k, P_{k+1}]$ and $f_{\bar{Q}}$ acts by the generator T_{4g-k} on the reflected interval $-[P_k, P_{k+1}] = [Q_{4g-k}, Q_{4g-k+1}]$, we have as a result that

$$f_{\bar{Q}}(x) = -f_{\bar{P}}(-x).$$

To prove (2.18), set $\widetilde{x} = \xi(x) := -\psi_{\bar{P}}(-x)$. Then $-\widetilde{x} = \psi_{\bar{P}}(-x)$ and $-x = \psi_{\bar{P}}^{-1}(-\widetilde{x})$, and then

$$\begin{aligned}\xi \circ f_{\bar{Q}} \circ \xi^{-1}(\widetilde{x}) &= \xi(f_{\bar{Q}}(x)) = \xi(-f_{\bar{P}}(-x)) = -\psi_{\bar{P}}(f_{\bar{P}}(-x)) \\ &= -\psi_{\bar{P}} \circ f_{\bar{P}} \circ \psi_{\bar{P}}^{-1}(-\widetilde{x}) = -\ell_{\bar{P}}(-\widetilde{x}).\end{aligned}$$

Since this is a function with constant slope λ, the claim (2.18) follows by uniqueness of the conjugacy.

Combining (2.18) with Theorem 11, we obtain $\psi_{\bar{P}}(-x) = -\psi_{\bar{P}}(x)$. Because $C_k = (k-2g)\alpha$, Theorem 13 implies

$$\psi_{\bar{P}}(C_k + x) = C_k + \psi_{\bar{P}}(x)$$

for all k. Therefore we compute that

$$\psi_{\bar{P}}(C_k + x) + \psi_{\bar{P}}(C_k - x) = C_k + \psi_{\bar{P}}(x) + C_k + \psi_{\bar{P}}(-x) = 2C_k. \qquad \square$$

A crucial observation for the proof of Theorem 1 is that $\psi_{\bar{P}}$ conjugates each T_k as a function on the circle $\mathbb{S}$, and the resulting function

$$S_k := \psi_{\bar{P}} \circ T_k \circ \psi_{\bar{P}}^{-1}$$

consists of two linear pieces, one with slope λ and the other with slope λ^{-1}. See Figure 2.5, where

$$P'_k := \psi_{\bar{P}}(P_k) \quad \text{and} \quad Q'_k := \psi_{\bar{P}}(Q_k).$$

Lemma 15. *The function $S_k : \mathbb{S} \to \mathbb{S}$ can be fully described as follows:*

(a) S_k is linear on $[P'_k, Q'_{k+1}]$ with slope λ;
(b) S_k is linear on $[Q'_{k+1}, P'_k]$ with slope λ^{-1}.

Proof. (a) By construction, S_k is linear on $[P'_k, P'_{k+1})$ since $[P_k, P_{k+1})$ is the interval where $f_{\bar{P}}$ acts as T_k. Given the central symmetry of T_k (Proposition 4) and $\psi_{\bar{P}}$ (Proposition 14) around C_k, the composition $S_k = \psi_{\bar{P}} \circ T_k \circ \psi_{\bar{P}}^{-1}$ must also be symmetric around C_k. The image of $[P'_k, P'_{k+1}]$ under the symmetry $C_k + x \mapsto C_k - x$ is $[Q'_k, Q'_{k+1}]$, and thus S_k is linear on $[Q'_k, Q'_{k+1}]$ with the same slope. Since the

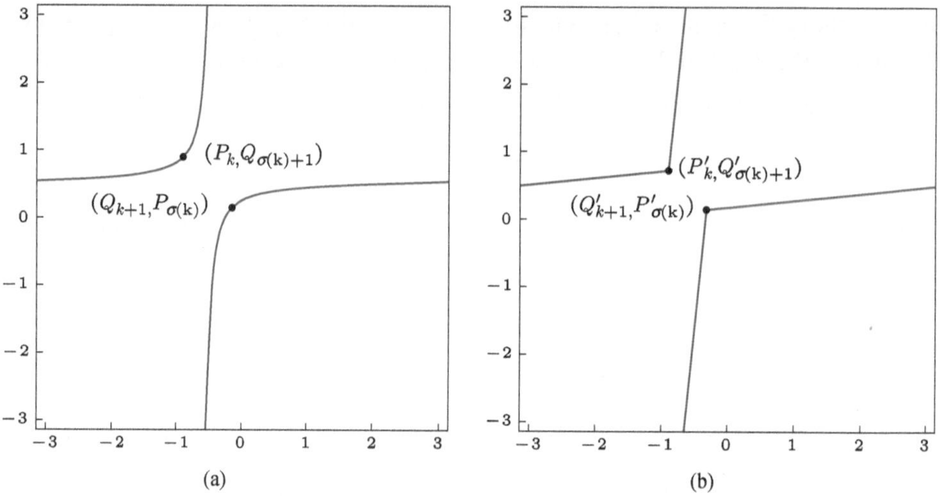

Figure 2.5 Plots of $T_k(x)$ (a) and $S_k(x)$ (b) with $k = 3$ and $g = 2$.

intervals of linearity $[P'_k, Q'_k]$ and $[Q'_k, Q'_{k+1}]$ overlap, there is no jump within their union, which is $[P'_k, Q'_{k+1}]$. We can in fact calculate

$$S_k(P'_k) = \psi_{\bar{P}}(T_k(P_k)) = \psi_{\bar{P}}(Q_{\sigma(k)+1}) = Q'_{\sigma(k)+1}$$
$$S_k(Q'_{k+1}) = \psi_{\bar{P}}(T_k(Q_{k+1})) = \psi_{\bar{P}}(P_{\sigma(k)}) = P'_{\sigma(k)}$$

directly using [15, Proposition 2.2].

(b) Because part (a) holds for all k, we know $S_{\sigma(k)}$ maps $[P'_{\sigma(k)}, Q'_{\sigma(k)+1}]$ linearly to $[Q'_{\sigma(\sigma(k))+1}, P'_{\sigma(\sigma(k))}] = [Q'_{k+1}, P'_k]$ with slope λ, and therefore $S^{-1}_{\sigma(k)}$ maps $[Q'_{k+1}, P'_k]$ linearly to $[P'_{\sigma(k)}, Q'_{\sigma(k)+1}]$ with slope $1/\lambda$. But

$$S^{-1}_{\sigma(k)} = (\psi_{\bar{P}} \circ T_{\sigma(k)} \circ \psi^{-1}_{\bar{P}})^{-1} = \psi_{\bar{P}} \circ T^{-1}_{\sigma(k)} \circ \psi^{-1}_{\bar{P}} = \psi_{\bar{P}} \circ T_k \circ \psi^{-1}_{\bar{P}}$$

is exactly S_k. □

2.5 Proof of Theorem 1

We can now prove the rigidity of topological entropy, that is, $h_{\text{top}}(f_{\bar{A}})$ is the same for all parameters $\bar{A}$ and for all fundamental polygons $\mathcal{F}$.

Regular Polygon

First, we prove Theorem 1 in the case where $f_{\bar{A}}$ is associated to a regular $(8g-4)$-gon. Let $\bar{A} = \{A_1, \ldots, A_{8g-4}\}$ consist of any points satisfying $A_k \in [P_k, Q_k]$. Because $S_k = \psi_{\bar{P}} \circ T_k \circ \psi^{-1}_{\bar{P}}$ is linear on all of $[P'_k, Q'_{k+1}]$ with slope λ by Theorem 15(1), the function $\psi_{\bar{P}} \circ f_{\bar{A}} \circ \psi_{\bar{P}}^{-1}$ (note the use of $f_{\bar{A}}$ with $\psi_{\bar{P}}$) is piecewise affine with constant slope λ, and so, by [18, Theorem 3′] applied to such maps, the topological entropy of $f_{\bar{A}}$ is $\log \lambda$.

Teichmüller Space

As explained in [3, Introduction], the Teichmüller space of a compact surface of genus g may be viewed as the space of marked $(8g-4)$-fundamental polygons, and the partitions of the boundary $\mathbb{S}$ for various polygons are related via a homeomorphism of $\mathbb{S}$ by the Fenchel–Nielsen Theorem.

Let $\widetilde{\Gamma}$ be a Fuchsian group such that $\widetilde{\Gamma}\backslash\mathbb{D}$ is a compact surface of genus g whose fundamental $(8g-4)$-gon $\widetilde{\mathcal{F}}$ is not regular. As explained in [16], there is a Fuchsian group Γ having a regular fundamental $(8g-4)$-gon $\mathcal{F}$ and an orientation-preserving homeomorphism $h\colon \overline{\mathbb{D}} \to \overline{\mathbb{D}}$ such that $\widetilde{\Gamma} = h \circ \Gamma \circ h^{-1}$. Side k of $\widetilde{\mathcal{F}}$ extends to a geodesic $\widetilde{P}_k\widetilde{Q}_{k+1}$ and is glued to side $\sigma(k)$ by the map $\widetilde{T}_k = h \circ T_k \circ h^{-1}$, where $\{T_k\}$ are generators of Γ identifying the sides of $\mathcal{F}$.

For any $\widetilde{A} = \{\widetilde{A}_1, ..., \widetilde{A}_{8g-4}\}$ with $\widetilde{A}_k \in [\widetilde{P}_k, \widetilde{Q}_k]$, we define

$$\widetilde{f}_{\widetilde{A}}(x) := \widetilde{T}_k(x) \quad \text{if } x \in [\widetilde{A}_k, \widetilde{A}_{k+1}).$$

Then the map $f_{\bar{A}}$ with $\bar{A} = \{h^{-1}(\widetilde{A}_1), ..., h^{-1}(\widetilde{A}_{8g-4})\}$ is associated to the regular fundamental polygon, and (correcting a typo in [16])

$$\widetilde{f}_{\widetilde{A}} = h \circ f_{\bar{A}} \circ h^{-1}.$$

Since $\widetilde{f}_{\widetilde{A}}$ is conjugate to $f_{\bar{A}}$, we conclude that $h_{\text{top}}(\widetilde{f}_{\widetilde{A}}) = h_{\text{top}}(f_{\bar{A}}) = \log \lambda$, and this completes the proof of the main theorem. (In fact, the map $\psi \circ h^{-1}$ with $\psi = \psi_{\bar{P}}$ from Section 2.4 will conjugate $\widetilde{f}_{\widetilde{A}}$ to a map of constant slope λ.)

Appendix A Proof of Theorem 12

We now prove Theorem 12, that is, that each cylinder interval $I_{\bar{P}}^{\omega}$ can be written as unions of cylinder intervals $I_{\bar{Q}}^{\eta}$ (see Figure A1 for the decompositions of $I_{\bar{P}}^{(1,16)}$ and $I_{\bar{P}}^{(1,17)}$). For the remainder of the section, we use the term "cylinder" (specifically, "$\bar{P}$-cylinder" and "$\bar{Q}$-cylinder") instead of "cylinder interval" for brevity.

From (2.14), we derive a common recursive description of cylinders:

$$I_{\bar{A}}^{(\omega_0,\omega_1,\ldots,\omega_n)} = I_{\omega_0} \cap f_{\bar{A}}^{-1}(I_{\bar{A}}^{(\omega_1,\ldots,\omega_n)}).$$

For our particular boundary maps, we have an alternative recursive relation: using the fact that each T_k is bijective on all of $\mathbb{S}$, we can compute

$$I_{\bar{P}}^{(\omega_0,\omega_1,\ldots,\omega_n)} = T_{\lceil \omega_0/2\rceil}^{-1}(I_{\bar{P}}^{(\omega_1,\ldots,\omega_n)})$$

for $\bar{P}$-admissible ω and

$$I_{\bar{Q}}^{(\omega_0,\omega_1,\ldots,\omega_n)} = T_{\lfloor \omega_0/2\rfloor}^{-1}(I_{\bar{Q}}^{(\omega_1,\ldots,\omega_n)})$$

for $\bar{Q}$-admissible ω without the need for an intersection. This is because $T_{\lceil \omega_0/2\rceil}^{-1}$ is contracting on $I_{\bar{P}}^{(\omega_1,\ldots,\omega_n)}$ and therefore $T_{\lceil \omega_0/2\rceil}^{-1}(I_{\bar{P}}^{(\omega_1,\ldots,\omega_n)})$ is already contained in I_{ω_0}. Note the use of ceiling $\lceil\ \rceil$ for $\bar{P}$ and floor $\lfloor\ \rfloor$ for $\bar{Q}$, owing to the fact that $f_{\bar{P}}$ acts by T_k on I_{2k-1} and I_{2k} while $f_{\bar{Q}}$ acts by T_k on I_{2k} and I_{2k+1}, and $k = \lceil \frac{2k-1}{2} \rceil = \lceil \frac{2k}{2} \rceil = \lfloor \frac{2k}{2} \rfloor = \lfloor \frac{2k+1}{2} \rfloor$. The preceding formulas can be extended recursively to

$$I_{\bar{P}}^{\omega} = T_{\lceil \omega_0/2\rceil}^{-1} \circ T_{\lceil \omega_1/2\rceil}^{-1} \circ \cdots \circ T_{\lceil \omega_{n-1}/2\rceil}^{-1}(I_{\omega_n}) \quad \text{if } \omega \text{ is } \bar{P}\text{-admissible},$$
$$I_{\bar{Q}}^{\omega} = T_{\lfloor \omega_0/2\rfloor}^{-1} \circ T_{\lfloor \omega_1/2\rfloor}^{-1} \circ \cdots \circ T_{\lfloor \omega_{n-1}/2\rfloor}^{-1}(I_{\omega_n}) \quad \text{if } \omega \text{ is } \bar{Q}\text{-admissible}.$$

Equations (2.5), (2.6), and (2.7) can be interpreted as statements about admissible pairs of symbols:

- In a $\bar{P}$-admissible sequence, an odd symbol $2k-1$ can only be followed by $2\sigma(k)+2$ or $2\sigma(k)+3$.
- In a $\bar{Q}$-admissible sequence, an odd symbol $2k-1$ can only be followed by $2\sigma(k)+8g-5$ or $2\sigma(k)+8g-4$.
- In a $\bar{P}$- or $\bar{Q}$-admissible sequence, an even symbol $2k$ can only be followed by a symbol from $\{2\sigma(k)+4, 2\sigma(k)+5, \ldots, 2\sigma(k)-4\}$.

In the final item above, and in Theorem 16 below, recall that these values are mod $16g-8$; see the explanation after (2.7) on page 27. The next lemma expands on the admissible pairs presented earlier and lists some longer admissible words used explicitly in the proof of Theorem 12.

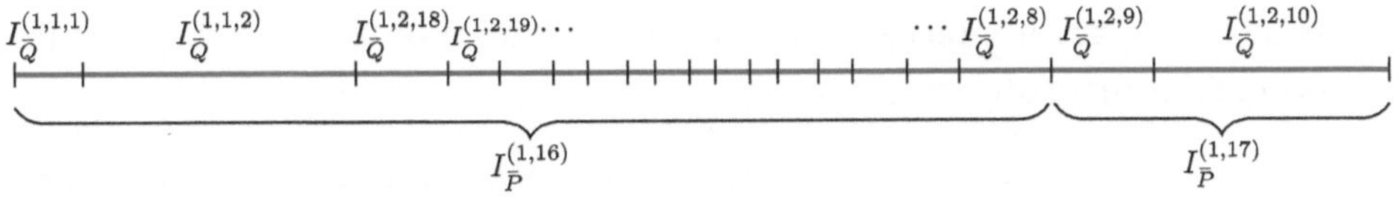

Figure A1 For $g=2$, $I_{\bar{P}}^{(1,16)}$ as a union of 17 $\bar{Q}$-cylinders (left) and $I_{\bar{P}}^{(1,17)}$ as a union of two $\bar{Q}$-cylinders (right).

Lemma 16.

(a) For all k, if $\ell \in \{2k+8g, 2k+8g+1, ..., 2k+8g-8\}$ *then* $(2k-1, 2\sigma(k)+8g-4, \ell)$ *is* $\bar{Q}$*-admissible.*
(b) For all k, if $\ell \in \{2\sigma(k)+4,\ 2\sigma(k)+5,\ ...,\ 2\sigma(k)-4\}$ *then* $(2k-1, 2\sigma(k)+8g-5, 2k, \ell)$ *is* $\bar{Q}$*-admissible.*
(c) For all k, if $\ell \in \{2k+8g-9, 2k+8g-8\}$ *then* $(2k-1, 2\sigma(k)+8g-5, 2k-1, 2\sigma(k)+8g-4, \ell)$ *is* $\bar{Q}$*-admissible.*

The proof of Lemma 16 consists of careful analysis of the transition matrix $M_{\bar{Q}}$ along with the useful identities

$$\sigma(k-1) = \sigma(k) - 4g + 3 \qquad \text{and} \qquad \sigma(k-2) = \sigma(k) + 2,$$

which follow by direct verification (see also [2, Lemma 3.2]).

The following two lemmas establish some relations among the generators $\{T_k\}$ which will be used in the proof of Theorem 12. We omit the composition notation (writing, e.g., $T_k^{-1} T_{\sigma(k)+1}^{-1}$, instead of $T_k^{-1} \circ T_{\sigma(k)+1}^{-1}$).

Lemma 17 ([15, Lemma 3.2]). $T_k^{-1} T_{\sigma(k)+1}^{-1} = T_{k-1}^{-1} T_{\sigma(k)+4g-2}^{-1}$.

Lemma 18. *For* $m \geq 1$, $T_k^{-1}\left(T_{\sigma(k)+1}^{-1} T_{k+4g-1}^{-1}\right)^m = \left(T_{k-1}^{-1} T_{\sigma(k)+4g-3}^{-1}\right)^m T_k^{-1}$.

Proof. The base case, $m = 1$, is proven using Lemma 17 twice, the second time for index $\sigma(k) + 4g - 2$:

$$T_k^{-1} T_{\sigma(k)+1}^{-1} T_{k+4g-1}^{-1} = T_{k-1}^{-1} T_{\sigma(k)+4g-2}^{-1} T_{k+4g-1}^{-1} = T_{k-1}^{-1} T_{\sigma(k)+4g-3}^{-1} T_k^{-1}.$$

Then $m > 1$ follows by induction:

$$\begin{aligned} T_k^{-1}\left(T_{\sigma(k)+1}^{-1} T_{k+4g-1}^{-1}\right)^m &= T_k^{-1}\left(T_{\sigma(k)+1}^{-1} T_{k+4g-1}^{-1}\right)^{m-1} T_{\sigma(k)+1}^{-1} T_{k+4g-1}^{-1} \\ &= \left(T_{k-1}^{-1} T_{\sigma(k)+4g-3}^{-1}\right)^{m-1} T_k^{-1} T_{\sigma(k)+1}^{-1} T_{k+4g-1}^{-1} \\ &= \left(T_{k-1}^{-1} T_{\sigma(k)+4g-3}^{-1}\right)^{m-1} T_{k-1}^{-1} T_{\sigma(k)+4g-3}^{-1} T_k^{-1} \\ &= \left(T_{k-1}^{-1} T_{\sigma(k)+4g-3}^{-1}\right)^m T_k^{-1}. \end{aligned}$$

□

We are now ready to proceed with an inductive proof of Theorem 12, with the following refinement of part (ii):

(a) If all ω_k are even, or if (ω_{n-1}, ω_n) are even but not of the form $(2m, 2\sigma(m)+4)$ for any m, then $I_{\bar{P}}^{\omega} = I_{\bar{Q}}^{(\eta_0, \ldots, \eta_n)}$ with η_n even, and therefore

$$I_{\bar{P}}^{\omega} = \bigcup_{i=0}^{16g-14} I_{\bar{Q}}^{(\eta_0, \ldots, \eta_n, 2\sigma(\eta_n/2)+4+i)}.$$

(b) If ω_n is even and either ω_{n-1} is odd or $(\omega_{n-1}, \omega_n) = (2m, 2\sigma(m)+4)$ for some m (but not all ω_i are even), then η_n is odd and

$$I_{\bar{P}}^{\omega} = I_{\bar{Q}}^{(\eta_0, \ldots, \eta_n, \eta_{n+1})} \cup I_{\bar{Q}}^{(\eta_0, \ldots, \eta_n, \eta_{n+1}+1)} \cup \bigcup_{i=0}^{16g-16} I_{\bar{Q}}^{(\eta_0, \ldots, \eta_n+1, 2\sigma(\frac{\eta_n+1}{2})+6+i)},$$

where $\eta_{n+1} = 2\sigma(\lfloor \eta_n/2 \rfloor) + 8g - 5$.

We begin with the base case $n=0$ for all parts. The original Markov partition sets I_i are both $\bar{P}$- and $\bar{Q}$-cylinders:

$$I_{\bar{P}}^{(\omega_0)} = I_{\omega_0} = I_{\bar{Q}}^{(\omega_0)}.$$

For $\omega_0 = 2k-1$ odd,

$$I_{\bar{P}}^{\omega} = I_{\bar{Q}}^{(2k-1)} = I_{\bar{Q}}^{(2k-1, 2\sigma(k)+8g-5)} \cup I_{\bar{Q}}^{(2k-1, 2\sigma(k)+8g-4)},$$

and for $\omega_0 = 2k$ even,

$$I_{\bar{P}}^{\omega} = I_{\bar{Q}}^{(2k)} = I_{\bar{Q}}^{(2k, 2\sigma(k)+4)} \cup I_{\bar{Q}}^{(2k, 2\sigma(k)+5)} \cup \cdots \cup I_{\bar{Q}}^{(2k, 2\sigma(k)-5)} \cup I_{\bar{Q}}^{(2k, 2\sigma(k)-4)}$$

by (2.6) and (2.7) with $f_{\bar{A}} = f_{\bar{Q}}$.

For some parts of the proof, $n=0$ is a sufficient base case, but we do at times implicitly assume $n \geq 1$, so we also provide here a "base case" with $n=1$. If ω_0 is even, then

$$I_{\bar{P}}^{(\omega_0, \omega_1)} = I_{\bar{Q}}^{(\omega_0, \omega_1)},$$

and Equations (2.15) and (2.16) follow immediately when ω_1 is odd, or, respectively, even. If $\omega_0 = 2k-1$ is odd, then ω_1 can be either $2\sigma(k)+2$ or $2\sigma(k)+3$. We investigate the interval $I_{\bar{P}}^{(2k-1, 2\sigma(k)+3)}$. For that, notice that $I_{\bar{P}}^{(2k-1, 2\sigma(k)+3)} = T_k^{-1}(I_{2\sigma(k)+3})$. From relation (2.6) written for index $\sigma(k)+2$, the interval $I_{2\sigma(k)+3}$ itself can be expressed as

$$\begin{aligned} I_{2\sigma(k)+3} &= T^{-1}_{\sigma(k)+1}([P_{\sigma(\sigma(k)+1))-1}, P_{\sigma(\sigma(k)+1))}]) = T^{-1}_{\sigma(k)+1}[P_{(k-2)+4g-2}, P_{(k-2)+4g-1}] \\ &= T^{-1}_{\sigma(k)+1}(I_{2(k-2)+8g-5} \cup I_{2(k-2)+8g-4}) = T^{-1}_{\sigma(k)+1}(I_{2k+8g-9} \cup I_{2k+8g-8}). \end{aligned}$$

Now we use Lemma 17 to write

$$\begin{aligned} I_{\bar{P}}^{(2k-1, 2\sigma(k)+3)} &= T_k^{-1} T^{-1}_{\sigma(k)+1}(I_{2k+8g-9} \cup I_{2k+8g-8}) \\ &= T^{-1}_{k-1} T^{-1}_{\sigma(k-1)-1}(I_{2k+8g-9} \cup I_{2k+8g-8}) \\ &= T^{-1}_{k-1} T^{-1}_{\sigma(k)+4g-2}(I_{2k+8g-9} \cup I_{2k+8g-8}) \\ &= T^{-1}_{k-1}\big(I_{\bar{Q}}^{(2\sigma(k)+8g-4, 2k+8g-9)} \cup I_{\bar{Q}}^{(2\sigma(k)+8g-4, 2k+8g-8)}\big) \\ &= I_{\bar{Q}}^{(2k-1, 2\sigma(k)+8g-4, 2k+8g-9)} \cup I_{\bar{Q}}^{(2k-1, 2\sigma(k)+8g-4, 2k+8g-8)}, \end{aligned}$$

which proves (2.15), that is, part (i), for $n = 1$.

The other $\bar{P}$-cylinder interval $I_{\bar{P}}^{(2k-1, 2\sigma(k)+2)} = I_{\bar{P}}^{(2k-1)} \setminus I_{\bar{P}}^{(2k-1, 2\sigma(k)+3)}$. Since

$$\begin{aligned} I_{\bar{P}}^{(2k-1)} = I_{\bar{Q}}^{(2k-1)} &= I_{\bar{Q}}^{(2k-1, 2\sigma(k)+8g-5)} \cup I_{\bar{Q}}^{(2k-1, 2\sigma(k)+8g-4)} \\ &= I_{\bar{Q}}^{(2k-1, 2\sigma(k)+8g-5, 2k-1)} \cup I_{\bar{Q}}^{(2k-1, 2\sigma(k)+8g-5, 2k)} \\ &\quad \cup \bigcup_{\ell=2k+8g}^{2k+8g-8} I_{\bar{Q}}^{(2k-1, 2\sigma(k)+8g-4, \ell)} \end{aligned}$$

by Theorem 16(a) and, rewriting $I_{\bar{P}}^{(2k-1, 2\sigma(k)+3)}$ as the union of two $\bar{Q}$-cylinders in the preceding equations, we have

$$\begin{aligned} I_{\bar{P}}^{(2k-1, 2\sigma(k)+2)} &= I_{\bar{Q}}^{(2k-1, 2\sigma(k)+8g-5, 2k-1)} \cup I_{\bar{Q}}^{(2k-1, 2\sigma(k)+8g-5, 2k)} \\ &\quad \cup \bigcup_{\ell=2k+8g}^{2k+8g-10} I_{\bar{Q}}^{(2k-1, 2\sigma(k)+8g-4, \ell)}, \end{aligned}$$

proving (2.17) for $n = 1$.

We proceed now with induction for $n \geq 2$. We say that a cylinder $I_{\bar{A}}^{(\omega_0, \ldots, \omega_n)}$ has *rank* $n+1$. Assume $I_{\bar{P}}^{\omega}$ of rank $\leq n$ is a union of $\bar{Q}$-cylinders as desired; we want $I_{\bar{P}}^{(\omega_0, \ldots, \omega_n)}$ to be a union of $\bar{Q}$-cylinders of rank $n+2$.

When ω_0 is even, the induction argument is straightforward for both parts. We demonstrate it for part (i), that is, when ω_n is odd. Using the induction hypothesis for $I_{\bar{P}}^{(\omega_1, \ldots, \omega_n)}$, we have

$$
\begin{aligned}
I_{\bar{P}}^{(\omega_0,\omega_1,\ldots,\omega_n)} &= T_{\omega_0/2}^{-1}\big(I_{\bar{P}}^{(\omega_1,\ldots,\omega_n)}\big)\\
&= T_{\omega_0/2}^{-1}\big(I_{\bar{Q}}^{(\eta_1,\ldots,\eta_n,\eta_{n+1})} \cup I_{\bar{Q}}^{(\eta_1,\ldots,\eta_n,\eta_{n+1}+1)}\big) \text{ by induction}\\
&= T_{\omega_0/2}^{-1} I_{\bar{Q}}^{(\eta_1,\ldots,\eta_n,\eta_{n+1})} \cup T_{\omega_0/2}^{-1} I_{\bar{Q}}^{(\eta_1,\ldots,\eta_n,\eta_{n+1}+1)}\\
&= I_{\bar{Q}}^{(\omega_0,\eta_1,\ldots,\eta_n,\eta_{n+1})} \cup I_{\bar{Q}}^{(\omega_0,\eta_1,\ldots,\eta_n,\eta_{n+1}+1)},
\end{aligned}
$$

where the final substitution uses the fact that $\eta_1 = \omega_1$ (from induction) and that the pair $(\omega_0,\eta_1) = (\omega_0,\omega_1)$ is $\bar{P}$-admissible if and only if it is $\bar{Q}$-admissible (because ω_0 is even, and the even rows of $M_{\bar{P}}$ and $M_{\bar{Q}}$ are identical). Part (ii) can be treated similarly.

We now prove parts (i) and (ii) separately when ω_0 is odd.

(i) From the induction hypothesis,

$$
I_{\bar{P}}^{(\omega_1,\omega_2,\ldots,\omega_n)} = I_{\bar{Q}}^{(\xi_1,\xi_2,\ldots,\xi_n,\xi_{n+1})} \cup I_{\bar{Q}}^{(\xi_1,\xi_2,\ldots,\xi_n,\xi_{n+1}+1)}
$$

with $\xi_1 = \omega_1$ and ξ_{n+1} odd. (We use ξ here instead of η because the terms ξ_i will not necessarily be η_i for $I_{\bar{P}}^{(\omega_0,\ldots,\omega_n)}$ from the statement of Theorem 12.) Thus

$$
\begin{aligned}
I_{\bar{P}}^{\omega} &= T_{\lceil\omega_0/2\rceil}^{-1}\big(I_{\bar{P}}^{(\omega_1,\omega_2,\omega_3,\ldots,\omega_n)}\big)\\
&= T_{\lceil\omega_0/2\rceil}^{-1}\Big(I_{\bar{Q}}^{(\omega_1,\xi_2,\xi_3,\ldots,\xi_{n+1})} \cup I_{\bar{Q}}^{(\omega_1,\xi_2,\xi_3,\ldots,\xi_{n+1}+1)}\Big)\\
&= T_{\lceil\omega_0/2\rceil}^{-1} T_{\lfloor\omega_1/2\rfloor}^{-1}\Big(I_{\bar{Q}}^{(\xi_2,\xi_3,\ldots,\xi_{n+1})} \cup I_{\bar{Q}}^{(\xi_2,\xi_3,\ldots,\xi_{n+1}+1)}\Big).
\end{aligned}
$$

Let $\omega_0 = 2k-1$. Then ω_1 must be $2\sigma(k)+2$ or $2\sigma(k)+3$, and either way $\lfloor\omega_1/2\rfloor = \sigma(k)+1$, giving

$$
I_{\bar{P}}^{\omega} = T_k^{-1} T_{\sigma(k)+1}^{-1}\Big(I_{\bar{Q}}^{(\xi_2,\xi_3,\ldots,\xi_{n+1})} \cup I_{\bar{Q}}^{(\xi_2,\xi_3,\ldots,\xi_{n+1}+1)}\Big). \tag{2.19}
$$

There are now several cases and sub-cases to consider; these are summarized in Figure A2.

If $\xi_1 = 2\sigma(k)+2$, then $\xi_2 \in \{2k+8g-2, 2k+8g-1, \ldots, 2k+8g-10\}$, and if $\xi_1 = 2\sigma(k)+3$, then ξ_2 is $2k+8g-9$ or $2k+8g-8$. Other than when $\xi_2 \in \{2k+8g-2, 2k+8g-1\}$, we can apply Theorem 17 to (2.19) to get

$$
I_{\bar{P}}^{\omega} = T_{k-1}^{-1} T_{\sigma(k)+4g-2}^{-1}\Big(I_{\bar{Q}}^{(\xi_2,\xi_3,\ldots,\xi_{n+1})} \cup I_{\bar{Q}}^{(\xi_2,\xi_3,\ldots,\xi_{n+1}+1)}\Big),
$$

and then Theorem 16(a) implies

$$
\begin{aligned}
I_{\bar{P}}^{\omega} &= T_{\left\lfloor\frac{2k-1}{2}\right\rfloor}^{-1} T_{\left\lfloor\frac{2\sigma(k)+8g-4}{2}\right\rfloor}^{-1}\Big(I_{\bar{Q}}^{(\xi_2,\xi_3,\ldots,\xi_{n+1})} \cup I_{\bar{Q}}^{(\xi_2,\xi_3,\ldots,\xi_{n+1}+1)}\Big)\\
&= I_{\bar{Q}}^{(2k-1,2\sigma(k)+8g-4,\xi_2,\xi_3,\ldots,\xi_{n+1})} \cup I_{\bar{Q}}^{(2k-1,2\sigma(k)+8g-4,\xi_2,\xi_3,\ldots,\xi_{n+1}+1)}.
\end{aligned}
$$

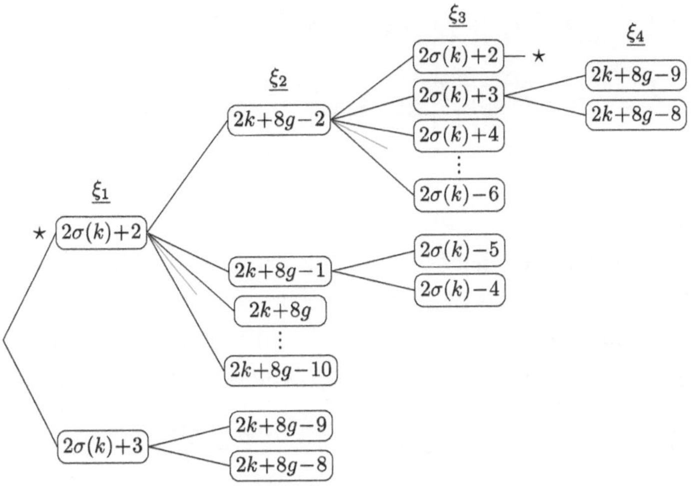

Figure A2 Relevant cases when $\omega_0 = 2k - 1$ is odd (all indices mod $16g - 8$).

We are left with analyzing the cases $\xi_2 = 2k + 8g - 2$ and $\xi_2 = 2k + 8g - 1$, with $\xi_1 = 2\sigma(k) + 2$. Here $\lfloor \xi_2/2 \rfloor = k + 4g - 1$ and so we proceed from (2.19) as

$$\begin{aligned} I_{\bar{P}}^{\omega} &= T_k^{-1} T_{\sigma(k)+1}^{-1} T_{k+4g-1}^{-1} \left(I_{\bar{Q}}^{(\xi_3,\dots,\xi_{n+1})} \cup I_{\bar{Q}}^{(\xi_3,\dots,\xi_{n+1}+1)} \right) \\ &= T_{k-1}^{-1} T_{\sigma(k)+4g-3}^{-1} T_k^{-1} \left(I_{\bar{Q}}^{(\xi_3,\dots,\xi_{n+1})} \cup I_{\bar{Q}}^{(\xi_3,\dots,\xi_{n+1}+1)} \right), \end{aligned}$$

using Lemma 18 with $m = 1$. If $\xi_2 = 2k + 8g - 2$, then $\xi_3 \in \{2\sigma(k) + 2, 2\sigma(k) + 3, \dots, 2\sigma(k) - 6\}$, and if $\xi_2 = 2k + 8g - 1$, then ξ_3 is $2\sigma(k) - 5$ or $2\sigma(k) - 4$. For all possible pairs (ξ_2, ξ_3) *except* $(\xi_2, \xi_3) = (2k + 8g - 2, 2\sigma(k) + 2)$ and $(\xi_2, \xi_3) = (2k + 8g - 2, 2\sigma(k) + 3)$, Lemma 16(2) implies precisely that

$$\begin{aligned} I_{\bar{P}}^{\omega} &= T_{k-1}^{-1} T_{\sigma(k)+4g-3}^{-1} T_k^{-1} \left(I_{\bar{Q}}^{(\xi_3,\dots,\xi_{n+1})} \cup I_{\bar{Q}}^{(\xi_3,\dots,\xi_{n+1}+1)} \right) \\ &= T^{-1}_{\left\lfloor \frac{2k-1}{2} \right\rfloor} T^{-1}_{\left\lfloor \frac{2\sigma(k)+8g-5}{2} \right\rfloor} T^{-1}_{\left\lfloor \frac{2k}{2} \right\rfloor} \left(I_{\bar{Q}}^{(\xi_3,\dots,\xi_{n+1})} \cup I_{\bar{Q}}^{(\xi_3,\dots,\xi_{n+1}+1)} \right) \\ &= I_{\bar{Q}}^{(2k-1,\, 2\sigma(k)+8g-5,\, 2k,\, \xi_3,\dots,\xi_{n+1})} \cup I_{\bar{Q}}^{(2k-1,\dots,\xi_{n+1}+1)}. \end{aligned}$$

Now we only need to analyze the cases $\xi_3 = 2\sigma(k) + 2$ and $\xi_3 = 2\sigma(k) + 3$, where we have already set $\xi_2 = 2k + 8g - 2$ and $\xi_1 = 2\sigma(k) + 2$.

If $\xi_3 = 2\sigma(k) + 3$, then ξ_4 is either $2k + 8g - 8$ or $2k + 8g - 9$, and so

$$\begin{aligned} I_{\bar{P}}^{\omega} &= T_{k-1}^{-1} T_{\sigma(k)+4g-3}^{-1} T_k^{-1} \left(I_{\bar{Q}}^{(\xi_3,\dots,\xi_{n+1})} \cup I_{\bar{Q}}^{(\xi_3,\dots,\xi_{n+1}+1)} \right) \\ &= T_{k-1}^{-1} T_{\sigma(k)+4g-3}^{-1} T_k^{-1} T_{\sigma(k)+1}^{-1} \left(I_{\bar{Q}}^{(\xi_4,\dots,\xi_{n+1})} \cup I_{\bar{Q}}^{(\xi_4,\dots,\xi_{n+1}+1)} \right) \end{aligned}$$

$$= T_{k-1}^{-1}T_{\sigma(k)+4g-3}^{-1}T_{k-1}^{-1}T_{\sigma(k)+4g-2}^{-1}\left(I_{\bar{Q}}^{(\xi_4,\dots,\xi_{n+1})} \cup I_{\bar{Q}}^{(\xi_4,\dots,\xi_{n+1}+1)}\right) \quad \text{by Lemma 17}$$
$$= T^{-1}_{\left\lfloor\frac{2k-1}{2}\right\rfloor}T^{-1}_{\left\lfloor\frac{2\sigma(k)+8g-5}{2}\right\rfloor}T^{-1}_{\left\lfloor\frac{2k-1}{2}\right\rfloor}T^{-1}_{\left\lfloor\frac{2\sigma(k)+8g-4}{2}\right\rfloor}\left(I_{\bar{Q}}^{(\xi_4,\dots,\xi_{n+1})} \cup I_{\bar{Q}}^{(\xi_4,\dots,\xi_{n+1}+1)}\right)$$
$$= I_{\bar{Q}}^{(2k-1,2\sigma(k)+8g-5,2k-1,2\sigma(k)+8g-4,\xi_4,\dots,\xi_{n+1})} \cup I_{\bar{Q}}^{(2k-1,\dots,\xi_{n+1}+1)}$$

by Theorem 16(c).

If $\xi_3 = 2\sigma(k)+2$, notice that $\xi_3 = \xi_1$, so we now analyze the situation when the sequence $(\xi_1,\dots,\xi_{n+1})$ consists of several alternating entries $(2\sigma(k)+2, 2k+8g-2)$ until some $\xi_j \notin \{2\sigma(k)+2, 2k+8g-2\}$ (this situation is denoted by $\star$ in Figure A2). Notice that $j < n+1$: otherwise, all $\xi_1,\dots,\xi_n$ would be even, and then

$$I_{\bar{Q}}^{(\xi_1,\xi_2,\dots,\xi_{n-1},\xi_n,\xi_{n+1})} \subset I_{\bar{Q}}^{(\xi_1,\xi_2,\dots,\xi_{n-1},\xi_n)} = I_{\bar{P}}^{(\xi_1,\xi_2,\dots,\xi_{n-1},\xi_n)}$$

would imply $I_{\bar{P}}^{(\xi_1,\xi_2,\dots,\xi_n)} = I_{\bar{P}}^{(\omega_1,\omega_2,\dots,\omega_n)}$, which is not possible since ω_n is odd.

We assume j is odd (the case of even j can be treated similarly). Then

$$(\xi_1,\xi_2,\dots,\xi_{n+1}) = (2\sigma(k)+2, 2k+8g-2,\dots,2\sigma(k)+2,2k+8g-2,\xi_j,\dots,\xi_{n+1}),$$

where ξ_j is one of $\{2\sigma(k)+3,\dots,2\sigma(k)-6\}$. Thus

$$I_{\bar{P}}^{\omega} = T_k^{-1}\left(I_{\bar{Q}}^{(\xi_1,\xi_2,\xi_3,\dots,\xi_{n+1})} \cup I_{\bar{Q}}^{(\xi_1,\dots,\xi_{n+1}+1)}\right)$$
$$= T_k^{-1}\left(T_{\sigma(k)+1}^{-1}T_{k+4g-1}^{-1}\right)^{(j-1)/2}\left(I_{\bar{Q}}^{(\xi_j,\dots,\xi_{n+1})} \cup I_{\bar{Q}}^{(\xi_j,\dots,\xi_{n+1}+1)}\right)$$
$$= \left(T_{k-1}^{-1}T_{\sigma(k)+4g-3}^{-1}\right)^{(j-1)/2}T_k^{-1}\left(I_{\bar{Q}}^{(\xi_j,\dots,\xi_{n+1})} \cup I_{\bar{Q}}^{(\xi_j,\dots,\xi_{n+1}+1)}\right) \tag{2.20}$$

by Lemma 18. For $\xi_j \neq 2\sigma(k)+3$, Lemma 16(2) implies that

$$I_{\bar{P}}^{\omega} = \left(T^{-1}_{\left\lfloor\frac{2k-1}{2}\right\rfloor}T^{-1}_{\left\lfloor\frac{2\sigma(k)+8g-5}{2}\right\rfloor}\right)^{(j-1)/2}T^{-1}_{\left\lfloor\frac{2k}{2}\right\rfloor}\left(I_{\bar{Q}}^{(\xi_j,\dots,\xi_{n+1})} \cup I_{\bar{Q}}^{(\xi_j,\dots,\xi_{n+1}+1)}\right)$$
$$= I_{\bar{Q}}^{(2k-1,2\sigma(k)+8g-5,\dots,2k-1,2\sigma(k)+8g-5,2k,\xi_j,\dots,\xi_{n+1})} \cup I_{\bar{Q}}^{(2k-1,\dots,\xi_{n+1}+1)},$$

and for $\xi_j = 2\sigma(k)+3$ we proceed from (2.20) with

$$I_{\bar{P}}^{\omega} = \left(T_{k-1}^{-1}T_{\sigma(k)+4g-3}^{-1}\right)^{(j-1)/2}T_k^{-1}\left(I_{\bar{Q}}^{(\xi_j,\dots,\xi_{n+1})} \cup I_{\bar{Q}}^{(\xi_j,\dots,\xi_{n+1}+1)}\right)$$
$$= \left(T_{k-1}^{-1}T_{\sigma(k)+4g-3}^{-1}\right)^{(j-1)/2}T_k^{-1}T_{\sigma(k)+1}^{-1}\left(I_{\bar{Q}}^{(\xi_{j+1},\dots,\xi_{n+1})} \cup I_{\bar{Q}}^{(\xi_j,\dots,\xi_{n+1}+1)}\right)$$
$$= \left(T_{k-1}^{-1}T_{\sigma(k)+4g-3}^{-1}\right)^{(j-1)/2}T_{k-1}^{-1}T_{\sigma(k)+4g-2}^{-1}\left(I_{\bar{Q}}^{(\xi_{j+1},\dots,\xi_{n+1})} \cup I_{\bar{Q}}^{(\xi_j,\dots,\xi_{n+1}+1)}\right)$$
$$= I_{\bar{Q}}^{(2k-1,2\sigma(k)+8g-5,\dots,2k-1,2\sigma(k)+8g-4,\xi_{j+1},\dots,\xi_{n+1})} \cup I_{\bar{Q}}^{(2k-1,\dots,\xi_{n+1}+1)},$$

using Lemma 17 for a substitution and then Lemma 16(a) for the final line since, following $\xi_j = 2\sigma(k)+3$, we know ξ_{j+1} is either $2k-8g-9$ or $2k-8g-8$. Having followed all paths in Figure A2, this completes the proof of part (i).

(ii) When ω_n is even, there are two possible structures, Equations (2.16) and (2.17), for the decomposition of $I_{\bar{P}}^{\omega}$, corresponding to the two cases (a) and (b) on page 40.

(a) First, if all ω_i are even, then $I_{\bar{P}}^{\omega} = I_{\bar{Q}}^{\omega}$ because the even rows of $M_{\bar{P}}$ and $M_{\bar{Q}}$ coincide. Then $I_{\bar{P}}^{\omega}$ can be trivially decomposed into $16g-15$ cylinders of higher rank as in (2.16).

If (ω_{n-1}, ω_n) are both even and not of the form $(2m, 2\sigma(m)+4)$ for any m, then, from the induction hypothesis for case (a), there exists a $\bar{Q}$-admissible sequence $(\xi_1, \xi_2, ..., \xi_n)$ such that

$$I_{\bar{P}}^{(\omega_1, \omega_2, ..., \omega_n)} = I_{\bar{Q}}^{(\xi_1, \xi_2, ..., \xi_n)},$$

with ξ_n even. Now the analogue of relation (2.19) is

$$I_{\bar{P}}^{\omega} = T_k^{-1} T_{\sigma(k)+1}^{-1} \big(I_{\bar{Q}}^{(\xi_2, ..., \xi_n)}\big).$$

The sequence $(\xi_1, \xi_2, ..., \xi_n)$ cannot consist entirely of alternating even entries $(2\sigma(k)+2, 2k+8g-2)$: if this were the case, then

$$I_{\bar{Q}}^{(\xi_1, \xi_2, ..., \xi_n)} = I_{\bar{P}}^{(\xi_1, \xi_2, ..., \xi_n)}, \text{ so } (\omega_1, \omega_2, ..., \omega_n) = (\xi_1, \xi_2, ..., \xi_n),$$

which is impossible since the last two entries (ω_{n-1}, ω_n) are not of the form $(2m, 2\sigma(m)+4)$. One can then proceed as in case (i) and express $I_{\bar{P}}^{\omega}$ as a single $\bar{Q}$-cylinder of rank $n+1$, which is then a union of $16g-15$ $\bar{Q}$-cylinders of rank $n+2$ as desired.

(b) If ω_{n-1} is odd or the final pair $(\omega_{n-1}, \omega_n) = (2m, 2\sigma(m)+4)$ for some m, then we have (2.17), as will we now show.

We follow the proof of (1), where a stricter key step $j < n$ will now follow from the new assumptions. Indeed, from the induction hypothesis for case (b),

$$I_{\bar{P}}^{(\omega_1, \omega_2, ..., \omega_n)} = I_{\bar{Q}}^{(\xi_1, ..., \xi_n, \xi_{n+1})} \cup I_{\bar{Q}}^{(\xi_1, ..., \xi_n, \xi_{n+1}+1)} \cup \bigcup_{i=0}^{16g-16} I_{\bar{Q}}^{(\xi_1, ..., \xi_n+1, 2\sigma(\frac{\xi_n+1}{2})+6+i)},$$

where $\xi_1 = \omega_1$, ξ_n is odd, and $\xi_{n+1} = 2\sigma(\lfloor \xi_n/2 \rfloor)+8g-5$. The analogous statement to (2.19) is now

$$I_{\bar{P}}^{\omega} = T_k^{-1} T_{\sigma(k)+1}^{-1} \Big(I_{\bar{Q}}^{(\xi_2, ..., \xi_{n+1})} \cup I_{\bar{Q}}^{(\xi_2, ..., \xi_{n+1}+1)} \cup \bigcup_{i=0}^{16g-16} I_{\bar{Q}}^{(\xi_2, ..., \xi_n+1, 2\sigma(\frac{\xi_n+1}{2})+6+i)}\Big).$$

Notice that it is not possible for the sequence $(\xi_1, \xi_2, ..., \xi_n)$ to consist entirely of alternating entries $(2\sigma(k)+2, 2k+8g-2)$ because ξ_n is odd. Nor can the sequence $(\xi_1, \xi_2, ..., \xi_n + 1)$ consist entirely of such alternating even entries because then (ξ_{n-1}, ξ_n) would not be $\bar{Q}$-admissible.

Therefore, there exists $j < n$ such that the sequence $(\xi_1, ..., \xi_j)$ stops alternating between $2\sigma(k) + 2$ and $2k + 8g - 2$. We can then express each $T_k^{-1} T_{\sigma(k)+1}^{-1} \big(I_{\bar{Q}}^{(\xi_2, ...)}\big)$ in the preceding equation as a $\bar{Q}$-cylinder of rank $n + 2$, thus making $I_{\bar{P}}^{\omega}$ a union of $2 + (16g - 17) = 16g - 15$ $\bar{Q}$-cylinders of rank $n + 2$. This does not affect the last two entries of the $\bar{Q}$-cylinders from the induction hypothesis, so the structure of the decomposition is as needed.

References

[1] A. Abrams. Extremal parameters and their duals for boundary maps associated to Fuchsian groups, *Illinois J. Math.* **65** (2021) No. 1, 153–179.

[2] A. Abrams, S. Katok. Adler and Flatto revisited: cross-sections for geodesic flow on compact surfaces of constant negative curvature. *Stud. Math.* **246** (2019), 167–202.

[3] A. Abrams, S. Katok, I. Ugarcovici. Flexibility of measure-theoretic entropy of boundary maps associated to Fuchsian groups. *Ergod. Theory Dyn. Syst.* **42** (2) 2021, 389–401.

[4] R. Adler, L. Flatto. Geodesic flows, interval maps, and symbolic dynamics. *Bull. Amer. Math. Soc.* **25** (1991), No. 2, 229–334.

[5] R. Adler, A. Konheim, M. McAndrew. Topological entropy. *Trans. Amer. Math. Soc.* **114** (1965), 309–319.

[6] L. Alsedà, J. Llibre, M. Misiurewicz. Combinatorial Dynamics and Entropy in Dimension One, Second Edition, *Advanced Series in Nonlinear Dynamics Volume 5*, World Scientific, River Edge, NJ, 2000.

[7] L. Alsedà, M. Misiurewicz. Semiconjugacy to a map of a constant slope. *Discrete Continuous Dyn. Syst. Ser. B.* **20** (10) 2015, 3403–3413.

[8] R. Bowen. Entropy for group endomorphisms and homogeneous spaces. *Trans. American Math. Society.* **153** (1971), 401–414; erratum, **181** (1973), 509–510.

[9] R. Bowen, C. Series. Markov maps associated with Fuchsian groups. *Inst. Hautes Études Sci. Publ. Math.* No. 50 (1979), 153–170.

[10] M. Denker, G. Keller, M. Urbański. On the uniqueness of equilibrium states for piecewise monotone mappings. *Stud. Math.* **97** (1990), 27–36.

[11] E. Dinaburg. The relation between topological entropy and metric entropy. *Soviet Math. Dokl.* **11** (1970), 13–16.

[12] F. Hofbauer. On intrinsic ergodicity of piecewise monotonic transformations with positive entropy II, *Israel J. Math.* **38** (1981), 107–115.

[13] I. Gelfand. Normierte ringe (German). *Rec. Math. [Mat. Sbornik] N.S.*, **9** (1941), 3–24.

[14] S. Katok. *Fuchsian Groups.* University of Chicago Press, Chicago, 1992.

[15] S. Katok, I. Ugarcovici. Structure of attractors for boundary maps associated to Fuchsian groups. *Geom. Dedicata* **191** (2017), 171–198.

[16] S. Katok, I. Ugarcovici. Correction to: Structure of attractors for boundary maps associated to Fuchsian groups. *Geom. Dedicata* **198** (2019), 189–191.
[17] J. Milnor, W. Thurston. On iterated maps of the interval, in J. C. Alexander, *Dynamical Systems*, Lecture Notes in Mathematics, 1342, Springer, Berlin, 1988, 465–563.
[18] M. Misiurewicz, W. Szlenk. Entropy of piecewise monotone mappings. *Stud. Math.* **67** (1980), 45–63.
[19] M. Misiurewicz, K. Ziemian. Horseshoes and entropy for piecewise continuous piecewise monotone maps, in G. Gyorgi et al. (ed.), *From Phase Transitions to Chaos*, World Scientific Publishing, River Edge, NJ, 1992, 489–500.
[20] W. Parry. Intrinsic Markov chains. *Trans. Amer. Math. Soc.* **112** (1964), 55–66.
[21] W. Parry. Symbolic dynamics and transformations of the unit interval. *Trans. Amer. Math. Soc.* **122** (1966), 368–378.
[22] P. Walters, *An Introduction to Ergodic Theory*, Graduate Texts in Mathematics, **79**, Springer, New York, 1975.

3

Flexibility of Entropies for Piecewise Expanding Unimodal Maps

LLUÍS ALSEDÀ, MICHAŁ MISIUREWICZ, AND RODRIGO A. PÉREZ*

We investigate the *flexibility* of the entropy (topological and metric) for the class of piecewise expanding unimodal maps. We show that the only restrictions for the values of the topological and metric entropies in this class are that both are positive, the topological entropy is at most log 2, and the metric entropy is not larger than the topological entropy.

In order to have a better control on the metric entropy, we work mainly with topologically mixing piecewise expanding skew tent maps, for which there are only two different slopes. For those maps, there is an additional restriction that the topological entropy is larger than $\frac{1}{2}\log 2$.

Moreover, we generalize and give a different interpretation of the Milnor–Thurston formula connecting the topological entropy and the kneading determinant for unimodal maps.

3.1 Introduction

Recently an important program in dynamical systems was initiated by Anatole Katok. It concerns *flexibility*, that is, the idea that for a given class of dynamical systems, dynamical invariants (for instance entropies) can take arbitrary values, subject only to natural restrictions. Various results in this direction were obtained for instance in papers [E, EK, BKRH].

Here we investigate the family of piecewise expanding unimodal maps. While they are not smooth, they are piecewise smooth (in fact, the maps that we consider are piecewise linear). For those maps, by [LaY], there exists an absolutely continuous invariant probability measure. By [LiY], this measure is unique. Therefore we

* This work was partially supported by grant number 426602 from the Simons Foundation to Michał Misiurewicz and by the Spanish grant MTM2017-86795-C3-1-P from the "Agencia Estatal de Investigación" (AEI). Michał Misiurewicz also acknowledges the support and hospitality of the Centre de Recerca Matemàtica, Barcelona, Spain. We are indebted to the anonymous referee for finding a minor error in the first version of the chapter.

can consider its metric entropy (which is also equal to its Lyapunov exponent), as well as the topological entropy of the map. Both entropies are positive, topological entropy is at most $\log 2$, and by the Variational Principle, the metric entropy is not larger than the topological entropy. We will show (Theorem B) that those are the only restrictions for the values of those entropies.

In order to have a better control on the metric entropy, we will work mainly with piecewise expanding skew tent maps, for which there are only two different slopes. In particular, for the topologically mixing expanding skew tent maps we prove a version of a theorem on flexibility of entropies. For those maps, there is the additional restriction that the topological entropy is larger than $\frac{1}{2}\log 2$. Again, it turns out that there are no additional restrictions (Theorem A).

There are two basic things that we have to prove in order to get Theorem A. One is continuity of the density of the absolutely continuous invariant probability measure as a function of a map, and the other one is existence of maps with small metric entropy. For this, we need strong estimates on the density of this measure. Classical methods, initiated in [LaY], using the variation and integral, are difficult and give too weak estimates. In particular, in [BK] continuity of the density as a function of the map is proved only at maps for which the turning point is not periodic.

The problem with this classical approach is that it is difficult to trace the trajectory of a density under the iterates of the Frobenius–Perron operator. This is due to the fact that most points have two preimages and the value of the image of the density at a given point depends on the values (and derivatives, that are usually different) at those preimages. However, for skew tent maps there is an alternative to this procedure. Instead of looking at the whole density, we look only at its jumps (discontinuities). Those jumps propagate along one trajectory of the map, and it is easy to keep track of them.

As the old saying goes, *nihil sub sole novum*,[1] and this method was employed by Ito, Tanaka and Nakada [ITN] over 40 years ago. They obtained a simple formula for the densities of absolutely continuous invariant measures for skew tent maps. Their results are not as widely known as they deserve, probably due to the fact that the term "skew tent map" was not used at that time.[2]

Besides proving our main theorems about flexibility and theorems that lead to them, we make an interesting observation. Namely, the formula of Milnor and Thurston [MT], connecting for unimodal maps the kneading sequence to the topological entropy, can be reinterpreted easily as the fact that for a tent map the sum of reciprocals of derivatives of all iterates of the map at the critical value is zero. It

[1] Eccles. 1:10 (Vulg.).

[2] A search in the MathSciNet suggests that it was used for the first time in [MV].

turns out that this is also true for skew tent maps. We wonder whether this can be translated back into a language involving some entropy-like quantities.

The paper is organized as follows. In Section 3.2 we give the basic definitions. In Section 3.3 we prove that if the positive slope of a mixing expanding skew tent map is large then the density of the absolutely continuous invariant probability measure is close to 1. In Section 3.4 we show that an additional assumption in the preceding section is necessary. In Section 3.5 we prove continuous dependence of this density on the map, while in Section 3.6 we prove continuous dependence of the metric entropy on the map. We do it for a larger class of maps, namely, we do not assume mixing. In Section 3.7 we modify the standard square root construction (basically inverse of the renormalization process) in order to stay in the class of piecewise expanding maps. Then, in Section 3.8 we prove our main theorems, and in Section 3.9 we make the observation we mentioned.

3.2 Definitions

An interval map $f\colon [0,1] \to [0,1]$ is called *piecewise expanding* if there is a finite partition of $[0,1]$ into smaller intervals, and on the closure of each of those smaller intervals f is of class C^2 and $|f'| \geq T$ for some constant $T > 1$. If f is unimodal, this partition can be finer than the partition into pieces of monotonicity (laps).

For a unimodal map $f\colon [0,1] \to [0,1]$ we assume that f is increasing on the left lap and decreasing on the right one. If c is the turning (critical) point for f then the *core* of f is the interval $[f^2(c), f(c)]$.

A *skew tent* map is a unimodal map which is linear (we will use this term in the sense of "affine"; this is a common terminology in real analysis) on each lap (see Figure 3.1). There are three popular models for skew tent maps. In the first one the map f is defined on some interval containing 0 in its interior, 0 is the turning point, and $f(0) = 1$ ([MV]). In the second and third ones f maps $[0,1]$ to itself. In the second model, $f(0) = f(1) = 0$ ([BK]). In the third model, $[0,1]$ is the core of f ([ITN]). We will use the third model. Thus, in particular, we will have $f(c) = 1$ and $f(1) = 0$. The slopes of f will be denoted by s (the left slope) and $-t$ (the right slope). The condition that f maps $[0,1]$ to itself translates to the condition $\frac{1}{s} + \frac{1}{t} \geq 1$.

The two main spaces of skew tent maps that we consider are the space $\mathfrak{Y}$ of all piecewise expanding skew tent maps and its subspace $\mathfrak{X}$ consisting of topologically mixing piecewise expanding skew tent maps. For computations it is good to remember that $s, t > 1$, that $c = (t-1)/t$, and that the fixed point is $t/(t+1)$ (there is a unique fixed point, except when $f(0) = 0$, and then we mean the other fixed point); see again Figure 3.1. In terms of slopes, each map of $\mathfrak{Y}$ is determined by slopes $s, t > 1$, subject to $\frac{1}{s} + \frac{1}{t} \geq 1$.

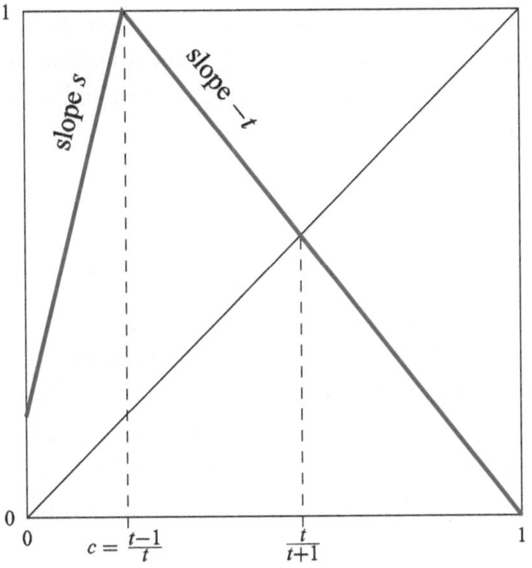

Figure 3.1 A skew tent map.

Lemma 2.1. *For a map $f \in \mathfrak{Y}$ the following properties are equivalent:*

(i) $f \in \mathfrak{X}$,

(ii) $f(0)$ *is to the left of the fixed point of* f,

(iii) $t > \frac{1}{s} + \frac{1}{t}$,

(iv) *topological entropy of f is larger than* $\frac{1}{2}\log 2$.

Proof. By Theorem 4.70 of [Ru], (i) implies (iv). To show that (iv) implies (ii), suppose that $f(0)$ is equal to or to the right of the fixed point. Then every point to the left of the fixed point is mapped to the right of the fixed point, and vice versa. The map f^2 restricted to the interval between 0 and the fixed point is unimodal, so its topological entropy is at most $\log 2$. Thus, the entropy of f is at most $\frac{1}{2}\log 2$.

To show that (ii) implies (i), note that $f(0) = 1 - \frac{s(t-1)}{t} = \frac{s+t-st}{t}$. Thus, by the formula (13) of [ITN] (in that paper our s, t are called a, b), our space $\mathfrak{X}$ is the same as the space D^* from [ITN] (formally, in that paper the authors speak about parameters, while we speak about maps). By Theorem 3.5 of [ITN], any map from that space is topologically exact, so it is topologically mixing. This proves that the properties (i), (ii), and (iv) are equivalent.

To show equivalence of (ii) and (iii), observe that $1 - \frac{s(t-1)}{t} < \frac{t}{t+1}$ is equivalent to $\frac{1}{t+1} < \frac{s(t-1)}{t}$, which is equivalent to $st^2 > t + s$, which is equivalent to (iii). □

We consider the spaces $\mathfrak{Y}$ and $\mathfrak{X}$ with the uniform (sup) topology. Parametrization via the absolute values of the slopes gives the same topology.

By [LaY], piecewise expanding maps have an absolutely continuous invariant probability measure (*acip* in short). By [LiY], if the map is unimodal, this measure is unique. We will use the notation μ for this measure and ϱ for its density. For μ multiplied by a positive constant (that is, an absolutely continuous invariant measure that is finite, but not necessarily normed) we will use the acronym *acim*. Of course, by "absolutely continuous" we mean absolutely continuous with respect to the Lebesgue measure λ.

We will use notation from *kneading theory* (see, e.g., [CE] or [MT]). The *itinerary* of a point x is the sequence $A_0, A_1, \ldots$, where $A_i = L$ if $f^i(x) < c$, $A_i = R$ if $f^i(x) > c$, and $A_i = C$ if $f^i(x) = c$. If there is no symbol C in this sequence, the itinerary is infinite. If there is a C, the sequence terminates at the first C. The central notion is the *kneading sequence* of the map f. There are two versions of it. The simpler one, which we will use in the proof of Lemma 3.3 and in Section 3.4, defines the kneading sequence as the itinerary of 1 (the image of the turning point). The other version, which we will use in all other places, defines the kneading sequence as the limit of the itineraries of x as x goes to 1 from the left. In this version, the symbol C does not appear in the kneading sequence.

3.3 Density Is Close to 1

The following theorem is proved in [ITN] (see Figure 3.2 for some examples of normalized densities obtained in this way).

Theorem 3.1. *For $f \in \mathfrak{Y}$, the function*

$$\widehat{\varrho} = \sum_{k=0}^{\infty} \frac{1}{(f^k)'(0)} \chi_{[f^k(0),1]} \tag{3.1}$$

is the density of an acim for f.

Remark 3.2. We have to explain how to understand the derivatives in (3.1) if 0 is periodic of period $p > 1$, because formally f^k is not differentiable at 0 for $k > p-2$. In [ITN], the authors use the chain rule and pretend that $f'(c) = -t$, so we can use the same convention in (3.1). However, it really does not matter. We have

$$\sum_{k=rp}^{rp+p-1} \frac{1}{(f^k)'(0)} \chi_{[f^k(0),1]} = \frac{1}{(f^{rp})'(0)} \sum_{k=0}^{p-1} \frac{1}{(f^k)'(0)} \chi_{[f^k(0),1]} = \frac{1}{(f^{rp})'(0)} \sum_{k=0}^{p-2} \frac{1}{(f^k)'(0)} \chi_{[f^k(0),1]}.$$

We could remove $\frac{1}{(f^{p-1})'(0)}\chi_{[f^{p-1}(0),1]}$, because $f^{p-1}(0) = 1$ and $[1,1]$ has Lebesgue measure zero. Therefore,

$$\sum_{k=0}^{\infty} \frac{1}{(f^k)'(0)}\chi_{[f^k(0),1]} = \left(\sum_{r=0}^{\infty} \frac{1}{(f^{rp})'(0)}\right) \sum_{k=0}^{p-2} \frac{1}{(f^k)'(0)}\chi_{[f^k(0),1]}.$$

Thus, $\sum_{k=0}^{p-2} \frac{1}{(f^k)'(0)}\chi_{[f^k(0),1]}$ is the density of an acim for f. This expression is well defined, because it does not involve $f'(c)$. Therefore,

$$\left(\sum_{r=0}^{\infty} \frac{1}{(f^{rp})'(0)}\right) \sum_{k=0}^{p-2} \frac{1}{(f^k)'(0)}\chi_{[f^k(0),1]} \tag{3.2}$$

is also an invariant density, *no matter how we define the values of* $(f^{rp})'(0)$. If we use the chain rule, we can define that always $f'(c) = -t$, or that always $f'(c) = s$, but also we can define that sometimes it is $-t$ and sometimes s.

Theoretically, there is a danger that the sum $\sum_{r=0}^{\infty} \frac{1}{(f^{rp})'(0)}$ is zero. However, the ratio of the rth and $(r+1)$st terms is equal to $(f^p)'(0)$ with $f'(c)$ defined as s or $-t$. We will see in the next lemma that this ratio is larger than 2. Consequently, our infinite sum is nonzero.

Lemma 3.3. *Assume that* $f \in \mathfrak{Y}$ *and* 0 *is periodic for* f *of period* $p > 1$. *Then*

$$\prod_{k=0}^{p-1} |f'(f^k(0))| \geq 1 + \min(s, t^2) > 2, \tag{3.3}$$

no matter whether $f'(c)$ *is defined as* s *or* $-t$.

Proof. Assume first that $f \in \mathfrak{X}$. In this case we prove that

$$\prod_{k=0}^{p-1} |f'(f^k(0))| \geq 1 + s > 2. \tag{3.4}$$

If the orbit of 0 includes a point from $(c, 1)$, then the left-hand side of (3.4) is at least st^2, which, by Lemma 2.1, is larger than $t + s$, which is larger than $1 + s$. The remaining possibility is that the kneading sequence of f is $RL^{p-2}C$. Then $f(0) \leq c$, so $s \geq \frac{1-c}{c} = \frac{1}{t-1}$. Therefore, $st \geq 1 + s$, and (3.4) follows.

Let us now consider the general situation of $f \in \mathfrak{Y}$. As we noticed in the proof of Lemma 2.1, if f is not mixing, then $f([0, c]) \subset [c, 1]$ and vice versa. Then f^2 restricted to $[0, c]$ maps $[0, c]$ to itself, and is a piecewise linear unimodal map. Thus, if we restrict it further, to its core I (see Section 3.2), we get a map which is almost a skew tent map, except that its domain is not $[0, 1]$ and that it has a minimum at the turning point. To remedy this, we rescale it by conjugating via the linear map from I onto $[0, 1]$ that reverses orientation. In such a way, we get a new skew tent map g, with slopes t^2 and $-st$. This process is called *renormalization* (see, e.g., [dMvS], Section II.5). If g is not mixing, we renormalize it again. This process has to stop, because the topological entropy of g is twice the topological entropy of f, and any map from $\mathfrak{Y}$ has positive entropy. Thus, by Lemma 2.1, maps from $\mathfrak{Y}$ which are renormalizable n times, but not $n+1$ times, have entropy larger than $\frac{1}{2^{n+1}} \log 2$ and smaller than or equal to $\frac{1}{2^n} \log 2$.

Now, if 0 is a periodic point of $f \in \mathfrak{Y}$ with period p and the nth renormalization of f (call it g) is mixing, then by the chain rule, the product of the derivatives of f along the orbit of 0 is equal to the product of the derivatives of g along the orbit of 0, but the latter one is, as we have already proved, at least 1 plus the left slope of g. If $n = 0$, then this slope is s; if $n = 1$, then it is t^2. If $n > 1$, then it is larger than s. This proves (3.3). □

By [K], $\widehat{\varrho}$ is bounded below by a positive constant on the support of the acim, which is the union of finitely many intervals. For $f \in \mathfrak{X}$, this support is just $[0, 1]$.

Let us estimate the variation of $\widehat{\varrho}$ for $f \in \mathfrak{X}$. Since $f(0)$ is to the left of the fixed point of f, there exists $m \geq 0$ such that the kneading sequence of f is $RLR^mL\ldots$. To see that, note that we allow $m = 0$. The first two symbols of the kneading sequence are always RL, so the only possibility that it does not start with RLR^mL is that it is RLR^∞, but then $f(0)$ is the fixed point. We will denote the space of all elements of $\mathfrak{X}$ with the kneading sequence $RLR^mL\ldots$, where $m \leq n$, by $\mathfrak{X}_n$. Then $\mathfrak{X}$ is the union of the ascending sequence of subsets $\mathfrak{X}_n$.

Lemma 3.4. *If $f \in \mathfrak{X}_n$ and $s \geq t$ then*

$$\operatorname{Var}(\widehat{\varrho}) \leq \frac{n+1}{s} + \frac{2s+1}{s^2}. \tag{3.5}$$

Proof. Since variation of the sum is not larger than the sum of variations, by (3.1) (Theorem 3.1), we have

$$\operatorname{Var}(\widehat{\varrho}) \leq \sum_{k=1}^{\infty} \frac{1}{|(f^k)'(0)|}. \tag{3.6}$$

This series converges because the map f is piecewise expanding.

If $f \in \mathfrak{X}_n$, then we have $|(f^k)'(0)| \geq s$ for $k = 1, 2, \ldots, n+1$ and $|(f^k)'(0)| \geq s^2 t^{k-2}$ for $k > n+1$, so

$$\operatorname{Var}(\widehat{\varrho}) \leq \frac{n+1}{s} + \frac{1}{s^2}\left(1 + \frac{1}{t} + \frac{1}{t^2} + \ldots\right) = \frac{n+1}{s} + \frac{1}{s^2} \cdot \frac{t}{t-1}. \tag{3.7}$$

By Lemma 2.1, $f \in \mathfrak{X}$ is equivalent to $st^2 - s - t \geq 0$. Solving this inequality for t, we get

$$t > \frac{1+\sqrt{1+4s^2}}{2s} \geq 1 + \frac{1}{2s}.$$

Since the function $t \mapsto \frac{t}{t-1}$ is decreasing, we get

$$\frac{t}{t-1} < \frac{1+\frac{1}{2s}}{\frac{1}{2s}} = 2s+1.$$

Therefore, for $f \in \mathfrak{X}_n$ we get the estimate (3.5). □

Now we normalize $\widehat{\varrho}$, that is, we set $\varrho = \widehat{\varrho} / \int \widehat{\varrho}$. Then ϱ is the density of the acip for f. Remember that by [LiY], the acip is unique.

Theorem 3.5. *For every n and $\varepsilon > 0$ there exists $s(\varepsilon, n)$ such that, if $f \in \mathfrak{X}_n$ has slope $s \geq s(\varepsilon, n)$, then $1 - \varepsilon \leq \varrho \leq 1 + \varepsilon$ and* $\mathrm{Var}(\varrho) \leq \varepsilon$.

To illustrate Theorem 3.5 we show in Figure 3.2 four densities for maps from $\mathfrak{X}_0$ with larger and larger slope s.

Proof of Theorem 3.5. Let us start by estimating the value of $\widehat{\varrho}(0)$. If 0 is periodic of period p, then

$$|\widehat{\varrho}(0) - 1| \leq \sum_{i=1}^{\infty} \frac{1}{|(f^{ip})'(0)|} \leq \sum_{i=1}^{\infty} \frac{1}{s^i} = \frac{1}{s-1}. \tag{3.8}$$

If 0 is not periodic, then $\widehat{\varrho}(0) = 1$, so (3.8) also holds.

Now, the existence of the required $s(\varepsilon, n)$ follows from the fact that the limit as $s \to \infty$ is 0 on the right-hand side of both (3.5) (Lemma 3.4) and (3.8). □

3.4 Necessity of Spaces $\mathfrak{X}_n$

We want to justify the assumption that $f \in \mathfrak{X}_n$ in Theorem 3.5. For this we show that, if we move through different classes $\mathfrak{X}_n$, then the limit density can be completely different (see Figure 3.3).

Let us fix $a \in (0, 1/2)$ and take the skew tent map f with $f(0) = a$ and the kneading sequence $RLR^{2n}C$ (then the orbit of the turning point is the Štefan periodic orbit of period $2n+3$). For $k = 0, 1, \ldots, 2n$, set

$$x_k = f^k(a) = \frac{t}{t+1} + \left(a - \frac{t}{t+1}\right)(-t)^k. \tag{3.9}$$

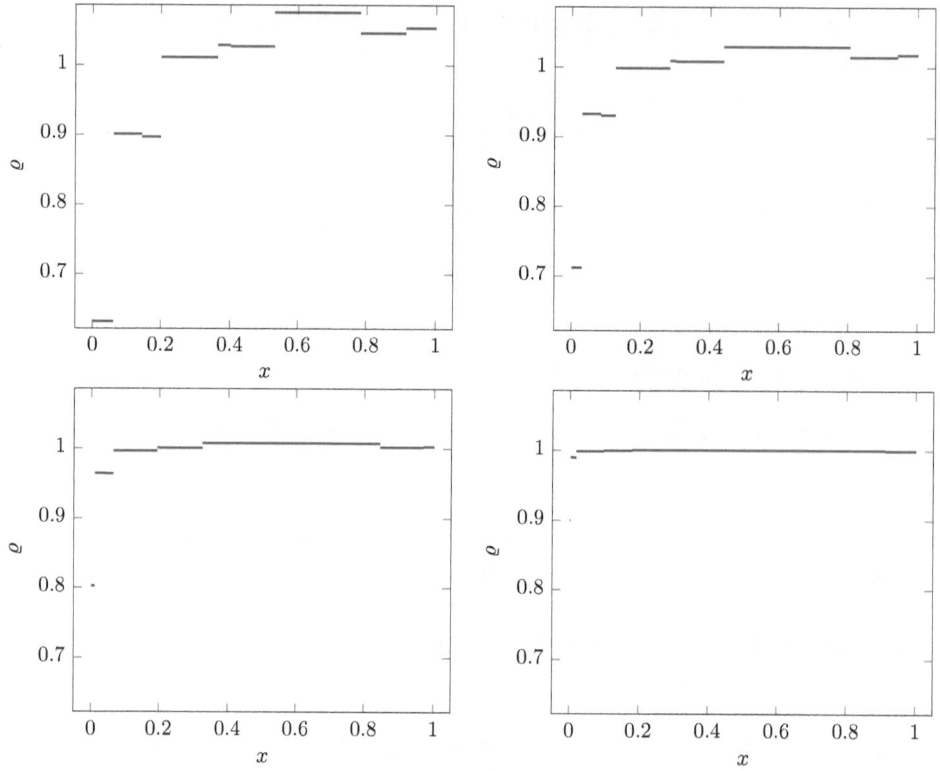

Figure 3.2 The graphs of the density ϱ for:
top left: $s = 2.35051\ldots$ and $t = \frac{10}{6}$;
top right: $s = 3.2339\ldots$ and $t = \frac{10}{7}$;
bottom left: $s = 4.9467\ldots$ and $t = \frac{10}{8}$;
bottom right: $s = 9.9837\ldots$ and $t = \frac{10}{9}$.

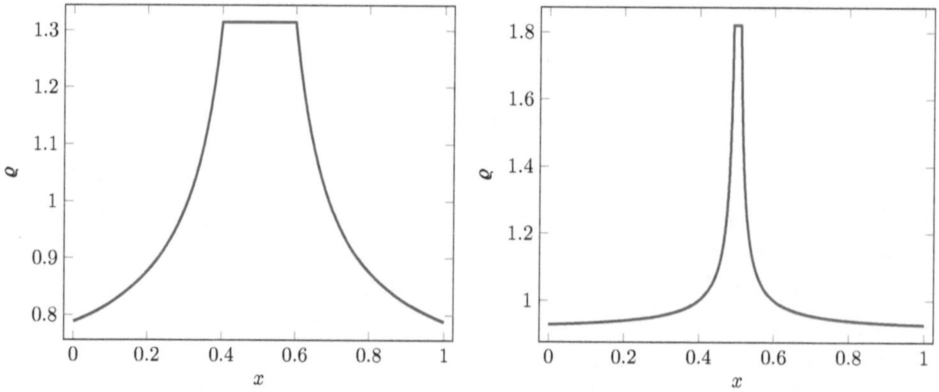

Figure 3.3 Two limit densities for $f(0) = a = 0.4$ (left) and $f(0) = a = 0.49$ (right).

Let $\overline{\varrho}$ be the density of the acim, normalized by $\overline{\varrho}(0) = 1$ (so, since the only preimage of 0 is 1, and $\overline{\varrho}$ is invariant under the Frobenius–Perron operator, we have $\overline{\varrho}(1) = t$). The function $\overline{\varrho}$ is constant on $[a, f(a)]$ and has jumps

$$\frac{1}{s(-t)^k} = \frac{t-1}{(1-a)(-1)^k t^{k+1}}$$

at x_k, since

$$s = \frac{1-a}{c} = \frac{1-a}{1-\frac{1}{t}} = \frac{t(1-a)}{t-1}.$$

Let us concentrate on the right part of the interval (the situation on the left one is very similar). If $\overline{p}$ is the value of $\overline{\varrho}$ on $[a, f(a)]$, then the value of $\overline{\varrho}$ at x_{2k+1} (more precisely, the limit from the left) is

$$\begin{aligned}\overline{\varrho}(x_{2k+1}) &= \overline{p} - \sum_{j=0}^{k-1} \frac{t-1}{(1-a)t^{2j+2}} \\ &= \overline{p} - \frac{t-1}{(1-a)t^2} \cdot \frac{1-1/t^{2k}}{1-1/t^2} = \overline{p} - \frac{1}{(1-a)(t+1)}\left(1 - \frac{1}{t^{2k}}\right).\end{aligned}$$

By (3.9),

$$\frac{1}{t^{2k}} = \frac{t\left(\frac{t}{t+1} - a\right)}{x_{2k+1} - \frac{t}{t+1}},$$

so

$$\overline{\varrho}(x_{2k+1}) = \overline{p} - \frac{1}{(1-a)(t+1)}\left(1 - \frac{t\left(\frac{t}{t+1} - a\right)}{x_{2k+1} - \frac{t}{t+1}}\right).$$

Since $x_{2n+1} = 1$, we have

$$\overline{p} = t + \frac{1}{(1-a)(t+1)}\left(1 - \frac{t\left(\frac{t}{t+1} - a\right)}{1 - \frac{t}{t+1}}\right) = \frac{1}{1-a}.$$

Now let us go with n to infinity, keeping a constant. Then t goes to $1, f(a)$ goes to $1-a$, and for every $\varepsilon > 0$ the points x_{2k+1} are ε-dense in $[1-a, 1]$ if n is large enough. Therefore, $\overline{\varrho}$ goes to the limit $\widetilde{\varrho}$, and for $x \geq 1-a$ we have

$$\widetilde{\varrho}(x) = \frac{1}{1-a} - \frac{1}{2(1-a)}\left(1 - \frac{\frac{1}{2} - a}{x - \frac{1}{2}}\right) = \frac{1}{2(1-a)} + \frac{1-2a}{4(1-a)} \cdot \frac{1}{x - \frac{1}{2}}.$$

Very similar computations give us

$$\widetilde{\varrho}(x) = \frac{1}{2(1-a)} + \frac{1-2a}{4(1-a)} \cdot \frac{1}{\frac{1}{2}-x}$$

for $x \le a$.

In such a way we get

$$\widetilde{\varrho}(x) = \begin{cases} \frac{1}{2(1-a)} + \frac{1-2a}{4(1-a)} \cdot \frac{1}{\frac{1}{2}-x} & \text{if } x < a, \\ \frac{1}{1-a} & \text{if } a \le x \le 1-a, \\ \frac{1}{2(1-a)} + \frac{1-2a}{4(1-a)} \cdot \frac{1}{x-\frac{1}{2}} & \text{if } x > 1-a, \end{cases} \tag{3.10}$$

so in particular, $\widetilde{\varrho}$ is symmetric with respect to $1/2$.

Now we have to normalize $\widetilde{\varrho}$ (that is, to divide it by its integral) in order to get the limit density ϱ. We have

$$\begin{aligned} \int \widetilde{\varrho}(x)\,dx &= 2\int_0^a \left(\frac{1}{2(1-a)} + \frac{1-2a}{4(1-a)} \cdot \frac{1}{\frac{1}{2}-x}\right) dx + \frac{1-2a}{1-a} \\ &= \frac{a}{1-a} + \frac{1-2a}{1-a} + \frac{1-2a}{2(1-a)} \int_0^a \frac{1}{\frac{1}{2}-x}\,dx \\ &= 1 - \frac{1-2a}{2-2a} \log\left(\frac{1}{2}-x\right)\Big|_0^a = 1 - \frac{1-2a}{2-2a}\log(1-2a). \end{aligned}$$

Thus the limit density ϱ of the acip is given by the function from formula (3.10) divided by $1-(1-2a)\log(1-2a)/(2-2a)$.

Observe that as a goes to $1/2$ then the maximal value of ϱ (at the plateau) increases to 2. This motivates us to make the following conjecture, with weaker and stronger versions.

Conjecture 4.1. *There exists a constant K such that for every $f \in \mathfrak{X}$ the density of the acip for f is bounded above by K.*

Conjecture 4.2. *For every $f \in \mathfrak{X}$ the density of the acip for f is bounded above by 2.*

3.5 Continuous Dependence of Densities

The next thing to show is that the density ϱ of the acip depends continuously on the map. In this section we will stress the dependence on the map, so we will write ϱ_f, $\widehat{\varrho}_f$, and so on.

For $\widehat{\varrho}_f = \widehat{\varrho}$ given by (3.1), let us look at its approximations $\widehat{\varrho}_{\ell,f}$, given by

$$\widehat{\varrho}_{\ell,f} = \sum_{k=0}^{\ell} \frac{1}{(f^k)'(0)} \chi_{[f^k(0),1]}. \tag{3.11}$$

We would like to prove that $\widehat{\varrho}_g$ as a function of g is continuous at any given f. Unfortunately, this is not true, because if 0 is periodic for f, the number $f'(c)$ is involved in the formula for $\widehat{\varrho}_f$, and it can change under arbitrarily small perturbations of f. We need to circumvent this using Remark 3.2. We will denote the ball of radius δ in $\mathfrak{Y}$ (in the sup metric), centered at f by $B(f,\delta)$. We will also denote the L^1 norm by $\|\cdot\|$.

Lemma 5.1. *Fix $f \in \mathfrak{Y}$ (with slopes s and $-t$) and $\varepsilon > 0$. Then there exists $\delta > 0$ such that for every $g \in B(f,\delta)$ there is a scalar $\gamma > \frac{\min(s,t^2)-1}{\min(s,t^2)+1}$ satisfying*

$$\|\widehat{\varrho}_g - \gamma\widehat{\varrho}_f\| < \varepsilon. \tag{3.12}$$

Proof. Suppose first that 0 is not periodic for f (or $f(0) = 0$). There exist constants $T > 1$ and $\delta > 0$ such that, if $g \in B(f,\delta)$, then the absolute values of the slopes of g are at least T. Then the sup distance between $\widehat{\varrho}_g$ and $\widehat{\varrho}_{\ell,g}$ is not larger than

$$\sum_{k=\ell+1}^{\infty} \frac{1}{T^k} = \frac{1}{T^\ell(T-1)}.$$

Let us choose ℓ so large that this is less than $\varepsilon/3$.

Observe that for $\alpha,\beta \in \mathbb{R}$ and $a,b \in [0,1]$ we have

$$\|\alpha\chi_{[a,1]} - \beta\chi_{[b,1]}\| \le |\alpha-\beta| + |a-b|\max(|\alpha|,|\beta|). \tag{3.13}$$

If δ is sufficiently small, then for each $k \le \ell$ the points $f^k(0)$ and $g^k(0)$ are as close to each other as we want, and the derivatives $(f^k)'(0)$ and $(g^k)'(0)$ are as close to each other as we want. Therefore, by (3.13), making δ sufficiently small yields

$$\sum_{k=1}^{\ell} |f^k(0) - g^k(0)| < \varepsilon/3,$$

so then $\|\widehat{\varrho}_{\ell,f} - \widehat{\varrho}_{\ell,g}\| < \varepsilon/3$. This proves that for a sufficiently small δ, if $g \in B(f,\delta)$, then $\gamma = 1$, and

$$\|\widehat{\varrho}_g - \widehat{\varrho}_f\| < \varepsilon. \tag{3.14}$$

Now we consider the case of 0 being periodic for f with period $p > 1$. In this case, it is not necessarily true that for a small δ, $(f^k)'(0)$ and $(g^k)'(0)$ are close to each other. However, this is true if we replace $\widehat{\varrho}_f$ and $\widehat{\varrho}_{\ell,f}$ by $\widehat{\widehat{\varrho}}_f$ and $\widehat{\widehat{\varrho}}_{\ell,f}$, where we change the values of $f'(f^{rp+p-2})$ to shadow the values of $g'(g^{rp+p-2})$ (look at the kneading sequence for g; if at the position $rp+p-2$ we see L, set $f'(f^{rp+p-2}(0)) = s$, otherwise set $f'(f^{rp+p-2})(0)) = -t$). Here we are using Remark 3.2. While always $f^{rp+p-2}(0) = c$, we are allowed to choose values for the derivative for different values of r. With this modification, instead of (3.14) we get (3.12) for some γ.

By Lemma 3.3, we have

$$1-\sum_{r=1}^{\infty}\frac{1}{(1+\min(s,t^2))^r}\leq\sum_{r=0}^{\infty}\frac{1}{(f^{rp})'(0)}\leq 1+\sum_{r=1}^{\infty}\frac{1}{(1+\min(s,t^2))^r},$$

independently of whether our choices for $f'(f^{rp+p-2}(0))$ were as for $\widehat{\varrho}_f$ or as for $\widehat{\widehat{\varrho}}_f$. Since the right and left expressions in the preceding equations equal $1-\frac{1}{\min(s,t^2)}$ and $1+\frac{1}{\min(s,t^2)}$ respectively, we obtain the uniform estimate $\gamma>\frac{\min(s,t^2)-1}{\min(s,t^2)+1}$. □

Now we can prove the main result of this section.

Theorem 5.2. *The map $g\mapsto\varrho_g$ from $\mathfrak{Y}$ to $L^1([0,1])$ is continuous.*

Proof. We use Lemma 5.1, but we have to switch from $\widehat{\varrho}$ to $\varrho=\widehat{\varrho}/\int\widehat{\varrho}=\widehat{\varrho}/\|\widehat{\varrho}\|$. Given $f\in\mathfrak{Y}$ and $\varepsilon>0$, let δ, g, and γ be as in Lemma 5.1. We have

$$\begin{aligned}\|\varrho_f-\varrho_g\| &= \left\|\frac{\widehat{\varrho}_f}{\|\widehat{\varrho}_f\|}-\frac{\widehat{\varrho}_g}{\|\widehat{\varrho}_g\|}\right\| \leq \left\|\frac{\gamma\widehat{\varrho}_f}{\gamma\|\widehat{\varrho}_f\|}-\frac{\widehat{\varrho}_g}{\gamma\|\widehat{\varrho}_f\|}\right\|+\left\|\frac{\widehat{\varrho}_g}{\gamma\|\widehat{\varrho}_f\|}-\frac{\widehat{\varrho}_g}{\|\widehat{\varrho}_g\|}\right\|\\ &= \frac{\|\gamma\widehat{\varrho}_f-\widehat{\varrho}_g\|}{\gamma\|\widehat{\varrho}_f\|}+\|\widehat{\varrho}_g\|\frac{\big|\|\widehat{\varrho}_g\|-\gamma\|\widehat{\varrho}_f\|\big|}{\gamma\|\widehat{\varrho}_f\|\cdot\|\widehat{\varrho}_g\|}\\ &= \frac{\|\gamma\widehat{\varrho}_f-\widehat{\varrho}_g\|+\big|\|\widehat{\varrho}_g\|-\gamma\|\widehat{\varrho}_f\|\big|}{\gamma\|\widehat{\varrho}_f\|}\leq\frac{2\|\gamma\widehat{\varrho}_f-\widehat{\varrho}_g\|}{\gamma\|\widehat{\varrho}_f\|}.\end{aligned}$$

Thus, by Lemma 5.1,

$$\|\varrho_f-\varrho_g\|\leq\frac{2\varepsilon(\min(s,t^2)+1)}{(\min(s,t^2)-1)\|\widehat{\varrho}_f\|}.$$

This shows continuity of the map $g\mapsto\varrho_g$ at f. Since $f\in\mathfrak{Y}$ was arbitrary, this map is continuous on $\mathfrak{Y}$. □

3.6 Continuous Dependence of Metric Entropy

For $g\in\mathfrak{Y}$ denote by μ_g the acip for g.

Theorem 6.1. *The map $g\mapsto h_{\mu_g}(g)$ from $\mathfrak{Y}$ to $\mathbb{R}$ is continuous.*

Proof. By the Rohlin Lemma ([P, R]), we have

$$h_{\mu_f}(f)=\int\log|f'|\,d\mu_f=\int\log|f'|\,\varrho_f\,d\lambda,$$

where λ is the Lebesgue measure.

Fix $f \in \mathfrak{Y}$. The density ϱ_f is bounded above by some constant M. There are also constants N, η such that, for every $g \in B(f, \eta)$, the function $\log |g'|$ is bounded above by N.

On the other hand, in view of Theorem 5.2 and since the map $g \mapsto \log |g'|$ from $\mathfrak{Y}$ to $L^1([0, 1])$ is continuous, for every $\varepsilon > 0$ there exists $\delta \in (0, \eta)$ such that

$$\| \log |f'| - \log |g'| \| < \frac{\varepsilon}{2M}, \text{ and}$$
$$\|\varrho_f - \varrho_g\| < \frac{\varepsilon}{2N},$$

whenever $g \in B(f, \delta)$. Thus,

$$\begin{aligned} |h_{\mu_f}(f) - h_{\mu_g}(g)| &= \left| \int \log |f'| \varrho_f \, d\lambda - \int \log |g'| \varrho_g \, d\lambda \right| \\ &\le \int \big| \log |f'| - \log |g'| \big| \varrho_f \, d\lambda + \int \log |g'| \, |\varrho_f - \varrho_g| \, d\lambda \\ &\le M \| \log |f'| - \log |g'| \| + N \|\varrho_f - \varrho_g\| < \varepsilon. \end{aligned}$$

□

3.7 Rectangular Root

To switch from a map g to a map with the topological and metric entropies halved, we need a square root procedure. This procedure is a kind of reverse to the renormalization procedure (see the proof of Lemma 3.3). The traditional square root procedure works as follows. Remember that when we say "linear" we mean what in algebra is called "affine."

Suppose we have a unimodal map g with an acip ν. Then we define a unimodal map G in the following way. On $[0, 1/3]$, $G(x) = (2 + g(1 - 3x))/3$; on $[2/3, 1]$, $G(x) = 1 - x$; on $[1/3, 2/3]$ G is linear to make it continuous. Figure 3.4 shows g, G, and G^2. For G^2 the intervals $[0, 1/3]$ and $[2/3, 1]$ are invariant, and on each of them G^2 is linearly conjugate to g. The map G maps each of them to the other one. All points of $[1/3, 2/3]$, except the fixed point, are eventually mapped by iterates of G to $[0, 1/3] \cup [2/3, 1]$ (except in the case when all those points are fixed points of G^2). Therefore $h(G) = (1/2)h(g)$ and $h_\varkappa(G) = (1/2)h_\nu(g)$, where $\varkappa$ is the acip for G obtained from ν (we will define it rigorously in a moment).

The problem with this procedure is that even if we start with a piecewise expanding map, we end up with a map that has intervals on which the absolute value of the slope is 1. To modify this procedure in such a way that the resulting map is also piecewise expanding, let us look at the original procedure from the point of view of compositions of maps.

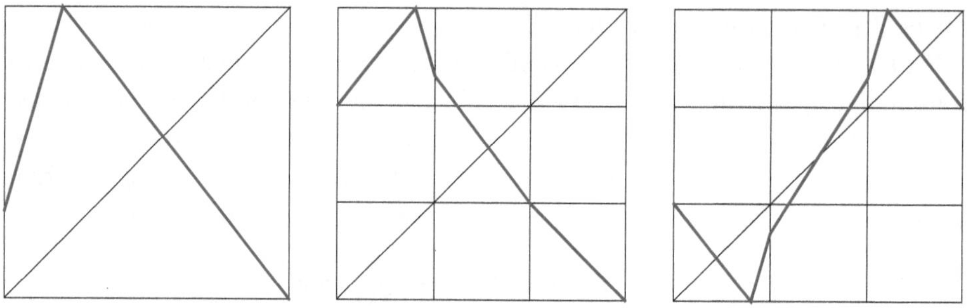

Figure 3.4 The square root procedure. Maps g, G, and G^2.

Let φ be the linear orientation-reversing map that sends the interval $[0, 1/3]$ onto the interval $[0, 1]$. Then $\varphi(x) = 1 - 3x$ and $\varphi^{-1}(x) = (1 - x)/3$. Similarly, let ψ be the linear orientation-preserving map, sending the interval $[2/3, 1]$ onto the interval $[0, 1]$. Then $\psi(x) = 3x - 2$ and $\psi^{-1}(x) = (2 + x)/3$. Now we see that on $[0, 1/3]$ we have $G = \psi^{-1} \circ g \circ \varphi$ and on $[2/3, 1]$ we have $G = \varphi^{-1} \circ \psi$. Since G sends $[0, 1/3]$ onto $[2/3, 1]$ and vice versa, on $[0, 1/3]$ we have $G^2 = \varphi^{-1} \circ g \circ \varphi$ and on $[2/3, 1]$ we have $G^2 = \psi^{-1} \circ g \circ \psi$. This explains why on both intervals G^2 is linearly conjugate to g.

This also allows us to give a precise definition of the measure $\varkappa$. Namely, $\varkappa$ is the average of the images of the measure ν under the maps φ^{-1} and ψ^{-1}. The equalities from the preceding paragraph show that since ν is invariant for g, $\varkappa$ is invariant for G.

If you look at Figure 3.4, you see nine small squares in a big square. If we change some of them into rectangles, our procedure will work better. Therefore, we will call this procedure the *rectangular root*.

If $g(0) > 0$, then we define the rectangular root procedure by changing in the definition of φ the interval $[0, 1/3]$ to $[0, (1+\varepsilon)/3]$ for some sufficiently small $\varepsilon > 0$ (see Figure 3.5). The slopes of G on $[0, (1 + \varepsilon)/3]$ are now equal to the slopes of g divided by $1 + \varepsilon$, so if ε is sufficiently small, their absolute values are still larger than 1. The slope of G on $[2/3, 1]$ is $-(1 + \varepsilon)$. The interval $[(1 + \varepsilon)/3, 2/3]$ is mapped by G onto a larger interval, so the absolute value of the slope is larger than 1 as well.

If $g(0) = 0$, then we simply remove the middle interval. That is, we take as φ the linear orientation-reversing maps, sending the interval $[0, (1 + \varepsilon)/2]$ onto the interval $[0, 1]$, and as ψ the linear orientation-preserving map, sending the interval $[(1+\varepsilon)/2, 1]$ onto the interval $[0, 1]$. Then we set $G = \psi^{-1} \circ g \circ \varphi$ on $[0, (1+\varepsilon)/2]$ and $G = \varphi^{-1} \circ \psi$ on $[(1 + \varepsilon)/2, 1]$ (see Figure 3.6). As in the first case, if $\varepsilon > 0$ is sufficiently small, then G is piecewise expanding.

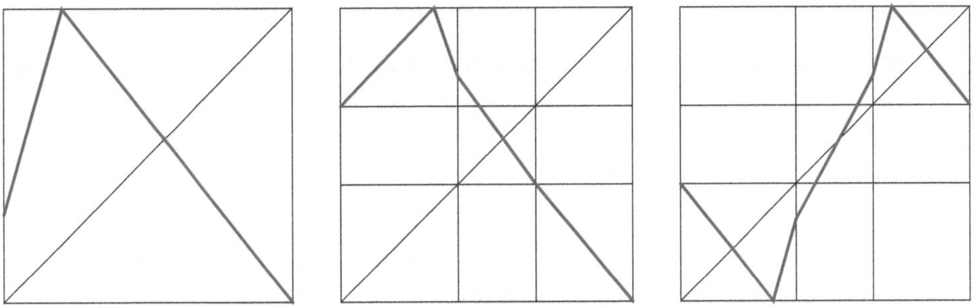

Figure 3.5 The rectangular root procedure when $g(0) > 0$. Maps g, G, and G^2.

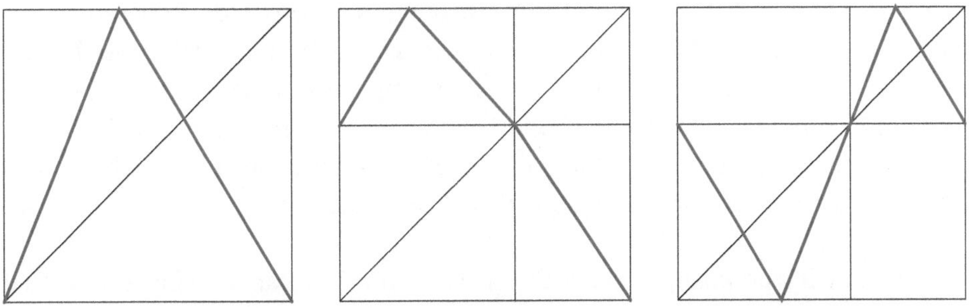

Figure 3.6 The rectangular root procedure when $g(0) = 0$. Maps g, G and G^2.

3.8 Main Theorems

The first flexibility result is about mixing skew tent maps (the space $\mathfrak{X}$).

Theorem A. *For every pair $a, b \in \mathbb{R}$ with $\frac{1}{2}\log 2 < a \le \log 2$ and $0 < b \le a$ there exists a piecewise expanding mixing skew tent map f for which $h(f) = a$ and $h_\mu(f) = b$, where μ is the acip for f.*

Proof. If $\log s_a = a$, then $f \in \mathfrak{X}$ with slopes s_a and $-s_a$ has topological entropy a. By [MV, Theorem C] and the subsequent remark, there exists a number $\gamma > 1$ and a continuous decreasing function $\beta\colon (1,\gamma] \to [1,\infty)$ such that $\lim_{t\searrow 1}\beta(t) = \infty$ and for $g \in \mathfrak{X}$ we have $h(g) = a$ if and only if the slopes of g are $\beta(t), -t$ for $t \in (1,\gamma]$. This in particular implies that $s_a \le \gamma$ and $\beta(s_a) = s_a$. Moreover, by [MV, Theorem C], the skew tent maps with the slopes $\beta(t), -t$ have the same kneading sequence for all t, so we will be able to use Theorem 3.5 for them.

Let $f_t \in \mathfrak{X}$ be the function with slopes $\beta(t)$ and $-t$, and let μ_t be its acip. For $t = s_a$ this measure is also the measure with maximal entropy, so $h_{\mu_t}(f_t) = a$. As t goes to 1, then $\beta(t)$ goes to infinity and the turning point c_t goes to 0, so by Theorem 3.5 $\mu_t([0, c_t])$ goes to 0. Since the partition of $[0, 1]$ into $[0, c_t]$ and $[c_t, 1]$

is a generator, this implies that $h_{\mu_t}(f_t)$ goes to 0. Therefore, by Theorem 6.1 and continuity of the function β, there exists t such that $h_{\mu_t}(f_t) = b$. □

The next theorem shows flexibility of entropies for piecewise expanding unimodal maps.

Theorem B. *For every pair $a, b \in \mathbb{R}$ with $0 < a \leq \log 2$ and $0 < b \leq a$ there exists a piecewise expanding unimodal map f for which $h(f) = a$ and $h_\mu(f) = b$, where μ is the acip for f.*

Proof. If $a > \frac{1}{2}\log 2$, this follows from Theorem A. If $a \leq \frac{1}{2}\log 2$, there is $n \geq 1$ such that $\frac{1}{2}\log 2 < 2^n a \leq \log 2$. Then use Theorem A to find $g \in \mathfrak{X}$ with $h(g) = 2^n a$ and $h_\nu(g) = 2^n b$, where ν is the acip for g. Finally, use n times the rectangular root procedure (see Section 3.7) to get the desired f. Each time we use this procedure, both topological and metric entropy get divided by 2. Then f is a piecewise expanding unimodal map with $h(f) = a$ and $h_\mu(f) = b$. □

The following proposition shows that in Theorem B we cannot replace unimodal maps by skew tent maps.

Proposition 8.1. *There is no piecewise expanding skew tent map with topological entropy $\frac{1}{4}\log 2$ and metric entropy smaller than $\frac{1}{4}\log\frac{16}{15}$.*

Proof. Let $f \in \mathfrak{Y}$ have topological entropy $h(f) = \frac{1}{4}\log 2$. This means that f is twice renormalizable (see the proof of Lemma 3.3). If the slopes of f are s and $-t$, after two renormalizations we get a skew tent map g with slopes s^2t^2 and $-st^3$ and entropy $\log 2$, linearly conjugate to f^4 restricted to an invariant interval.

For a unimodal map, topological entropy being $\log 2$ is equivalent to this map having the kneading sequence RL^∞. So for a skew tent map this means that the sum of reciprocals of the absolute values of the slopes is 1. It is well known (and very easy to check) that then the Lebesgue measure is invariant, so the acip is the Lebesgue measure. Thus, we have $s + t = s^2t^3$. Remember that we assume $s, t > 1$.

Set $T = \max(s, t)$. We have $T^2 < s^2t^3 = s + t \leq 2T$, so $T < 2$. Hence, absolute values of both slopes of the second renormalization g are smaller than 16. Therefore $1/16 < c < 15/16$, so either $\lambda([0, c]) > 1/16$ or $\lambda((c, 1]) > 1/16$. Then,

$$h_\lambda(g) > -\frac{1}{16}\log\frac{1}{16} - \frac{15}{16}\log\frac{15}{16} = \log 16 - \frac{15}{16}\log 15 > \log 16 - \log 15 = \log\frac{16}{15}.$$

Thus, if μ is the acip for f, then $h_\mu(f) = \frac{1}{4}h_\lambda(g) > \frac{1}{4}\log\frac{16}{15}$. □

3.9 An Interesting Observation

Hidden in [ITN] is the formula called by the authors "f-expansion." Namely, for $f \in \mathfrak{Y}$ and $x \in [0, 1]$ we have (translating to our notation)

$$x = 1 - \frac{1}{t} \sum_{k=0}^{\infty} \frac{1}{(f^k)'(x)}. \tag{3.15}$$

Taking $y > c$ for which $f(y) = x$, we can reinterpret it as

$$f(y) = \sum_{k=0}^{\infty} \frac{1}{(f^k)'(y)}. \tag{3.16}$$

However, since (3.16) does not work for $y < c$, it does not look important. On the other hand, if we take in (3.15) $y = 1$, we get

$$\sum_{k=0}^{\infty} \frac{1}{(f^k)'(1)} = 0, \tag{3.17}$$

which is much more interesting.

If $s = t$, then the left-hand side of (3.17) is the value of the Milnor–Thurston kneading determinant at $1/s$. However, s is the exponential of the topological entropy of f. Thus, formula (3.17) is a generalization and a different interpretation of the Milnor–Thurston formula connecting the topological entropy and the kneading determinant for unimodal maps.

References

[BKRH] J. Bochi, A. Katok, and F. Rodriguez Hertz, *Flexibility of Lyapunov exponents*, arXiv preprint 1908.07891 [math.DS] (2019).

[BK] Z. Buczolich and G. Keszthelyi, *Isentropes and Lyapunov exponents*, Discrete Contin. Dyn. Syst. **40** (2020), 1989–2009.

[CE] P. Collet and J.-P. Eckmann, *Iterated maps on the interval as dynamical systems*, Progress in Physics **1**, Birkhäuser, Boston, MA, 1980.

[dMvS] W. de Melo and S. van Strien, *One-dimensional dynamics*, Springer, Berlin, 1993.

[E] A. Erchenko, *Flexibility of Lyapunov exponents for expanding circle maps*, Discrete Contin. Dyn. Syst. **39** (2019), 2325–2342.

[EK] A. Erchenko and A. Katok, *Flexibility of entropies for surfaces of negative curvature*, Israel J. Math. **232** (2019), 631–676.

[ITN] S. Ito, S. Tanaka, and H. Nakada, *On unimodal linear transformations and chaos II*, Tokyo J. Math. **2** (1979), 241–259.

[K] G. Keller, *Piecewise monotonic transformations and exactness*, Publications des séminaires de mathématiques et informatique de Rennes, no. 1 (1978), article no. 6, 32 pp.

[LaY] A. Lasota and J. A. Yorke, *On the existence of invariant measures for piecewise monotonic transformations*, Trans. Amer. Math. Soc. **186** (1973), 481–488

[LiY] T.-Y. Li and J. A. Yorke, *Ergodic transformations from an interval into itself*, Trans. Amer. Math. Soc. **235** (1978), 183–192.

[MT] J. Milnor and W. Thurston, *On iterated maps of the interval*, in "Dynamical Systems," pp. 465–563, Lecture Notes in Mathematics, **1342**, Springer, Berlin, 1988.

[MV] M. Misiurewicz and E. Visinescu, *Kneading sequences of skew tent maps*, Ann. Inst. H. Poincaré Probab. Statist. **27** (1991), 125–140.

[P] Ja. B. Pesin, *Characteristic Ljapunov exponents, and smooth ergodic theory* (Russian), Uspehi Mat. Nauk **32** (1977), 55–112.

[R] V. A. Rohlin, *Exact endomorphisms of a Lebesgue space* (Russian), Izv. Akad. Nauk SSSR Ser. Mat. **25** (1961), 499–530.

[Ru] S. Ruette, *Chaos on the interval*, University Lecture Series 67, American Mathematical Society, Providence, RI, 2017.

4

Recent Results and Open Questions on Pseudo-Rotations

BARNEY BRAMHAM AND ZHIYUAN ZHANG

Pseudo-rotations are rare amongst dynamical systems. Precisely what "rare" means, seems to lie at the cross-roads of several fundamental conjectures in symplectic topology. The smooth realization problem is another perspective from which pseudo-rotations are important objects. In this survey we compile some of the more recent results about pseudo-rotations. Theorem G and Corollary A are new. The pioneering work of Anosov and Katok [AK70] is fundamental throughout.

4.1 Introduction

The study of smooth realization is concerned with the basic problem, essentially formulated by von Neumann in 1932, of whether or not every finite entropy abstract measure-preserving transformation of a Lebesgue space is isomorphic to a smooth diffeomorphism preserving a volume element on a compact manifold. We can also ask what obstructions there could be for such a smooth realization. Many interesting results started to appear around the 1970s. We refer the readers to [LT77, Cha66, Kat67, Kat79, Shk67] for some classical results. One of the pioneering works [AK70], and crucial to the topics in this article, is due to Anosov and Katok.

4.1.1 The Anosov–Katok Construction

In [AK70] Anosov and Katok constructed the first example of a smooth area-preserving diffeomorphism of the 2-disc $\mathbb{D}$ which is ergodic with respect to the area. Put another way, ergodicity on the disc is not an obstruction to smooth realization.[1]

[1] In contrast, ergodicity is easy to realize by smooth hyperbolic diffeomorphisms on the torus. These maps always have positive entropy; in contrast, the construction in [AK70] has zero entropy, because the mechanism that enables orbit travel is fundamentally different.

To construct the diffeomorphism in [AK70] they developed a new technique, now called "Approximation by Conjugation," or the Anosov–Katok construction. The influence of their ideas goes far beyond this single example. For a detailed survey on this method see [FK04] (see also [Kun21] in this volume and [HCP19] and [LS20] in a symplectic context). Roughly the construction goes as follows:

1. At each step we have an area-preserving diffeomorphism f_n of the form $H_n R_{P_n/Q_n} H_n^{-1}$, where R_α denotes the standard rotation on $\mathbb{D}$ through angle $2\pi\alpha$;
2. We pick a smooth area-preserving diffeomorphism h_n of $\mathbb{D}$, which commutes with R_{P_n/Q_n}, such that the image under h_n of every circle centered at 0 wiggles drastically, and the mass distribution tends to the Lebesgue measure as n grows;
3. We take some extremely large integer q_{n+1} and set

$$f_{n+1} = H_{n+1} R_{P_{n+1}/Q_{n+1}} H_{n+1}^{-1},$$

where $H_{n+1} = H_n h_n$ and $P_{n+1}/Q_{n+1} = P_n/Q_n + 1/q_{n+1}$. Notice that by commutativity we have $f_{n+1} = H_n(h_n R_{1/q_{n+1}} h_n^{-1}) R_{P_n/Q_n} H_n^{-1}$ and so f_{n+1} converges to f_n with all derivatives as $q_n \to +\infty$. Thus by taking q_{n+1} sufficiently large, depending on H_n and h_n, we can ensure by Arzelà–Ascoli that the sequence f_n converges in the smooth category to some area-preserving smooth diffeomorphism f. Moreover, since we can control the mass distribution of each f_n-orbit of length Q_n, we can control "almost all" iterates of f by letting the convergence of f_n be sufficiently fast.

The limit map is an area-preserving diffeomorphism of $\mathbb{D}$ without any periodic points besides the origin. However, unlike a rotation through an irrational angle, it can have dense orbits or invariant pseudo-circles. In short, the following objects can have non-trivial dynamics:

DEFINITION 1.1. A *pseudo-rotation* on $\mathbb{D}$ is a non-wandering homeomorphism $f: \mathbb{D} \to \mathbb{D}$ that is isotopic to the identity, and has no periodic points except at the origin which is a fixed point.

This definition was introduced in [BCLP04, BCL06], where the name "pseudo-rotation" was coined, or rather *irrational pseudo-rotation*. See also the notes [Cro06]. However, the study of pseudo-rotations can be traced back to Birkhoff [Bir50] – before they had a name.

There are at least two other ways to characterize area-preserving pseudo-rotations: By the work of Franks [Fra92], an equivalent definition would be those area-preserving homeomorphisms (isotopic to the identity) having a *finite number* of periodic points. Another equivalent definition would be that every orbit (or even just almost every orbit) has a well-defined asymptotic rotation number about the fixed point at the origin, and that this rotation number is the *same irrational value*,

$$\rho(f) \in \mathbb{R}/\mathbb{Z},$$

for all trajectories, see [Fra88a, Fra89, Fra88b, Fra92]. This value, $\rho(f) \in \mathbb{R}/\mathbb{Z}$, is called *the rotation number* of the pseudo-rotation f and coincides with the rotation number of f restricted to $\partial\mathbb{D}$. These reformulations are thus highly non-trivial. As we will see in what follows, many of the results about pseudo-rotations depend on the nature of the rotation number, and some parallels to the theory of circle diffeomorphisms arise.

Some other surfaces admit a definition of pseudo-rotation: On the closed annulus $\mathbb{A} = \mathbb{R}/\mathbb{Z} \times [0, 1]$, the open annulus, the two-sphere $\mathbb{S}^2$, and the two-torus $\mathbb{T}^2$, a pseudo-rotation f can be defined as a homeomorphism that is isotopic to the identity, which is either area-preserving or non-wandering, and which has the minimum number of periodic points; none for $\mathbb{A}$ and $\mathbb{T}^2$, and two for $\mathbb{S}^2$ which are then always fixed points[2]. On closed higher genus surfaces all Hamiltonian homeomorphisms have infinitely many periodic points, which falls into the framework of the Conley Conjecture; see [LeC06] and also [FH03]. We will mostly refer to the disc and closed annulus.

The article is structured as follows. In Section 4.2 two fundamental results of opposite nature are recalled, in which the rotation number provides a sharp dividing line[3] between opposing dynamical behavior; one result establishes rigidity or integrability phenomena for Diophantine rotation numbers, while the other shows strongly non-integrable behavior for Liouvillean rotation numbers. In Section 4.3 we discuss further results and questions on integrable features. We also present two new results in [BZ23], Theorem 6 and Corollary A. In Section 4.4 we look at a rigidity phenomenon that surprisingly is strongest for Liouvillean rotation numbers. In Sections 4.1–4.4 the discussion is confined to surfaces. In Section 4.5 we summarize some recent symplectic developments in higher dimensions. Section 4.6 lists some further open questions.

Finally, the perspective that we have taken is neither exhaustive nor broad. There are many other interesting related topics, such as a detailed study of the rotation set in [KPS16, LT18, Koc16] and the world of C^0-pseudo-rotations, in which even the entropy can be positive; see [Cro06].

For the remainder of this chapter, unless stated otherwise, **all pseudo-rotations considered will be C^∞-smooth and preserve a smooth volume form.**

[2] These conditions can be equivalently formulated in terms of a constant irrational rotation number on $\mathbb{A}$, resp. on $\mathbb{S}^2$ minus the two fixed points, resp. a constant irrational rotation vector $\rho(f) \in (\mathbb{R}^2\backslash\mathbb{Q}^2)/\mathbb{Z}^2$ on $\mathbb{T}^2$. See, for instance, Crovisier's note [Cro06].

[3] At least in the C^∞-category. In, for example, the real analytic category the dividing line might be different.

4.2 Exotic Pseudo-Rotations

From the formulation of a pseudo-rotation in terms of rotation number one expects the dynamics of a pseudo-rotation to be quite rigid. But, as mentioned earlier, the Anosov–Katok method permits moderate orbit divergence and the construction of exotic, namely non-integrable, pseudo-rotations with, for example, dense orbits, weak mixing, or invariant pseudo-circles. Currently all constructions build on the Anosov–Katok method in some way. In this section we describe the results more precisely.

The first constructions in [AK70] resulted in pseudo-rotations with a restricted (highly Liouvillean) rotation number. This was necessary to get the convergence of the approximating sequence described in the introduction. Recall that $\alpha \in \mathbb{R}$, resp. in $\mathbb{R}/\mathbb{Z}$, is Liouville if it is irrational and for all $C > 0$ there exists $Q \in \mathbb{N}$ so that

$$d(Q\alpha, \mathbb{Z}) < \frac{1}{Q^C}.$$

Equivalently if it is irrational and not Diophantine. The Liouville numbers form a Baire set of measure zero. So, Liouville numbers are abundant from a topological perspective while Diophantine numbers are abundant in measure.

It was unclear for which rotation numbers ergodicity or weak mixing could be achieved, until Fayad and Saprykina refined the approach in [AK70] and showed:

THEOREM 1 ([FS05]). *For every Liouville number $\alpha \in \mathbb{R}/\mathbb{Z}$ there exists a weakly mixing pseudo-rotation on $\mathbb{D}$, resp. $\mathbb{A}$, with rotation number α.*

This result is remarkably sharp. Indeed, the set of rotation numbers in Theorem 1 is optimal due to the following result of Fayad and Krikorian:

THEOREM 2 ([FK09]). *Every area-preserving C^∞ diffeomorphism of $\mathbb{A}$, isotopic to the identity, with Diophantine rotation number on one boundary circle, has smooth invariant circles filling up a set of positive measure in any neighborhood of that boundary circle, on which the dynamics is smoothly conjugate to rotation maps.*

This result, proven using KAM methods, is a key step toward *Herman's last geometric theorem*; see Theorem 4. Remarkably there is no "local twist" assumption near the boundary, so that more conventional KAM methods do not apply.

It is unknown whether weakly mixing can be upgraded to mixing; not a single example is known and many negative results are discussed in Section 4.4. Another natural question concerns minimizing the number of ergodic invariant measures:

THEOREM 3 ([FK04], Theorem 3.3). *There exists a pseudo-rotation on $\mathbb{D}$, resp. $\mathbb{A}$, having only three ergodic invariant Borel probability measures. The ergodic measures are the* *area form ω and the Dirac measures supported on the fixed point and $\partial\mathbb{D}$, resp. the two boundary components of $\mathbb{A}$.*

More generally, for each $\ell \geq 3$ they construct pseudo-rotations with exactly ℓ ergodic invariant Borel probability measures.

Despite this result, a pseudo-rotation can never be minimal on the complement of its fixed points. Indeed a very general result of Le Calvez–Yoccoz [LY97] states that no homeomorphism of the plane minus a finite number of points can be minimal. Franks and Misiurewicz re-proved this, using Conley theory [Fra99, FM02], showing, moreover, that for a pseudo-rotation on the sphere, neither of the two fixed points can be isolated as invariant sets. In other words, each arbitrarily small punctured neighborhood of each fixed point of a pseudo-rotation contains an invariant set.

In [Cro06, Theorem 3.4], Crovisier has constructed smooth pseudo-rotations on $\mathbb{S}^2$ having a minimal compact invariant set with positive Lebesgue measure and which is connected, but not locally connected at any point (a pseudo-circle).

4.3 Integrability and Integrable Approximations

We begin with two remarkable questions of Birkhoff and Herman concerning the actual *integrability* of certain pseudo-rotations.

In [Bir50, page 712] (see also [Bir41]) Birkhoff conjectured that every real analytic area-preserving pseudo-rotation on the 2-sphere is integrable in the following sense:

CONJECTURE 1 (Birkhoff's sphere conjecture). *Is it true that every real-analytic and area-preserving pseudo-rotation on the 2-sphere is conjugate to a rigid irrational rotation?*

Very recently, Pierre Berger has announced a counterexample to the above conjecture; see [Ber22]. It has been known for a long time that such a statement cannot be true in the smooth category, at least for Liouvillean rotation numbers, because of Theorem 1. However, in [Her98] Herman posed the following:

QUESTION 1 (Herman[4]). *Is it true that every C^∞-smooth and area-preserving pseudo-rotation on $\mathbb{D}$ or $\mathbb{A}$ or $\mathbb{S}^2$ with Diophantine rotation number is smoothly conjugate to a rigid rotation?*

[4] Actually Herman stated the question for the closed annulus, but the dynamics of a C^1 map near a fixed point are in any case more restrictive than those near a boundary circle. Moreover, instead of area-preserving he asks the question more generally for maps $f\colon \mathbb{A} \to \mathbb{A}$ with the so-called intersection property: $f(\gamma) \cap \gamma \neq \emptyset$ for every essential closed curve γ.

The Diophantine condition is necessary because of Theorem 1. There is some evidence in support of a positive answer to Question 1. This is the so-called Herman's last geometric theorem[5] which handles the perturbative setting:

THEOREM 4 ([FK09], Herman's last geometric theorem). *Every smooth and area-preserving pseudo-rotation of the annulus with Diophantine rotation number and which is sufficiently close to a rigid rotation is smoothly conjugate to a rigid rotation.*

Here is a version of their result described more precisely: Let $\alpha \in \mathbb{R}/\mathbb{Z}$ be Diophantine, and denote by $R_\alpha : \mathbb{A} \to \mathbb{A}$ the rigid rotation $(\theta, r) \mapsto (\theta + \alpha, r)$. Then there exist $k \in \mathbb{N}$ and $\varepsilon > 0$, both depending on α, so that every smooth area-preserving pseudo-rotation $f : \mathbb{A} \to \mathbb{A}$ with rotation number α and satisfying

$$\|f - R_\alpha\|_{C^k(\mathbb{A})} < \varepsilon,$$

can be C^∞-conjugated to R_α. In fact, they assume only that f satisfies the intersection condition in place of area-preserving (see Question 1), and moreover k and ε only depend on the Diophantine class that α belongs to.

A more recent perturbative result is announced by Avila–Krikorian for area-preserving pseudo-rotations in an open C^∞ neighborhood of the rigid rotations on $\mathbb{D}$. Compared to the result in [FK09], the closeness of the pseudo-rotation to the rigid rotation here is independent of the rotation number. This is a significant improvement.

Since it is a feature of the Anosov–Katok method that it produces maps as limits of integrable systems, more precisely as limits of maps that can be conjugated to rotations, it is natural to ask whether *all* pseudo-rotations have this property.

QUESTION 2. *Is every smooth and area-preserving pseudo-rotation f on the disc or annulus or sphere a C^∞-limit of area-preserving maps f_k that are conjugate to rigid rotations?*

In contrast to the preceding discussion of genuine integrability, it is possible that the rotation number of the pseudo-rotation plays no decisive role here. The first positive result in this direction was proven using methods from symplectic geometry:

THEOREM 5 ([Bra15a]). *Any smooth area-preserving pseudo-rotation f on the closed disc is a C^0-limit of maps f_k of the form $h_k R_{p_k/q_k} h_k^{-1}$, where $h_k \in \mathrm{Diff}^\infty(\mathbb{D})$ is orientation preserving and fixes the origin, and $(p_k, q_k) \in \mathbb{Z} \times \mathbb{N}$.*

[5] Apparently Katok suggested this name, and the authors of [FK09] also use this.

Each approximating map f_k is constructed as the holonomy map of a foliation $\mathcal{F}_k$ of a mapping torus for f, where the leaves come from foliations by pseudo-holomorphic curves, a theory developed in [HWZ98, HWZ02, HWZ03]; see also [BH12]. As $k \to \infty$ the leaves collapse onto the trajectories of the mapping torus. See also [LeC16] for an approach using generating functions, using machinery developed in [Cha84, LeC91a, LeC91b, LeC04], that applies to pseudo-rotations of class C^1, but which requires conjugacy to a rotation on the boundary. Let us also note an interesting older result from [BCLP04, Corollary 1.2] of a reverse nature. They show that R_α can be approximated by C^0-conjugates of any fixed C^0-pseudo-rotation with rotation number α, in the C^0-topology. They mention that their methods do not seem to yield the reverse approximations discussed here.

Note that there are no restrictions on the rotation number in Theorem 5. On the other hand, the convergence is only in a C^0-sense and, moreover, the approximating maps are probably not area-preserving.

Very recently, the authors have obtained the following result[6] using a completely different method, which achieves both of these improvements, albeit for generic rotation numbers.

THEOREM 6 ([BZ23]). *There is a Baire generic subset $\mathcal{R} \subset \mathbb{R}/\mathbb{Z}$ such that any C^∞ area-preserving pseudo-rotation on $\mathbb{A}$ with rotation number $\alpha \in \mathcal{R}$ is a C^∞-limit of maps of the form $hR_\alpha h^{-1}$ where $h \in \mathrm{Diff}^\infty(\mathbb{A}, \omega)$.*

Theorem 6 can also be viewed as a partial generalization of the following well-known result for circle diffeomorphisms of Herman–Yoccoz:

THEOREM 7. *For any irrational $\alpha \in \mathbb{R}/\mathbb{Z}$, denote by $F^\infty(\alpha)$ the set of C^∞ circle diffeomorphisms with rotation number α, and denote by $O^\infty(\alpha)$ the set of C^∞ circle diffeomorphisms which are C^∞-conjugate to the standard rotation R_α. Then*

$$F^\infty(\alpha) = \begin{cases} \overline{O^\infty(\alpha)} = O^\infty(\alpha) & \text{if } \alpha \text{ is Diophantine} \\ \\ \overline{O^\infty(\alpha)} \neq O^\infty(\alpha) & \text{if } \alpha \text{ is Liouville} \end{cases}$$

where the closures are taken in the C^∞-topology.

The Diophantine part of Theorem 7 was first established in the perturbative setting near rotations by Arnold [Arn61], and globalized by Herman in [Her79] and Yoccoz in [Yoc84]. The Liouville part was a conjecture of Herman in [Her79]. In the same paper, Herman proved his conjecture for a Baire generic α by considering the variation of rotation numbers. The full conjecture was solved by

[6] Even more recently, Avila and Krikorian have announced an extension of our result to any non-Brjuno rotation number.

Yoccoz in [Yoc95b] by considering appropriate renormalizations of the circle diffeomorphism.

Our proof of Theorem 6 is somewhat similar to the proof of Yoccoz in [Yoc95b], with a new key ingredient on smooth norm estimation coming from [AFLXZ20]. We consider a certain renormalization scheme of a pseudo-rotation that is inspired by the aforementioned work of Avila–Krikorian. Though we are unable to transfer the strong estimates for circle diffeomorphisms to general pseudo-rotations, we are able to extract some useful information from a sequence of suitably renormalized pseudo-rotations using the smooth norm bound and a strong (but generic) arithmetic condition on the rotation number.

There is an interesting Corollary of Theorem 6 as follows.

COROLLARY A ([BZ23]). *For a Baire generic $\alpha \in \mathbb{R}/\mathbb{Z}$, the set of smooth weakly mixing and area-preserving pseudo-rotations on the closed annulus with rotation number α forms a Baire generic subset with empty interior, with respect to the C^∞-topology.*

This follows immediately from Theorem 6 and the Anosov-Katok construction.

The construction of dynamically interesting pseudo-rotations in [AK70] seems to have been at least a partial motivation for a question of Katok [Kat09] that shifts the focus onto the vanishing of the entropy of sufficiently smooth pseudo-rotations:

QUESTION 3 (Katok). *Is every smooth area and orientation-preserving surface diffeomorphism with zero topological entropy a limit of integrable systems?*

Since smooth pseudo-rotations have vanishing topological entropy, the area-preserving ones can be seen as the simplest examples in the domain of this question. Indeed, on the disc, annulus, or sphere, the only possible zero entropy Hamiltonian systems with a dense orbit are pseudo-rotations. This follows from the following decomposition theorem of Franks and Handel:

THEOREM 8 ([FH12]). *Let f be a smooth Hamiltonian diffeomorphism of the two-sphere with zero topological entropy, infinite order, and at least three periodic points. Then the complement of the fixed points decomposes as a countable union of pairwise disjoint open f-invariant annuli whose closure is the entire space.*

The decomposition says much more, in particular that on each invariant annulus $U \subset \mathbb{S}^2$ there is a well-defined rotation number for trajectories in U giving rise to a f-invariant function $\rho_U : U \to \mathbb{R}/\mathbb{Z}$ having two surprising features: ρ_U is defined on all of U (not just almost everywhere), and ρ_U is everywhere continuous. Moreover, each point in the boundary of U has α- and ω-limit sets singletons $\{p\}$ and $\{q\}$ that depend only on the boundary component of U that the initial point lies in. With these stronger conclusions one can see that zero entropy and a dense orbit forces the system to be a pseudo-rotation: Indeed, if f is not a pseudo-rotation then f has at least three periodic points. So either f is of finite order, or the Franks–Handel decomposition of f has either more than one invariant annulus or an invariant annulus on which the rotation number is non-constant. In all these cases, f cannot have a dense orbit.

To progress further on the above question of Katok, it appears natural to study in the next stage zero entropy surface diffeomorphisms with only finitely many periodic points. This complicated problem goes beyond the scope of this survey, but let us mention recent related work of Le Calvez [LeC20].

4.4 A Liouvillean Rigidity

In Problem 2.2 in [FK04] they ask whether there exists a mixing pseudo-rotation on the disc or annulus. The following comment [FK04, p. 1505] suggests that mixing examples were considered a likely possibility at the time: "It is an open problem whether there exists a circle action on a compact manifold and α such that the corresponding space $\mathcal{A}_\alpha$ (essentially of objects with rotation number α) contains a non-rigid transformation. This would be a first step toward proving the existence of mixing maps in $\mathcal{A}_\alpha$." The terminology "rigid" as used here is taken from [FK04], although it seems to be older. The meaning is: f is C^k-rigid if it has a sequence of iterates f^{n_j} which converges in a C^k-sense to the identity. So it is an extremely strong and synchronized form of recurrence. However, even C^∞-rigidity does not preclude weakly mixing, because it says only something about a very particular "thin" sequence of iterates. Nevertheless it is easily seen that even C^0-rigidity precludes strong mixing. In this light the following statement seems somewhat unexpected:

THEOREM 9 ([Bra15b]). *Any smooth area-preserving pseudo-rotation f on the closed disc having super Liouvillean rotation number is C^0-rigid. In particular, f cannot be topologically mixing.*

A number $\alpha \in \mathbb{R}/\mathbb{Z}$ is called super-Liouville if, for all $C > 0$, there exists $Q \in \mathbb{N}$ so that

$$d(Q\alpha, \mathbb{Z}) < \frac{1}{C^Q}. \tag{4.1}$$

A rather surprising aspect of Theorem 9 is that it shows a kind of "Liouvillean rigidity," in contrast to Theorem 1. This was proven using holomorphic curve methods building on the proof of Theorem 5. The generating functions approach in [LeC16] also recovered this result in the regularity and boundary setting explained before.

Later a completely different and much more elementary argument was found by Avila, Fayad, Le Calvez, Xu, and the second author in [AFLXZ20] that applied also with low regularity. The key observation to recover Theorem 9 is the following lemma.

LEMMA 4.1 ([AFLXZ20], Free Discs are Small). *Let f be a continuous area-preserving pseudo-rotation of $\mathbb{D}$. If $\rho(f) = \varepsilon \in \mathbb{R}/\mathbb{Z}$, then any topological disc D such that $\omega(D) > \varepsilon$ satisfies $f(D) \cap D \neq \emptyset$. Here ω is the normalized Lebesgue measure, namely, $\omega(\mathbb{D}) = 1$.*

This lemma also allowed to prove rigidity of any Hölder pseudo-rotation with sufficiently Liouvillean rotation number.

Using much more sophisticated methods, they extended the rigidity result from super-Liouville to non-Brjuno rotation numbers:

THEOREM 10 ([AFLXZ20]). *If f is an area-preserving pseudo-rotation on $\mathbb{D}$ of class C^k, $k \geq 2$, with non-Brjuno rotation number, then f is C^{k-1}-rigid. In particular, f cannot be topologically mixing.*

See Question 10. Recall that an irrational number $\alpha \in \mathbb{R}/\mathbb{Z}$ is Brjuno if

$$\sum_{n\geq 0} \frac{\ln q_{n+1}}{q_n} < \infty,$$

where q_n is the sequence of denominators of the best rational approximations to α. For comparison, the super-Liouville condition can be reformulated as $\limsup \frac{\ln q_{n+1}}{q_n} = +\infty$. The set of Diophantine numbers is properly contained in the set of Brjuno numbers. Building on Theorem 10, they established the following in the real analytic category:

THEOREM 11 ([AFLXZ20]). *A real analytic area-preserving disc diffeomorphism with zero topological entropy is not topologically mixing.*

The proof of Theorem 10 goes as follows. By Lemma 4.1, the C^0-distance to the identity $d_{C^0}(f^{q_n}, \mathrm{Id})$ can be bounded from above in terms of the rotation number of f^{q_n}. From the Liouville condition on $\rho(f)$, one finds a sequence $q_n \to \infty$ so that the rotation numbers $\rho(f^{q_n})$ decay to zero faster than the inverse of any polynomial function of q_n. Although this is not yet enough to show rigidity, the non-Brjuno condition gives a tiny bit more, but nothing as strong as the exponential decay that would result from an exponential Liouville condition. The next key step is the theorem below, which can be thought of as an effective version of the well-known Katok's closing lemma in [Kat80] restricted to the conservative setting. A similar effective closing lemma with an extra angle condition was obtained earlier by Climenhaga–Pesin [CP16]. The readers can find other applications of this line of ideas in [FZ17] and [CDP16].

THEOREM 12 ([AFLXZ20]). *Let S be a compact surface. There exist constants* $A_0 > 0$, $\theta_0 \in (0,1)$, $H_0 > 0$ *such that for any* $g \in \mathrm{Diff}^\infty(S,\omega)$, $q \in \mathbb{N}$, $A > A_0$, $D \geq A$, $q \geq D^{H_0}$, *if*

$$\|Dg\| \geq A,$$
$$\|D^2 g\| \geq D,$$
$$\|Dg^q\| \leq A^{\theta_0 q},$$

then g has a hyperbolic periodic point.

Indeed, Theorem 12 gives an explicit subexponential upper bound on $\|Df^{q_n}\|$, which in turn yields an explicit subexponential upper bound on $\|D^r f^{q_n}\|$, for each $r \geq 1$. Then by Hadamard's convexity theorem, we can show that $\|D^r f^{q_n}\|$ decays as n tends to infinity.

The proof of Theorem 11 goes as follows. By [FH12], as described after Theorem 8, a smooth area-preserving disc diffeomorphism with zero topological entropy and strictly more than two periodic points cannot be topologically transitive. Thus it suffices to study a pseudo-rotation $f \in \mathrm{Diff}^\omega(\mathbb{D},\omega)$. By Theorem 10 we may further suppose that the rotation number of f is Brjuno. On the other hand, it follows from a result of Rüssmann [Rüs67] that any real analytic area-preserving pseudo-rotation on $\mathbb{D}$ with *Brjuno* rotation number has invariant curves near the boundary and thus cannot be topologically mixing.

One is led to speculate that the remaining rotation numbers in the middle, more precisely those that are Brjuno and Liouville, either admit smooth mixing examples, or they are non-mixing for different reasons than those produced with KAM or [AFLXZ20, Bra15b].

Note that Lemma 4.1 provides an upper bound on the C^0 distance of a C^1 pseudo-rotation to the identity in terms of its C^1 norm and its rotation number. In [WZ18], Wang and the second author generalized Lemma 4.1 to pseudo-rotations on the torus $\mathbb{T}^2$ under a condition called *Bounded mean motion* (see [Jäg09] for an important consequence of this condition). This leads to an answer (see also [WY21]) to the following question of Norton–Sullivan in [NS96]: is there a natural geometric condition (e.g., smoothness) on a diffeomorphism $f\colon \mathbb{T}^2 \to \mathbb{T}^2$ which is semi-conjugate to a translation, that forces f to be topologically conjugate to the translation?

Finally, an interesting corollary to Theorem 12 is that not every C^∞ circle diffeomorphism can be realized as the boundary map of a pseudo-rotation on $\mathbb{D}$. Indeed, Yoccoz constructed in [Yoc95b] a non-rigid C^∞ circle diffeomorphism with non-Brjuno rotation number. This means that the following question must have an interesting answer:

QUESTION 4. *Which orientation-preserving circle diffeomorphisms can be realized as the boundary map of a pseudo-rotation on $\mathbb{D}$?*

This question is intended for circle maps with Liouvillean rotation number, since Diophantine circle maps can be smoothly linearized [Her79, Yoc84] and every orientation-preserving circle diffeomorphism is the boundary of an area preserving diffeomorphism. It also means that there is a non-empty class of orientation-preserving circle diffeomorphisms (with Liouvillean rotation number) with the remarkable property, that every area-preserving diffeomorphism of the disc with this boundary condition must have infinitely many interior periodic points.

We will return to results on rigidity later in Section 4.5.

4.5 Pseudo-Rotations in Higher Dimensions

There are many important recent developments in higher dimensions, most notably in [GG18a, HCP19, LS20].

4.5.1 Definition in Higher Dimensions

Several interesting ways to generalize the two-dimensional definition have emerged. We discuss some of these now.

Pseudo-rotations as Hamiltonian diffeomorphisms: This is probably the most natural from a symplectic perspective and is especially interesting because of its connection to various conjectures in symplectic topology, as discussed in Section 4.5.2:

DEFINITION 5.1. Let (M, ω) be a smooth manifold with a symplectic structure. A *Hamiltonian pseudo-rotation* is a Hamiltonian diffeomorphism $\phi \colon M \to M$ with a finite and minimal number of periodic points, and these periodic points are all fixed points.

This is the definition introduced in [GG18a]. See also Definition 5.3, where we interpret the word *minimal* more liberally. The word minimal may be unnecessary: In all known examples in arbitrary dimensions, Definition 5.1 coincides with the so-called perfect Hamiltonian diffeomorphisms introduced in [CGG12].

DEFINITION 5.2. Let (M, ω) be a smooth manifold with a symplectic structure. A Hamiltonian diffeomorphism $\phi \colon M \to M$ is called *perfect* if it has a finite number of periodic points.

In [CGG20] they work with a different Floer theoretic definition of pseudo-rotation that could also turn out to be equivalent to Definition 5.1.

As an example, on $\mathbb{C}P^n$ Definition 5.1 can be made more explicit since the theoretic minimal number of periodic points $n+1$ is known to be achieved; see Section 4.5.3. Thus, a Hamiltonian pseudo-rotation on $\mathbb{C}P^n$ is any Hamiltonian diffeomorphism with precisely $n+1$ fixed points and no other periodic points.

Pseudo-rotations as symplectic diffeomorphisms: If one replaces Hamiltonian diffeomorphism with symplectic diffeomorphism in Definition 5.1 above, there may be no periodic points at all, for example an irrational rotation on $\mathbb{T}^2$. Hernández-Corbato and Presas show in [HCP19] that if a symplectic manifold (M, ω) admits a locally free S^1-action by symplectomorphisms – see [FGM91] for examples – then there exists a symplectic diffeomorphism f, symplectically isotopic to the identity, that is minimal. That is, each orbit $\{f^k(x)\}_{k \in \mathbb{Z}}$, $x \in M$, is dense in M. In particular, f has no periodic points at all. They construct f using the Anosov–Katok method, following more closely the approach in [FH77].

Pseudo-rotations in odd dimensions: In [Kat73], Katok adapted the approximation by conjugation method to construct autonomous Hamiltonians on various spaces including $\mathbb{R}^{2n}$ equipped with the standard symplectic form ω_0, whose constant energy levels are ergodic and have precisely n periodic orbits. This includes many convex energy surfaces and therefore Reeb flows on each sphere $(S^{2n+1}, \xi_{\mathrm{st}})$ equipped with its standard contact structure. In [AGZ22] they realize a large class of pseudo-rotations on $\mathbb{D}$ as the return map of a Reeb flow on $(\mathbb{S}^3, \xi_{\mathrm{st}})$.

If a contact manifold (M, ξ) admits a locally free S^1-action by contactomorphisms, then in [HCP19] they use the Anosov–Katok method to construct a contactomorphism isotopic to the identity (through contactomorphisms) that is strictly ergodic (equivalently, uniquely ergodic and minimal).

Pseudo-rotations defined via rotation vectors: In the realm of volume-preserving dynamical systems there is so much flexibility that just requiring a minimal number of periodic orbits seems to be too mild by itself to expect a fruitful comparison with two-dimensional pseudo-rotations without restricting the rotation vectors (even in dimension two; see [Koc16]). The definition of an (irrational) pseudo-rotation in [BCLP04] is that the rotation set is a single irrational number.

There is a body of important works in this direction. We merely mention in higher dimensions that Fayad–Katok in [FK04] construct volume-preserving flows on the 3-torus $\mathbb{T}^3$ using an extension of the Anosov–Katok method that requires a $\mathbb{T}^k$-action with $k \geq 3$ (rather than the more usual $k = 1$), that are volume-preserving and whose set of rotation vectors forms a single point set and thus should be considered a pseudo-rotation. This construction is particularly interesting, in that it is mixing (rather than say weakly mixing). This is the only source of mixing pseudo-rotations that we are aware of; see also [Fay02]. Using rotation vectors to define pseudo-rotations seems to work well on tori as they have a large first homology group. But it is unclear how to adapt this to say simply connected spaces such as $\mathbb{C}P^n$.

For the rest of Section 4.5 all pseudo-rotations will mean Hamiltonian pseudo-rotations as in Definition 5.1. We will just refer to these as pseudo-rotations, that is, drop the word Hamiltonian.

The following is conjectured in [GG18a]:

QUESTION 5. *Are the fixed points of every pseudo-rotation all non-degenerate? If so, are they always elliptic?*

Where non-degenerate means that all eigenvalues of the linearization at each fixed point are different from 1, and elliptic means they, moreover, all lie on the unit circle. Of course, if the answer is positive in both cases, applying the statement to iterates implies that all eigenvalues must have irrational argument. This is the case in dimension 2; see [Bra15a, GG18a], where the former is a simple application of a deep result of Franks' [Fra88a] and the Poincaré–Birkhoff theorem, while the latter prove it using Floer homology.

4.5.2 Pseudo-Rotations in a Broader Symplectic Context

Definition 5.1 is extremely compelling from the perspective of symplectic geometry as it sits on the frontier of several fundamental conjectures in the subject (some of which are now Theorems); the Hamiltonian Arnold conjecture, the Conley conjecture, the Hofer–Zehnder conjecture, and the more recent Chance–McDuff conjecture. We recall what this means.

As remarked before, pseudo-rotations are rare. This is the assertion of the Conley conjecture, which says that for a broad class of closed symplectic manifolds every Hamiltonian diffeomorphism has infinitely many periodic points. Thus pseudo-rotations can only exist on symplectic manifolds for which the Conley conjecture "fails."[7]

[7] The state-of-the-art result is that the Conley conjecture holds for (M, ω) unless there exists $A \in \pi_2(M)$ for which $\langle c_1(TM), A\rangle > 0$ and $\langle \omega, A\rangle > 0$, where c_1 denotes the first Chern class associated to ω; see [GG17, CG18] and former developments [SZ92, Hin09, Gin10].

Then there is the issue of the difference between having finitely many versus a minimal number of periodic points. The Arnold conjecture [Arn78] predicts a lower bound for the minimal number of fixed points, in terms of the topology of M. In the case of non-degenerate Hamiltonians the conjecture is a theorem,[8] and the lower bound is the sum of the Betti numbers of M. A weaker lower bound is predicted for degenerate Hamiltonians.

Now the Hofer–Zehnder conjecture [HZ94] comes in and says that if a Hamiltonian map has more than the minimum number of fixed points predicted by the Arnold conjecture, then it must have infinitely many periodic points. For example, on $\mathbb{C}P^n$ the Arnold conjecture says that any Hamiltonian diffeomorphism f must have at least $n+1$ fixed points (in both the degenerate and non-degenerate settings), and so the Hofer–Zehnder conjecture predicts that if f has at least $n+2$ periodic points, then it has infinitely many periodic points. On $\mathbb{C}P^1$ this is a celebrated result of Franks [Fra92], while on $\mathbb{C}P^n$ this is a recent breakthrough result of Shelukhin [She22] under a mild non-degeneracy assumption; see also [CGG21, Ala20, Ala22].

In short: The Hofer–Zehnder conjecture asserts that every perfect Hamiltonian diffeomorphism is a Hamiltonian pseudo-rotation and that it has precisely the number of fixed points required by Arnold's conjecture.

Finally, the Chance–McDuff conjecture [McD] states roughly that if the Conley conjecture fails for (M, ω), in particular if (M, ω) admits a pseudo-rotation, then some Gromov–Witten invariants for (M, ω) are non-vanishing. This suggests a fundamental relation between a dynamical phenomenon – a lack of periodic orbits – and the existence of stable solutions to a PDE, namely J-holomorphic spheres. In [CGG20] they confirm the conjecture for a large class of pseudo-rotations. See also [CGG22, She19b, She19c].

In light of the dichotomy in Arnold's conjecture, that the lower bound it provides deals separately with degenerate and non-degenerate systems, it is convenient to make the following definition:

DEFINITION 5.3. Let (M, ω) be a smooth manifold with a symplectic structure and denote by ℓ the sum of the Betti numbers. Let us refer to any Hamiltonian diffeomorphism $f : M \to M$ whose periodic points consist of ℓ non-degenerate fixed points, as a *non-degenerate (Hamiltonian) pseudo-rotation*.

In other words, such a map has the minimum number of periodic points amongst non-degenerate Hamiltonians. This nomenclature allows us to refer to all the examples in Section 4.5.3 as non-degenerate pseudo-rotations.[9]

[8] With contributions from many people, building on the work of Floer. See references in [AD14].

[9] This is convenient, but it is possible that this definition turns out to be weaker than a Hamiltonian pseudo-rotation that is also non-degenerate, since it may not be minimal in the sense of Definition 5.1. This would make the nomenclature unfortunate.

To emphasize: The state of the art is that all known perfect Hamiltonian diffeomorphisms are pseudo-rotations and are non-degenerate pseudo-rotations.

4.5.3 Existence of Hamiltonian Pseudo-Rotations in Higher Dimensions

Completely integrable examples: Consider a closed symplectic manifold (M, ω) that admits a Hamiltonian circle-action $R\colon \mathbb{S}^1 \to \mathrm{Ham}(M, \omega)$, $t \mapsto R_t$, having isolated fixed points $\mathrm{Fix}(R)$ and such that the action is locally free on $M\backslash\mathrm{Fix}(R)$. Then each element R_t, with $t \in \mathbb{R}/\mathbb{Z}$ irrational, will be a non-degenerate pseudo-rotation in the sense of Definition 5.3.

Alternatively, suppose (M^{2n}, ω) is toric. Recall this means that there exists an effective group action by a torus of half the dimension:

$$R\colon \mathbb{T}^n \to \mathrm{Ham}(M^{2n}, \omega),$$

where effective means that if $t \in \mathbb{T}^n\backslash\{0\}$ then $R_t \neq \mathrm{Id}$. Then each generic element of the action will be a non-degenerate pseudo-rotation. This is the source of many examples of non-degenerate pseudo-rotations in the sense of Definition 5.3; see also [CGG12]. For example, this includes on $\mathbb{C}P^n$ the Hamiltonian diffeomorphisms of the form

$$f\colon \mathbb{C}P^n \to \mathbb{C}P^n, \qquad [z_0 : \cdots : z_n] \mapsto [e^{2\pi i a_0} z_0 : \cdots : e^{2\pi i a_n} z_n],$$

where $a_0, \ldots, a_n$ are any real numbers. This f is a pseudo-rotation if and only if $a_i - a_j$ is irrational for each distinct i, j. The $n + 1$ fixed points of f are then $[1 : 0 : \cdots : 0], \ldots, [0 : \cdots : 0 : 1]$.

Similarly one can often construct pseudo-rotations on (M, ω) if it admits a Hamiltonian torus action $\mathbb{T}^k \to \mathrm{Ham}(M^{2n}, \omega)$ with isolated fixed points.

All of these examples that use circle or torus actions are in some sense integrable and have relatively simple dynamics.

Exotic examples: Non-integrable examples of pseudo-rotations arise easily by taking products of non-integrable pseudo-rotations on $\mathbb{S}^2$.

Much more interesting examples of pseudo-rotations are constructed by Le Roux–Seyfaddini [LS20] on all toric manifolds:

THEOREM 13 ([LS20]). *Suppose that (M^{2n}, ω) is a closed symplectic toric manifold. Then there exists a non-degenerate pseudo-rotation*[10] *f on M for which the symplectic volume form ω^n is ergodic. Moreover, f has no other ergodic invariant measures besides the volume form and the Dirac measures supported on its fixed points.*

Two non-trivial symplectic ingredients are used in the Anosov–Katok method here: The C^∞-Flux conjecture, proven by Ono [Ono06], is used to conclude from

[10] In the sense of Definition 5.3.

the sequence of converging Hamiltonian diffeomorphisms that the limit map is not just symplectic but actually Hamiltonian.[11] Secondly, the approximating maps are required to transport many small balls around the space. When $2n > 2$, there are often symplectic packing obstructions to doing this, even for a single ball, that do not arise in the volume-preserving category. To construct pseudo-rotations that are merely ergodic they could work around this difficulty by using sufficiently small balls. But to get a minimal number of ergodic measures this approach was unclear and they instead exploited certain global features of symplectic toric manifolds.

4.5.4 Liouvillean Rigidity

In [GG18a] Ginzburg–Gurel discovered a generalization of the C^0-rigidity result from Theorem 9 to pseudo-rotations on $\mathbb{C}P^n$. Instead of using foliations by holomorphic curves they used Floer theory.

In higher dimensions it is not at all obvious what should play the role of the rotation number. It turns out that one has to take into account the rotation numbers of all the fixed points simultaneously. Indeed, let f be a pseudo-rotation on $\mathbb{C}P^n$, which we recall from Section 4.5.1 means that f is a Hamiltonian diffeomorphism with precisely $n+1$ fixed points $x_0, x_1, \ldots, x_n$ and no other periodic points. After fixing a Hamiltonian isotopy to the identity, each fixed point has a well-defined mean Conley–Zehnder index $\hat{\mu}(x_i)$ with value in the circle $\mathbb{R}/2(n+1)\mathbb{Z}$ which can be thought of roughly as an asymptotic rotation number for the linearized flow at x_i. They call

$$\vec{\Delta}_f = \big(\hat{\mu}(x_0), \ldots, \hat{\mu}(x_n)\big) \in \mathbb{T}^{n+1}$$

the mean index vector of f, and say that it is exponentially Liouville if there exists $C > 0$ so that

$$\|q_j \vec{\Delta}_f\| \le e^{-Cq_j}$$

for some sequence $q_j \to +\infty$, where the left-hand side denotes the distance from $q_j\vec{\Delta}_f$ to the lattice $2(n+1)\mathbb{Z}^n$ with respect to the maximum norm. They prove the following:

THEOREM 14 ([GG18a]). *Let f be a pseudo-rotation on $\mathbb{C}P^n$ whose mean index vector $\vec{\Delta}_f$ is exponentially Liouville. Then there exists a sequence $q_j \to +\infty$ such that $f^{q_j} \to \mathrm{Id}$ in the C^0-topology. In particular, f cannot be topologically mixing.*

4.5.5 γ-Rigidity and Hofer-Rigidity

In [GG18a] Ginzburg–Gurel also showed a weaker form of rigidity which, however, holds independently of the rotation number, in particular, without any Liouville

[11] Being symplectic and isotopic to the identity is not enough in many symplectic manifolds.

condition. This is rigidity with respect to the so-called γ-norm, which they prove for pseudo-rotations on a class of symplectic manifolds that includes each $\mathbb{C}P^n$, and which was a new result even for $n = 1$, namely for pseudo-rotations on $\mathbb{S}^2$:

THEOREM 15 ([GG18a]). *Let f be a pseudo-rotation on $\mathbb{C}P^n$ with the Fubini–Study form. Then $\gamma(f^{q_j}) \to 0$ for some sequence of iterates $q_j \to +\infty$.*

The γ-norm was originally introduced in [Vit92] for compactly supported Hamiltonian diffeomorphisms of $\mathbb{R}^{2n}$ using generating functions; see also [HZ94]. The construction was then extended using Floer theory to successively broader classes of symplectic manifolds in [Sch00] and [Oh05]. It provides a map

$$\gamma : \mathrm{Ham}(M, \omega) \to [0, \infty)$$

that has many nice features, in particular it vanishes uniquely at the identity map (non-degeneracy) and is conjugation invariant: $\gamma(gfg^{-1}) = \gamma(f)$ for all f, g, so that $d_\gamma(f, g) := \|fg^{-1}\|_\gamma$ defines a biinvariant metric. It is also continuous with respect to the Hofer metric, another biinvariant metric on the group of Hamiltonian diffeomorphisms, which was introduced on $\mathbb{R}^{2n}$ by Hofer [Hof90, HZ94] and extended to increasingly larger classes of symplectic manifolds by Polterovich [Pol93] and Lalonde–McDuff [LM95]. There is an inequality between the associated norms: $\|f\|_\gamma \leq \|f\|_{\mathrm{Hof}}$.

In the course of the proof of Theorem 15 they establish the following inequality

$$\gamma(f^q) \leq C\|q\vec{\Delta}_f\| \qquad \forall q \in \mathbb{N} \tag{4.2}$$

relating the γ-norm and the mean rotation vector $\vec{\Delta}_f$ for pseudo-rotations on $\mathbb{C}P^n$, where C is a constant depending only on $\vec{\Delta}_f$ and not on q. The subject of γ-rigidity is also discussed in [GG19].

In [JS22] Joksimović and Seyfaddini show how to combine inequality (2.1) with the energy-capacity inequality due to Usher [Ush10] to give an alternative proof of the C^0-rigidity statement on $\mathbb{C}P^n$ under a super-Liouville condition.

In the same article [JS22] they also investigate pseudo-rotations that are well approximated by certain integrable Hamiltonian systems in a C^∞-sense. They refer to these as Anosov–Katok pseudo-rotations. Exploiting the special form of Anosov–Katok pseudo-rotations allows them to do without an inequality like (1.2) and to elegantly show that all Anosov–Katok pseudo-rotations are rigid with respect to the Hofer metric and hence also with respect to the γ-norm. Here they use their application of the energy-capacity inequality together with the biinvariance property of the Hofer metric, resp. the γ-norm.

Unlike C^0-rigidity, rigidity in these weaker senses does not appear to exclude mixing, at least not obviously so.

4.5.6 Invariant Sets

In [GG18a] a higher-dimensional counterpart to the results of Le Calvez–Yoccoz [LY97] and Franks–Misiurewicz [FM02], described at the end of Section 4.2, was discovered for certain symplectic manifolds, including the projective spaces:

THEOREM 16 ([GG18a]). *No fixed point of a pseudo-rotation of $\mathbb{C}P^n$ is isolated as an invariant set.*

This involves arguing with both the local Floer homology at the fixed points and also global algebraic properties of the quantum cohomology of (M, ω).

4.6 Further Questions

This final section contains questions that did not fit into the preceding sections. Many, but not all, concern the 2-dimensional case. Unless mentioned otherwise, all of the following questions concern the C^∞-smooth area-preserving case.

4.6.1 Integrable Approximations

The following question is about the possibility of generalizing Herman–Yoccoz's theory on circle diffeomorphisms to pseudo-rotations:

QUESTION 6. *Is every smooth and area-preserving pseudo-rotation on $\mathbb{A}$, $\mathbb{D}$, or $\mathbb{T}^2$ with Liouvillean rotation number α a C^∞-limit of smooth, area-preserving pseudo-rotations of the form $hR_\alpha h^{-1}$, where h is smooth and area-preserving?*

4.6.2 Almost-Reducibility

Another question that is motivated by the theory of circle diffeomorphisms is the following:

QUESTION 7. *Is it true that every pseudo-rotation f on the disc $\mathbb{D}$ is almost-reducible?*

Here a C^∞ diffeomorphism f on $\mathbb{D}$ is said to be almost-reducible if there exists a sequence of C^∞ area-preserving diffeomorphisms h_n such that $h_n^{-1} f h_n$ converges to $R_{\rho(f)}$ in the C^∞ topology. Avila and Krikorian have announced a result that says any C^∞ minimal circle diffeomorphism is almost-reducible.

The following question about almost-reducibility is asked in [WZ18], in connection to a question of Norton and Sullivan [NS96].

QUESTION 8. *Suppose that f is an area-preserving, C^∞ pseudo-rotation on $\mathbb{T}^2$ with a totally irrational rotation number, and semi-conjugate to a translation. Then is f almost-reducible?*

4.6.3 Wandering Domains

The following question also appears in [NS96]. Let us recall it here.

QUESTION 9. *Is there a C^∞ diffeomorphism of $\mathbb{T}^2$ (not necessarily area-preserving) that is semi-conjugate to a totally irrational translation and has a wandering domain?*

4.6.4 C^k-Rigidity

Motivated by the results in [Bra15b, AFLXZ20] on pseudo-rotations with Liouvillean rotation numbers, by Herman's conjecture concerning pseudo-rotations with Diophantine rotation number, and by the γ-rigidity in [GG18a], the following question is natural:

QUESTION 10. *Is every pseudo-rotation on the disc C^k-rigid for some $k \in \mathbb{N}$?*

4.6.5 Smooth Realization

The following question, asked in [AFLXZ20], is about the obstructions to smooth realization.

QUESTION 11. *Does there exist a smooth area-preserving diffeomorphism of the disc that is mixing with zero metric entropy or zero topological entropy?*

Another open question in the real analytic category is the following:

QUESTION 12. *Do there exist real analytic area-preserving diffeomorphisms of the disc or the sphere with zero metric entropy that is transitive or ergodic?*

Recall that the answer to Question 12 for zero topological rather than metric entropy is negative by Theorem 11, while for Question 11 it is still unclear, but any example would have to have Brjuno but Liouvillean rotation number.

4.6.6 Asymptotic Action, Linking Numbers, Calabi Invariant

Our final questions concern whether it is possible to compute these for pseudo-rotations in dimension 2, and to what extent they behave like rotations.

More precisely, let $f\colon \mathbb{D} \to \mathbb{D}$ be a pseudo-rotation of the disc with rotation number $\alpha \in \mathbb{R}/\mathbb{Z}$. Fix an isotopy to the identity or a generating Hamiltonian $H\colon \mathbb{R}/\mathbb{Z} \times \mathbb{D} \to \mathbb{R}$. Then the asymptotic action of f is defined by

$$\overline{\sigma}_f(p) := \lim_{n\to\infty} \frac{1}{n} \sum_{j=0}^{n-1} \sigma(f^j(p)),$$

where $\sigma : \mathbb{D} \to \mathbb{R}$ is the action of f, defined to be the unique solution to $d\sigma = f^*\lambda - \lambda$, and where λ is any primitive of $\omega = dx \wedge dy$, with boundary condition $f(p) = \int_{t\mapsto f_t(p)} \lambda$ for each $p \in \partial\mathbb{D}$. The limit function $\overline{\sigma}_f\colon \mathbb{D}_0 \subset \mathbb{D} \to \mathbb{R}$ is defined on a set of full measure by the ergodic theorem.

For any pair of distinct points $(p, q) \in \mathbb{D} \times \mathbb{D}$, the asymptotic linking number $\mathrm{lk}_f(p, q) \in \mathbb{R}$ also exists almost everywhere, and is the average amount that the path of vectors $t \mapsto f_t(p) - f_t(q) \in \mathbb{R}^2$ winds around as $t \to +\infty$.

The Calabi invariant of f can be defined as the average of the action, or equivalently by the ergodic theorem, as the average of the asymptotic action:

$$\mathrm{Cal}(f) := \int_{\mathbb{D}} \sigma_f(p)\,\omega_p = \int_{\mathbb{D}} \overline{\sigma}_f(p)\,\omega_p.$$

See [Bec21, MS17] for more details on these definitions. In the case that f is the rigid rotation R_α with isotopy $t \mapsto R_{t\alpha}$, the asymptotic action and linking functions are constant and one computes

$$\overline{\sigma}_{R_\alpha} = \pi\alpha, \qquad \mathrm{lk}_{R_\alpha} = 2\pi\alpha, \qquad \mathrm{Cal}(R_\alpha) = \pi^2\alpha.$$

Pirnapasov in [Pir21] has shown that this expression for Calabi indeed extends to all smooth pseudo-rotations: $\mathrm{Cal}(f) = \pi^2\rho(f)$. His approach made use of a deep result of Hutchings [Hut16]. More recently, Le Calvez [LeC22] has found an approach using generating functions, that has lower regularity requirements but requires the pseudo-rotation to be a rotation on the boundary of the disc. Special cases were also established by Joly [Jol21]. The following two questions are open:

QUESTION 13. *If f is a pseudo-rotation, is the asymptotic action of f defined everywhere (rather than just almost everywhere)? Is the asymptotic action of f constant (everywhere or almost everywhere)?*

Note that if the asymptotic action is constant almost everywhere, then by integrating and using the knowledge of the Calabi invariant just mentioned, it would follow that the constant indeed agrees with $\pi\rho(f)$.

QUESTION 14. *If f is a pseudo-rotation, is the asymptotic linking number of f defined everywhere off the diagonal in $\mathbb{D} \times \mathbb{D}$ (rather than just almost everywhere)? Is the asymptotic linking number of f constant (everywhere or almost everywhere)?*

By a formula due to Fathi [Fat80], see also [GG97, She15, Bec21] or rather its extension to non-compactly supported Hamiltonian disc maps [Bec22], one knows that if the asymptotic linking number is constant almost everywhere, then by integrating over $\mathbb{D} \times \mathbb{D}$, using our knowledge of the Calabi invariant again, it would follow that the constant indeed agrees with $2\pi\rho(f)$.

Acknowledgments

This article was initiated in 2019 while both authors were at the Institute for Advanced Study. We thank them for their hospitality and excellent working environment. B. B. would like to thank Viktor Ginzburg, Patrice Le Calvez, Frédéric Le Roux, and Sobhan Seyfaddini for helpful conversations. B. B. was partially supported by the SFB/TRR 191 "Symplectic Structures in Geometry, Algebra and Dynamics," funded by the DFG (B1 281071066 – TRR 191). Z. Z. was supported by the National Science Foundation under Grant No. DMS-1638352.

References

[AGZ22] P. Albers, H. Geiges, and K. Zehmisch, Pseudorotations of the 2-disc and Reeb flows on the 3-sphere, *Ergod. Theory Dyn. Syst.*, **42** (2022), 402–436.

[Ala20] S. Allais, On the Hofer–Zehnder conjecture on $\mathbb{C}P^d$ via generating functions (with an appendix by Egor Shelukhin), arXiv:2010.14172v2.

[Ala22] S. Allais, On the Hofer–Zehnder conjecture on weight projective spaces, arXiv:2201.13277.

[AK70] D. Anosov and A. B. Katok, New examples in smooth ergodic theory, *Trans. Moscow Math. Soc.*, **199** (1970), 1–35.

[Arn61] V. I. Arnold, Small Denominators I; on the Mapping of a Circle into itself, *Investijia Akad. Nauk, série Math*, **25** (1961), 21–96. *Transl. A. M. S. 2nd series*, **46**, 213–284.

[Arn78] V. I. Arnold, Mathematical methods of classical mechanics (Appendix 9), *Berline-Heidelberg-New York: Springer* 1978.

[AD14] M. Audin and M. Damian, Morse Theory and Floer Homology, *Universitext: Springer* (2014).

[AFLXZ20] A. Avila, B. Fayad, P. Le Calvez, D. Xu, and Z. Zhang, On mixing diffeomorphisms of the disc, *Invent. Math.*, **220** (2020), 673–714.

[Bec21] D. Bechara Senior, Asymptotic action and asymptotic winding number for area-preserving diffeomorphisms of the disk, *Ann. Math. Québec,* (2021), 1–17.

[Bec22] D. Bechara Senior, The Asymptotic Action and Generalizations of a Theorem of Fathi, Ph.D. thesis, Ruhr-Universität Bochum, 2022.

[Ber22] P. Berger, Analytic pseudo-rotations, arXiv:2210.03438.

[BCL06] F. Béguin, S. Crovisier, and F. Le Roux, Pseudo-rotations of the open annulus, *Bull. Braz. Math. Soc.*, **37** (2006), 275–306.

[BCLP04] F. Béguin, S. Crovisier, F. Le Roux, and A. Patou, Pseudo-rotations of the closed annulus: variation on a theorem of J. Kwapisz, *Nonlinearity*, **17** (2004), 1427–1453.

[Bir41] G. BIRKHOFF, Some unsolved problems of theoretical dynamics, *Science*, **94**(2452):598–600, 1941.

[Bir50] G. D. BIRKHOFF, Collected papers, Vol II, *American Mathematical Society*, New York City, 1950.

[Bra15a] B. BRAMHAM, Periodic approximations of irrational pseudo-rotations using pseudoholomorphic curves, *Ann. of Math.* **181** (2015), 1033–1086.

[Bra15b] B. BRAMHAM, Pseudo-rotations with sufficiently Liouvillean rotation number are C^0-rigid, *Invent. Math.*, **199** (2015), no. 2, 561–580.

[BH12] B. BRAMHAM AND H. HOFER, First steps towards a symplectic dynamics, *Surv. Differ. Geom.*, 17 (2012), 127–178.

[BZ23] B. BRAMHAM AND Z. ZHANG, On pseudo-rotations of the annulus with generic rotation number, arXiv:2301.04486.

[Cha66] R. V. CHACON, Transformations having continuous spectrum, *J. Math. Mech.*, **16**, 399–415 (1966).

[CGG12] M. CHANCE, V. GINZBURG, AND B. GUREL, Action-index relations for perfect Hamiltonian diffeomorphisms, *J. Symplectic Geom.*, 11 (2013), 449–474.

[Cha84] M. CHAPERON, Une idée du type "géodésiques brisées" pour les systèmes hamiltoniens, *C. R. Acad. Sc. Paris*, **298** (1984), 293–296.

[CG18] E. CINELI AND V. GINZBURG, Conley conjecture and local Floer homology, *Arch. Math. (Basel)*, **111** (2018), 647–656

[CGG20] E. CINELI, V. GINZBURG, AND B. GUREL, Pseudo-rotations and holomorphic curves, *Selecta Math. (N.S.)*, **26** (5) (2020).

[CGG21] E. CINELI, V. GINZBURG, AND B. GUREL, Another look at the Hofer–Zehnder conjecture, *J. Fixed Point Theory Appl.* (2022), 52–76.

[CGG22] E. CINELI, V. GINZBURG, AND B. GUREL, From pseudo-rotations to holomorphic curves via quantum Steenrod squares, *Int. Math. Res. Not.*, **3** (2022), 2274–2297.

[CP16] V. CLIMENHAGA AND Y. PESIN, Hadamard–Perron theorems and effective hyperbolicity, *Ergod. Theory Dyn. Syst.*, **36** (2016), 23–63.

[CDP16] V. CLIMENHAGA, D. DOLGOPYAT, AND Y. PESIN, Non-stationary non-uniform hyperbolicity: SRB measures for non-uniformly hyperbolic attractors, *Commun. Math. Phys.*, **346** (2016), 553–602.

[Cro06] S. CROVISIER, Exotic rotations, *Méthodes topologiques en dynamique des surfaces*, Grenoble (2006). www.math.u-psud.fr/~crovisie/grenoble2006 .pdf.

[Fat80] A. FATHI, Transformations et homéomorphismes préservant la mesure: systémes dynamiques minimaux, Ph.D. thesis, Orsay, 1980.

[FH77] A. FATHI AND M. HERMAN, Existence de diffeomorphismes minimaux, *Dynamical systems, Vol. I*, Astérisque **49** (1977), Société Mathématique de France, Paris, 37–59.

[Fay02] B. FAYAD, Analytic mixing reparametrizations of irrational flows, *Ergod. Theory Dyn. Syst.*, **22** (2002), 437–468.

[FK04] B. FAYAD AND A. B. KATOK, Constructions in elliptic dynamics, *Ergod. Theory Dyn. Syst.*, **24** (2004), 1477–1520.

[FK09] B. FAYAD AND R. KRIKORIAN, Herman's last geometric theorem, *Ann. Sci. Éc. Norm. Supér.* (4) **42** (2009), 193–219.

[FS05] B. FAYAD AND M. SAPRYKINA, weakly mixing disc and annulus diffeomorphisms with arbitrary Liouville rotation number on the boundary, *Ann. Sci. Éc. Norm. Supér.* (4) **38** (2005), 339–364.

[FZ17] B. FAYAD AND Z. ZHANG, An effective version of Katok's horseshoe theorem for conservative C^2 surface diffeomorphisms, *J. Mod. Dyn.*, **11** (2017), 425–445.

[FGM91] M. FERNANDEZ, A. GRAY, AND J. MORGAN, Compact symplectic manifolds with free circle actions, and Massey products, *Michigan Math. J.*, **38** (1991), 271–283.

[Fra88a] J. FRANKS, Generalizations of the Poincaré–Birkhoff theorem, *Annals of Math. (2)*, **128** (1988), 139–151; Erratum to "Generalizations of the Poincaré–Birkhoff theorem," *Ann. Math. (2)*, **164** (2006), 1097–1098.

[Fra88b] J. FRANKS, Recurrence and fixed points of surface homeomorphisms, *Ergod. Theory Dyn. Syst.*, **8** (1988), Charles Conley Memorial Issue, 99–107.

[Fra89] J. FRANKS, Realizing rotation vectors for torus homeomorphisms, *Trans. Amer. Math. Soc.*, **311** (1989), 107–115.

[Fra92] J. FRANKS, Geodesics on S^2 and periodic points of annulus homeomorphisms, *Invent. Math.*, **108** (1992), 403–418.

[Fra99] J. FRANKS, The Conley index and non-existence of minimal homeomorphisms, *Proceedings of the Conference on Probability, Ergodic Theory, and Analysis (Evanston, IL, 1997)*, *Illinois J. Math.*, **43** (1999), 457–464.

[FH03] J. FRANKS AND M. HANDEL, Periodic points of Hamiltonian surface diffeomorphisms, *Geom. Topol.*, **7** (2003), 713–756.

[FH12] J. FRANKS AND M. HANDEL, Entropy zero area-preserving diffeomorphisms of S^2, *Geom. Topol.*, **16** (2012), 2187–2284.

[FM02] J. FRANKS AND M. MISIUREWICZ, Topological methods in dynamics, *Handbook of Dynamical Systems*, Vol. 1A, 547–598, North-Holland, 2002.

[GG97] J-M. GAMBAUDO AND E. GHYS, Enlacements asymptotiques, *Topology,* **36** (1997) 1355–1379.

[Gin10] V. L. GINZBURG, The Conley conjecture, *Ann. Math.*, **172** (2010), 1127–1180.

[GG17] V. L. GINZBURG AND B. GUREL, Conley conjecture revisited, *Int. Math. Res. Not.*, **2019** (2017), 761–798.

[GG18a] V. L. GINZBURG AND B. GUREL, Hamiltonian pseudo-rotations of projective spaces, *Invent. Math.*, **214** (2018), 1081–1130.

[GG19] V. L. GINZBURG AND B. GUREL, Approximate identities and Lagrangian Poincaré recurrence, *Arnold Math. J.*, 5 (2019), 5–14.

[Her79] M. R. HERMAN, Sur la conjugasion différentiable des difféomorphismes du cercle a des rotations, *Inst. Hautes Études Sci. Publ. Math.*, **49** (1979), 5–233.

[Her98] M. R. HERMAN, Some open problems in dynamical systems, *Proceedings of the International Congress of Mathematicians*, vol. II (Berlin 1998), Extra vol. II, 797–808.

[HCP19] L. HERNÁNDEZ-CORBATO AND F. PRESAS, The conjugation method in symplectic dynamics, *Rev. Mat. Iberoam*, **35** (2019), 1–22.

[Hin09] N. HINGSTON, Subharmonic solutions of Hamiltonian equations on tori, *Ann. of Math.*, **170** (2009), 525–560.

[Hof90] H. HOFER, On the topological properties of a symplectic map, *Proc. Roy. Soc. Edinburgh Sect.* A **115** (1990), 25–38.

[HWZ98] H. HOFER, K. WYSOCKI, AND E. ZEHNDER, The dynamics on three-dimensional strictly convex energy surfaces, *Ann. Math.* **148** (1998), 197–289.

[HWZ02] H. HOFER, K. WYSOCKI, AND E. ZEHNDER, Pseudoholomorphic curves and dynamics in three dimensions, *Handbook of Dynamical Systems*, Vol. 1A, North-Holland, (2002), 1129–1188.

[HWZ03] H. HOFER, K. WYSOCKI, AND E. ZEHNDER, Finite energy foliations of tight three-spheres and Hamiltonian dynamics, *Ann. Math.* **157** (2003), 125–255.

[HZ94] H. HOFER AND E. ZEHNDER, Symplectic invariants and Hamiltonian dynamics, Birkhäuser.

[Hut16] M. HUTCHINGS, Mean action and the Calabi invariant, *J. Mod. Dyn.*, **10** (2016) 511–539.

[Jäg09] T. JÄGER, Linearization of conservative toral homeomorphisms, *Invent. Math.*, **176** (2009), 601–616.

[JS22] D. JOKSIMOVIĆ AND S. SEYFADDINI, A Hölder-type inequality for the C^0-distance and Anosov-Katok pseudo-rotations, arXiv:2207.11813.

[Jol21] B. JOLY, The Calabi invariant for Hamiltonian diffeomorphisms of the unit disk, arXiv:2102.09352.

[Kat67] A. B. KATOK, Spectral properties of dynamical systems with an integral invariant on the torus, *Funct. Anal. Appl.*, **1** (1967), 46–56.

[Kat73] A. B. KATOK, Ergodic perturbations of degenerate integrable Hamiltonian systems, *Izv. Akad. Nauk SSSR Ser. Mat.*, **37** (1973), 539–576.

[Kat79] A. B. KATOK, Bernoulli diffeomorphisms on surfaces, *Ann. Math.*, **110** (1979), 529–547.

[Kat80] A. B. KATOK, Lyapunov exponents, entropy and periodic orbits for diffeomorphisms, *Publ. Mathématiques*, **51** (1980), 137–174.

[Kat09] A. B. KATOK, Open questions in elliptic dynamics, Preprint www.personal.psu.edu/axk29/elliptic.pdf.

[Koc16] A. KOCSARD, On the dynamics of minimal homeomorphisms of $\mathbb{T}^2$ which are not pseudorotations, *Ann. Sci. Éc. Norm. Supér.* (4) **54** (2021), 991–1034.

[KPS16] A. KOROPECKI, A. PASSEGGI, AND M. SAMBARINO: The Franks–Misiurewicz conjecture for extensions of irrational rotations, *Ann. Sci. Éc. Norm. Supér.* (4) **54** (2021), 1035–1049 (2021).

[Kun21] P. KUNDE, On the smooth realization problem and the AbC method, in this volume.

[LM95] F. LALONDE AND D. MCDUFF, The geometry of symplectic energy, *Ann. Math.* **141** (1995), 349–371.

[LeC91a] P. LE CALVEZ, Propriétés dynamiques de l'anneau et du tore, *Astérisque*, **204** (1991), Société mathématique de France.

[LeC91b] P. LE CALVEZ, Décomposition des difféomorphismes du tore en applications déviant la verticale, *Mémoires Soc. Math. France*, **79** (1991), 1–122.

[LeC04] P. LE CALVEZ, Ensembles invariants non enlacés des difféomorphismes du tore et de l'anneau, *Invent. Math.*, **155** (2004), 561–603.

[LeC06] P. LE CALVEZ, Periodic orbits of Hamiltonian homeomorphisms of surfaces, *Duke Math. J.* **133** (2006), 125–184.

[LeC16] P. LE CALVEZ, A finite dimensional approach to Bramham's approximation theorem, *Annales de l'Institut Fourier*, **66** (2016), 2169–2202.

[LeC20] P. LE CALVEZ, Conservative surface homeomorphisms with finitely many periodic points, *J. Fixed Point Theory Appl.*, **24** (2022).

[LeC22] P. LE CALVEZ, A finite dimensional proof of a result of Hutchings about irrational pseudo-rotations, arXiv:2207.07319.

[LT18] P. LE CALVEZ AND F. TAL, Forcing theory for transverse trajectories of surface homeomorphisms, *Invent. Math.*, **212** (2018), 619–729.

[LY97] P. LE CALVEZ AND J.-C. YOCCOZ, Un théoreme d'indice pour les homéomorphismes du plan au voisinage d'un point fixe, *Ann. Math.*, **146** (1997), 241–293.

[LS20] F. LE ROUX AND S. SEYFADDINI, The Anosov–Katok method and pseudorotations in symplectic dynamics, *J. Fixed Point Theory Appl.* (to appear). arXiv:2010.06237, 2020.

[LT77] D. LIND AND J. -P. THOUVENOT, Measure-preserving homeomorphisms of the torus represent all finite entropy ergodic transformations, *Math. Systems Theory*, Vol. **11**, 1977, 275–285.

[McD] D. MCDUFF, Hamiltonian S^1-manifolds are uniruled, *Duke Math. J.*, **146** (2009), 449–507.

[MS17] D. MCDUFF AND D. SALAMON, Introduction to symplectic topology, *Oxford Graduate Texts in Mathematics*, **27**, Oxford University Press, third edition, 2017.

[NS96] A. NORTON AND D. SULLIVAN, Wandering domains and invariant conformal structures for mappings of the 2-torus, *Ann. Acad. Sci. Fenn. Math.*, **21** (1996), 51–68.

[Oh05] Y.-G. OH, Spectral invariants, analysis of the Floer moduli space, and geometry of the Hamiltonian diffeomorphism group, *Duke Math. J.*, **130** (2005), 199–295.

[Ono06] K. ONO, Floer–Novikov cohomology and the flux conjecture, *Geom. Funct. Anal.*, **16**(5) (2006) 981–1020.

[Pir21] A. PIRNAPASOV, Mean action and the Calabi invariant revisited and an application to pseudo-rotations, arXiv:2102.09533.

[Pol93] L. POLTEROVICH, Symplectic displacement energy for Lagrangian submanifolds, *Ergod. Theory Dyn. Syst.*, **13** (1993), 357–367.

[Rüs67] H. RÜSSMANN, Über die Normalform analytischer Hamiltonscher Differentialgleichungen in der Nähe einer Gleichgewichtslösung, *Math. Ann.*, **169** (1967), 55–72.

[SZ92] D. SALAMON AND E. ZEHNDER, Morse theory for periodic solutions of Hamiltonian systems and the Maslov index, *Comm. Pure Appl. Math.*, **45** (1992), 1303–1360.

[Sch00] M. SCHWARZ, On the action spectrum for closed symplectically aspherical manifolds, *Pacific J. Math.*, **193** (2000), 419–461.

[She15] E. SHELUKHIN, "enlacements asymptotiques" revisited, *Annales mathématiques du Québec,* **39** (2015), 205–208.

[She19b] E. SHELUKHIN, Pseudo-rotations and Steenrod squares, *J. Mod. Dyn.* (to appear). arXiv:1905.05108.

[She19c] E. SHELUKHIN, Pseudo-rotations and Steenrod squares revisited, *Math. Res. Lett.* (to appear). arXiv:1909.12315, 2019.

[She22] E. SHELUKHIN, On the Hofer–Zehnder conjecture, *Ann. Math.*, (2) **195** (2022), 775–839.

[Shk67] M. D. SHKLOVER, On dynamical systems on the torus with continuous spectrum, *Izv. Vuzov*, **10** (1967), 113–124.

[Ush10] M. USHER, The sharp energy-capacity inequality, *Commun. Contemp. Math.*, **12** (2010), 457–473.

[Vit92] C. VITERBO, Symplectic topology as the geometry of generating functions, *Math. Ann.*, **292** (1992), 685–710.

[WY21] J. WANG AND H. YANG, A question of Norton–Sullivan in the analytic case, *International Mathematics Research Notices*, **2021** (22), 17201–17219.

[WZ18] J. WANG AND Z. ZHANG, The rigidity of pseudo-rotations on the two-torus and a question of Norton-Sullivan, *Geom. Funct. Anal.*, **28** (2018), 1487–1516.

[Yoc84] J.-C. YOCCOZ, Conjugaison différentiable des difféomorphismes du cercle dont le nombre de rotation vérifie une condition diophantienne. (French) [Differentiable conjugation of diffeomorphisms of the circle whose rotation number satisfies a Diophantine condition], *Ann. Sci. École Norm. Sup.*, **17** (1984), 333–359.

[Yoc95b] J.-C. YOCCOZ, Centralisateurs et conjugaison différentiable des difféomorphismes du cercle, *Astérisque*, **231** (1995), 89–242.

5

SRB and Equilibrium Measures via Dimension Theory

VAUGHN CLIMENHAGA*

For Tolya, who taught me to learn things a second time.

It is well known that SRB (Sinai–Ruelle–Bowen) and equilibrium measures for uniformly hyperbolic flows admit a product structure in terms of measures on stable and unstable leaves with scaling properties given by the potential function. We describe a construction of these leaf measures analogous to the definition of Hausdorff measure, relying on the Pesin–Pitskel's description of topological pressure as a dimensional characteristic using Bowen balls. These leaf measures were constructed for discrete-time systems by the author, Ya. Pesin, and A. Zelerowicz. In the continuous-time setting here, the description of the scaling properties is more complete, and we use a similar procedure with two-sided Bowen balls to directly produce the equilibrium measure itself.

5.1 Introduction

5.1.1 Overview

Our story takes place where dynamical systems meets two other major theories: thermodynamics and dimension theory. In dimension theory, Hausdorff dimension is defined via a family of measures with prescribed scaling behavior when we zoom in geometrically; for a given set, exactly one of these measures has the correct scaling properties. In thermodynamic formalism, one can use a similar approach to define topological entropy [Bow73] and topological pressure [PP84] via a family of measures that have prescribed scaling behavior when we zoom in dynamically. Exactly one of these measures has the correct scaling properties.

The rest of this chapter explores the following idea, introduced in [CPZ19, CPZ20]: *for a uniformly hyperbolic system, the measure produced by this approach*

* The author was partially supported by NSF grant DMS-1554794.

from dimension theory can be used to describe the equilibrium measure in thermodynamic formalism and its conditional measures on stable and unstable leaves.

We emphasize that the leaf measures we study, and their relationship to the equilibrium measure, are by no means new; see Section 5.1.3. The novelty here consists of the following two aspects:

- the observation here and in [CPZ19, CPZ20] that these measures can be constructed using tools from dimension theory (though this was foreshadowed by [Ham89, Has89], which studied the measure of maximal entropy);
- the adaptation of this technique to give a direct formula for the equilibrium measure itself, without working on the leaves.

To illustrate this second point, we give a new construction for the Sinai–Ruelle–Bowen (SRB) measure on a hyperbolic attractor; see Section 5.2 for full definitions, and Section 5.3.5 for an extension to equilibrium measures. The construction uses a version of "two-sided Bowen balls" $B^*_{s,t}(x,r)$ whose definition requires the following local product structure of a hyperbolic attractor Λ for a smooth flow $(f_t)_{t\in\mathbb{R}}$: there is a constant $L > 0$ such that for all sufficiently small $r > 0$, any $x \in \Lambda$, and any $z \in B(x,r)$; there is a point y on the local unstable manifold of z such that $f_\beta(y)$ is on the local stable manifold of x for some $\beta = \beta(x,z) \in [-Lr, Lr]$. For large values of $s, t > 0$, we will use the sets

$$B^*_{s,t}(x,r) := \left\{ z \in \Lambda\colon \sup_{\tau\in[-s,t]} d(f_\tau z, f_\tau x) < r \text{ and } |\beta(x,z)| < \frac{r}{s+t} \right\}.$$

Theorem 1.1. *Let M be a Riemannian manifold and $\Lambda \subset M$ a hyperbolic attractor for a smooth flow $(f_t)_{t\in\mathbb{R}}$. Then for all sufficiently small $r > 0$, there is $c > 0$ such that the SRB measure μ on Λ is given by*

$$\mu(Z) = \lim_{T\to\infty} \inf \sum_{(x,s,t)\in\mathcal{E}} \frac{c}{s+t} \det(Df_{s+t}|_{E^{\mathrm{u}}_{f_{-s}(x)}})^{-1}$$

for all Borel $Z \subset \Lambda$, where the infimum is over all $\mathcal{E} \subset \Lambda \times [T,\infty)^2$ such that $Z \subset \bigcup_{(x,s,t)\in\mathcal{E}} B^*_{s,t}(x,r)$.

5.1.2 Topological Pressure and Equilibrium Measures

Throughout the chapter, M is a smooth compact Riemannian manifold, and $F = (f_t)_{t\in\mathbb{R}}$ is a smooth flow on M. We will consider a compact F-invariant set $\Lambda \subset M$, which in all our results is assumed to be a topologically transitive locally maximal hyperbolic set. Given a continuous function $\varphi\colon \Lambda \to \mathbb{R}$, which we call a *potential*, the *topological pressure* $P(\varphi)$ of (Λ, F, φ) can be defined in three ways.

Definition 1.2 (Variational principle).

$$P(\varphi) = \sup \Big(h_\mu(f_1) + \int \varphi \, d\mu \Big), \tag{5.1}$$

where $h_\mu(f_1)$ is measure-theoretic entropy and the supremum is over all Borel F-invariant probability measures μ on Λ.

Definition 1.3 (Growth rate). Writing $\Phi(x,t) = \int_0^t \varphi(f_s x)\, ds$,

$$P(\varphi) = \lim_{r\to 0} \varlimsup_{t\to\infty} \frac{1}{t} \log \inf \sum_{x\in E} e^{\Phi(x,t)}, \tag{5.2}$$

where the infimum is over all sets $E \subset \Lambda$ such that Λ is covered by the collection $\{B_t(x,r) : x \in E\}$ of *Bowen balls* given by

$$B_t(x,r) = \{ y \in M : d(f_s x, f_s y) < r \text{ for all } s \in [0,t] \}.$$

Definition 1.4 (Dimension via a critical value). Fixing $r > 0$, define for each $\alpha \in \mathbb{R}$ an outer measure m^α on Λ by

$$m^\alpha(Z) = \lim_{T\to\infty} \inf \sum_{(x,t)\in\mathcal{E}} e^{\Phi(x,t) - t\alpha}, \tag{5.3}$$

where the infimum is over all finite or countable sets $\mathcal{E} \subset \Lambda \times [T, \infty)$ such that the Bowen balls $\{B_t(x,r) : (x,t) \in \mathcal{E}\}$ cover Z.[1] Then there is a critical value of α given by

$$P(\varphi, r) = \sup\{\alpha : m^\alpha(\Lambda) = \infty\} = \inf\{\alpha : m^\alpha(\Lambda) = 0\}, \tag{5.4}$$

and the topological pressure is $P(\varphi) = \lim_{r\to 0} P(\varphi, r)$.

A measure achieving the supremum in the variational principle is called an *equilibrium measure*; this includes measures of maximal entropy and SRB measures. Questions of existence, uniqueness, and properties of equilibrium measures lie at the heart of the study of thermodynamic formalism for dynamical systems.

When Λ is a topologically transitive locally maximal hyperbolic set and $\varphi : \Lambda \to \mathbb{R}$ is Hölder continuous, it is well known that there is a unique equilibrium measure μ, which is also the unique probability measure satisfying the following *Gibbs property*: for all sufficiently small $r > 0$ there is $Q = Q(r)$ such that[2]

$$\mu(B_t(x,r)) = Q^{\pm 1} e^{\Phi(x,t) - tP(\varphi)} \text{ for all } x \in \Lambda \text{ and } t > 0. \tag{5.5}$$

[1] A crucial difference between this and the previous definition is that here the order t may vary from one Bowen ball to the next.

[2] We write $A = Q^{\pm 1} B$ to mean $Q^{-1} B \le A \le QB$.

In this setting we have $P(\varphi) = P(\varphi, r)$ for all sufficiently small r, and choosing $\alpha = P(\varphi) = P(\varphi, r)$ we see that the weight in (5.3) associated to each (x, t) is exactly $e^{\Phi(x,t)-tP(\varphi)}$ as in (5.5).

In this chapter we describe three related constructions of μ based on dimension theory, and in particular on the outer measure in (5.3).

(1) *Pushforward and average.* With $\alpha = P(\varphi)$, (5.3) defines a finite Borel measure m_x^{u} on each local unstable manifold. The averaged pushforwards $\nu_t = \frac{1}{t}\int_0^t (f_s)_* m_x^{\mathrm{u}}\, ds$ converge in the weak* topology to a scalar multiple of the unique equilibrium measure μ. See Theorem 3.1.
(2) *Construct a product.* Reversing time gives an analogous construction of Borel measures m_x^{s} on stable leaves. Using the product structure (unstable) $\times$ (stable) $\times$ (flow direction), one can construct a product measure on neighborhoods in Λ by $m_x^{\mathrm{u}} \times m_x^{\mathrm{s}} \times \mathrm{Leb}$, which is equivalent to μ. See Theorem 3.10.
(3) *A two-sided definition.* Theorem 1.1 is a specific case of a more general result that adapts Definition 1.4, using two-sided Bowen balls to refine in both the stable and unstable directions, and refining "manually" in the flow direction, thus directly obtaining a Borel measure on Λ, which is a scalar multiple of μ. See Theorem 3.15.

In the first two constructions involving the leaf measures, an important role is played by the scaling properties of these measures under the flow and under holonomy maps; see Theorem 3.4. In fact, these scaling properties characterize the leaf measures up to a scalar; see Corollaries 3.11 and 3.12, where this is deduced from uniqueness of μ together with the fact that the density function in Theorem 3.10 can be written down explicitly, which allows us to recover the leaf measures from the conditional measures of μ. In Corollary 3.13, we show how these results give a direct construction of the conditional measures of the SRB measure along *stable* leaves.

5.1.3 History and Related Results

The author is not aware of any analogues in the literature of the third construction just mentioned, via two-sided Bowen balls as in Theorem 1.1 and Section 5.3.5. For the first two constructions, the history is richer.

5.1.3.1 Pushforward and Average

Applying the "pushforward and average" procedure to an initial reference measure is a standard way of proving the Krylov–Bogolubov theorem on existence of an invariant measure. For uniformly hyperbolic systems, Sinai [Sin68] and Ruelle [Rue76] proved that taking volume as the initial reference measure leads to a limiting measure whose conditionals on unstable leaves are equivalent to leaf volume, which in turn shows that this *Sinai–Ruelle–Bowen (SRB) measure* governs the asymptotic behavior of volume-typical trajectories. Pesin and Sinai [PS82] used leaf volume as the original reference measure and generalized the result on unstable conditionals to the partially hyperbolic setting. See [CLP17] for an overview of this approach and of SRB measures more generally.

This approach was extended beyond the SRB case to more general equilibrium measures in recent joint work of the author with Ya. Pesin and A. Zelerowicz [CPZ19, CPZ20]. This showed that for uniformly (and some partially) hyperbolic diffeomorphisms, (5.3) defines a finite Borel measure on each unstable leaf whose averaged pushforwards converge to a multiple of the unique equilibrium measure. Theorem 3.1 uses this to deduce the result for flows.

5.1.3.2 Construct a Product

The "construct a product" procedure has been extensively explored for the measure of maximal entropy (MME), where it was carried out by Sinai for Anosov diffeomorphisms [Sin68] and Margulis for Anosov flows [Mar70]; this was extended to the Axiom A setting by Ruelle and Sullivan for diffeomorphisms [RS75], and Bowen and Marcus for flows [BM77]. Margulis uses a functional analytic approach, while the other papers rely on Markov partitions.

Let us highlight two important precursors to the present work, which constructed the leaf measures for the MME as the Hausdorff measures associated to a dynamically defined metric. This was done for geodesic flows in negative curvature by Hamenstädt [Ham89] and for more general Anosov flows by Hasselblatt [Has89]. It is reasonable to view Theorem 3.10 as an extension of these results to equilibrium measures associated to nonconstant potential functions; see Section 5.4.3.3.

For nonconstant potential functions, the product structure of the corresponding equilibrium measure was described by Haydn [Hay94] for flows, and by Leplaideur [Lep00] for diffeomorphisms. Both of these papers established the scaling properties of the leaf measures under the dynamics and under holonomy; Haydn referred to these properties as *Margulis' cocycle equations*. Haydn's proof uses Markov partitions and the correspondence between equilibrium measures for a map and its suspension flow [BR75]. Leplaideur's proof does not use a full-fledged Markov partition or symbolic dynamics, but does require a proper rectangle with the Markov property, on which the first return map is considered. Both proofs use the Perron–Frobenius operator and obtain the leaf measures via its eigendata.

In light of the previous paragraph, it is worth pointing out that the construction of an equilibrium state over a subshift of finite type using Ruelle's Perron–Frobenius operator [Bow08] can be viewed as a product construction, even when the underlying system is non-invertible; one can interpret the eigenmeasure as a measure on local unstable leaves, and the eigenfunction as giving the total weight of local stable leaves in the natural extension.

5.1.3.3 Reliance on the Literature

The proofs we give in Section 5.4 involve a certain amount of "cheating" by relying on known results from the literature, up to and including the fact that there is a unique equilibrium measure and that this is also the unique invariant Gibbs probability measure. With a little more effort (and a longer chapter), the arguments here could be made self-contained. This would require us to provide proofs of the following facts.

(1) Uniform counting bounds for leafwise partition sums: these play a key role in the proof that the measures we construct are positive and finite. See [CPZ20, §6] for details of this argument in the discrete-time case, using ideas that go back to Bowen's work using the specification property [Bow75].
(2) Ergodicity of the invariant measure built using any of the three constructions. In [CPZ19, CPZ20] this is done for the discrete-time case using the Hopf argument.
(3) An ergodic Gibbs measure must be the unique equilibrium measure; this argument is standard and can be found in [Bow75].

We have chosen to shorten the proofs in this chapter by relying on known results, but for potential applications to non-uniform hyperbolicity (see the next section) it would likely be productive to seek direct proofs of these facts along the lines in [CPZ19, CPZ20].

5.1.4 Non-uniform Hyperbolicity

For uniformly hyperbolic systems, thermodynamic formalism is very well understood, and in this setting the results in this chapter are mostly intended to illuminate a well-known subject from a slightly different angle. It may be hoped, however, that the approach presented here will eventually bear fruit for non-uniformly hyperbolic systems, where the story is much less complete [CP17].

One well-established approach to thermodynamic formalism for non-uniformly hyperbolic systems uses a Markov structure in one guise or another, via countable-state Markov partitions, Young towers, or inducing schemes; see [CP17] for an overview of the relevant literature. More recently, the author and D. J. Thompson have introduced an alternate approach based on Bowen's proof of uniqueness for uniformly hyperbolic systems using expansivity and specification [Bow75]. There are non-uniform versions of these properties [CT12, CT16] that have found applications in examples such as geodesic flow in nonpositive curvature [BCFT18] and no conjugate points [CKW21]; see the forthcoming survey [CT21] for a more complete description of this approach.

If one wants to go beyond uniqueness of the equilibrium measure and establish strong statistical properties, local product structure, and so on, then the well-established symbolic approach gives stronger results than the more recent specification-based approach.[3] Given that the approach in the present chapter shares much in common with the specification-based approach, such as a reliance on uniform counting bounds and bare-hands constructions, it is worth asking whether the present approach can also be generalized to non-uniform hyperbolicity, and then used to establish some of these stronger results.

One example of this symbiosis is found in recent work by the author, Gerhard Knieper, and Khadim War [CKW21] that used the specification approach to establish uniqueness of the MME for geodesic flows on surfaces with no conjugate points, and then used the Patterson–Sullivan approach to obtain a product structure for this measure; this in turn can be used to establish the Margulis asymptotics for the number of closed geodesics on such surfaces [CKW22]. The Patterson–Sullivan approach can be viewed as a leafwise construction analogous to the present one [Kai90], and thus one may reasonably hope to carry out a similar procedure in other non-geometric examples, including more general non-uniformly hyperbolic systems, hyperbolic systems with singularities such as billiards, or singular hyperbolic flows such as the Lorenz attractor. Of course, one would need to first carry out the specification approach in these settings, or at least adapt enough of it to prove the uniform counting bounds necessary for the construction here, and so for the time being this is all rather aspirational.

Outline of the Chapter

In Section 5.2, we recall some background definitions from the literature on hyperbolic sets for flows and on Carathéodory dimension characteristics. In Section 5.3, we state the main results that were briefly described earlier. We give the proofs in Section 5.4.

5.2 Background Definitions

The reader who is familiar with hyperbolic dynamics can likely skip this section on a first reading and proceed directly to the main results in Section 5.3. This section contains basic definitions and standard facts about hyperbolic sets for flows (Section 5.2.1), including some lemmas that must be stated before our main results

[3] However, it is worth mentioning recent work of Call and Thompson [CT22, Cal22] that uses the specification approach to establish the K property, and even Bernoullicity in some cases.

in order to establish the scale at which various constructions are made; it also describes the general notion of Carathéodory dimension characteristic (Section 5.2.2) that we use when we define various measures in Section 5.3. We refer to the recent book of Fisher and Hasselblatt [FH19, Chapters 5 and 6] for proofs of facts about hyperbolic flows that are omitted here, and to Pesin's book [Pes97] for a more in-depth treatment of dimension theory in dynamical systems.

5.2.1 Hyperbolic Sets

Throughout, recall that M is a smooth compact Riemannian manifold, and $F = (f_t)_{t\in\mathbb{R}}$ is a smooth flow on M.

Definition 2.1 (Hyperbolic sets). A *hyperbolic set* for F is a compact F-invariant set $\Lambda \subset M$ over which the tangent bundle admits a DF-invariant splitting $T_\Lambda M = E^{\mathrm{s}} \oplus E^{\mathrm{u}} \oplus E^0$ such that E^0 is the flow direction,[4] E^{s} is uniformly contracting, and E^{u} is uniformly expanding: there are $C_0 \geq 1$ and $\chi \in (0,1)$ such that

$$\|Df_t|_{E^{\mathrm{s}}}\| \leq C_0\chi^t \quad \text{and} \quad \|Df_{-t}|_{E^{\mathrm{u}}}\| \leq C_0\chi^t \quad \text{for all } t \geq 0. \tag{5.6}$$

By passing to an *adapted metric* and increasing χ slightly,[5] we assume without loss of generality that $C_0 = 1$. In the adapted metric, we can also assume that $E^{\mathrm{s}}, E^{\mathrm{u}}, E^0$ are all orthogonal, and that the flow has unit speed so that $\|Df_t|_{E^0}\| = 1$ for all $t \in \mathbb{R}$.

We write d for the distance function on M induced by the (adapted) Riemannian metric, and $B(x,r)$ for the ball centered at $x \in M$ with radius $r > 0$. More generally, given an injectively immersed connected submanifold $W \subset M$, we will write d_W for the distance function on W induced by the restriction of the Riemannian metric, and

$$B(x,r,W) = \{y \in W : d_W(y,x) < r\} \tag{5.7}$$

for the corresponding ball in W centered at $x \in W$ with radius $r > 0$.

Definition 2.2 (Cone fields on Λ). Given $\kappa > 0$, the *stable and unstable cone fields of width κ on Λ* are

$$\begin{aligned} C^u_x = C^u_x(\kappa) &:= \{v + w : v \in E^{\mathrm{u}}_x,\ w \in E^{\mathrm{s}}_x \oplus E^0_x,\ \|w\| < \kappa\|v\|\}, \\ C^s_x = C^s_x(\kappa) &:= \{v + w : v \in E^{\mathrm{s}}_x,\ w \in E^{\mathrm{u}}_x \oplus E^0_x,\ \|w\| < \kappa\|v\|\}. \end{aligned} \tag{5.8}$$

[4] We do not allow Λ to contain fixed points for the flow, so E^0 is always a one-dimensional subspace.

[5] If we do not pass to an adapted metric, then our constructions give equivalent measures, but we do not have a formula for the Radon–Nikodym derivatives and thus cannot write down the product construction explicitly, or recover the leaf measures from the conditional measures; see Section 5.4.3.3. In Section 5.3.5 we will drop the assumption that the metric is adapted.

For every $t > 0$ and $x \in \Lambda$ we have

$$\overline{Df_t(x)(C^u_x)} \subset C^u_{f_tx} \quad \text{and} \quad \overline{Df_{-t}(x)(C^s_x)} \subset C^s_{f_{-t}x}. \tag{5.9}$$

Moreover, for every $\theta, \kappa > 0$ there is $t_0 \in \mathbb{R}$ such that given any $x \in \Lambda$ and $v \in T_xM$ with $\angle(v, E^s_x \oplus E^0_x) \geq \theta$, we have

$$Df_t(x)(v) \in C^u_{f_tx}(\kappa) \quad \text{for all } t \geq t_0. \tag{5.10}$$

Fixing $\bar{\chi} \in (\chi, 1)$, by (5.6) and (5.9) we can choose $\kappa > 0$ sufficiently small that for all $x \in \Lambda$ and $t \geq 0$, we have

$$\begin{aligned} \|Df_t(x)(v)\| &\geq \bar{\chi}^{-t}\|v\| \text{ for all } v \in C^u_x(\kappa), \\ \|Df_{-t}(x)(v)\| &\geq \bar{\chi}^{-t}\|v\| \text{ for all } v \in C^s_x(\kappa). \end{aligned} \tag{5.11}$$

From now on, C^u_x and C^s_x will be defined with this value of κ unless otherwise specified.

Definition 2.3 (Extension to a neighborhood). Extend the distributions E^s and E^u continuously to a neighborhood of Λ (note that they need not be invariant off of Λ). Define cone families C^s_x, C^u_x on this neighborhood by (5.8), with κ as in (5.11). Fixing $\lambda \in (\bar{\chi}, 1)$, there is $r_0 > 0$ such that if $x \in M$ and $T > 0$ have the property that $f_t(x) \in B(\Lambda, r_0) := \bigcup_{y \in \Lambda} B(y, r_0)$ for every $t \in [0, T]$, then for all $t \in (0, T]$ we have

$$\overline{Df_t(x)(C^u_x)} \subset C^u_{f_tx} \text{ and } \|Df_t(x)(v)\| \geq \lambda^{-t}\|v\| \text{ for all } v \in C^u_x, \tag{5.12}$$

and similarly for the stable cones with time reversed.

Definition 2.4 (Admissible manifolds). An embedded connected submanifold $W \subset B(\Lambda, r_0)$ is *u-admissible* if $T_xW \subset C^u_x$ for all $x \in W$,[6] and *s-admissible* if $T_xW \subset C^s_x$ for all $x \in W$.

It follows from (5.12) that

$$\begin{aligned} &\text{if } W \text{ is } u\text{-admissible and } f_\tau(W) \subset B(\Lambda, r_0) \text{ for all } \tau \in [0, t], \\ &\qquad \text{then } d_W(x, y) \leq \lambda^t d_{f_tW}(f_tx, f_ty) \text{ for all } x, y \in W. \end{aligned} \tag{5.13}$$

A similar result holds for s-admissible manifolds if we replace f_τ and f_t by $f_{-\tau}$ and f_{-t}.

[6] This definition does not require W to have the same dimension as E^u, but in practice we will only consider u-admissible manifolds of maximal dimension.

Definition 2.5 (Global manifolds $W^{\mathrm{s}}(x)$, $W^{\mathrm{u}}(x)$). Given a point $x \in \Lambda$, the *global stable and unstable manifolds* of x are

$$W^{\mathrm{s}}(x) = \{y \in M : d(f_t y, f_t x) \to 0 \text{ as } t \to \infty\},$$
$$W^{\mathrm{u}}(x) = \{y \in M : d(f_t y, f_t x) \to 0 \text{ as } t \to -\infty\}.$$

These are injectively immersed manifolds such that $T_x W^*(x) = E^*_x$ for every $x \in \Lambda$ and $* \in \{\mathrm{s}, \mathrm{u}\}$. They are clearly flow-invariant in the sense that $f_t(W^*(x)) = W^*(f_t x)$ for every $x \in \Lambda$, $* \in \{\mathrm{s}, \mathrm{u}\}$, and $t \in \mathbb{R}$.

Definition 2.6 (Local manifolds $W^{\mathrm{s}}(x,\delta)$, $W^{\mathrm{u}}(x,\delta)$). Given $x \in \Lambda$ and $\delta > 0$, let $W^{\mathrm{s}}(x,\delta) := B(x,\delta,W^{\mathrm{s}}(x))$, and similarly for W^{u}. There exists $\delta_0 > 0$ such that for all $x \in \Lambda$ and $\delta \in (0,\delta_0]$, the set $W = W^{\mathrm{s}}(x,\delta)$ has the following properties:

- it is s-admissible;[7]
- there is a topological ball $B \subset E^{\mathrm{s}}_x$ containing the origin and a function $\psi : B \to E^{\mathrm{u}}_x$ such that $W = \exp_x(\operatorname{graph}\psi)$.

We refer to any $W \subset W^{\mathrm{s}}(x)$ with these properties as a *local stable manifold*, or a *local stable leaf.* Local unstable manifolds are defined analogously, reversing the roles of s, u. These local leaves are C^1 and depend continuously on x in the C^1 topology.

If W is a local stable manifold of x, then $f_t(W)$ is a local stable manifold of $f_t(x)$ for all $t \geq 0$, and similarly for unstable manifolds with $t \leq 0$. Thus (5.13) and its analogue for stable manifolds gives

$$\begin{aligned} d(f_t y, f_t z) &\leq \lambda^t d(y,z) \text{ for all } y,z \in W^{\mathrm{s}}(x,\delta_0) \text{ and } t \geq 0,\\ d(f_{-t} y, f_{-t} z) &\leq \lambda^t d(y,z) \text{ for all } y,z \in W^{\mathrm{u}}(x,\delta_0) \text{ and } t \geq 0, \end{aligned} \tag{5.14}$$

where here and throughout the remainder of the paper $\delta_0 > 0$ is fixed as in Definition 2.6.

Definition 2.7 (Weak manifolds $W^{cs,cu}$). Given $x \in \Lambda$ and $\delta \in (0,\delta_0]$, we consider the following *local weak stable and unstable manifolds*:

$$W^{\mathrm{cs}}(x,\delta) = \{f_t(y) : y \in W^{\mathrm{s}}(x,\delta), |t| < \delta\},$$
$$W^{\mathrm{cu}}(x,\delta) = \{f_t(y) : y \in W^{\mathrm{u}}(x,\delta), |t| < \delta\}.$$

7 Observe that in the Anosov case $\Lambda = M$, every submanifold of $W^{\mathrm{s}}(x)$ is s-admissible, but that when $\Lambda \neq M$ there may be submanifolds of $W^{\mathrm{s}}(x)$ that are not s-admissible because they contain points that are too far from Λ.

We also consider the *global weak stable and unstable manifolds*

$$W^{\mathrm{cs}}(x) = \{f_t y : y \in W^{\mathrm{s}}(x), t \in \mathbb{R}\} = \bigcup_{t\in\mathbb{R}} W^{\mathrm{s}}(f_t x),$$
$$W^{\mathrm{cu}}(x) = \{f_t y : y \in W^{\mathrm{u}}(x), t \in \mathbb{R}\} = \bigcup_{t\in\mathbb{R}} W^{\mathrm{u}}(f_t x).$$

Definition 2.8 (Weak-stable transversals). A C^1 injectively immersed connected submanifold $W \subset M$ is *transverse to the weak-stable direction* if $T_x M = T_x W \oplus T_x W^{\mathrm{cs}}(x) = T_x W \oplus E^{\mathrm{s}}_x \oplus E^0_x$ for all $x \in W \cap \Lambda$. Write $\mathcal{W}^u$ for the set of all such W.

Given $\theta > 0$, let $\mathcal{W}^u_\theta$ denote the set of $W \in \mathcal{W}^u$ such that $\angle(w, v) \geq \theta$ for any $x \in W \cap \Lambda$, $w \in T_x W$, and $v \in T_x W^{\mathrm{cs}}(x) = E^{\mathrm{s}}_x \oplus E^0_x$. Reversing the roles of u and s, we define $\mathcal{W}^s$ and $\mathcal{W}^s_\theta$ analogously.

Using (5.10), there exists $r_1 \in (0, r_0]$ such that the following is true.

Lemma 2.9. *Given any $W \in \mathcal{W}^u$ and $x \in W \cap \Lambda$, we have*

$$\lim_{t\to\infty} d_{C^1}(B(f_t x, r_1, f_t W), W^{\mathrm{u}}(f_t x, r_1)) = 0. \tag{5.15}$$

Given $\theta > 0$, this convergence is uniform over all $W \in \mathcal{W}^u_\theta$ and $x \in W \cap \Lambda$. In particular, there is $t_1 = t_1(\theta) \geq 0$ such that for any $W \in \mathcal{W}^u_\theta$, $x \in W \cap \Lambda$, and $t \geq t_1$, the manifold $B(f_t x, r_1, f_t W)$ is u-admissible.

Our main results will be given for a hyperbolic set satisfying the following two additional properties.

- *Local maximality*: $\Lambda = \bigcap_{t\in\mathbb{R}} f_t(U)$ for some open $U \supset \Lambda$.
- *Topological transitivity*: Λ contains a dense orbit.

Definition 2.10 (Basic sets). A topologically transitive locally maximal hyperbolic set Λ will be referred to as a *basic set*.

Definition 2.11 (Local product structure). There are $L, \delta_1 > 0$ such that if $x, y \in \Lambda$ have $d(x, y) < \delta_1$, then the intersection $W^{\mathrm{cs}}(x, L\delta_1) \cap W^{\mathrm{u}}(y, L\delta_1)$ consists of a single point, which we denote $[x, y]$ and refer to as the *bracket* of x and y.[8] Moreover, $[x, y]$ lies in both of $W^{\mathrm{cs}}(x, Ld(x, y))$ and $W^{\mathrm{u}}(y, Ld(x, y))$. A hyperbolic set Λ has *local product structure* if $[x, y] \in \Lambda$ for every $x, y \in \Lambda$ with $d(x, y) < \delta_1$.

Given a hyperbolic set Λ, local product structure is equivalent to local maximality. In particular, basic sets have local product structure.

8 This is sometimes called the *Bowen bracket* or the *Smale bracket*.

Definition 2.12 (Attractors)**.** A basic set Λ is an *attractor* if there is an open set $U \supset \Lambda$ such that $\overline{f_t(U)} \subset U$ for all $t > 0$.

5.2.2 Carathéodory Dimension and Measure

We recall a construction from [Pes97, §10] generalizing Hausdorff dimension and measure.

Definition 2.13. A *Carathéodory dimension structure*, or *C-structure*, on a set X is given by the following data.

(1) An indexed collection $\mathcal{F} = \{U_s \subset X : s \in \mathcal{S}\}$ of subsets of X.
(2) Functions $\xi, \eta, \psi : \mathcal{S} \to [0, \infty)$ such that
   (H1) $\eta(s) = 0$ if and only if $U_s = \emptyset$, and similarly for ψ;[9]
   (H2) for every $\delta > 0$, there exists $\varepsilon > 0$ such that $\eta(s) \leq \delta$ for any $s \in \mathcal{S}$ with $\psi(s) \leq \varepsilon$;
   (H3) for every $\varepsilon > 0$, there is a finite or countable $\mathcal{G} \subset \mathcal{S}$ such that $\bigcup_{s\in\mathcal{G}} U_s \supset X$ and $\psi(\mathcal{G}) := \sup\{\psi(s) : s \in \mathcal{S}\} \leq \varepsilon$.

No conditions are placed on ξ.

Given a C-structure $(\mathcal{S}, \mathcal{F}, \xi, \eta, \psi)$, a set $Z \subset X$, and $\varepsilon > 0$, let

$$\mathbb{E}(Z, \varepsilon) = \Big\{\mathcal{G} \subset \mathcal{S} : Z \subset \bigcup_{s\in\mathcal{G}} U_s \text{ and } \psi(\mathcal{G}) \leq \varepsilon\Big\}.$$

By [Pes97, Proposition 1.1], the following defines an outer measure on X for each $\alpha \in \mathbb{R}$:

$$m_C^\alpha(Z) := \lim_{\varepsilon\to 0} \inf_{\mathcal{G}\in\mathbb{E}(Z,\varepsilon)} \sum_{s\in\mathcal{G}} \xi(s)\eta(s)^\alpha, \tag{5.16}$$

where $m_C^\alpha(\emptyset) := 0$. Then by [Pes97, Proposition 1.2] there is a unique $\alpha_C \in [-\infty, \infty]$ such that $m_C^\alpha(X) = \infty$ for all $\alpha < \alpha_C$, and $m_C^\alpha(X) = 0$ for all $\alpha > \alpha_C$. Writing $\dim_C(X) = \alpha_C$ for this *Carathéodory dimension* of X, one is naturally led to the question of whether $m_C^{\alpha_C}(X)$ is positive and finite. This requires further information about the C-structure, which we will explore more in the next section.

When X is a metric space, one may ask whether $m_C := m_C^{\alpha_C}$ gives a Borel measure on X. Recall that an outer measure m on a metric space X is a *metric outer measure* if $m(A \cup B) = m(A) + m(B)$ whenever $d(A, B) := \inf\{d(x, y) : x \in A, y \in B\} > 0$. Every metric outer measure has the property that all Borel sets are measurable; see [Roy88, Proposition 12.41] or [Fed69, §2.3.2(9)].

9 In [Pes97] there is an extra requirement that the case $U_s = \emptyset$ occur, but this can be safely omitted by defining $m_C^\alpha(\emptyset) = 0$ in (5.16).

Lemma 2.14. *Suppose $\{\mathcal{F}, \xi, \eta, \psi\}$ is a C-structure with the property that for every $\rho > 0$, there is $\varepsilon > 0$ such that $\psi(s) \leq \varepsilon$ implies* $\operatorname{diam}(U_s) < \rho$. *Then $m_C = m_C^{\alpha C}$ is a metric outer measure, and hence gives a Borel measure on X.*

Proof. If $d(A,B) > 0$, there is $\varepsilon > 0$ such that if U_s intersects A, U_t intersects B, and $\psi(s), \psi(t) \leq \varepsilon$, then $U_s \cap U_t = \emptyset$, which means that every $\mathcal{G} \in \mathbb{E}(A \cup B, \varepsilon)$ splits into two disjoint subsets, one covering A and the other covering B. This implies that $m_C(A \cup B) = m_C(A) + m_C(B)$. □

5.3 Main Results

All the results in this section will be proved in §5.4.

5.3.1 Reference Measures on Unstable Leaves and Transversals

From now on we assume that Λ is a basic set for a flow $F = (f_t)_{t \in \mathbb{R}}$ and $\varphi \colon \Lambda \to \mathbb{R}$ is a Hölder continuous potential function: that is, there is $\sigma > 0$ such that

$$|\varphi|_\sigma := \sup \left\{ \frac{|\varphi(x) - \varphi(y)|}{d(x,y)^\sigma} : x, y \in \Lambda, x \neq y \right\} < \infty. \tag{5.17}$$

Given $x \in \Lambda$ and $t \in \mathbb{R}$, we write[10]

$$\Phi(x,t) := \int_0^t \varphi(f_s x)\, ds. \tag{5.18}$$

Our first task is to construct a *system of W^{cs}-transversal measures*: by this we mean a family of Borel measures $\{m_W^{\mathrm{u}} \colon W \in \mathcal{W}^u\}$ with the following properties.

- For every $W \in \mathcal{W}^u$, the measure m_W^{u} is supported on $W \cap \Lambda$ in the sense that $m_W^{\mathrm{u}}(M \setminus (W \cap \Lambda)) = 0$.
- If $W_1, W_2 \subset \mathcal{W}^u$, and $W_1 \subset W_2$, then $m_{W_1}^{\mathrm{u}}(Z) = m_{W_2}^{\mathrm{u}}(Z)$ for all Borel $Z \subset W_1$.

We will see in Section 5.3.2 that the system of measures we construct also has important *continuity* and *conformality* properties, and in Section 5.3.4 that these properties actually characterize the system of measures up to a scalar multiple.

Let $\delta_0 > 0$ be the size of the local manifolds as in Definition 2.6, and let $r_1 > 0$ be as in Lemma 2.9. We now fix once and for all a scale $0 < r < \min(r_1, \delta_0/3)$. Given $W \in \mathcal{W}^u$, define the W-Bowen ball centered at $x \in W$ with radius r and order $t \geq 0$ by[11]

$$B_t(x,r,W) = \{y \in W \colon d_{f_\tau W}(f_\tau y, f_\tau x) < r \text{ for all } \tau \in [0,t]\}. \tag{5.19}$$

[10] Observe that for $t < 0$ this is the negative of the Birkhoff integral along the orbit segment from $f_t x$ to x: we have $\Phi(x,t) = -\Phi(f_t x, -t)$.

[11] This definition uses the *intrinsic* metric on W and thus differs slightly from [CPZ19, CPZ20], where the leaf measures were defined using Bowen balls in the extrinsic metric from M.

Given $W \in \mathcal{W}^u$, define a C-structure by $X = W \cap \Lambda$, $\mathcal{S} = X \times [0, \infty)$, $U_{(x,t)} = B_t(x, r, W) \cap \Lambda$, $\xi(x, t) = e^{\Phi(x,t)}$, and $\eta(x, t) = \psi(x, t) = e^{-t}$. It turns out (for example, as a consequence of (5.22)) that the Carathéodory dimension $\dim_C(W \cap \Lambda)$ for this C-structure is equal to the topological pressure $P(\varphi)$, and we will consider the corresponding outer measure on $W \cap \Lambda$ defined by (5.16) with $\alpha = P(\varphi)$, which can be described as follows. Given $Z \subset W \cap \Lambda$ and $T > 0$, let

$$\mathbb{E}(Z, T) = \Big\{\mathcal{E} \subset (W \cap \Lambda) \times [T, \infty) \colon Z \subset \bigcup_{(x,t)\in\mathcal{E}} B_t(x, r, W)\Big\}. \tag{5.20}$$

Members of $\mathbb{E}(Z, T)$ can be thought of as collections of Bowen balls "enveloping" the set Z. Then we define m^{u}_W by

$$m^{\mathrm{u}}_W(Z) = \lim_{T\to\infty} \inf_{\mathcal{E}\in\mathbb{E}(Z,T)} \sum_{(x,t)\in\mathcal{E}} e^{\Phi(x,t)-tP(\varphi)}. \tag{5.21}$$

We will prove in Section 5.4.1 that for every $W \in \mathcal{W}^u_\theta$, $x \in W \cap \Lambda$, and $t \geq t_2 = t_2(\theta)$, we have $\operatorname{diam} B_t(x, r, W) \leq 2r\lambda^{t-t_2}$; see (5.66). Thus this C-structure satisfies the hypothesis of Lemma 2.14, and we conclude that m^{u}_W is a Borel measure.[12] Observe that different values of r give different (but equivalent) measures m^{u}_W, but since we fix r throughout the chapter, we will suppress this dependence from the notation.

Now we restrict to the case of local unstable leaves and assume that $W = W^{\mathrm{u}}(x, \delta)$ for some $x \in \Lambda$ and $\delta \in (0, \delta_0]$. In this case we write $m^{\mathrm{u}}_x = m^{\mathrm{u}}_W$, and we have the following.[13]

Theorem 3.1. *Let $F = (f_t)_{t\in\mathbb{R}}, \Lambda, \varphi, r$ be as previously. Given $\delta > 0$, there is $K > 0$ such that for every local unstable leaf $W = W^{\mathrm{u}}(x, \delta)$ with $x \in \Lambda$, the Borel measure m^{u}_x on W defined by* (5.21) *satisfies*

$$K^{-1} \leq m^{\mathrm{u}}_x(W) \leq K. \tag{5.22}$$

Moreover, for every $x \in \Lambda$, the measures

$$\nu_t := \frac{1}{t}\int_0^t (f_s)_* m^{\mathrm{u}}_x \, ds = \frac{1}{t}\int_0^t m^{\mathrm{u}}_x \circ f_{-s} \, ds \tag{5.23}$$

are weak convergent as $t \to \infty$ to a limiting measure, whose normalization μ is independent of x. This measure μ is the unique equilibrium measure for φ on Λ. It is ergodic, gives positive weight to every (relatively) open set in Λ, and has the Gibbs property* (5.5).

12 This is the step where it is crucial that we work on $W \cap \Lambda$ and not on Λ itself; the outer measure defined by (5.21) on Λ is not a Borel measure. We will return to this point in Section 5.3.5.

13 Theorem 3.1 holds for $W \in \mathcal{W}^u$ as well, but in that case cannot be deduced directly from [CPZ20], which considered only local unstable leaves. Rather, it follows from the case here together with the scaling properties in the next section. For $W \in \mathcal{W}^u$, the lower bound in (5.22) depends not on the diameter of W itself, but on the diameter of its projection to a local unstable leaf under holonomy.

We will prove Theorem 3.1 in Section 5.4.2 using the corresponding results in [CPZ20], which cover partially hyperbolic diffeomorphisms and can be applied to the time-τ map f_τ. A more expository overview of the principles driving the proof is given in [CPZ19, §6].

5.3.2 Scaling Properties

An important role in [CPZ19, CPZ20] is played by the behavior of the measures m_x^{u} under transformation by the dynamics and by holonomy maps. In our present setting we can strengthen the results proved there. Note that now we return to the general case $W \in \mathcal{W}^u$ instead of restricting our attention to local unstable leaves.

Definition 3.2. We say that a system of W^{cs}-transversal measures $\{\bar{m}_W^{\mathrm{u}} : W \in \mathcal{W}^u\}$ is *φ-conformal* if given any $t \in \mathbb{R}$ and $W, f_t(W) \in \mathcal{W}^u$, the measures $(f_t)_*\bar{m}_W^{\mathrm{u}}$ and $\bar{m}_{f_t(W)}^{\mathrm{u}}$ are equivalent, and for each Borel $Z \subset W$ we have

$$\bar{m}_{f_t(W)}^{\mathrm{u}}(f_t Z) = \int_Z e^{tP(\varphi)-\Phi(z,t)}\, d\bar{m}_W^{\mathrm{u}}(z). \tag{5.24}$$

In terms of the Radon–Nikodym derivatives this is equivalent to

$$\frac{d((f_{-t})_*\bar{m}_{f_t(W)}^{\mathrm{u}})}{d\bar{m}_W^{\mathrm{u}}}(z) = e^{tP(\varphi)-\Phi(z,t)}, \tag{5.25}$$

$$\frac{d((f_t)_*\bar{m}_W^{\mathrm{u}})}{d\bar{m}_{f_t(W)}^{\mathrm{u}}}(f_t z) = e^{\Phi(z,t)-tP(\varphi)}. \tag{5.26}$$

Definition 3.3. Given $W_1, W_2 \in \mathcal{W}^u$, a *weak-stable holonomy* between W_1 and W_2 is a homeomorphism $\pi : W_1 \cap \Lambda \to W_2 \cap \Lambda$ such that $\pi(z) \in W^{\mathrm{cs}}(z)$ for all $z \in W_1 \cap \Lambda$. If $\pi(z) \in W^{\mathrm{cs}}(z, \delta)$ for all $z \in W_1 \cap \Lambda$, then we say that π is a *weak-stable δ-holonomy.*[14] Replacing W^{cs} by W^{s} gives the definition of *strong-stable holonomy,*[15] and reversing the roles of u, s defines *weak-unstable* and *strong-unstable holonomies.*

We say that a system of W^{cs}-transversal measures $\{\bar{m}_W^{\mathrm{u}} : W \in \mathcal{W}^u\}$ is *continuous* if for every $\varepsilon > 0$ there is $\delta > 0$ such that if W_1, W_2 are u-admissible manifolds with $d_{C^1}(W_1, W_2) < \delta$ that are related by a weak-stable δ-holonomy π, then for every Borel $Z \subset W_1 \cap \Lambda$, we have

$$\bar{m}_{W_2}^{\mathrm{u}}(\pi Z) = e^{\pm\varepsilon}\bar{m}_{W_1}^{\mathrm{u}}(Z). \tag{5.27}$$

[14] Given $\delta > 0$, it follows from the local product structure that if $x \in W_1$ and $y \in W_2$ are sufficiently close, then there are neighborhoods of x, y in $W_1 \cap \Lambda$ and $W_2 \cap \Lambda$ that are related by a weak-stable δ-holonomy.

[15] It is worth noting that strong-stable holonomies do not always exist, for example between W and $f_t(W)$.

Theorem 3.4. *Let F, Λ, φ, r be as in Section 5.3.1, and let $W \in \mathcal{W}^u$ and $t \in \mathbb{R}$. Then the system of W^{cs}-transversal measures $\{m_W^u : W \in \mathcal{W}^u\}$ defined in* (5.20)–(5.21) *is φ-conformal and continuous.*

Remark 3.5. In the discrete-time setting of [CPZ19, CPZ20], the natural analogue of (5.20)–(5.21) defines a system of transversal measures that is φ-conformal, but it is not known whether or not the system defined there is also continuous in the sense of (5.27); all that is proved is that this relationship holds for *some* $\varepsilon > 0$, which suffices for the analogue of Theorem 3.1, but not for the remaining results that we prove here. See Section 5.4.3.3 for more details.

Remark 3.6. We will eventually see in Corollary 3.12 that up to a scalar multiple, the measures defined in (5.20)–(5.21) give the *only* system of W^{cs}-transversal measures that is φ-conformal and continuous. Two special cases are worth highlighting.

- When $\varphi \equiv 0$, we have $P(\varphi) = h_{top}(f) =: h$, the topological entropy, and (5.24) gives $m_{f_t(W)}^u = e^{th}(f_t)_* m_W^u$, which is the scaling property satisfied by the Margulis measures on unstable leaves.
- When φ^u is the *geometric potential*

$$\varphi^u(x) = -\lim_{t \to 0} \frac{1}{t} \log \det(Df_t|_{E_x^u}) \tag{5.28}$$

and Λ is an attractor, it can be shown that $P(\varphi^u) = 0$ and that the unique equilibrium measure is the SRB measure on Λ [FH19, §7.4]. Then $e^{\Phi^u(x,t)} = \det(Df_t|_{E_x^u})^{-1}$ and (5.24) gives

$$m_{f_t(W)}^u(f_t Z) = \int_Z \det(Df_t|_{E_z^u})\, dm_W^u(z), \tag{5.29}$$

which is the same scaling property satisfied by leaf volume.

Since the Margulis measures and leaf volumes also satisfy the continuity property, it will follow from the uniqueness result in Corollary 3.12 that in these two cases these systems of measures are in fact given by m_W^u (up to a global scaling constant).

It turns out that the properties of φ-conformality and continuity are enough to completely describe how a system of W^{cs}-transversal measures transforms under weak-stable holonomies. We will restrict our attention to δ_0-holonomies, since this is all we need later on.[16]

[16] In fact, we will prove the following results separately for global holonomies along flow lines and along strong-stable manifolds, but to combine these results into a formula for weak-stable holonomies requires some care because t in (5.30) may not be uniquely determined by x and y unless we work locally.

Given $x \in \Lambda$ and $y \in W^{cs}(x, \delta_0) \cap \Lambda$, there is a unique $t = t(x,y) \in (-\delta_0, \delta_0)$ such that $f_t(x) \in W^s(y, \delta_0)$. We define

$$\omega^+(x,y) = \Phi(x,t) - tP(\varphi) + \int_0^\infty (\varphi(f_{\tau+t}x) - \varphi(f_\tau y))\, d\tau. \tag{5.30}$$

The improper integral converges absolutely because $d(f_{\tau+t}x, f_\tau y) \to 0$ exponentially quickly and φ is Hölder continuous.

Theorem 3.7. *Let F, Λ, φ be as in Section 5.3.1, and let $\{\bar{m}^u_W : W \in \mathcal{W}^u\}$ be any continuous and φ-conformal system of W^{cs}-transversal measures. Suppose that $W_1, W_2 \in \mathcal{W}^u$ are related by a weak-stable δ_0-holonomy $\pi : W_1 \cap \Lambda \to W_2 \cap \Lambda$. Then the measures $\pi_* \bar{m}^u_{W_1}$ and $\bar{m}^u_{W_2}$ are equivalent, and we have*

$$\frac{d(\pi_* \bar{m}^u_{W_1})}{d\bar{m}^u_{W_2}}(\pi(z)) = e^{\omega^+(z, \pi z)}. \tag{5.31}$$

It is worth noting the special cases

$$\omega^+(x,y) = \int_0^\infty (\varphi(f_\tau x) - \varphi(f_\tau y))\, d\tau \qquad \text{when } x \in W^s(y), \tag{5.32}$$

and for points on the same orbit,

$$\omega^+(x, f_t x) = \Phi(x,t) - tP(\varphi). \tag{5.33}$$

In particular, when $W_2 = f_t(W_1)$ and $\pi = f_t$, (5.31) reduces to the φ-conformality property (5.26).

We will also use the cocycle relation

$$\omega^+(x,y) = \omega^+(x,z) + \omega^+(z,y) \quad \text{for } y, z \in W^{cs}(x, \delta_0/2), \tag{5.34}$$

which can be proved by first observing that the definition (5.30) gives

$$\begin{aligned} \omega^+(x,y) &= \lim_{T\to\infty} \big(\Phi(x, T + t(x,y)) - \Phi(y, T)\big) - t(x,y)P(\varphi) \\ &= \lim_{T\to\infty} \big(\Phi(x, T) - \Phi(y, T - t(x,y))\big) - t(x,y)P(\varphi). \end{aligned} \tag{5.35}$$

Thus the two terms on the right-hand side of (5.34) can be written as

$$\omega^+(x,z) = \lim_{T\to\infty} \big(\Phi(x, T + t(x,z)) - \Phi(z, T)\big) - t(x,z)P(\varphi),$$
$$\omega^+(z,y) = \lim_{T\to\infty} \big(\Phi(z, T) - \Phi(y, T - t(z,y))\big) - t(z,y)P(\varphi);$$

adding these and using the identity $t(x,y) = t(x,z) + t(z,y)$, we see that the right-hand side of (5.34) is equal to

$$\lim_{T\to\infty} \big(\Phi(x, T + t(x,z)) - \Phi(y, T - t(z,y))\big) - t(x,y)P(\varphi).$$

Writing $T' = T - t(z,y)$ so that $T + t(x,z) = T' + t(x,y)$, this becomes

$$\lim_{T'\to\infty} \big(\Phi(x, T' + t(x,y)) - \Phi(y, T')\big) - t(x,y)P(\varphi),$$

which is equal to $\omega^+(x,y)$ by (5.35), thus proving (5.34).

5.3.3 Reference Measures on Stable Leaves and Transversals

Recall from Definition 2.8 that $\mathcal{W}^s$ denotes the set of C^1 injectively immersed submanifolds $W \subset M$ that are transverse to the weak-unstable direction. Given $W \in \mathcal{W}^s$, we construct a Borel measure m^{s}_W on $W \cap \Lambda$ by applying the construction in Section 5.3.1 to the time-reversed flow. In terms of the original flow, this can be defined using *backwards Bowen balls*: given $x \in \Lambda$ and $t \geq 0$, let

$$B_t^-(x,r,W) = \{z \in M \colon d_{f_{-\tau}W}(f_{-\tau}z, f_{-\tau}x) < r \text{ for all } \tau \in [0,t]\}.$$

Then, given $Z \subset W \cap \Lambda$ and $T > 0$, let

$$\mathbb{E}^-(Z,T) = \Big\{\mathcal{E} \subset (W \cap \Lambda) \times [T,\infty) \colon Z \subset \bigcup_{(x,t)\in\mathcal{E}} B_t^-(x,r,W)\Big\}. \tag{5.36}$$

Finally, define m^{s}_W by

$$m^{\mathrm{s}}_W(Z) = \lim_{T\to\infty} \inf_{\mathcal{E}\in\mathbb{E}^-(Z,T)} \sum_{(x,t)\in\mathcal{E}} e^{\Phi(f_{-t}x,t)-tP(\varphi)}. \tag{5.37}$$

Observe that the Birkhoff integral in (5.37) evaluates as

$$\Phi(f_{-t}x,t) = \int_0^t \varphi(f_{\tau-t}x)\,d\tau = \int_{-t}^0 \varphi(f_s x)\,ds = -\Phi(x,-t);$$

care must be taken with the signs because the Φ notation favors the initial point of the orbit segment, rather than the final one. Theorem 3.1 shows that this defines a positive finite Borel measure on each local stable leaf,[17] and that pushing *backwards* and averaging gives a family of measures converging to a scalar multiple of the unique equilibrium measure.

Versions of the scaling properties in Section 5.3.2 hold for the system of W^{cu}-transversal measures $\{m^{\mathrm{s}}_W \colon W \in \mathcal{W}^s\}$.

Definition 3.8. We say that a system of W^{cu}-transversal measures $\{\bar{m}^{\mathrm{s}}_W \colon W \in \mathcal{W}^s\}$ is *φ-conformal* if, given any $t \in \mathbb{R}$ and $W, f_t(W) \in \mathcal{W}^s$, the measures $(f_t)_* m^{\mathrm{s}}_W$ and $m^{\mathrm{s}}_{f_t(W)}$ are equivalent, and for each Borel $Z \subset W$ we have

$$\bar{m}^{\mathrm{s}}_{f_t(W)}(f_t Z) = \int_Z e^{-tP(\varphi)+\Phi(z,t)}\,d\bar{m}^{\mathrm{s}}_W(z). \tag{5.38}$$

[17] As with m^{u}, positivity and finiteness on more general compact $W \in \mathcal{W}^s$ will follow from this together with the results on scaling under holonomies.

The system is *continuous* if, for every $\varepsilon > 0$, there is $\delta > 0$ such that if W_1, W_2 are s-admissible manifolds with $d_{C^1}(W_1, W_2) < \delta$ that are related by a weak-unstable δ-holonomy π, then for every Borel $Z \subset W_1 \cap \Lambda$, we have

$$\bar{m}^{\mathrm{s}}_{W_2}(\pi Z) = e^{\pm\varepsilon} m^{\mathrm{s}}_{W_1}(Z). \tag{5.39}$$

Given $x \in \Lambda$ and $y \in W^{\mathrm{cu}}(x, \delta_0) \cap \Lambda$, there is a unique $t = t(x,y) \in (-\delta_0, \delta_0)$ such that $f_t(x) \in W^{\mathrm{u}}(y, \delta_0)$. We define

$$\omega^-(x,y) := -\Phi(x,t) + tP(\varphi) + \int_0^\infty (\varphi(f_{t-\tau}x) - \varphi(f_{-\tau}y))\, d\tau \tag{5.40}$$

and observe that ω^- satisfies the cocycle equation (5.34). Note that when $y = f_t x$, we have

$$\omega^-(x, f_t x) = -\Phi(x,t) + tP(\varphi) = -\omega^+(x, f_t x). \tag{5.41}$$

The following result is just Theorem 3.7 with time reversed.

Theorem 3.9. *Let F, Λ, φ be as in Section 5.3.1, and let $\{\bar{m}^{\mathrm{s}}_W : W \in \mathcal{W}^s\}$ be any continuous φ-conformal system of W^{cu}-transversal measures. Suppose that $W_1, W_2 \in \mathcal{W}^s$ are related by a weak-unstable δ_0-holonomy $\pi : W_1 \cap \Lambda \to W_2 \cap \Lambda$. Then the measures $\pi_* \bar{m}^{\mathrm{s}}_{W_1}$ and $\bar{m}^{\mathrm{s}}_{W_2}$ are equivalent, and we have*

$$\frac{d(\pi_* \bar{m}^{\mathrm{s}}_{W_1})}{d\bar{m}^{\mathrm{s}}_{W_2}}(\pi(z)) = e^{\omega^-(z,\pi z)}. \tag{5.42}$$

By Theorem 3.4 with time reversed, the system of W^{cu}-transversal measures given by (5.36) and (5.37) is continuous and φ-conformal, so it satisfies the conclusion of Theorem 3.9.

5.3.4 A Product Construction

In this section we once again restrict our attention to local stable and unstable leaves. Let $\overline{\mathcal{W}}^s \subset \mathcal{W}^s$ denote the collection of local stable leaves, and $\overline{\mathcal{W}}^u \subset \mathcal{W}^u$ the collection of local unstable leaves.[18] With $\delta_0 > 0$ as in Definition 2.6, we denote the collection of local weak-stable leaves by

$$\overline{\mathcal{W}}^{cs} := \Big\{W_\delta := \bigcup_{t\in(-\delta,\delta)} f_t(W) : W \in \overline{\mathcal{W}}^s, \delta \in (0, \delta_0]\Big\}, \tag{5.43}$$

and similarly for the collection $\overline{\mathcal{W}}^{cu}$ of local weak-unstable leaves.

[18] Recall from Definition 2.6 that $W \in \overline{\mathcal{W}}^s$ need not be of the form $W^{\mathrm{s}}(x, \delta)$: in general, any subset of a local stable leaf that is a topological ball of maximal dimension is itself a local stable leaf, and similarly for unstable leaves.

By a *system of measures on local unstable leaves* we mean a family of Borel measures $\{\bar{m}^{\mathrm{u}}_W : W \in \overline{\mathcal{W}}^u\}$ satisfying the support and consistency properties from the beginning of §5.3.1, and similarly with "unstable" replaced by stable, weak-unstable, and weak-stable. The definitions of continuity and φ-conformality for these systems remain the same as for the corresponding systems on transversals, and Theorems 3.7 and 3.9 still hold for such systems of measures with identical proofs.

Now suppose $\{\bar{m}^{\mathrm{s}}_W : W \in \overline{\mathcal{W}}^s\}$ and $\{\bar{m}^{\mathrm{u}}_W : W \in \overline{\mathcal{W}}^u\}$ are continuous and φ-conformal systems of measures on local stable and unstable leaves. Define corresponding systems of measures on local weak-stable and weak-unstable leaves by

$$\bar{m}^{\mathrm{cs}}_{W_\delta} = \int_{-\delta}^{\delta} \bar{m}^{\mathrm{s}}_{f_t W}\, dt \qquad \text{and} \qquad \bar{m}^{\mathrm{cu}}_{W_\delta} = \int_{-\delta}^{\delta} \bar{m}^{\mathrm{u}}_{f_t W}\, dt. \tag{5.44}$$

Let $L, \delta_1 > 0$ be as in Definition 2.11, so that given any $x, y \in \Lambda$ with $d(x,y) \in \delta_1$, the bracket $[x,y]$ is well-defined as the unique point in $W^{\mathrm{cs}}(x, L\delta_1) \cap W^{\mathrm{u}}(y, L\delta_1)$. Thus, given $q \in \Lambda$ and $\delta \in (0, \delta_1/2)$ and writing

$$R^{\mathrm{u}}_q := W^{\mathrm{u}}(q,\delta) \cap \Lambda, \qquad R^{\mathrm{cs}}_q := W^{\mathrm{cs}}(q,\delta) \cap \Lambda,$$

we can define

$$R_q := \{[x,y] : x \in R^{\mathrm{u}}_q, y \in R^{\mathrm{cs}}_q\}.$$

Given $x \in R^{\mathrm{u}}_q$ and $y \in R^{\mathrm{cs}}_q$, we write $\Psi_q(x,y) = [x,y]$, so that Ψ_q is a homeomorphism between $R^{\mathrm{u}}_q \times R^{\mathrm{cs}}_q$ and R_q.

From now on we fix δ sufficiently small that $\operatorname{diam} R_q < \delta_1$, and thus the bracket is well-defined for every pair of points in R_q. For this fixed value of δ, we write

$$\bar{m}^{\mathrm{u}}_q := \bar{m}^{\mathrm{u}}_{W^{\mathrm{u}}(q,\delta)} \qquad \text{and} \qquad \bar{m}^{\mathrm{cs}}_q := \bar{m}^{\mathrm{cs}}_{W^{\mathrm{cs}}(q,\delta)}. \tag{5.45}$$

Given $z = \Psi_q(x,y) \in R_q$, we write

$$R^{\mathrm{u}}_q(z) := W^{\mathrm{u}}(z, L\delta) \cap R_q, \qquad R^{\mathrm{cs}}_q(z) := W^{\mathrm{cs}}(z, L\delta) \cap R_q,$$

and we observe that $z \in R^{\mathrm{cs}}_q(x) \cap R^{\mathrm{u}}_q(y)$, so that $x \in R^{\mathrm{cs}}_q(z) \cap R^{\mathrm{u}}_q(q)$ and $y \in R^{\mathrm{cs}}_q(q) \cap R^{\mathrm{u}}_q(z)$. In other words,

$$z = [x,y], \qquad x = [z,q], \qquad y = [q,z]. \tag{5.46}$$

Now we can formulate the main result on the product construction of the equilibrium measure.

Theorem 3.10. *Let $F, \Lambda, \varphi, \delta$ be as in Section 5.3.1, and let $\{\bar{m}^{\mathrm{u}}_x, \bar{m}^{\mathrm{cs}}_x : x \in \Lambda\}$ be given by (5.44) and (5.45) from any continuous φ-conformal systems of measures on local stable* *and unstable leaves. Then for every $q \in \Lambda$, the following equations all define the same measure on R_q: given a Borel set $Z \subset R_q = \Psi_q(R_q^{\mathrm{u}} \times R_q^{\mathrm{cs}})$, put*

$$\mu_q(Z) = \int_Z e^{\omega^+(z,[z,q])+\omega^-(z,[q,z])}\, d((\Psi_q)_*(\bar{m}_q^{\mathrm{u}} \times \bar{m}_q^{\mathrm{cs}}))(z), \tag{5.47}$$

$$\mu_q(Z) = \int_{R_q^{\mathrm{cs}}} \int_{R_q^{\mathrm{u}}} e^{\omega^+([x,y],x)+\omega^-([x,y],y)} \mathbf{1}_Z([x,y])\, d\bar{m}_q^{\mathrm{u}}(x)\, d\bar{m}_q^{\mathrm{cs}}(y), \tag{5.48}$$

$$\mu_q(Z) = \int_{R_q^{\mathrm{cs}}} \int_{R_q^{\mathrm{u}}(y)\cap Z} e^{\omega^-(z,y)}\, d\bar{m}_y^{\mathrm{u}}(z)\, d\bar{m}_q^{\mathrm{cs}}(y), \tag{5.49}$$

$$\mu_q(Z) = \int_{R_q^{\mathrm{u}}} \int_{R_q^{\mathrm{cs}}(x)\cap Z} e^{\omega^+(z,x)}\, d\bar{m}_x^{\mathrm{cs}}(z)\, d\bar{m}_q^{\mathrm{u}}(x). \tag{5.50}$$

Moreover, there is a unique Borel measure μ on Λ such that $\mu(Z) = \mu_q(Z)$ for all $q \in \Lambda$ and $Z \subset R_q$. The measure μ is flow-invariant, nonzero, and finite. It has the Gibbs property and thus is a scalar multiple of the unique equilibrium measure.

If one redefines the bracket using W^{s} and W^{cu} instead of W^{u} and W^{cs}, then one obtains an analogue of Theorem 3.10 with *u* replaced by *cu*, and *cs* replaced by *s*, everywhere they appear.

The following is an immediate consequence of (5.49) and the definition of conditional measures (see [Cou16, §14.2] or [VO16, §§5.1–5.2]).

Corollary 3.11. *Given $q \in \Lambda$, consider the partition of R_q into local unstable leaves: let $\hat{\mu}_q$ be the transversal measure on R_q^{cs} defined by $\hat{\mu}_q(y) = \mu(\bigcup_{y\in Y} R_q^{\mathrm{u}}(y))$, and let $\{\mu_y^{\mathrm{u}} : y \in R_q^{\mathrm{cs}}\}$ denote the conditional measures of μ characterized (for $\hat{\mu}_q$-a.e. y) by*

$$\mu(Z) = \int_{R_q^{\mathrm{cs}}} \mu_y^{\mathrm{u}}(Z \cap R_q^{\mathrm{u}}(y))\, d\hat{\mu}_q(y). \tag{5.51}$$

Then $\hat{\mu}_q$ and $\bar{m}_q^{\mathrm{cs}}$ are equivalent, and for $\hat{\mu}_q$-a.e. y (hence $\bar{m}_q^{\mathrm{cs}}$-a.e. y) the measures μ_y^u and $\bar{m}_y^{\mathrm{u}}$ are equivalent as well. Moreover, defining $h\colon R_q^{\mathrm{cs}} \to (0,\infty)$ by

$$h(y) := \int_{R_q^{\mathrm{u}}} e^{\omega^+([x,y],x)}\, d\bar{m}_q^{\mathrm{u}}(x) = \bar{m}_y^{\mathrm{u}}(R_q^{\mathrm{u}}(y)), \tag{5.52}$$

where the two expressions are equal by Theorem 3.7, we have

$$\frac{d\hat{\mu}_q}{d\bar{m}_q^{\mathrm{cs}}}(y) = h(y) \quad \text{and} \quad \frac{d\mu_y^u}{d\bar{m}_y^{\mathrm{u}}}(z) = \frac{e^{\omega^-(z,y)}}{h(y)}. \tag{5.53}$$

Corresponding results hold for the transversal and conditional measures associated to the partitions of R_q by local stable, weak-stable, and weak-unstable leaves.

We prove the following corollary in Section 5.4.5.2.

Corollary 3.12. *If $\{\bar{m}^{\mathrm{u}}_W : W \in \overline{\mathcal{W}}^u\}$ is any continuous φ-conformal system of measures on local unstable leaves, then there is a real number $c > 0$ such that $\bar{m}^{\mathrm{u}}_W = c m^{\mathrm{u}}_W$ for every $W \in \overline{\mathcal{W}}^u$, where m^{u}_W is the leaf measure defined in* (5.20)–(5.21). *Similar results hold for local stable, weak-stable, and weak-unstable leaves.*

We return to the case when Λ is an attractor and $\varphi = \varphi^{\mathrm{u}}$ is the geometric potential as in (5.28), so that $P(\varphi) = 0$, $e^{\Phi^u(y,t)} = \det(Df_t|_{E^{\mathrm{u}}_y})^{-1}$, and, given $y \in \Lambda$, $z \in W^{\mathrm{u}}(y,\delta)$, we have

$$e^{\omega^-(z,y)} = \lim_{t\to-\infty} \frac{e^{\Phi^u(z,t)}}{e^{\Phi^u(y,t)}} = \lim_{t\to-\infty} \frac{\det(Df_t|_{E^{\mathrm{u}}_y})}{\det(Df_t|_{E^{\mathrm{u}}_z})}. \tag{5.54}$$

Let ν^{u}_y be the Borel measure on $W^{\mathrm{u}}(y,\delta)$ given by

$$\frac{d\nu^{\mathrm{u}}_y}{d\,\mathrm{Leb}_{W^{\mathrm{u}}(y,\delta)}}(z) = \lim_{t\to-\infty} \frac{\det(Df_t|_{E^{\mathrm{u}}_y})}{\det(Df_t|_{E^{\mathrm{u}}_z})}, \tag{5.55}$$

where $\mathrm{Leb}_{W^{\mathrm{u}}(y,\delta)}$ denotes leaf volume. Given $q \in \Lambda$, define a Borel measure ν^{s}_q on $W^{\mathrm{s}}(q,\delta) \cap \Lambda$ by

$$\nu^{\mathrm{s}}_q(Z) = \lim_{T\to\infty} \inf_{\mathcal{E}\in\mathbb{E}^-(Z,T)} \sum_{(y,t)\in\mathcal{E}} \det(Df_{-t}|_{E^{\mathrm{u}}_y}), \tag{5.56}$$

where $\mathbb{E}^-(Z,T)$ is as in (5.36). Using this, define a measure ν^{cs}_q on $R^{\mathrm{cs}}_q = W^{\mathrm{cs}}(q,\delta) \cap \Lambda$ by

$$\nu^{\mathrm{cs}}_q(Z) = \int_{-\delta}^{\delta} \nu^{\mathrm{s}}_q(f_t Z)\,dt. \tag{5.57}$$

Then (5.49) gives the following.

Corollary 3.13. *Let Λ be a hyperbolic attractor and μ its associated SRB measure. Given $q \in \Lambda$, there exists $c = c(q) > 0$ such that*

$$\mu(Z) = c \int_{R^{\mathrm{cs}}_q} \nu^{\mathrm{u}}_y(Z \cap R^{\mathrm{u}}_q(y))\,d\nu^{\mathrm{cs}}_q(y) \tag{5.58}$$

for every Borel $Z \subset R_q$, where ν^{u}_y, ν^{cs}_q are given by (5.55)–(5.57).

One can write similar analogues of (5.47), (5.48), and (5.50). The fact that the measures ν^{u}_y are the conditional measures of the SRB measure is well known [Sin68, Lemma 2.3]. To the best of the author's knowledge, the explicit description of the transverse measure ν^{cs}_q in (5.56)–(5.57) is new, as is the corresponding description of the stable conditional measures in terms of ν^{s}_q that can be obtained using Corollary 3.11 (though we do not write out these details).

5.3.5 Refining in All Directions

We conclude our main results by giving one more description of the unique equilibrium measure via a Carathéodory dimension structure, this time on Λ itself rather than on a local stable or unstable leaf. In what follows we do *not* assume that the metric is adapted.

To define a C-structure with $X = \Lambda$ and obtain a Borel measure via Lemma 2.14, we cannot use the one-sided Bowen balls we have used so far, because $\bigcap_{t>0} B_t(x,r)$ is a local weak-stable manifold, rather than a single point; similarly, $\bigcap_{t>0} B_t^-(x,r)$ is a local weak-unstable manifold. We can make some progress by using *two-sided Bowen balls*

$$\begin{aligned} B_{s,t}^{\pm}(x,r) &:= B_t(x,r) \cap B_s^-(x,r) \\ &= \{y \in M : d(f_\tau x, f_\tau y) < r \text{ for all } \tau \in [-s,t]\}, \end{aligned}$$

for which taking an intersection over all $s,t > 0$ gives a small piece of the orbit of x. Thus given a sufficiently small embedded disc D that is transverse to the flow direction, we have $\bigcap_{s,t>0} B_{s,t}^{\pm}(x,r) \cap D = \{x\}$ for each $x \in D \cap \Lambda$, and so we could use two-sided Bowen balls to define a C-structure on $D \cap \Lambda$ that gives a Borel measure; then we could flow this measure forward to get an invariant measure on Λ.

Rather than describing the details of this, we go one step further: provided we choose $r \in (0, \delta_1]$, we can use the local product structure in Definition 2.11 whenever $x, y \in \Lambda$ satisfy $d(x,y) < r$ to obtain $z = [x,y] \in W^{\mathrm{cs}}(x, Lr) \cap W^{\mathrm{u}}(y, Lr)$ and $z' = f_\beta(z) \in W^{\mathrm{s}}(x, Lr)$ for some $\beta = \beta(x,y) \in (-Lr, Lr)$. Then we make the following definition for $x \in \Lambda$ and $s,t > 0$:

$$B_{s,t}^*(x,r) := \{y \in B_{s,t}^{\pm}(x,r) \cap \Lambda : |\beta(x,y)| < r/(s+t)\}. \tag{5.59}$$

Now we consider the C-structure defined by $X = \Lambda$, $\mathcal{S} = X \times [1,\infty)^2$, and $U_{(x,s,t)} = B_{s,t}^*(x,r)$, together with

$$\xi(x,s,t) = \frac{e^{\Phi(f_{-s}x, s+t)}}{s+t}, \quad \eta(x,s,t) = e^{-(s+t)}, \quad \psi(x,s,t) = e^{-\min(s,t)}.$$

The conditions of Lemma 2.14 are satisfied, so this C-structure produces a Borel measure, which we describe more carefully in what follows.

Remark 3.14. As motivation for the formula for ξ, observe that $B_{s,t}^*(x,r)$ is roughly the "product" of $B_t(y,r)$ in the unstable direction, $B_s^-(x,r)$ in the stable direction, and an interval in the flow direction: we want to produce a measure m with a product structure whose conditionals give weight $e^{\Phi(x,t)}$ along the unstable, $e^{\Phi(f_{-s}x,s)}$ along the stable, and $1/(s+t)$ along the flow; the product of these weights gives ξ. Note that the factor of $1/(s+t)$ has no thermodynamic meaning and in particular does not depend on the potential; it is roughly proportional to the amount of time a trajectory takes to cross $B^*_{s,t}(x,r)$, and is present to guarantee that the measure m defined by (5.61) is flow-invariant.

The formula for η is as in our previous C-structures: $e^{-\ell}$, where ℓ is the length of the orbit segment that defines the Bowen ball we use. The formula for ψ guarantees that both s and t are large when ψ is small, which mimics our previous constructions.

We remark that it would be tempting to simplify this C-structure by requiring that $s = t$, so that we use symmetric two-sided Bowen balls $B_t(x,r) \cap B_t^-(x,r)$; however, our proof of flow-invariance requires us to allow the case $s \neq t$.

This C-structure defines a Borel measure m on Λ as follows:

$$\mathbb{E}^*(Z,T) := \Big\{\mathcal{E} \subset \Lambda \times [T,\infty)^2 : Z \subset \bigcup_{(x,s,t)\in\mathcal{E}} B^*_{s,t}(x,r)\Big\}, \tag{5.60}$$

$$m(Z) := \lim_{T\to\infty} \inf_{\mathcal{E}\in\mathbb{E}^*(Z,T)} \sum_{(x,s,t)\in\mathcal{E}} \frac{1}{s+t} e^{\Phi(f_{-s}x,s+t)-(s+t)P(\varphi)}. \tag{5.61}$$

Theorem 3.15. *Let F, Λ, φ, r be as in Section 5.3.1. Then the Borel measure defined by (5.60)–(5.61) is nonzero, finite, flow-invariant, and a scalar multiple of the unique equilibrium measure.*

Observe that Theorem 1.1 is a direct consequence of Theorem 3.15 in the case when Λ is an attractor and $\varphi = \varphi^{\mathrm{u}}$ is the geometric potential.

Remark 3.16. Returning to the approach in Section 5.3.4, one could adapt the construction here to use "one-sided Bowen balls" on weak-unstable leaves that also include a restriction on $|\beta(x,y)|$ as in (5.59), and thus obtain the measures m_x^{cu} (and similarly m_x^{cs}) directly from a C-structure rather than first obtaining m_x^{u} (or m_x^{s}) and pushing under the flow.

Remark 3.17. The precise form of the factor $(s+t)^{-1}$ in the definition of B^* and of m is not essential: one could use any function $\zeta(s+t)$ where $\lim_{T\to\infty} \zeta(T) = 0$ and $\underline{\lim}_{T\to\infty} \zeta(T+\eta)/\zeta(T) > 0$ for every $\eta > 0$ – the latter bound is needed at the end of the proof of Lemma 4.9. Replacing $r/(s+t)$ in (5.59) with $\zeta(s+t)$, and defining $\xi(x,s,t) = e^{\Phi(f_{-s}x,s+t)}\zeta(s+t)$, the proofs still go through.

One could avoid using such a function by introducing an extra parameter in the definition of the C-structure and putting

$$\begin{aligned} \mathcal{S} &= \Lambda \times [1,\infty)^2 \times (0,1), \\ U_{(x,s,t,\delta)} &= \{y \in B^{\pm}_{s,t}(x,r) \cap \Lambda : |\beta(x,y)| < \delta\}, \end{aligned}$$

together with the functions

$$\xi(x,s,t,\delta) = \delta e^{\Phi(f_{-s}x,s+t)},$$
$$\eta(x,s,t,\delta) = e^{-(s+t)},$$
$$\psi(x,s,t,\delta) = \max(e^{-s}, e^{-t}, \delta).$$

It is a matter of taste which route one takes; here we have opted for the definition with fewer parameters and the explicit choice $\zeta(T) = 1/T$.

5.4 Proofs

5.4.1 Bowen Balls on Leaves

Recall from (5.19) that we define Bowen balls on transversals according to the intrinsic metric:

$$B_t(x,r,W) = \{y \in W : d_{f_\tau W}(f_\tau y, f_\tau x) < r \text{ for all } \tau \in [0,t]\}.$$

Now (5.13) immediately gives the following.

Lemma 4.1. *If W is a u-admissible manifold and $r \in (0, r_0]$, then for all $x \in W \cap \Lambda$ and $t \geq 0$ we have*

$$B_t(x,r,W) = f_{-t}B(f_t x, r, f_t W) \subset B(x, r\lambda^t, W). \tag{5.62}$$

We also get a version of this lemma for $W \in \mathcal{W}^u$. Let r_1 and $t_1 = t_1(\theta)$ be as in Lemma 2.9, and let

$$C_1 = C_1(\theta) := \sup\{\|Df_t(x)\| : x \in M, t \in [-t_1, t_1]\}. \tag{5.63}$$

Lemma 4.2. *For every $\theta > 0$, there is $t_2 \geq 0$ such that given any $r \in (0, r_1]$, $W \in \mathcal{W}^u_\theta$, any $x \in W \cap \Lambda$, and any $t \geq t_2$, we have*

$$B_t(x,r,W) = f_{-t}B(f_t x, r, f_t W), \tag{5.64}$$

and $B(f_t x, r, f_t W)$ is u-admissible. Moreover, given $\eta > 0$, we have

$$\begin{aligned} B_{t+\eta}(x,r,W) &= f_{-t}B_\eta(f_t x, r, f_t W) \\ &\subset f_{-t}B(f_t x, r\lambda^\eta, f_t W) = B_t(x, r\lambda^\eta, W), \end{aligned} \tag{5.65}$$

and in particular

$$\operatorname{diam} B_{t+\eta}(x,r,W) \leq 2r\lambda^\eta. \tag{5.66}$$

Proof. Clearly the left-hand side of (5.64) is contained in the right-hand side, so it suffices to prove the other inclusion. By Lemma 2.9, there is $t_1 \geq 0$ such that for all r, W, x as in the lemma, and any $t \geq t_1$, $B(f_t x, r, f_t W)$ is u-admissible. Then it follows from Lemma 4.1 that, for all such t and every $\tau \in [t_1, t]$, we have

$$f_{-(t-\tau)}B(f_t x, r, f_t W) \subset B(f_\tau x, r\lambda^{t-\tau}, f_\tau W).$$

In particular, we have

$$f_{-(t-t_1)}B(f_t x, r, f_t W) \subset B(f_{t_1} x, r\lambda^{t-t_1}, f_{t_1} W),$$

and using this together with (5.63) we see that for every $\tau \in [0, t_1]$ we have

$$f_{-(t-\tau)}B(f_t x, r, f_t W) \subset B(f_\tau x, C_1 r\lambda^{t-t_1}, f_\tau W).$$

Thus, by choosing t_2 sufficiently large that $C_1\lambda^{t_2-t_1} \leq 1$, we guarantee that

$$f_\tau(f_{-t}B(f_t x, r, f_t W)) \subset B(f_\tau x, r, f_\tau W)$$

for all $t \geq t_2$ and $\tau \in [0, t]$. This shows that the right-hand side of (5.64) is contained in the left-hand side.

For (5.65), we apply (5.64) as follows:

$$B_{t+\eta}(x, r, W) = \bigcap_{\tau\in[0,\eta]} B_t(f_\tau x, r, W) = \bigcap_{\tau\in[0,\eta]} f_{-t}B(f_\tau(f_t x), r, f_{\tau+t} W)$$

and observe that the final intersection is $f_{-t}B_\eta(f_t x, r, f_t W)$, which proves the first equality in (5.65). The inclusion and the final equality follow from Lemma 4.1. Then (5.66) follows immediately upon observing that $B_t(x, r\lambda^\eta, W) \subset B(x, r\lambda^\eta)$. □

5.4.2 Finiteness and Pushforwards: Proof of Theorem 3.1

We will deduce Theorem 3.1 from the corresponding results in [CPZ20], which deal with partially hyperbolic diffeomorphisms. Thus the proof here should be read less with the intent of understanding the mechanisms driving Theorem 3.1, and more with the intent of understanding the relationship between the continuous- and discrete-time settings. For a description of the general mechanisms, we refer to [CPZ19, §6].

5.4.2.1 Relationship to Discrete-Time Results and Uniform Bounds

Before proving the bounds in (5.22), we describe the relationship to [CPZ20], observing that every time-τ map f_τ ($\tau > 0$) of a flow on a hyperbolic set Λ is partially hyperbolic. The results in [CPZ20] apply to a diffeomorphism f with a compact invariant set Λ on which f is partially hyperbolic and topologically transitive, with a local product structure between unstable and center-stable foliations, and for which the center-stable foliation satisfies a certain "Lyapunov stability" condition that is automatic for the time-τ map of a flow. In order to apply the results in [CPZ20] to f_τ we only need to check topological transitivity of f_τ. For this we use the following lemma, which may be of independent interest (though we do not claim that it is new).

Lemma 4.3. *Let X be a compact metric space and $(f_t)_{t\in\mathbb{R}}$ a continuous flow on X that is topologically transitive: for every open $U, V \subset X$ there is $t > 0$ such that $f_t(U) \cap V \neq \emptyset$. Then there is a residual set of $\tau \in (0,\infty)$ such that the time-τ map f_τ is topologically transitive.*

Proof. Note that, given any open $U, V \subset X$, the set

$$I(U,V) := \{\tau > 0 \colon f_{k\tau}(U) \cap V \neq \emptyset \text{ for some } k \in \mathbb{N}\}$$

is open. We claim that it is dense in $(0,\infty)$. Indeed, given any $(a,b) \subset (0,\infty)$, if $k \geq b/(b-a)$ then $kb - ka \geq b > a$, so $(k+1)a < kb$ and therefore $(ka,kb) \cap ((k+1)a,(k+1)b) \neq \emptyset$; this implies that

$$\bigcup_{k\in\mathbb{N}} (ka,kb) \supset \left[\frac{ab}{b-a},\infty\right). \tag{5.67}$$

Since the flow is transitive, there exists $t \geq ab/(b-a)$ such that $f_t(U) \cap V \neq \emptyset$ (note that t can be chosen arbitrarily large by iterating the definition of transitivity). By (5.67) there is $k \in \mathbb{N}$ such that $t/k \in (a,b)$. Since $t/k \in I(U,V)$, this shows that $I(U,V)$ is open and dense in $(0,\infty)$.

Now let $\{U_n\}_{n\in\mathbb{N}}$ be a basis for the topology on X. Any τ in the residual set $\bigcap_{k,n} I(U_k,U_n)$ has the property that f_τ is topologically transitive. Note that this set is dense in $(0,\infty)$ by the Baire category theorem. □

Returning to the setting of Theorem 3.1, use Lemma 4.3 to choose $\tau > 0$ such that the time-τ map f_τ is topologically transitive, and thus writing $\psi(y) := \Phi(y,\tau) = \int_0^\tau \varphi(f_t y)\,dt$, the conditions of [CPZ20] are satisfied by (Λ, f_τ, ψ). Observe that

$$S_n^{f_\tau}\psi(y) := \sum_{k=0}^{n-1} \psi(f_\tau^k y) = \int_0^{\tau n} \varphi(f_t y)\,dt = \Phi(y,\tau n), \tag{5.68}$$

and the topological pressures of the map f_τ and the flow F are related by [Wal75, Corollary 4.12(iii)]:

$$P(f_\tau,\psi) = \tau P(\varphi). \tag{5.69}$$

(We continue to write $P(\varphi) = P(F, \varphi)$ for the pressure on the flow.) The construction in [CPZ20, §3.2] defines a Borel measure $m_x^{\tau,r}$ on $W = W^{\mathrm{u}}(x, \delta)$ as follows: we work with discrete-time Bowen balls[19]

$$B_n^{f_\tau}(y, r) := \{z \colon d(f_\tau^k y, f_\tau^k z) < r \text{ for all } 0 \le k < n\},$$

for which the corresponding collection of enveloping sets is

$$\mathbb{E}_\tau^r(Z, N) := \Big\{\mathcal{E} \subset (W \cap \Lambda) \times \{N, N+1, \dots\} \colon Z \subset \bigcup_{(y,n)\in\mathcal{E}} B_n^{f_\tau}(y, r)\Big\},$$

and then produce a measure by

$$m_x^{\tau,r}(Z, N) := \inf_{\mathcal{E}\in\mathbb{E}_\tau^r(Z,N)} \sum_{(y,n)\in\mathcal{E}} e^{S_n^{f_\tau}\psi(y) - nP(f_\tau,\psi)}, \tag{5.70}$$

$$m_x^{\tau,r}(Z) := \lim_{N\to\infty} m_x^{\tau,r}(Z, N). \tag{5.71}$$

By (5.68) and (5.69), we have

$$e^{\Phi(N\tau) - n\tau P(\varphi)} = e^{S_n^{f_\tau}\psi(y) - nP(f_\tau,\psi)}, \tag{5.72}$$

which will let us relate the definitions of m_x^{u} in (5.21) and $m_x^{\tau,r}$ in (5.71) by comparing $\mathbb{E}(Z, T)$ and $\mathbb{E}_\tau^r(Z, N)$. To avoid confusion we will write $B_t^F(y, r, W)$ for the continuous-time Bowen ball from (5.19) in the remainder of this argument.

First observe that, for each $n \in \mathbb{N}$, we have $B_{\tau n}^F(y, r, W) \subset B_n^{f_\tau}(y, r)$. Thus given $\mathcal{E} \in \mathbb{E}(Z, \tau N)$, we have

$$Z \subset \bigcup_{(y,t)\in\mathcal{E}} B_t^F(y, r, W) \subset \bigcup_{(y,t)\in\mathcal{E}} B_{\lfloor t/\tau\rfloor\tau}^F(y, r, W) \subset \bigcup_{(y,t)\in\mathcal{E}} B_{\lfloor t/\tau\rfloor}^{f_\tau}(y, r),$$

and thus by (5.72),

$$m_x^{\tau,r}(Z, N) \le \sum_{(y,t)\in\mathcal{E}} e^{S_{\lfloor t/\tau\rfloor}^{f_\tau}\psi(y) - \lfloor t/\tau\rfloor P(f_\tau,\psi)} = \sum_{(y,t)\in\mathcal{E}} e^{\Phi(y,\lfloor t/\tau\rfloor\tau) - \lfloor t/\tau\rfloor\tau P(\varphi)}.$$

Observe that, given $t \in [n\tau, (n+1)\tau]$, we have

$$\Phi(y, n\tau) - \Phi(y, t) = -\Phi(f_{n\tau} y, t - n\tau) \le \tau\|\varphi\|,$$

and similarly

$$-n\tau P(\varphi) + tP(\varphi) = (t - n\tau)P(\varphi) \le \tau|P(\varphi)|,$$

so we obtain

$$m_x^{\tau,r}(Z, N) \le \sum_{(y,t)\in\mathcal{E}} e^{\tau(\|\varphi\| + |P(\varphi)|)} e^{\Phi(y,t) - tP(\varphi)}.$$

[19] Observe that these Bowen balls are defined with respect to the metric on M, rather than the leaf metric on W as in (5.19).

Taking an infimum over all $\mathcal{E} \in \mathbb{E}(Z, \tau N)$ and sending $N \to \infty$ gives

$$m_x^{\tau,r}(Z) \leq C_2^\tau m_x^{\mathrm{u}}(Z), \quad \text{where } C_2 = e^{\|\varphi\|+|P(\varphi)|}. \tag{5.73}$$

For a bound in the other direction, observe that by uniform continuity of the flow there is $\rho > 0$ such that

$$d(y,z) < \rho \text{ implies } d_{f_t W}(f_t y, f_t z) < r \text{ for all } t \in [0, \tau]. \tag{5.74}$$

Then $B_n^{f_\tau}(y, \rho) \subset B_{\tau n}^F(y, r, W)$, so, given any $\mathcal{E} \in \mathbb{E}_\tau^\rho(Z, N)$, we have

$$Z \subset \bigcup_{(y,n)\in\mathcal{E}} B_n^{f_\tau}(y, \rho) \subset \bigcup_{(y,n)\in\mathcal{E}} B_{\tau n}^F(y, r, W),$$

and thus defining $m_x^{\mathrm{u}}(Z, T)$ as the infimum in (5.21) (without taking the limit), we have

$$m_x^{\mathrm{u}}(Z, N\tau) \leq \sum_{(y,n)\in\mathcal{E}} e^{\Phi(y,n\tau)-n\tau P(\varphi)} = \sum_{(y,n)\in\mathcal{E}} e^{S_n^{f_\tau}\psi(y)-nP(f_\tau,\psi)}.$$

Taking an infimum over all $\mathcal{E} \in \mathbb{E}_\tau^\rho(Z, N)$ and sending $N \to \infty$ gives $m_x^{\mathrm{u}}(Z) \leq m_x^{\tau,\rho}(Z)$, and combining this with (5.73) we obtain

$$C_2^{-\tau} m_x^{\tau,r}(Z) \leq m_x^{\mathrm{u}}(Z) \leq m_x^{\tau,\rho}(Z) \text{ for all } Z \subset W^{\mathrm{u}}(x, \delta). \tag{5.75}$$

Now the bounds in (5.22) follow immediately from the corresponding bounds for $m_x^{\tau,r}$ and $m_x^{\tau,\rho}$ in [CPZ20, Theorem 4.2], together with (5.75).

We record the following consequence of (5.75) for use in the next section: for every $x \in \Lambda$ and $\tau \in (0, 1]$ such that f_τ is transitive, writing $W = W^{\mathrm{u}}(x, \delta)$, we have

$$m_x^{\tau,r}(W) \leq C_2^\tau m_x^{\mathrm{u}}(W) \leq C_2 m_x^{\mathrm{u}}(W) =: C_3(r) < \infty. \tag{5.76}$$

We will use this in what follows with r replaced by ρ.

5.4.2.2 *Averaged Pushforwards Converge to Equilibrium*

As shown earlier, there is a residual set of $\tau \in (0, \infty)$ such that the discrete-time system $(\Lambda, f_\tau, \psi = \int_0^\tau \varphi \circ f_s \, ds)$ satisfies the hypotheses of [CPZ20]. By [CPZ20, Theorem 4.7], the measures $m_x^{\tau,r}$ defined in (5.71) have the property that

$$\nu_n^{\tau,r} := \frac{1}{n} \sum_{k=0}^{n-1} (f_{k\tau})_* m_x^{\tau,r} \xrightarrow{\text{weak*}} m_x^{\tau,r}(W)\mu, \tag{5.77}$$

where μ is the unique equilibrium measure for (Λ, f_τ, ψ). Moreover, μ is ergodic, fully supported on Λ, and has the Gibbs property.

Every equilibrium measure for the continuous-time system (Λ, F, φ) is also an equilibrium measure for the discrete-time system. A measure that is ergodic, fully supported, and Gibbs for the discrete-time system has the same properties for the continuous-time system. Thus to prove Theorem 3.1, it suffices to show that the measures ν_t defined in (5.23) converge to the same limit as the measures $\nu_n^{\tau,r}$, up to a scalar. Equivalently, we must prove that if $\mu_1 = \lim_{j\to\infty} \nu_{t_j}$ is any limit point of the measures ν_t, then μ_1 is a scalar multiple of μ. Since μ is ergodic and μ_1 is invariant, it will suffice to prove that $\mu_1 \ll \mu$.

We will do this using (5.75) and the observation that if m_n, μ_n are any sequences of measures on a compact metric space that are weak* convergent to m_0, μ_0 respectively, and that have the property $m_n(Z) \leq \mu_n(Z)$ for all measurable Z, then the same is true of m_0, μ_0. Indeed, this property is equivalent to $\int \zeta \, dm_n \leq \int \zeta \, d\mu_n$ for all nonnegative continuous functions ζ, and this inequality survives the limit by the definition of weak* convergence. We summarize this as

$$m_n \xrightarrow{\text{weak*}} m_0,\ \mu_n \xrightarrow{\text{weak*}} \mu_0,\ m_n \leq \mu_n \quad \Rightarrow \quad m_0 \leq \mu_0. \tag{5.78}$$

To prove that $\mu_1 = \lim \nu_{t_j} \ll \mu$, start by observing that, given any $t, \eta > 0$ and $Z \subset \Lambda$, we have

$$\begin{aligned}|\nu_{t+\eta}(Z) - \nu_t(Z)| &= \left| \frac{1}{t+\eta} \int_0^{t+\eta} (f_s)_* m_x^{\mathrm{u}}(Z)\, ds - \frac{1}{t} \int_0^t (f_s)_* m_x^{\mathrm{u}}(Z)\, ds \right| \\ &\leq \frac{1}{t+\eta} \int_t^{t+\eta} (f_s)_* m_x^{\mathrm{u}}(Z)\, ds + \left| \frac{1}{t+\eta} - \frac{1}{t} \right| \int_0^t (f_s)_* m_x^{\mathrm{u}}(Z)\, ds \\ &\leq \frac{1}{t+\eta} \cdot \eta m_x^{\mathrm{u}}(W) + \frac{\eta}{t(t+\eta)} \cdot t m_x^{\mathrm{u}}(W) = \frac{2\eta m_x^{\mathrm{u}}(W)}{t+\eta}.\end{aligned}$$

Thus, for every $\zeta \in C(\Lambda)$, any $\tau > 0$, and any $t \in [(n-1)t\tau, n\tau]$ for some $n \in \mathbb{N}$, we can apply this with $\eta = n\tau - t$ (and thus $\frac{\eta}{t+\eta} \leq \frac{\tau}{n\tau} = \frac{1}{n}$) to get

$$\left| \int \zeta \, d\nu_t - \int \zeta \, d\nu_{n\tau} \right| \leq \frac{2}{n} \|\zeta\| m_x^{\mathrm{u}}(W). \tag{5.79}$$

This implies that for every $\tau > 0$, the sequence $n_j = \lfloor t_j/\tau \rfloor \to \infty$ has the property that $\nu_{n_j\tau} \to \mu_1$. Now observe that

$$\nu_{n\tau} = \frac{1}{n\tau} \int_0^\tau \sum_{k=0}^{n-1} (f_{k\tau+s})_* m_x^{\mathrm{u}}\, ds = \frac{1}{\tau} \int_0^\tau (f_s)_* \left(\frac{1}{n} \sum_{k=0}^{n-1} (f_{k\tau})_* m_x^{\mathrm{u}} \right) ds.$$

Fix $\rho > 0$ small enough that (5.74) is satisfied with $\tau = 1$, and thus for all $\tau \in (0, 1]$ as well. By (5.75), we have

$$\nu_{n\tau} \leq \frac{1}{\tau} \int_0^\tau (f_s)_* \nu_n^{\tau,\rho}\, ds =: \bar{\nu}_n^{\tau,\rho}. \tag{5.80}$$

Let $C_3 = C_3(\rho)$ be as in (5.76) (note that we replace r by ρ) and consider the space $\mathcal{M}$ of all finite Borel measures on Λ with total weight $\leq C_3$; this space is compact in the weak* topology, which is induced by some metric D. The flow $\{(f_t)_*\}_{t\in\mathbb{R}}$ is continuous on $\mathcal{M}$, hence uniformly continuous, and thus for every $\ell \in \mathbb{N}$ there is $\tau \in (0,1)$ such that $D((f_t)_*\nu, \nu) \leq 1/\ell$ whenever $|t| \leq \tau$ and f_τ is topologically transitive (by Lemma 4.3).

Let $c := m_x^{\tau,\rho}(W)$; by (5.76) this is at most C_3, and thus $\nu_n^{\tau,\rho} \in \mathcal{M}$ for all n. Observe that $\nu_n^{\tau,\rho} \to c\mu$ by (5.77). Thus there is N such that $D(\nu_n^{\tau,\rho}, c\mu) \leq 1/\ell$ for all $n \geq N$. By our choice of τ, we conclude that $D(\bar{\nu}_n^{\tau,\rho}, c\mu) \leq 2/\ell$ for all such n. By passing to a subsequence if necessary, we can assume that

$$\nu_{n_j\tau} \to \mu_1 \text{ and } \bar{\nu}_{n_j}^{\tau,\rho} \to \bar{\nu}_\ell \text{ with } D(\bar{\nu}, c\mu) \leq 2/\ell.$$

Then (5.78) and (5.80) give $\mu_1 \leq \bar{\nu}_\ell$, and since $\bar{\nu}_\ell \to c\mu$ as $\ell \to \infty$, applying (5.78) again gives $\mu_1 \leq c\mu$. In particular, $\mu_1 \ll \mu$, so ergodicity of μ and flow-invariance of μ_1 imply that μ_1 is a scalar multiple of μ, which completes the proof.

5.4.3 *Conformality and Continuity: Proof of Theorem 3.4*

5.4.3.1 *Scaling under the dynamics*

In this section we prove that the system $\{m_W^{\mathrm{u}} : W \in \mathcal{W}^u\}$ is φ-conformal in the sense of (5.24) from Definition 3.2, using the following.

Lemma 4.4. *Given $W \in \mathcal{W}^u$, suppose that $Z \subset W \cap \Lambda$, $t, c \in \mathbb{R}$, and $\varepsilon > 0$ are such that for all $y \in Z$, we have $|\Phi(y,t) - c| < \varepsilon$. Then*

$$m_{f_t(W)}^{\mathrm{u}}(f_t Z) = e^{\pm 3\varepsilon} \int_Z e^{tP(\varphi) - \Phi(z,t)}\, dm_W^{\mathrm{u}}(z). \tag{5.81}$$

Once the lemma is established, we can deduce (5.24) as follows: fixing $\varepsilon > 0$, write $Z = \bigsqcup_i Z_i$ where each Z_i satisfies the condition of the lemma with some c_i, and thus satisfies (5.81). Summing over i gives (5.81) for Z as well, and since $\varepsilon > 0$ was arbitrary this proves (5.24). Thus it suffices to prove the lemma.

Proof of Lemma 4.4. Throughout the following, we fix Z, W, t, c, ε as in the statement. The function $x \mapsto \Phi(x,t)$ is continuous on Λ, hence uniformly continuous, so there is $\delta > 0$ such that given any $x \in Z$ and $y \in B(x,\delta) \cap \Lambda$, we have

$$\Phi(y,t) = \Phi(x,t) \pm \varepsilon = c \pm 2\varepsilon.$$

By (5.66), there is T_0 such that for all $\tau \geq T_0$ and $x \in Z$, we have $B_\tau(x, r, W) \subset B(x, \delta, W)$, and thus

$$\Phi(y,t) = c \pm 2\varepsilon \text{ for every } y \text{ such that } B_\tau(y, r, W) \cap Z \neq \emptyset. \tag{5.82}$$

Let $\mathbb{E}'(Z,T) = \{\mathcal{E} \subset \mathbb{E}(Z,T) \colon B_\tau(y,r,W) \cap Z \neq \emptyset$ for all $(y,\tau) \in \mathcal{E}\}$, and observe that in the definition of m_W^{u} it suffices to take an infimum over $\mathbb{E}'(Z,T)$. Now consider the map

$$S \colon (W \cap \Lambda) \times [T,\infty) \to (f_t(W) \cap \Lambda) \times [T-t,\infty),$$
$$(y,\tau) \mapsto (f_t y, \tau - t).$$

Given t_1 as in Lemma 4.2 and $\tau \geq \max(t_1, t_1 + t)$, that lemma gives

$$\begin{aligned} f_t(B_\tau(y,r,W)) &= f_t(f_{-\tau}(f_\tau y, r, f_\tau W)) \\ &= f_{-(\tau-t)}(f_{\tau-t}(f_t y), r, f_{\tau-t}(f_t W)) = B_{\tau-t}(f_t y, r, f_t W). \end{aligned}$$

Thus for $T \geq \max(t_1, t_1 + t)$, we see that $\mathcal{E} \subset (W \cap \Lambda) \times [T,\infty)$ lies in $\mathbb{E}'(Z,T)$ if and only if $S(\mathcal{E})$ lies in $\mathbb{E}'(f_t Z, T-t)$; in other words, S gives a bijection between $\mathbb{E}'(Z,T)$ and $\mathbb{E}'(f_t Z, T-t)$. Moreover, (5.82) gives the following estimate for each $(y,\tau) \in \mathbb{E}'(Z,T)$:

$$\Phi(y,\tau) = \Phi(y,t) + \Phi(f_t y, \tau - t) = c \pm 2\varepsilon + \Phi(f_t y, \tau - t).$$

Now we conclude that

$$\begin{aligned} m_W^{\mathrm{u}}(Z) &= \lim_{T\to\infty} \inf_{\mathcal{E}\in\mathbb{E}'(Z,T)} \sum_{(y,\tau)\in\mathcal{E}} e^{\Phi(y,\tau)-\tau P(\varphi)} \\ &= e^{c\pm 2\varepsilon} \lim_{T\to\infty} \inf_{\mathcal{E}\in\mathbb{E}'(Z,T)} \sum_{(y,\tau)\in\mathcal{E}} e^{\Phi(f_t y,\tau-t)-\tau P(\varphi)} \\ &= e^{c\pm 2\varepsilon} \lim_{T\to\infty} \inf_{\mathcal{E}'\in\mathbb{E}'(f_t Z,T-t)} \sum_{(y',\tau')\in\mathcal{E}} e^{\Phi(y',\tau')-(\tau'+t)P(\varphi)}, \end{aligned}$$

which yields

$$m_W^{\mathrm{u}}(Z) = e^{c\pm 2\varepsilon - tP(\varphi)} m_{f_t(W)}^{\mathrm{u}}(f_t Z). \tag{5.83}$$

At the same time, we have $\Phi(z,t) = c \pm \varepsilon$ on Z, so

$$\int_Z e^{tP(\varphi)-\Phi(z,t)}\, dm_W^{\mathrm{u}}(z) = e^{tP(\varphi)-c\pm\varepsilon} m_W^{\mathrm{u}}(Z). \tag{5.84}$$

Combining (5.83) and (5.84) gives (5.81) and proves the lemma. □

5.4.3.2 *Continuity*

In this section we prove that the system $\{m_W^{\mathrm{u}} \colon W \in \mathcal{W}^u\}$ is continuous in the sense of Definition 3.3. Let W_1, W_2, π be as in Definition 3.3. Given $t \geq 0$, there is a weak-stable δ-holonomy $f_t W_1 \cap \Lambda \to f_t W_2 \cap \Lambda$ given by

$$\pi_t = f_t \circ \pi \circ f_{-t}. \tag{5.85}$$

Lemma 4.5. *For every $\varepsilon > 0$, there is $\delta > 0$ such that if W_1, W_2, π are as in Definition 3.3, then for all $t \geq 0$ and $x, y \in W_1 \cap \Lambda$ such that $d(f_t x, f_t y) \leq r_0$, we have*

$$d_{f_t W_2}(\pi_t(f_t x), \pi_t(f_t y)) = e^{\pm\varepsilon} d_{f_t W_1}(f_t x, f_t y). \tag{5.86}$$

Proof. Choose $\delta > 0$ such that (5.86) holds when $t = 0$ for every pair of admissibles W_1, W_2 that are related by a δ-holonomy and satisfy $d_{C^1}(W_1, W_2) < \delta$. Then by cone-invariance, $B(f_t x, r_0, f_t W_1)$ and its image under π_t are admissibles containing $f_t x, f_t y$ and $\pi_t(f_t x), \pi_t(f_t y)$, respectively. Moreover, the C^1 distance between these admissibles is nonincreasing in t due to uniform expansion along E^{u} and nonexpansion along $E^{\mathrm{s}} \oplus E^0$, and thus is at most δ for every $t \geq 0$, so (5.86) continues to hold for all such t. □

Lemma 4.6. *For every $r \in (0, r_0)$ and $\eta > 0$, there exists $\delta > 0$ such that if W_1, W_2, π are as in Definition 3.3, then, for every $x \in W_1 \cap \Lambda$ and $t \geq \eta$, we have*

$$\pi(B_t(x, r, W_1) \cap \Lambda) \subset B_{t-\eta}(\pi x, r, W_2). \tag{5.87}$$

Proof. Choose $\varepsilon > 0$ such that $re^{\varepsilon} < \min(r_0, r\lambda^{-\eta}) =: r'$, and let $\delta > 0$ be given by Lemma 4.5. Then given W_1, W_2, π, x, t as in the statement, it follows from Lemma 4.5 that for every $y \in B_t(x, r, W_1) \cap \Lambda$ and $\tau \in [0, t]$, we have

$$d_{f_\tau W_2}(f_\tau(\pi x), f_\tau(\pi y)) \leq e^{\varepsilon} d_{f_\tau W_1}(f_\tau x, f_\tau y) < e^{\varepsilon} r < r'.$$

Since this holds for all $\tau \in [0, t]$, we have $\pi y \in B_t(\pi x, r', W_2)$, and we conclude that

$$\pi(B_t(x, r, W_1) \cap \Lambda) \subset B_t(\pi x, r', W_2) \subset B_{t-\eta}(\pi x, r'\lambda^{\eta}, W_2), \tag{5.88}$$

where the last inclusion uses (5.65). Since $r'\lambda^{\eta} \leq r$, this proves the lemma. □

Now we prove continuity. Given $\varepsilon > 0$, choose $\eta > 0$ such that

$$3\eta\|\varphi\| + \frac{|\varphi|_\sigma \eta^\sigma}{\sigma|\log\lambda|} \leq \frac{\varepsilon}{2} \quad \text{and} \quad \eta|P(\varphi)| \leq \frac{\varepsilon}{2}. \tag{5.89}$$

Then choose $\delta \in (0, \eta]$ such that Lemma 4.6 holds. Given W_1, W_2, π as in Definition 3.3 for this choice of δ, it follows from Lemma 4.6 that if $\mathcal{E} \in \mathbb{E}(Z, T)$ for some $Z \subset W_1 \cap \Lambda$, then

$$\pi(Z) \subset \bigcup_{(x,t)\in\mathcal{E}} \pi(B_t(x, r, W_1) \cap \Lambda) \subset \bigcup_{(x,t)\in\mathcal{E}} B_{t-\eta}(\pi x, r, W_2).$$

From this we see that $\{(\pi x, t-\eta) : (x, t) \in \mathcal{E}\} \subset \mathbb{E}(\pi Z, T-\eta)$, and we deduce that

$$m^{\mathrm{u}}_{W_2}(\pi Z) \leq \lim_{T\to\infty} \inf_{\mathcal{E}\in\mathbb{E}(Z,T)} \sum_{(x,t)\in\mathcal{E}} e^{\Phi(\pi x, t-\eta)-(t-\eta)P(\varphi)}. \tag{5.90}$$

To proceed we must compare $\Phi(\pi x, t-\eta)$ and $\Phi(x,t)$. There is $\tau \in (-\delta, \delta)$ such that $f_t(x) \in W^s(\pi x, \delta)$, and thus for all $\hat{t} > 0$ we have

$$\begin{aligned} |\Phi(f_t x, \hat{t}) - \Phi(\pi x, \hat{t})| &\leq \int_0^\infty |\varphi(f_s(f_t x)) - \varphi(f_s(\pi x))|\, ds \\ &\leq \int_0^\infty |\varphi|_\sigma (\delta\lambda^s)^\sigma\, ds = \frac{|\varphi|_\sigma \delta^\sigma}{\sigma |\log\lambda|}. \end{aligned} \tag{5.91}$$

Since $\delta \leq \eta$, we deduce that $|\tau - \eta| \leq 2\eta$, so

$$\begin{aligned} \Phi(\pi x, t-\eta) &= \Phi(\pi x, t-\tau) \pm 2\eta\|\varphi\|, \\ \Phi(x,t) &= \Phi(f_\tau x, t-\tau) \pm \eta\|\varphi\|, \end{aligned}$$

which together with (5.89) and (5.91) gives

$$|\Phi(\pi x, t-\eta) - \Phi(x,t)| \leq \varepsilon/2.$$

Using this estimate in (5.90) gives

$$m^{\mathrm{u}}_{W_2}(\pi Z) \leq \lim_{T\to\infty} \inf_{\mathcal{E}\in\mathbb{E}(Z,T)} e^{\Phi(x,t)+\varepsilon/2-tP(\varphi)} e^{\eta P(\varphi)} \leq e^{\varepsilon} m^{\mathrm{u}}_{W_1}(Z),$$

where the last inequality uses the second half of (5.89). By symmetry, this proves that $m^{\mathrm{u}}_{W_2}(\pi Z) = e^{\pm\varepsilon} m^{\mathrm{u}}_{W_1}(Z)$, which completes the proof of Theorem 3.4.

5.4.3.3 *Alternate Definitions for Discrete-Time and Non-Adapted Metrics*

In discrete time, the system of measures $\{m^{\mathrm{u}}_x\}_{x\in\Lambda}$ was introduced in [CPZ19, CPZ20], which also proved that this system is φ-conformal and has an absolute continuity property under stable holonomies. However, it remains open whether this construction produces a continuous system in the sense of Definition 3.3; see [CPZ19, Theorem 4.10, §6.2.2] and [CPZ20, Theorem 4.6, §7.3] for the results on holonomy maps that take the place of this property, and note that no specific formula for the Radon–Nikodym derivative is given there (compare to Theorem 3.7, which we prove in the next section).

The key step in the proof of continuity in Section 5.4.3.2 is to observe that under a δ-holonomy, Bowen balls of radius r on W_1 are mapped into Bowen balls of a slightly larger radius r' on W_2, and that these can in turn be covered by Bowen balls of the original radius by making a small increase in the continuous time parameter $t \in \mathbb{R}^+$ and then using (5.65). In discrete time, this step does not work because there is no way to make an arbitrarily small increase in the time parameter $n \in \mathbb{N}$.

One way to obtain continuity as in Definition 3.3 for the discrete-time construction would be to allow the radius r to vary, and to introduce an extra factor into the "weight" assigned each ball in the C-structure. For example, one might fix a continuous function $\zeta(r)$ and then redefine the C-structure on $X = W^{\mathrm{u}}(x,\delta) \cap \Lambda$ by

$$\mathcal{S} = X \times \mathbb{N} \times (0, \infty), \qquad U_{(x,n,r)} = B_n(x, r),$$

with ξ, η, ψ given by

$$\xi(x, n, r) = \zeta(r) e^{\sum_{k=0}^{n-1} \varphi(f^k x)}, \qquad \eta(x, n, r) = \psi(x, n, r) = e^{-n}.$$

With an appropriate choice of ζ – say constant and positive near 0, then increasing quickly on a small interval of r so as to guarantee that all "optimal" enveloping sets use only values of r from this interval – it should be possible to guarantee that in the discrete-time uniformly hyperbolic setting, the system of leaf measures defined by these C-structures satisfies the continuity property in Definition 3.3, and that they are uniformly equivalent to the original leaf measures defined in [CPZ19].

It is also worth pointing out that our proofs of φ-conformality and continuity relied on the fact that we work in an adapted metric. Working in a different metric would produce uniformly equivalent systems of measures, following arguments similar to those in Section 5.4.2.1. It is not clear whether these new systems of measures would satisfy φ-conformality and continuity: the arguments at the end of [Has89] suggest that one might expect "almost-everywhere" results instead of "everywhere" results in this case.

Because we work in an adapted metric, the leafwise Bowen balls are exactly the metric balls in the Hamenstädt–Hasselblatt metric from [Ham89, Has89], and then the main thing differentiating our construction from theirs is that we assign weights according to the potential function.[20] An approach that might perhaps be more natural, and which is done in [CPZ19, CPZ20], is to use the extrinsic metric from M on each leaf, instead of the intrinsic leaf metric; doing this here would once again lead to uniformly equivalent systems of measures (again following the arguments in Section 5.4.2.1) but we would not be able to carry out our proofs of φ-conformality or continuity, since we would not have Lemma 4.2.

5.4.4 Weak-Stable Holonomy Formula: Proof of Theorem 3.7

To prove Theorem 3.7, we start by using the chain rule for Radon–Nikodym derivatives to reduce to the case of two u-admissible manifolds that are extremely close; then we will use the continuity property.

Lemma 4.7. *Suppose $W_1, W_2 \in \mathcal{W}^u$ are related by a weak-stable holonomy π. Given $s, t \geq 0$, let $\pi_{s,t} = f_t \circ \pi \circ f_{-s}$ be the corresponding weak-stable holonomy between $f_s W_1 \cap \Lambda$ and $f_t W_2 \cap \Lambda$. Suppose, moreover, that $(\pi_{s,t})_* \bar{m}^{\mathrm{u}}_{f_s W_1} \ll \bar{m}^{\mathrm{u}}_{f_t W_2}$. Then $\pi_* \bar{m}^{\mathrm{u}}_{W_1} \ll \bar{m}^{\mathrm{u}}_{W_2}$, and we have*

[20] Note that [Has89, Lemma 6] gives the proof of continuity for the zero potential using that construction.

$$\frac{d(\pi_* \bar{m}^{\mathrm{u}}_{W_1})}{d\bar{m}^{\mathrm{u}}_{W_2}}(\pi z) = e^{\Phi(z,s)-\Phi(\pi z,t)-(s-t)P(\varphi)} \frac{d(\pi_{s,t})_* \bar{m}^{\mathrm{u}}_{f_s W_1}}{d\bar{m}^{\mathrm{u}}_{f_t W_2}}(\pi_{s,t} f_s z). \tag{5.92}$$

Proof. We use the chain rule for Radon–Nikodym derivatives in the following form: given measures $\nu_0, \nu_1, \nu_2, \nu_3$ and maps g^1, g^2, g^3 such that $g^i_* \nu_{i-1} \ll \nu_i$ for $i = 1, 2, 3$, and writing $x_i = g_i(x_{i-1})$, we have

$$\begin{aligned}\frac{d(g^3_* g^2_* g^1_* \nu_0)}{d\nu_3}(x_3) &= \frac{d(g^3_* g^2_* g^1_* \nu_0)}{d(g^3_* g^2_* \nu_1)}(x_3) \frac{d(g^3_* g^2_* \nu_1)}{d(g^3_* \nu_2)}(x_3) \frac{d(g^3_* \nu_2)}{d\nu_3}(x_3) \\ &= \frac{d(g^1_* \nu_0)}{d\nu_1}(x_1) \frac{d(g^2_* \nu_1)}{d\nu_2}(x_2) \frac{d(g^3_* \nu_2)}{d\nu_3}(x_3).\end{aligned} \tag{5.93}$$

With $g^1 = f_s$, $g^2 = \pi_t$, and $g^3 = f_{-t}$, (5.93) gives

$$\frac{d(\pi_* \bar{m}^{\mathrm{u}}_{W_1})}{d\bar{m}^{\mathrm{u}}_{W_2}}(\pi z) = \frac{d(f_s)_* \bar{m}^{\mathrm{u}}_{W_1}}{d\bar{m}^{\mathrm{u}}_{f_s W_1}}(f_s z) \frac{d(\pi_{s,t})_* \bar{m}^{\mathrm{u}}_{f_s W_1}}{d\bar{m}^{\mathrm{u}}_{f_t W_2}}(f_t \pi z) \frac{d(f_{-t})_* \bar{m}^{\mathrm{u}}_{f_t W_2}}{d\bar{m}^{\mathrm{u}}_{W_2}}(\pi z).$$

From φ-conformality, the first and third factors simplify as

$$\frac{d(f_s)_* \bar{m}^{\mathrm{u}}_{W_1}}{d\bar{m}^{\mathrm{u}}_{f_s W_1}}(f_s z) \frac{d(f_{-t})_* \bar{m}^{\mathrm{u}}_{f_t W_2}}{d\bar{m}^{\mathrm{u}}_{W_2}}(\pi z) = e^{\Phi(z,s)-sP(\varphi)} e^{tP(\varphi)-\Phi(\pi z,t)},$$

completing the proof of Lemma 4.7. □

Now as in Theorem 3.7 we assume that there is a weak-stable δ_0-holonomy $\pi : W_1 \cap \Lambda \to W_2 \cap \Lambda$. Given $x \in W_1 \cap \Lambda$, there is $\tau(x) \in (-\delta_0, \delta_0)$ such that $f_{\tau(x)} \in W^{\mathrm{s}}(\pi x, \delta_0)$. Fix $\varepsilon > 0$ and let $\delta > 0$ be given by continuity (Definition 3.3). For each $x \in W_1 \cap \Lambda$ there are $t > 0$ and a relatively open set $W_1' \subset W_1$ containing x such that writing $W_2' = \pi(W_1')$, we have the following, using Lemma 2.9.

- $f_{t+\tau(x)}(W_1')$ and $f_t(W_2')$ are u-admissible.
- $d_{C^1}(f_{t+\tau(x)}(W_1'), f_t(W_2')) < \delta$.
- $\pi_{t+\tau,t} : f_{t+\tau(x)}(W_1') \cap \Lambda \to f_t(W_2') \cap \Lambda$ is a weak-stable δ-holonomy.
- $|\tau(z) - \tau(x)|(\|\varphi\| + |P(\varphi)|) < \varepsilon$ for all $z \in W_1'$.
- $\Phi(z, t + \tau(z)) - \Phi(\pi z, t) - \tau(z)P(\varphi) = \omega^+(z, \pi z) \pm \varepsilon$ for all $z \in W_1'$.

Using the first three properties, continuity, and Lemma 4.7 (with $s = t+\tau(x)$) gives

$$\frac{d(\pi_* \bar{m}^{\mathrm{u}}_{W_1'})}{d\bar{m}^{\mathrm{u}}_{W_2'}}(\pi z) = e^{\Phi(z,t+\tau(x))-\Phi(\pi z,t)-\tau(x)P(\varphi)\pm\varepsilon}.$$

Then the fourth property gives

$$\Phi(z, t + \tau(x)) - \tau(x)P(\varphi) = P(z, t + \tau(z)) - \tau(z)P(\varphi) \pm \varepsilon,$$

from which we deduce that

$$\frac{d(\pi_* \bar{m}^{\mathrm{u}}_{W_1'})}{d\bar{m}^{\mathrm{u}}_{W_2'}}(\pi z) = e^{\Phi(z,t+\tau(z))-\Phi(\pi z,t)-\tau(z)P(\varphi)\pm 2\varepsilon} = e^{\omega^+(z,\pi z)\pm 3\varepsilon},$$

where the last equality uses the fifth property. Since $\bar{m}^{\mathrm{u}}_{W_j'}$ and $\bar{m}^{\mathrm{u}}_{W_j}$ agree on W_j', we get the same result with W_j' replaced by W_j. Finally, since $\varepsilon > 0$ was arbitrary, this proves (5.31).

5.4.5 Product Construction: Proof of Theorem 3.10

5.4.5.1 Proof of the Theorem

In this section we prove Theorem 3.10. There are four things to prove.

(1) The equations (5.47)–(5.50) all agree.
(2) There is a unique measure μ such that $\mu|_{R_q} = \mu_q$ for all $q \in \Lambda$.
(3) The measure μ is flow-invariant, nonzero, and finite.
(4) μ is a scalar multiple of the unique equilibrium measure.

For the first, start by defining a function $h_q \colon R_q^{\mathrm{u}} \times R_q^{\mathrm{cs}} \to (0,\infty)$ by

$$h_q(x,y) := e^{\omega^+([x,y],x)+\omega^-([x,y],y)}. \tag{5.94}$$

Then, recalling (5.46), we can write (5.47) as

$$\mu_q(Z) = \int_Z h_q(\Psi_q^{-1}(z))\, d((\Psi_q)_*(\bar{m}_q^{\mathrm{u}} \times \bar{m}_q^{\mathrm{cs}}))(z),$$

while (5.48) becomes

$$\mu_q(Z) = \int_{\Psi_q^{-1}(Z)} h_q(x,y)\, d(\bar{m}_q^{\mathrm{u}} \times \bar{m}_q^{\mathrm{cs}})(x,y).$$

These integrals agree by the definition of pushforward, so (5.47) and (5.48) agree. To compare (5.48) and (5.49) it suffices to observe that given any $y \in R_q^{\mathrm{cs}}$, there is a weak-stable δ-holonomy $\pi \colon R_q^{\mathrm{u}} \to R_q^{\mathrm{u}}(y)$ given by $\pi(x) = [x,y]$, and Theorem 3.7 shows that for every continuous $\zeta \colon R_q^{\mathrm{u}}(y) \to \mathbb{R}$, we have

$$\begin{aligned}\int_{R_q^{\mathrm{u}}} \zeta(\pi x)\mathbf{1}_Z([x,y])\, d\bar{m}_q^{\mathrm{u}}(x) &= \int_{R_q^{\mathrm{u}}(y)\cap Z} \zeta(z)\, d(\pi_*\bar{m}_q^{\mathrm{u}})(z) \\ &= \int_{R_q^{\mathrm{u}}(y)\cap Z} \zeta(z) e^{\omega^+([z,q],z)}\, d\bar{m}_y^{\mathrm{u}}(z).\end{aligned} \tag{5.95}$$

We use this with

$$\zeta(z) = h_q(\Psi_q^{-1}(z)) = e^{\omega^+(z,[z,q])+\omega^-(z,y)}$$

so that

$$\zeta(\pi x) = e^{\omega^+([x,y],x)+\omega^-([x,y],y)} \quad \text{and} \quad \zeta(z)e^{\omega^+([z,q],z)} = e^{\omega^-(z,y)}.$$

Then the first integral in (5.95) is the inner integral in (5.48), while the last integral in (5.95) becomes the inner integral in (5.49), so that (5.48) and (5.49) agree. A similar argument shows the equivalence of (5.48) and (5.50).

To prove that the family of measures $\{\mu_q \colon q \in \Lambda\}$ defines a unique Borel measure μ on Λ, it suffices to show that they are consistent in the sense that

$$\mu_q(Z) = \mu_p(Z) \text{ whenever } Z \subset R_q \cap R_p; \tag{5.96}$$

then one defines μ by $\mu(Z) = \mu_q(Z)$ whenever $Z \subset R_q$, and extends by additivity.[21] We observe that when $p \in R_q^{\mathrm{cs}}$, (5.96) follows immediately from (5.49) and the fact that $\bar{m}_q^{\mathrm{cs}}$ and $\bar{m}_p^{\mathrm{cs}}$ agree on $R_q^{\mathrm{cs}} \cap R_p^{\mathrm{cs}}$. Similarly, (5.50) establishes (5.96) when $p \in R_q^{\mathrm{u}}$.

Now given any $q, p \in \Lambda$ such that $R_q \cap R_p \neq \emptyset$, we can take x to be any point in the intersection and then observe that the measures μ_q, $\mu_{[x,q]}$, μ_x, $\mu_{[x,p]}$, μ_p all agree on $R_q \cap R_p$ by the previous paragraph. This proves (5.96) and thus establishes existence of a unique μ as claimed.

The fact that μ is nonzero and finite follows from the corresponding properties for each μ_q, together with the fact that Λ can be covered with finitely many sets R_q by compactness. For flow-invariance, it suffices to show that given $q \in \Lambda$, $Z \subset R_q$, and $t \in \mathbb{R}$ such that $f_t Z \subset R_q$, we have $\mu_q(f_t Z) = \mu_q(Z)$. To show this, we observe that on the leaf $W^{\mathrm{cs}}(x)$, we can use (5.38) and (5.44), together with the observation that $-tP(\varphi) + \Phi(z,t) = \omega^+(z, f_t z)$, to write

$$\begin{aligned}
\bar{m}_{f_t x}^{\mathrm{cs}}(f_t Z) &= \int_{-\delta}^{\delta} \bar{m}_{f_\tau(f_t x)}^{\mathrm{s}}(f_t Z)\,d\tau = \int_{-\delta}^{\delta} \bar{m}_{f_t(f_\tau x)}^{\mathrm{s}}(f_t Z)\,d\tau \\
&= \int_{-\delta}^{\delta} \int_Z e^{\omega^+(z,f_t z)}\,d\bar{m}_{f_\tau x}^{\mathrm{s}}(z)\,d\tau = \int_Z e^{\omega^+(z,f_t z)}\,d\bar{m}_x^{\mathrm{cs}}(z)\,d\tau.
\end{aligned}$$

We will use this with t replaced by $-t$ to obtain

$$\frac{d((f_t)_*\bar{m}_{f_{-t}x}^{\mathrm{cs}})}{d\bar{m}_x^{\mathrm{cs}}}(y) = e^{\omega^+(y,f_{-t}y)} \quad \text{and} \quad \frac{d\bar{m}_x^{\mathrm{cs}}}{d((f_t)_*\bar{m}_{f_{-t}x}^{\mathrm{cs}})}(y) = e^{\omega^+(f_{-t}y,y)}.$$

Using the second of these together with the definition of μ_q in (5.50), we have

$$\begin{aligned}
\mu_q(f_t Z) &= \int_{R_q^{\mathrm{u}}} \int_{R_q^{\mathrm{cs}}(x)} e^{\omega^+(y,x)} \mathbf{1}_{f_t Z}(y)\,d\bar{m}_x^{\mathrm{cs}}(y)\,d\bar{m}_q^{\mathrm{u}}(x) \\
&= \int_{R_q^{\mathrm{u}}} \int_{R_q^{\mathrm{cs}}(x)} e^{\omega^+(y,x)} \mathbf{1}_Z(f_{-t}y) e^{\omega^+(f_{-t}y,y)}\,d((f_t)_*\bar{m}_{f_{-t}x}^{\mathrm{cs}})(y)\,d\bar{m}_q^{\mathrm{u}}(x).
\end{aligned}$$

[21] See [Mar70, §4] and [Mar04, §4] for more details of this standard procedure.

Writing $y' = f_{-t}y$ and using the cocycle relation (5.34), this becomes

$$\mu_q(f_t Z) = \int_{R_q^{\mathrm{u}}} \int_{R_q^{\mathrm{cs}}(x)} e^{\omega^+(y',x)} \mathbf{1}_Z(y')\, d\bar{m}^{\mathrm{cs}}_{f_{-t}x}(y')\, d\bar{m}^{\mathrm{u}}_q(x).$$

Finally, since for all $y' \in Z$ we also have $f_t(y') \in R_q$ (by our assumption that $f_t Z \subset R_q$), we can use the fact that $\bar{m}^{\mathrm{cs}}_{f_{-t}x}$ and $\bar{m}^{\mathrm{cs}}_x$ agree on the overlap of the center-stable leaves to obtain

$$\mu_q(f_t Z) = \int_{R_q^{\mathrm{u}}} \int_{R_q^{\mathrm{cs}}(x)} e^{\omega^+(y',x)} \mathbf{1}_Z(y')\, d\bar{m}^{\mathrm{cs}}_x(y')\, d\bar{m}^{\mathrm{u}}_q(x) = \mu_q(Z),$$

establishing the desired flow-invariance.

Finally, to prove that μ is a scalar multiple of the unique equilibrium measure, it suffices to show that there is $C_4 > 0$ such that the upper Gibbs bound $\mu_q(B_t(q,r)) \leq C_4 e^{\Phi(q,t) - tP(\varphi)}$ holds for all $q \in \Lambda$ and $t > 0$, since then μ is absolutely continuous with respect to the unique invariant Gibbs probability measure, which is ergodic.

To prove the upper Gibbs bound, observe that for δ sufficiently small, for each $q \in \Lambda$ and $y \in W^{\mathrm{cs}}(q,\delta) \cap \Lambda$, writing $W = W^{\mathrm{u}}(y, L\delta)$ we have

$$B_t(q,\delta) \cap W \subset B_t(y, 2\delta, W) = f^{-t} W^{\mathrm{u}}(f_t y, 2\delta),$$

where the last equality uses Lemma 4.1. Thus φ-conformality gives

$$\bar{m}^{\mathrm{u}}_y(B_t(q,\delta)) \leq \sup_{z \in B_t(q,\delta)} e^{\Phi(z,t) - tP(\varphi)} \bar{m}^{\mathrm{u}}_{f_t y}(W^{\mathrm{u}}(f_t y, 2\delta)).$$

Since φ is Hölder continuous, it satisfies the Bowen property

$$\sup_{q \in \Lambda} \sup_{t > 0} \sup_{z \in B_t(q,\delta)} |\Phi(z,t) - \Phi(q,t)| < \infty.$$

Using the previous two estimates and Theorem 3.1, we get

$$\bar{m}^{\mathrm{u}}_y(B_t(q,\delta)) \leq C_5 e^{\Phi(q,t) - tP(\varphi)},$$

where $C_5 > 0$ does not depend on q, y, or t. Now the upper Gibbs bound for μ_q follows immediately from (5.49) and the fact that $e^{\omega^-(x,y)}$ and $\bar{m}^{\mathrm{cs}}_q(R^{\mathrm{cs}}_q)$ are uniformly bounded above independently of $q \in \Lambda$, $y \in R^{\mathrm{u}}_q$, and $x \in R^{\mathrm{cs}}_q(y)$. This completes the proof of Theorem 3.10.

5.4.5.2 *Proof of Corollary 3.12*

To prove Corollary 3.12, we start by observing that, given $q \in \Lambda$, it follows from (5.53) that, for $\hat{\mu}_q$-a.e. $y \in W^{\mathrm{cs}}_q$, we have

$$\frac{\bar{m}^{\mathrm{u}}_y(Z)}{m^{\mathrm{u}}_y(Z)} = \frac{\bar{m}^{\mathrm{u}}_y(R^{\mathrm{u}}_q(y))}{m^{\mathrm{u}}_y(R^{\mathrm{u}}_q(y))} =: g(q,y) \text{ for all Borel } Z \subset R^{\mathrm{u}}_q(y). \tag{5.97}$$

Observe that $g(q, y)$ is independent of δ (even though $R^{\mathrm{u}}_q(y)$ depends on δ) because one obtains the same ratio for every Borel Z. Since both systems $\{\bar{m}^{\mathrm{u}}_y\}$ and $\{m^{\mathrm{u}}_y\}$ scale according to Theorem 3.7 under weak-stable holonomies, we conclude that g is, in fact, independent of y. More precisely, choosing any $y \in W^{\mathrm{cs}}_q$ such that (5.97) holds, let $\pi : R^{\mathrm{u}}_q \to R^{\mathrm{u}}_q(y)$ be a weak-stable δ_0-holonomy: then by (5.31), for every Borel $Z \subset R^{\mathrm{u}}_q$ we have

$$\frac{\bar{m}^{\mathrm{u}}_q(Z)}{m^{\mathrm{u}}_q(Z)} = \frac{\pi_* \bar{m}^{\mathrm{u}}_q(\pi Z)}{\pi_* m^{\mathrm{u}}_q(\pi Z)} = \frac{\int_{R^{\mathrm{u}}_q(y)} e^{\omega^+(\pi^{-1}z, z)} \, d\bar{m}^{\mathrm{u}}_y(z)}{\int_{R^{\mathrm{u}}_q(y)} e^{\omega^+(\pi^{-1}z, z)} \, dm^{\mathrm{u}}_y(z)} = g(q, y) =: g(q).$$

Now given any $t \in \mathbb{R}$, since $\bar{m}^{\mathrm{u}}_q$ and m^{u}_q are fully supported on R^{u}_q, we can choose a Borel set $Z \subset R^{\mathrm{u}}_q$ such that $f_t Z \subset R^{\mathrm{u}}_{f_t q}$, and use the φ-conformality property (5.24) to deduce that

$$g(f_t q) = \frac{\bar{m}^{\mathrm{u}}_{f_t q}(f_t Z)}{m^{\mathrm{u}}_{f_t q}(f_t Z)} = \frac{\int_Z e^{tP(\varphi) - \Phi(z,t)} \, d\bar{m}^{\mathrm{u}}_q(z)}{\int_Z e^{tP(\varphi) - \Phi(z,t)} \, dm^{\mathrm{u}}_q(z)} = g(q).$$

Thus g is flow-invariant, and since the unique equilibrium measure μ is ergodic, g is constant μ-a.e.: there is $c > 0$ such that $g(q) = c$ for μ-a.e. $q \in \Lambda$. Moreover, since $\operatorname{supp} \mu = \Lambda$ and g is continuous (because the systems $\{\bar{m}^{\mathrm{u}}_q\}$ and $\{m^{\mathrm{u}}_q\}$ are), we conclude that $g(q) = c$ for every $q \in \Lambda$, which completes the proof of Corollary 3.12.

5.4.6 Refining in All Directions: Proof of Theorem 3.15

Let m be the measure on Λ defined by (5.60)–(5.61). Observe that it is a Borel measure by Lemma 2.14 because (5.66) and the local product structure give

$$\operatorname{diam} B^*_{s,t}(x, r) \le 2rL(\lambda^{s-t_0} + \lambda^{t-t_0}) + \frac{C_6 r}{s+t},$$

where

$$C_6 := \max_{x \in M} \left\| \frac{d}{dt} f_t(x)|_{t=0} \right\| \tag{5.98}$$

is the maximum speed of the flow, so that $d(x, f_t x) \le C_6 |t|$ for all x, t.

We prove Theorem 3.15 by showing that m is flow-invariant and has the Gibbs property (5.5); we will use the fact that (Λ, F, φ) is already known to have a unique flow-invariant Gibbs probability measure.

5.4.6.1 Flow-Invariance

To prove that $m(f_\tau Z) = m(Z)$ for all $Z \subset \Lambda$ and $\tau \in \mathbb{R}$, we first observe that, given $x \in \Lambda$ and $s, t \ge |\tau|$, we have

$$z \in B^{\pm}_{s,t}(x, r) \quad \Leftrightarrow \quad f_\tau z \in B^{\pm}_{s+\tau, t-\tau}(f_\tau x, r).$$

Moreover, given $z \in B^{\pm}_{s,t}(x,r)$, we have

$$\beta(f_\tau x, f_\tau z) = \beta(x,z),$$

from which it follows that

$$f_\tau B^*_{s,t}(x,r) = B^*_{s+\tau,t-\tau}(f_\tau x, r). \tag{5.99}$$

Now given $Z \subset \Lambda$ and $T \geq |\tau|$, consider the map

$$S\colon \Lambda \times [T,\infty)^2 \to \Lambda \times [T-|\tau|,\infty)^2,$$
$$(x,s,t) \mapsto (f_\tau x, s+\tau, t-\tau).$$

By (5.99), given $Z \subset \Lambda$ and $\mathcal{E} \in \mathbb{E}^*(Z,T)$, we have

$$f_\tau Z \subset f_\tau\Big(\bigcup_{(x,s,t)\in\mathcal{E}} B^*_{s,t}(x,r)\Big) \subset \bigcup_{(x,s,t)\in\mathcal{E}} B^*_{s+\tau,t-\tau}(f_\tau x, r),$$

so that $S(\mathcal{E}) \subset \mathbb{E}^*(f_\tau Z, T-|\tau|)$. Since $(s+\tau)+(t-\tau) = s+t$, it follows that

$$m(f_\tau Z) \leq \lim_{T\to\infty} \inf_{\mathcal{E}\in\mathbb{E}^*(Z,T)} \sum_{(x,s,t)\in\mathcal{E}} \frac{e^{\Phi(f_{-(s+\tau)}(f_\tau x), s+t)-(s+t)P(\varphi)}}{s+t}$$
$$= \lim_{T\to\infty} \inf_{\mathcal{E}\in\mathbb{E}^*(Z,T)} \sum_{(x,s,t)\in\mathcal{E}} \frac{e^{\Phi(f_{-s}(x), s+t)-(s+t)P(\varphi)}}{s+t} = m(Z).$$

Since $m(f_\tau Z) \leq m(Z)$ for all $\tau \in \mathbb{R}$ and Borel $Z \subset \Lambda$, we conclude that m is flow-invariant.

5.4.6.2 Gibbs Property

We start with a general lemma.

Lemma 4.8. *There are $C_7, T_1, \eta > 0$ such that, for every flow-invariant measure ν on Λ and every $(x,s,t) \in \Lambda \times [T_1,\infty)^2$, we have*

$$\nu(B^*_{s,t}(x,r)) \leq C_7(s+t)^{-1}\nu(B^{\pm}_{s-\eta,t-\eta}(x,r)), \tag{5.100}$$
$$\nu(B^{\pm}_{s,t}(x,r)) \leq C_7(s+t)\nu(B^*_{s-\eta,t-\eta}(x,r)). \tag{5.101}$$

Proof. Given $z \in B^{\pm}_{s,t}(x,r) \cap \Lambda$, the local product structure gives $y := [x,z] \in W^{\mathrm{u}}(z, rL)$ and $y' = f_{\beta(x,z)}(y) \in W^{\mathrm{s}}(x, rL)$ with $f_{-s}(y') \in W^{\mathrm{s}}(f_{-s}x, rL)$ and $f_t(y) \in W^{\mathrm{u}}(f_t z, rL)$. Given $\eta \in [0, \min(s,t)]$ and $\tau \in [-s+\eta, t-\eta]$, we deduce from (5.14) that[22]

$$d(f_\tau z, f_\tau y) \leq C_0 r L \lambda^\eta \quad \text{and} \quad d(f_\tau y', f_\tau x) \leq C_0 r L \lambda^\eta.$$

[22] In fact we are using the analogue of (5.14) for the original metric, which is not necessarily adapted; this is the reason for the inclusion of the constant C_0.

Since we also have $d(f_\tau y', f_\tau y) \leq C_6|\beta(x,z)|$, the triangle inequality gives

$$d(f_\tau x, f_\tau z) \leq 2C_0 rL\lambda^\eta + C_6|\beta(x,z)| \tag{5.102}$$

for all $z \in B^{\pm}_{s,t}(x,r)$ and $\tau \in [-s+\eta, t-\eta]$.

Now we prove (5.100). By (5.102) there are $T_1 > \eta > 0$ such that for all $z \in B^*_{s,t}(x,r)$ and $\tau \in [-s+\eta, t-\eta]$, we have

$$d(f_\tau x, f_\tau z) \leq 2C_0 rL\lambda^\eta + C_6 T_1^{-1} < r/2.$$

In other words, we have

$$B^*_{s,t}(x,r) \subset \{z \in B^{\pm}_{s-\eta,t-\eta}(x,r/2) \colon |\beta(x,z)| < r/(s+t)\}. \tag{5.103}$$

Now let $\xi > 0$ be such that $d(z, f_\tau z) < r/2$ for all $z \in \Lambda$ and $|\tau| < \xi$; then we have $f_\tau(x) \in B^*_{s,t}(x,r)$ for all $|\tau| < \xi$, and

$$f_\tau(z) \in B^{\pm}_{s-\eta,t-\eta}(x,r) \text{ for all } z \in B^*_{s,t}(x,r) \text{ and } |\tau| < \xi. \tag{5.104}$$

Consider $(x,s,t) \in \Lambda \times [T_1,\infty)^2$ and let $\gamma = \frac{2r}{s+t}$. Then for all $0 \leq k < N := \lfloor \frac{\xi}{\gamma} \rfloor = \lfloor \frac{\xi}{2r}(s+t) \rfloor$, we have $f_{k\gamma} B^*_{s,t}(x,r) \subset B^{\pm}_{s-\eta,t-\eta}(x,r)$ by (5.104), and, moreover, these N sets are disjoint because $\beta(x, f_{k\gamma} z) = k\gamma + \beta(x,z) = k\gamma \pm \frac{r}{s+t}$ for all $z \in B^*_{s,t}(x,r)$. These sets all have the same measure by flow-invariance, and thus $\nu(B^{\pm}_{s-\eta,t-\eta}(x,r)) \geq N\nu(B^*_{s,t}(x,r))$, which proves (5.100).

Now we prove (5.101). There is $C_8 > 0$ such that $\beta(x,z) \leq C_8 r$ for all $x \in \Lambda$ and $z \in B(x,r)$. Given $(x,s,t) \in \Lambda \times [T_1,\infty)^2$, let $\gamma = \frac{r}{s+t}$, $N = \lceil C_8 r/\gamma \rceil = \lceil C_8(s+t) \rceil$, and $x_k = f_{k\gamma}(x)$ for all $|k| \leq N$. Then for every $z \in B^{\pm}_{s,t}(x,r)$, there is $|k| \leq N$ such that $|\beta(x_k,z)| < \gamma = \frac{r}{s+t}$. For every $\tau \in [-s+\eta, t-\eta]$, (5.102) gives

$$d(f_\tau x_k, f_\tau z) \leq 2C_0 rL\lambda^\eta + C_6/(s+t) < r/2,$$

and thus $z \in B^*_{s-\eta,t-\eta}(x_k, r/2) \subset f_{k\gamma} B^*_{s-\eta,t-\eta}(x,r)$. By flow-invariance we have

$$\nu(B^{\pm}_{s,t}(x,r)) \leq \nu\Big(\bigcup_{k=-N}^{N} f_{k\gamma} B^*_{s-\eta,t-\eta}(x,r)\Big) = (2N+1)\nu(B^*_{s-\eta,t-\eta}(x,r)),$$

which proves (5.101). □

Lemma 4.9. *Let μ be the unique flow-invariant Gibbs probability measure for (Λ, F, φ). Then there is $C_9 > 0$ such that, for every $(x,s,t) \in \Lambda \times [T_1,\infty)^2$, we have*

$$\mu(B^*_{s,t}(x,r)) = C_9^{\pm 1}(s+t)^{-1} e^{\Phi(f_{-s}x, s+t) - (s+t)P(\varphi)}. \tag{5.105}$$

Proof. Since μ is a Gibbs measure, there is $Q > 0$ such that for all $(x,s,t) \in \Lambda \times [0,\infty)^2$, we have

$$\mu(B^{\pm}_{s,t}(x,r)) = \mu(f_{-s}B^{\pm}_{s,t}(x,r)) = \mu(B_{s+t}(f_{-s}x,r))$$
$$= Q^{\pm 1}e^{\Phi(f_{-s}x,s+t)-(s+t)P(\varphi)}. \tag{5.106}$$

Combining (5.100) and the upper bound in (5.106) gives

$$\mu(B^*_{s,t}(x,r)) \le C_7Q(s+t)^{-1}e^{\Phi(f_{-s+\eta}x,s+t-2\eta)-(s+t-2\eta)P(\varphi)}$$
$$\le C_7Q(s+t)^{-1}e^{\Phi(f_{-s}x,s+t)-(s+t)P(\varphi)}e^{2\eta(\|\varphi\|+|P(\varphi)|)},$$

which proves the upper bound in (5.105). For the lower bound we use (5.101) and the lower bound in (5.106) to get

$$\mu(B^*_{s,t}(x,r)) \ge C_7^{-1}(s+t+2\eta)^{-1}\mu(B^{\pm}_{s+\eta,t+\eta}(x,r))$$
$$\ge \frac{(s+t)e^{\Phi(f_{-s}x,s+t)-(s+t)P(\varphi)}}{C_7Q(s+t+2\eta)(s+t)e^{2\eta(\|\varphi\|+|P(\varphi)|)}},$$

which completes the proof since $(s+t)/(s+t+2\eta) \ge T_1/(T_1+2\eta)$. □

Now we can prove that m has the Gibbs property. For the lower bound, consider any Borel set $Z \subset \Lambda$, any $T \ge T_1$, and any $\mathcal{E} \in \mathbb{E}^*(Z,T)$: by (5.105), we have

$$\sum_{(x,s,t)\in\mathcal{E}} \frac{e^{\Phi(f_{-s}(x),s+t)-(s+t)P(\varphi)}}{s+t} \ge \sum_{(x,s,t)\in\mathcal{E}} \frac{1}{C_9}\mu(B^*_{s,t}(x,r)) \ge \frac{\mu(Z)}{C_9}.$$

Taking an infimum over all such $\mathcal{E}$ and sending $T \to \infty$ gives $m(Z) \ge \mu(Z)/C_9$. Since μ has the Gibbs property, this proves the lower Gibbs bound for m.

To prove the upper Gibbs bound, fix $x \in \Lambda$ and $\tau > 0$, and let $Z = B_\tau(x,r/2)\cap\Lambda$. Given $T > \tau$, let $E_T \subset Z$ be maximal with the property that for any $y,z \in E_T$ with $z \ne y$, we have $z \notin B^*_{T,T}(y,r)$; then $Z \subset \bigcup_{y\in E_T} B^*_{T,T}(y,r)$, while at the same time the sets $B^*_{T,T}(y,r/2)$ are disjoint, and we have

$$m(Z) \le \varlimsup_{T\to\infty} \sum_{y\in E_T} \frac{1}{s+t}e^{\Phi(f_{-T}y,2T)-2TP(\varphi)}$$
$$\le \varlimsup_{T\to\infty} \sum_{y\in E_T} C_9(r/2)\mu(B^*_{T,T}(y,r/2))$$
$$\le \varlimsup_{T\to\infty} C_9(r/2)\mu\Big(\bigcup_{y\in E_T} B^*_{T,T}(y,r/2)\Big).$$

Given $y \in E_T$, observe that $B^*_{T,T}(y,r/2) \subset B_\tau(y,r/2) \subset B_\tau(x,r)$, and thus the above computation shows that

$$m(B_\tau(x,r/2)) \le C_9(r/2)\mu(B_\tau(r)).$$

Thus the upper Gibbs bound for m (at scale $r/2$) follows from the upper Gibbs bound for μ (at scale r).

References

[BCFT18] K. Burns, V. Climenhaga, T. Fisher, and D. J. Thompson, *Unique equilibrium states for geodesic flows in nonpositive curvature*, Geom. Funct. Anal. **28** (2018), no. 5, 1209–1259. MR 3856792

[BM77] Rufus Bowen and Brian Marcus, *Unique ergodicity for horocycle foliations*, Israel J. Math. **26** (1977), no. 1, 43–67. MR 451307

[Bow73] Rufus Bowen, *Topological entropy for noncompact sets*, Trans. Amer. Math. Soc. **184** (1973), 125–136. MR 0338317

[Bow08] Rufus Bowen, *Equilibrium states and the ergodic theory of Anosov diffeomorphisms*, revised ed., Lecture Notes in Mathematics, vol. 470, Springer-Verlag, Berlin, 2008, With a preface by David Ruelle, edited by Jean-René Chazottes. MR 2423393

[Bow75] Rufus Bowen, *Some systems with unique equilibrium states*, Math. Systems Theory **8** (1974/75), no. 3, 193–202. MR 0399413

[BR75] Rufus Bowen and David Ruelle, *The ergodic theory of Axiom A flows*, Invent. Math. **29** (1975), no. 3, 181–202. MR 0380889

[Cal22] Benjamin Call, *The K-property for some unique equilibrium states in flows and homeomorphisms*, Ergod. Theory Dyn. Syst. **42** (2022), no. 8, 2509–2532.

[CKW21] Vaughn Climenhaga, Gerhard Knieper, and Khadim War, *Uniqueness of the measure of maximal entropy for geodesic flows on certain manifolds without conjugate points*, Adv. Math. **376** (2021), 107452, 44. MR 4178924

[CKW22] Vaughn Climenhaga, Gerhard Knieper, and Khadim War, *Closed geodesics on surfaces without conjugate points*, Commun. Contemp. Math. **24** (2022), no. 6, Paper No. 2150067, 35. MR 4466167

[CLP17] Vaughn Climenhaga, Stefano Luzzatto, and Yakov Pesin, *The geometric approach for constructing Sinai-Ruelle-Bowen measures*, J. Stat. Phys. **166** (2017), no. 3–4, 467–493. MR 3607577

[Cou16] Yves Coudène, *Ergodic theory and dynamical systems*, Universitext, Springer-Verlag London, Ltd., London; EDP Sciences, [Les Ulis], 2016. Translated from the 2013 French original [MR3184308] by Reinie Erné. MR 3586310

[CP17] Vaughn Climenhaga and Yakov Pesin, *Building thermodynamics for non-uniformly hyperbolic maps*, Arnold Math. J. **3** (2017), no. 1, 37–82. MR 3646530

[CPZ19] Vaughn Climenhaga, Yakov Pesin, and Agnieszka Zelerowicz, *Equilibrium states in dynamical systems via geometric measure theory*, Bull. Amer. Math. Soc. (N.S.) **56** (2019), no. 4, 569–610. MR 4007162

[CPZ20] Vaughn Climenhaga, Yakov Pesin, and Agnieszka Zelerowicz, *Equilibrium measures for some partially hyperbolic systems*, J. Mod. Dyn. **16** (2020), 155–205.

[CT12] Vaughn Climenhaga and Daniel J. Thompson, *Intrinsic ergodicity beyond specification: β-shifts, S-gap shifts, and their factors*, Israel J. Math. **192** (2012), no. 2, 785–817. MR 3009742

[CT16] Vaughn Climenhaga and Daniel J. Thompson, *Unique equilibrium states for flows and homeomorphisms with non-uniform structure*, Adv. Math. **303** (2016), 745–799. MR 3552538

[CT21] Vaughn Climenhaga and Daniel J. Thompson, *Beyond Bowen's specification property*, Thermodynamic Formalism: CIRM Jean-Morlet Chair, Fall 2019 (Mark Pollicott and Sandro Vaienti, eds.), Lecture Notes in Mathematics, vol. 2290, Springer, 2021, arXiv:2009.09256.

[CT22] Benjamin Call and Daniel J. Thompson, *Equilibrium states for self-products of flows and the mixing properties of rank 1 geodesic flows*, J. Lond. Math. Soc. (2) **105** (2022), no. 2, 794–824. MR 4400937

[Fed69] Herbert Federer, *Geometric measure theory*, Die Grundlehren der mathematischen Wissenschaften, Band 153, Springer-Verlag New York Inc., New York, 1969. MR 0257325

[FH19] Todd Fisher and Boris Hasselblatt, *Hyperbolic flows*, Zurich Lectures in Advanced Mathematics, vol. 25, European Mathematical Society, 2019.

[Ham89] Ursula Hamenstädt, *A new description of the Bowen–Margulis measure*, Ergod. Theory Dyn. Syst. **9** (1989), no. 3, 455–464. MR 1016663

[Has89] Boris Hasselblatt, *A new construction of the Margulis measure for Anosov flows*, Ergod. Theory Dyn. Syst. **9** (1989), no. 3, 465–468. MR 1016664

[Hay94] Nicolai T. A. Haydn, *Canonical product structure of equilibrium states*, Random Comput. Dynam. **2** (1994), no. 1, 79–96. MR 1265227

[Kai90] Vadim A. Kaimanovich, *Invariant measures of the geodesic flow and measures at infinity on negatively curved manifolds*, Ann. Inst. H. Poincaré Phys. Théor. **53** (1990), no. 4, 361–393, Hyperbolic behaviour of dynamical systems (Paris, 1990). MR 1096098

[Lep00] Renaud Leplaideur, *Local product structure for equilibrium states*, Trans. Amer. Math. Soc. **352** (2000), no. 4, 1889–1912. MR 1661262

[Mar70] Grigory A. Margulis, *Certain measures that are connected with U-flows on compact manifolds*, Funkcional. Anal. i Priložen. **4** (1970), no. 1, 62–76. MR 0272984

[Mar04] Grigoriy A. Margulis, *On some aspects of the theory of Anosov systems*, Springer Monographs in Mathematics, Springer-Verlag, Berlin, 2004, With a survey by Richard Sharp: Periodic orbits of hyperbolic flows, Translated from the Russian by Valentina Vladimirovna Szulikowska. MR 2035655

[Pes97] Yakov B. Pesin, *Dimension theory in dynamical systems*, Chicago Lectures in Mathematics, University of Chicago Press, Chicago, IL, 1997, Contemporary views and applications. MR 1489237

[PP84] Yakov B. Pesin and B. S. Pitskel', *Topological pressure and the variational principle for noncompact sets*, Funktsional. Anal. i Prilozhen. **18** (1984), no. 4, 50–63, 96. MR 775933

[PS82] Yakov B. Pesin and Yakov G. Sinaĭ, *Gibbs measures for partially hyperbolic attractors*, Ergod. Theory Dyn. Syst. **2** (1982), no. 3–4, 417–438 (1983). MR 721733

[Roy88] Halsey L. Royden, *Real analysis*, third ed., Macmillan Publishing Company, New York, 1988. MR 1013117

[RS75] David Ruelle and Dennis Sullivan, *Currents, flows and diffeomorphisms*, Topology **14** (1975), no. 4, 319–327. MR 415679

[Rue76] David Ruelle, *A measure associated with axiom-A attractors*, Amer. J. Math. **98** (1976), no. 3, 619–654. MR 0415683

[Sin68] Yakov G. Sinaĭ, *Markov partitions and U-diffeomorphisms*, Funkcional. Anal. i Priložen **2** (1968), no. 1, 64–89. MR 0233038

[VO16] Marcelo Viana and Krerley Oliveira, *Foundations of ergodic theory*, Cambridge Studies in Advanced Mathematics, vol. 151, Cambridge University Press, Cambridge, 2016. MR 3558990

[Wal75] Peter Walters, *A variational principle for the pressure of continuous transformations*, Amer. J. Math. **97** (1975), no. 4, 937–971. MR 390180

6

Abelian Livshits Theorems and Geometric Applications

ANDREY GOGOLEV AND FEDERICO RODRIGUEZ HERTZ*

Dedicated to the memory of Anatole Katok, our mentor and friend.

We introduce a notion of abelian cohomology in the context of smooth flows. This is an equivalence relation which is weaker than the standard cohomology equivalence relation for flows. We develop Livshits theory for abelian cohomology over transitive Anosov flows. In particular, we prove an abelian Livshits theorem for homologically full Anosov flows. Then we apply this theorem to strengthen marked length spectrum rigidity for negatively curved surfaces. We also present an application to rigidity of contact Anosov flows. Some new results on homologically full Anosov flows are also given.

6.1 Introduction

Cohomological equations play a crucial role in dynamical systems theory. In the setting when the dynamical system is a smooth flow $X^t: M \to M$ generated by the vector field X the equation

$$\varphi = L_X u$$

is called a *cohomological equation*. Here $u: M \to \mathbb{R}$ is an unknown function and L_X is the Lie derivative in the direction of X. If the preceding equation has a solution, then the function $\varphi: M \to \mathbb{R}$ is called an *X^t-coboundary.*[1] Accordingly, two continuous functions φ and ψ are said to be *X^t-cohomologous* if $\varphi - \psi$ is a coboundary. We recommend A. Katok's book for a general introduction of cohomology for dynamical systems [K03].

The basic question is to decide whether the space of X^t-coboundaries $\mathcal{B}_X$ is closed in an appropriate topology (Hölder, smooth, etc.). If this is the case, then $\mathcal{B}_X$

* The authors were partially supported by NSF grants DMS-1823150 and DMS-1500947 & DMS-1900778, respectively

1 We use the term X^t-cohomologous rather than just "cohomologous" to avoid confusion when we consider cohomology of the manifold.

is given by the intersection of kernels of all X^t-invariant bounded linear functionals. Even better is to explicitly describe a family of functionals which would yield $\mathcal{B}_X$. If $X^t\colon M \to M$ is a transitive Anosov flow on a compact manifold M, then this problem was solved by A. Livshits in the classical paper [Liv72]. He proved that in this case the space $\mathcal{B}_X$, as a subspace of the space of Hölder continuous functions, is characterized by countably many functionals which are given by integration over periodic orbits of X^t.

Much later, A. Katok and A. Kononenko introduced a new type of functionals called periodic cycle functionals to handle the case of partially hyperbolic dynamical systems. They proved that the space of coboundaries is closed for partially hyperbolic diffeomorphisms and flows which satisfy certain local accessibility condition on its stable and unstable foliation [KK96]. The local accessibility was later relaxed to a more general accessibility condition by Wilkinson [W13].

In this paper we revisit the idea of Katok–Kononenko and define a version of periodic cycle functionals. We prove several versions of an *abelian Livshits Theorem* for transitive Anosov flows, where abelian cohomology is a certain weaker equivalence relation than cohomology.[2] Roughly speaking, this relation is given by the standard cohomology on the universal abelian cover. Then we give an application to marked length spectrum rigidity. More specifically, we improve the Croke–Otal marked length spectrum rigidity theorem ([Cr90, Ot90]) in the following way. If, for a given pair of negatively curved compact surfaces, the marked lengths spectra on homologically trivial orbits coincide, then, in fact, the full marked length spectra match and, by the Croke–Otal [Cr90, Ot90] rigidity theorem, the surfaces are isometric. We also apply our machinery to the rigidity problem of contact Anosov flows. Namely, under a technical assumption, if two contact Anosov flows are smoothly orbit equivalent, then, in fact, the first flow is conjugate to a very special reparametrization of the second flow. Both of these applications also strongly rely on R. Sharp's work on homologically full Anosov flows [Sh93].

Organization

In the next section we recall the definition of homologically full Anosov flows and recall some results of Sharp. We provide a new characterization of homologically full Anosov flows in terms of transitivity on the universal abelian cover. We also prove that contact Anosov flows are homologically full. In Section 6.3 we state and prove two abelian Livshits theorems: one for general Anosov flows in terms of periodic cycle functionals and one for homologically full Anosov flows in terms of homologically trivial periodic data. We also give an "integer Livshits theorem."

[2] To avoid confusion we stress that the term "abelian" does not refer to the group of values. All cocycles in this paper are $\mathbb{R}$-valued.

In Section 6.4 we establish several properties of reparametrized flows which will be needed for applications. In particular, we study how equilibrium states and Sharp's minimizers for homologically full Anosov flows behave under reparametrization. In Section 6.5 we apply an abelian Livshits Theorem to the conjugacy problem of homologically full Anosov flows similarly to the way that the classical Livshits Theorem applies to the conjugacy problem of transitive Anosov flows. Then, in Section 6.6, we use the preceding results to weaken the assumption in the Croke–Otal rigidity theorem to matching of homologically trivial spectra only. In Section 6.7 we also give an application to the conjugacy problem for contact Anosov flows. Finally, in Section 6.8, we present an instructive example showing that our results from Sections 6.6 and 6.7 are optimal in certain ways. We also pose two open questions, one at the end of Section 6.2 and one at the end of Section 6.3.

Acknowledgments

The authors would like to thank all the referees for their careful reading of the paper and suggestions for improvement.

Notation

We will denote by $[\alpha]$ the homology class of a periodic orbit or, more generally, a loop α. Similarly by $[\omega]$ we will denote the cohomology class of a closed 1-form ω.

6.2 Homologically Full Anosov Flows

Here we recall some notions and results due to R. Sharp on homologically full Anosov flows [Sh93]. We also establish some results about such flows which will be needed for our abelian Livshits theory and applications, but may have some independent interest. We begin with a general definition based on the idea that "homologically full" should mean that points can move asymptotically in any direction in homology of the manifold.

Let M be a compact smooth manifold and let $X^t\colon M \to M$ be a smooth flow. Denote by $\mathcal{M}(X)$ the space of X^t-invariant probability measures. Recall that $\mathcal{M}(X) \neq \varnothing$ by the Krylov–Bogolyubov theorem [KB37]. Following Schwartzman [Sch57], define the asymptotic cycle $\rho\colon \mathcal{M}(X) \to H_1(M,\mathbb{R}) \simeq Hom(H^1(M,\mathbb{R}),\mathbb{R})$ by

$$\rho_\mu([\omega]) = \int_M \omega(X)d\mu$$

where $\omega(X)$ is the evaluation of the 1-from ω on the generating vector field X. Recall that this is indeed well defined, namely $\rho_\mu([\omega])$ is independent of the 1-form representative of the cohomology class $[\omega]$. We define *the Schwartzman simplex* as the set $\mathcal{S}(X) = \{\rho_\mu : \mu \in \mathcal{M}(X)\} \subset H_1(M, \mathbb{R})$ which is a compact convex set.

Definition 2.1. A smooth flow $X^t : M \to M$ is called *homologically full* if the origin is contained in the interior of $\mathcal{S}(X)$.

Definition 2.2. Denote by $\mathcal{V}(X) \subset H_1(M, \mathbb{R})$ the linear span of $\mathcal{S}(X)$. Then $X^t : M \to M$ is called *homologically ample* if the origin is contained in the interior of $\mathcal{S}(X)$ as a subset of $\mathcal{V}(X)$.

For example, the geodesic flow on a flat 2-torus is homologically ample, but not homologically full. On the other hand, if X^t is a transitive homologically ample Anosov flow, then it is homologically full. Indeed, this follows from the fact that homology classes of periodic orbits of X^t span $H_1(M, \mathbb{Z})$ which is due to Parry and Pollicott [PP86]. (Plante proved the same for volume-preserving flows earlier, using a much simpler argument [Pl73].)

Remark 2.3. Let N be a compact Riemannian manifold. If $N \neq \mathbb{T}^2$, then the projection map $T^1N \to N$ induces the isomorphism $H_1(T^1N, \mathbb{R}) \simeq H_1(N, \mathbb{R})$. Let $X^t : T^1N \to T^1N$ be the geodesic flow. Then the Schwartzman simplex $\mathcal{S}(X) \subset H_1(T^1N, \mathbb{R})$ coincides with the unit ball in $H_1(N, \mathbb{R})$ with respect to so-called *stable norm* which is a very popular object of study in geometry [M96, Théorème 1.3.9].

Now we follow Sharp and recall the definition of the β-functional for Aņoov flows [Sh93]. Define $\beta : H^1(M, \mathbb{R}) \to \mathbb{R}$ using the formula

$$\beta([\theta]) = \sup_{\mu \in \mathcal{M}(X)} \left\{ h_\mu(X) + \int_M \theta(X) d\mu \right\}.$$

The definition is independent of a particular 1-form representative in the cohomology class. Indeed, if $d\alpha$ is an exact form, then $\int_M d\alpha(X) d\mu = \int_M L_X \alpha d\mu = 0$ and hence $\beta(\theta + d\alpha) = \beta(\theta)$. Sharp proved that β is convex.

Theorem 2.4 ([Sh93]). *Let $X^t : M \to M$ be a transitive Anosov flow. Then the following statements are equivalent.*

(i) *every homology class in $H_1(M, \mathbb{Z})$ is represented by a periodic orbit of X^t;*
(ii) *there exists a fully supported measure $\mu \in \mathcal{M}$ such that $\int_M \omega(X) d\mu = 0$ for all closed 1-forms ω;*
(iii) *the functional $\beta : H^1(M, \mathbb{R}) \to \mathbb{R}$ is bounded below and there exists unique $\xi_X \in H^1(M, \mathbb{R})$ for which the minimum is attained;*

(iv) *the convex hull of the set* $\{[\gamma]\colon \gamma$ *is a periodic orbit*$\} \subset H_1(M,\mathbb{R})$ *contains the origin in its interior.*

We note right away that, in the context of Anosov flows, (iv) is equivalent to our definition of homologically full because measures supported on periodic orbits are dense in $\mathcal{M}(X)$ [S72]. We call the cohomology class ξ_X from (iii) *Sharp's minimizer*. Further, Sharp established an asymptotic formula for the number of periodic orbits in a fixed homology class. When $\xi_X = 0$ the flow is "balanced" in the sense that asymptotic density of periodic orbits in different homology classes is the same.

The following result provides yet one more characterization of homologically full Anosov flows. We need it for our abelian Livshits theory, but it might be of some independent interest as well. Let $\hat{M}$ be *the universal abelian cover* of M, that is, the cover which corresponds to the commutator subgroup $[\pi_1 M, \pi_1 M]$; its group of Deck transformation is given by $H_1(M,\mathbb{Z})$.

Theorem 2.5. *A transitive Anosov flow* $X^t\colon M \to M$ *is homologically full if and only if its lift to the universal abelian cover is transitive.*

Remark 2.6. Transitivity on the universal abelian cover is equivalent to absence of wandering points. Indeed, transitivity on the abelian cover clearly implies that the lift has no wandering points. In the other direction, by Anosov closing lemma, full non-wandering set implies that periodic points are dense. Then transitivity follows by the standard Smale's argument as in the following proof.

Proof. We begin with the following lemma.

Lemma 2.7. *If* $X^t\colon M \to M$ *is a homologically full transitive Anosov flow, then homologically trivial periodic orbits are dense in* M.

By [Sh93, Theorem 2], homologically trivial orbits equidistribute according to an equilibrium measure given by a potential $\xi(X)$, where ξ is a certain closed 1-form. (When X is a geodesic flow $\xi = 0$ and homologically trivial orbits equidistribute to the measure of maximal entropy.) In particular, because equilibrium states are fully supported, it follows that the homologically trivial orbits are dense. However, one can avoid using Sharp's machinery altogether and give a simpler proof by using shadowing. We briefly sketch this proof.

Proof of Lemma 2.7. For any $\varepsilon > 0$ we will construct an ε-dense homologically trivial periodic orbit. Begin with an ε-dense periodic geodesic γ_ε. Then by (i) of Theorem 2.4 there exist geodesics γ'_ε in the opposite homology class, that is, such that the sum $\gamma_\varepsilon + \gamma'_\varepsilon$ bounds a 2-cycle.

Consider a point $x \in M$ such that $\gamma'(t_\varepsilon) \to x$, $\varepsilon \to 0$. Pick a Markov partition for X^t in such a way that x is in the interior of a Markov rectangle R. Both γ_ε and γ'_ε intersect R at two points p and q, respectively, which are ε-close to each other. By concatenating symbolic periods of γ_ε and γ'_ε, we can find a periodic orbit η_ε which ε-shadows γ_ε first and then ε-shadows γ'_ε. Orbit η_ε intersects R very close to $[p, q]$ once and then intersect R very close to $[q, p]$. (Of course, there could be more points of intersection with R corresponding to other points of intersection of γ_ε and γ'_ε with R.) Applying Fried's construction [Fri83, pp. 300–301] to γ_ε, γ'_ε and η yields a 2-dimensional immersed surface whose boundary consists of these periodic orbits.[3] Moreover, η_ε is homologous to $\gamma_\varepsilon + \gamma'_\varepsilon$ and, hence, is homologically trivial. It remains to notice that γ_ε is contained in the ε-neighborhood of η. Hence η_ε is 2ε-dense in M. □

First assume that X^t is homologically full. Denote by $\hat{M}$ the universal abelian cover of M and by $\hat{X}^t \colon \hat{M} \to \hat{M}$ the lift of the flow X^t. Note that homologically trivial periodic orbits in M are precisely those periodic orbits which lift to periodic orbits of $\hat{X}^t$. Hence, by the preceding lemma, periodic orbits of $\hat{X}^t$ are dense in $\hat{M}$. Now we can apply a standard argument of Smale [Sm67, (7.5)] to conclude that $\hat{X}^t \colon \hat{M} \to \hat{M}$ is indeed a transitive flow. Namely, given two open sets U and V in $\hat{M}$, one can use local product structure and denseness of periodic orbits to connect U to V using a finite chain of local stable and unstable manifolds of periodic orbits. Then one can apply the λ-lemma to show that $\hat{X}^T(U) \cap V \neq \varnothing$ for some large T.[4]

It remains to check the converse implication. (This implication is not needed for Theorem 3.5.) Assume that $\hat{X}^t$ is transitive, namely $\{\hat{X}^t(\hat{x}) \colon t \in \mathbb{R}\}$ is dense in $\hat{M}$ for some $\hat{x} \in \hat{M}$.

The homology group $H_1(M, \mathbb{Z})$ is identified with the group of Deck transformations of the cover $\hat{M} \to M$. Take any $\gamma \in H_1(M, \mathbb{Z})$. Then for some $t_0 > 0$ the point $\hat{X}^{t_0}(\hat{x})$ is very close to $\gamma(\hat{x})$ so that $\hat{X}^{t_0}(\hat{x})$ and $\gamma(\hat{x})$ belong to the same small local product structure chart. Denote by x the image of $\hat{x}$ in M. Then $X^{t_0}(x)$ and x also belong to the same small local product structure chart. Hence, by Anosov closing lemma, the orbit segment $[x, X^{t_0}(x)]$ can be shadowed by a periodic orbit of a point y, $X^{t_1}(y) = y$, $t_1 \approx t_0$, which is very close to x. By the shadowing property, the orbit of y is homotopic to the orbit segment $[x, X^{t_0}(x)]$ concatenated with a short curve connecting $X^{t_0}(x)$ back to x. It follows that $\hat{X}^{t_1}(\hat{y}) = \gamma(\hat{y})$, where $\hat{y}$ is a lift of y. That is, γ is represented by a periodic orbit of X^t and, hence, X^t is homologically full. □

3 Fried considers 3-dimensional Anosov flows, but this particular construction works well in any dimension.
4 With some more care one could show the topological mixing property.

Remark 2.8. The latter implication (transitivity on the universal cover implies homologically full property) was also proved using a different argument in [DS19, Lemma 7.4].

The following theorem provides a natural class of homologically full Anosov flows. It is interesting whether any contact flow can be shown to be homologically ample.

Theorem 2.9. *Let $X^t\colon M \to M$ be a contact Anosov flow. Then X^t is homologically full.*

Proof. Let α be the positive contact form for X^t and let $m = \alpha \wedge (d\alpha)^k$ be the invariant volume form. Recall that by (ii) of Theorem 2.4: *X^t is homologically full if, for every $[\omega] \in H^1(M,\mathbb{R})$,*

$$\int_M \omega(X)m = 0.$$

So, let ω be a closed 1-form. Note that $\omega \wedge (d\alpha)^k$ is a top-dimensional form and, hence, $\omega \wedge (d\alpha)^k = \psi\alpha \wedge (d\alpha)^k$ for some smooth function ψ. Contraction with X yields $\omega(X)(d\alpha)^k = \psi\iota_X(\alpha \wedge (d\alpha^k)) = \psi(d\alpha)^k$. Hence $\psi = \omega(X)$. Now we have

$$\int_M \omega(X)m = \int_M \omega(X)\alpha\wedge(d\alpha)^k = \int_M \omega\wedge(d\alpha)^k = \int_M -d(\omega\wedge\alpha\wedge(d\alpha)^{k-1}) = 0.$$

□

Remark 2.10. It was pointed out to us by G. Paternain that the preceding result is well known; see, for example, [Pl72, Corollary 4.10].

Asaoka proved that any transitive codimension-1 Anosov flow is orbit equivalent to a volume-preserving Anosov flow [A08]. Also recall that Foulon and Hasselblatt developed contact surgery and created many examples of 3-dimensional contact Anosov flows [FH13].

Question 2.11. Is every 3-dimensional homologically full Anosov flow orbit equivalent to a contact Anosov flow?

6.3 Abelian Cohomology for Anosov Flows

Let $X^t\colon M \to M$ be a transitive Anosov flow on a closed compact manifold M of arbitrary dimension. A Hölder continuous function φ is called an *abelian coboundary* if there exists a smooth closed 1-form ω and a Hölder continuous function u, which is continuously differentiable along X, such that

$$\varphi = \omega(X) + L_X u. \tag{6.1}$$

Recall that $\omega(X)$ stands for the contraction given by evaluation of ω on the generating vector field X. Accordingly, we say that the two functions φ and ψ are *abelian cohomologous* if $\varphi - \psi$ is an abelian coboundary.

Remark 3.1. Notice that the decomposition (6.1) is highly non-unique because we can change ω by any exact 1-form. Indeed, given any smooth function $v: M \to \mathbb{R}$, we can write a different decomposition:

$$\varphi = (\omega + dv)(X) + L_X(u - v).$$

However, one could make some canonical choice, for example, by asking ω to be harmonic with respect to a Riemannian metric. (Recall that, given a fixed Riemannian metric, there exists a unique harmonic representative in each cohomology class.)

We develop the counterpart of the standard Livshits theory [Liv72] for abelian cohomology. Specifically, we prove two *abelian Livshits Theorems* for transitive Anosov flows:

- General Livshits Theorem 3.3, which characterizes the space of abelian coboundaries as the intersection of kernels of periodic cycle functionals;
- Livshits Theorem 3.5 for homologically full Anosov flows, which characterizes abelian coboundaries via obstructions given by integration over homologically trivial periodic orbits.

At the end of this section we also give a similar proposition for functions whose periodic orbit obstructions take values in a rank one abelian subgroup of $\mathbb{R}$ and pose an open question for the case of finite rank.

Now we explain the term "abelian." Recall that the universal abelian cover $\hat{M} \to M$ is the cover which corresponds to the commutator subgroup $[\pi_1 M, \pi_1 M]$. Then the lift $\hat{\omega}$ of any closed 1-form ω on M is exact on $\hat{M}$. Hence a lift $\hat{\varphi}$ to $\hat{M}$ of an abelian coboundary $\varphi: M \to \mathbb{R}$ is a true coboundary for the lifted Anosov flow $\hat{X}$ because

$$\hat{\varphi} = \hat{\omega}(\hat{X}) + L_{\hat{X}}\hat{u} = d\alpha(\hat{X}) + L_{\hat{X}}\hat{u} = L_{\hat{X}}(\alpha + \hat{u}).$$

6.3.1 Katok-Kononenko Theory of Periodic Cycle Functionals Revisited

Given an Anosov flow $X^t: M \to M$ a *us-adapted path* is a piecewise smooth path $\gamma: [0,1] \to M$ such that each of its legs lies entirely in a stable or an unstable leaf of X^t. Analogously, an *Xus-adapted path* (or simply an *adapted path*) is a piecewise smooth path γ such that each of its legs is either a flow-line segment or lies entirely in a stable or an unstable leaf. An *Xus-adapted loop* is an *Xus*-adapted path which begins and ends at the same point.

Given an adapted path or loop γ, we proceed to define *periodic cycle functionals* $PCF_\gamma\colon C^\alpha(M,\mathbb{R}) \to \mathbb{R}$ as follows. If γ lies entirely in a stable leaf, then let

$$PCF_\gamma(\varphi) = \int_0^\infty \varphi(X^t(\gamma(0))) - \varphi(X^t(\gamma(1)))dt.$$

If γ lies entirely in an unstable leaf, then let

$$PCF_\gamma(\varphi) = \int_{-\infty}^0 \varphi(X^t(\gamma(1))) - \varphi(X^t(\gamma(0)))dt.$$

Note that convergence follows from exponential contraction/expansion and Hölder continuity of φ. If γ is a positively oriented orbit segment $\gamma = [x, \varphi^T(x)]$, $T > 0$, then let

$$PCF_\gamma(\varphi) = \int_0^T \varphi(X^t(x))dt,$$

and if $\gamma = [x, X^T(x)]$, $T < 0$, then let

$$PCF_\gamma(\varphi) = -\int_T^0 \varphi(X^t(x))dt.$$

Finally, for an adapted path γ define $PCF_\gamma(\varphi)$ as a the sum of values on each of the legs. The following properties are immediate from the definitions.

1. Any continuous path can be C^0 approximated by an *Xus*-adapted path.
2. The value of $PCF_\gamma(\varphi)$ only depends on the sequence of the endpoints of the legs of γ and is independent of the choice of leg between the endpoints.
3. If $\bar{\gamma}$ denotes the adapted path (or loop) with reversed orientation, then $PCF_{\bar{\gamma}}(\varphi) = -PCF_\gamma(\varphi)$.
4. Suppose α and β are adapted loops such that α contains a subpath γ and β contains $\bar{\gamma}$, the same subpath with the opposite orientation. Concatenating α with γ removed and β with $\bar{\gamma}$ removed results in a loop $\alpha * \beta$. Then we have the *additive property*,

$$PCF_{\alpha*\beta}(\varphi) = PCF_\alpha(\varphi) + PCF_\beta(\varphi). \tag{6.2}$$

Remark 3.2. As mentioned earlier, originally periodic cycle functionals were introduced by Katok and Kanonenko [KK96] as obstructions, given by adapted *us*-loops, to solving the cohomological equation in the setting of partially hyperbolic diffeomorphisms when periodic orbit obstructions are not readily available. Note that our definition is different from the original one, as we allow the flow direction in the definition of the adapted loop.

If φ is an X^t-coboundary, $\varphi = L_X u$, then it is easy to see that in all three cases (γ is contained in a stable leaf, unstable leaf, or an orbit segment) we have $PCF_\gamma(\varphi) = u(\gamma(1)) - u(\gamma(0))$. Hence periodic cycle functionals of Xus-adapted loops vanish. Similarly, if φ is an abelian coboundary, then periodic cycle functions vanish on homologically trivial Xus-adapted loops because these are the loops which can be lifted to the universal abelian cover where φ becomes a true coboundary. We prove that vanishing on homotopically (and even homologically) trivial Xus-adapted loops is also a sufficient condition for being an abelian coboundary.

Theorem 3.3. *Let $X^t\colon M \to M$ be a transitive Anosov flow and let $\varphi \in C^r(M)$, $r > 0$. Assume that $PCF_\gamma(\varphi) = 0$ for every homotopically trivial Xus-adapted loop γ. Then there exist a C^∞ smooth closed 1-form ω and $u \in C^{r_*}(M)$, such that*

$$\varphi = \omega(X) + L_X u.$$

In Theorem 3.3, $r_* = r$ if $r \notin \mathbb{N}$ and $r_* = r - 1 + Lip$ if $r \in \mathbb{N}$.

We immediately obtain the following corollary.

Corollary 3.4. *If function $\varphi\colon M \to \mathbb{R}$ is C^∞ smooth, then there is a C^∞ smooth closed 1-form ω such that*

$$\varphi = \omega(X).$$

Indeed we have $\varphi = \omega(X) + L_X u$, where $u \in C^\infty(M)$. Hence $\varphi = \omega(X) + du(X) = (\omega + du)(X)$.

Proof of Theorem 3.3. We lift all the objects to the universal cover $\tilde{M}$ and, by a light abuse of notation, we still denote by φ and X the lifts of φ and X to the universal cover. Pick a point $a \in \tilde{M}$ and define $\tilde{u}_a\colon \tilde{M} \to \mathbb{R}$ in the following way. For a given point $x \in \tilde{M}$, let γ be an adapted path starting at a and ending at x; we set

$$\tilde{u}_a(x) = PCF_\gamma(\varphi).$$

This definition is independent of the choice of γ because we have assumed that periodic cycle functionals vanish on homotopically trivial adapted loops. Note that, because we can assume that the last leg of γ is a flow segment, we have $L_X\tilde{u}_a = \varphi$ (though we will not use this last fact). It easily follows from the definition that, for any pair of points $a, b \in \tilde{M}$,

$$\tilde{u}_a - \tilde{u}_b = \tilde{u}_a(b). \tag{6.3}$$

Let $\mathcal{D} \simeq \pi_1 M$ be the group of deck transformations acting on $\tilde{M}$. The function $\tilde{u}_a$ solves the cohomological equation on $\tilde{M}$, but a priori is not $\mathcal{D}$-invariant, so it needs to be adjusted.

Let $T \in \mathcal{D}$ and let γ be an adapted path. Because $\varphi \circ T = \varphi$ we have

$$PCF_\gamma(\varphi) = PCF_{T(\gamma)}(\varphi).$$

Hence, for every $T \in \mathcal{D}$,

$$\tilde{u}_{T(a)}(T(x)) = \tilde{u}_a(x).$$

Define $c\colon \mathcal{D} \to \mathbb{R}$ in the following way:

$$\tilde{u}_a(T(x)) - \tilde{u}_a(x) = \tilde{u}_a(T(x)) - \tilde{u}_{T(a)}(T(x)) = \tilde{u}_a(T(a)) \stackrel{\text{def}}{=} c(T) \in \mathbb{R},$$

where we have used (6.3). Then c is a homomorphism. Indeed,

$$c(T \circ S) = \tilde{u}_a(T(S(a)) = \tilde{u}_a(T(S(a)) - \tilde{u}_a(S(a)) + \tilde{u}_a(S(a)) = c(T) + c(S).$$

Notice also that c does not depend on the choice of the base point a.

Now we use the isomorphism $Hom(\mathcal{D}, \mathbb{R}) \simeq H^1(M, \mathbb{R})$. Recall that the cohomology class corresponding to $c\colon \mathcal{D} \to \mathbb{R}$ is represented by a closed 1-form ω on M such that

$$c(T) = \int_{\gamma_T} \hat{\omega}(\dot{\gamma}_T(s))ds,$$

where γ_T is any curve starting at x and ending at $T(x)$ and $\hat{\omega}$ is the lift of ω to $\hat{M}$.

Note that $\bar{\varphi} = \varphi - \omega(X)$ is a function on M. Now take any periodic orbit γ in M and lift it to an orbit segment $\tilde{\gamma}$ in $\tilde{M}$. Then, of course, $PCF_{\tilde{\gamma}}(\bar{\varphi}) = PCF_\gamma(\bar{\varphi})$. Let x and $T(x)$, $T \in \mathcal{D}$, be the endpoints of $\tilde{\gamma}$. We have

$$\begin{aligned}\int_\gamma \bar{\varphi} = \int_{\tilde{\gamma}} \bar{\varphi} = \int_{\tilde{\gamma}} \varphi - \int_{\tilde{\gamma}} \omega(X) &= \tilde{u}_x(T(x)) - \tilde{u}_x(x) - c(T)\\ &= \tilde{u}_a(T(x)) - \tilde{u}_a(x) - c(T) = 0.\end{aligned}$$

Hence

$$\int_\gamma \bar{\varphi} = 0$$

for every closed orbit γ. Then by Livshits Theorem [Liv72] there exists a Hölder continuous u, continuously differentiable along X such that that $L_X u = \bar{\varphi}$, that is,

$$\varphi = L_X u + \omega(X).$$

Further, the Hölder exponent of u is the same as the Hölder exponent for φ. If φ is C^r with $r > 1$, then the de la Llave–Marco–Mariyón smooth Livshits Theorem [LMM86, Appendix A] applies and together with Journé's regularity lemma [Jou88] yields C^{r_*} regularity of u. □

6.3.2 Livshits Theorem for Homologically Trivial Orbits

Theorem 3.5. *Assume that $X^t : M \to M$ is a homologically full transitive Anosov flow and let $\varphi : M \to \mathbb{R}$ be a C^r, $r > 0$ function such that*

$$\int_\gamma \varphi = 0$$

for all homologically trivial periodic orbits γ. Then there is a C^∞ smooth closed 1-form ω on M and a function $u \in C^{r*}(M)$ such that

$$\varphi = \omega(X) + L_X u.$$

Remark 3.6. Notice that any homologically trivial periodic orbit bounds a 2-cycle. Hence, using (6.2), the integral $\int_\gamma \varphi$ can be decomposed into a sum of periodic cycle functionals of homotopically trivial *Xus*-adapted loops. Hence, it is easy to see that vanishing of periodic cycle functionals on homotopically trivial *Xus*-adapted loops implies vanishing on all homologically trivial periodic orbits. Thus Theorem 3.5 can be viewed as a strengthening of Theorem 3.3 in the setting of homologically full Anosov flows.

Proof. Let $\hat{M}$ be the universal abelian cover of M. We denote by $\hat{\varphi}$ the lift of φ to $\hat{M}$ and we still write X^t for the lift of the flow, as it won't cause any confusion. By Theorem 2.5 the lifted flow is transitive.

Let $x \in \hat{M}$ be a point with a dense orbit. Define

$$\hat{u}(X^t(x)) = \int_0^t \hat{\varphi}(X^\tau(x))d\tau.$$

Recall that homologically trivial periodic orbits of X^t are precisely those periodic orbits which lift to periodic orbits of the lifted flow on $\hat{M}$. Hence we have

$$\int_\gamma \hat{\varphi} = 0$$

for every periodic orbit γ on $\hat{M}$. Hence we can apply the classical argument of Livshits and conclude that u is Hölder continuous (with a uniform constant) on the orbit of x and hence extends to a Hölder function on $\hat{M}$. Further, $\hat{u}$ is continuously differentiable along the flow direction and solves the cohomological equation

$$L_X \hat{u} = \hat{\varphi}.$$

Let $\mathcal{D} \simeq H_1(M, \mathbb{Z})$ be the group of deck transformations of the covering $\hat{M} \to M$. Because $\hat{\varphi}$ is $\mathcal{D}$ invariant, we have $L_X(\hat{u} \circ T - \hat{u}) = 0$ for every $T \in \mathcal{D}$. And because X^t has a dense orbit, we conclude that $\hat{u} \circ T - \hat{u}$ is constant. Let $c(T) = \hat{u} \circ T - \hat{u}$. Then $c\colon \mathcal{D} \to \mathbb{R}$ is a homomorphism, indeed

$$c(T \circ S) = \hat{u} \circ (T \circ S) - \hat{u} = \hat{u} \circ (T \circ S) - \hat{u} \circ S + \hat{u} \circ S - \hat{u} = c(T) + c(S).$$

Now identify $\mathcal{D}$ with its orbit in $\hat{M}$. By the de Rham Theorem we can extend $c\colon \mathcal{D} \to \mathbb{R}$ to a smooth function $c\colon \hat{M} \to \mathbb{R}$ which is equivariant with respect to the $\mathcal{D}$ action, that is,

$$c \circ T - c = c(T).$$

Let $\hat{\omega} = dc$. Then $\hat{\omega}$ is an exact 1-form which is invariant under the action of $\mathcal{D}$. Hence it descends to a closed 1-form ω on M. (The form ω is a de Rham representative in the cohomology class given by $c \in \mathrm{Hom}(H_1(M, \mathbb{R}), \mathbb{R})$.) The function $\hat{u} - c$ is $\mathcal{D}$-invariant and hence descends to a function u on M. We have

$$L_X(\hat{u} - c) = \hat{\varphi} - \hat{\omega}(X).$$

Thus

$$L_X u = \varphi - \omega(X).$$

Finally, if φ is C^r, then, as in the proof of Theorem 3.3, smooth Livshits theory [LMM86, Appendix A], [Jou88] yields C^{r_*} regularity of u. □

6.3.3 Integer Periods Livshits Theorem

We will prove the following proposition using the circle valued Livshits Theorem.

Proposition 3.7. *Let $X^t\colon M \to M$ be a transitive Anosov flow and let $\varphi\colon M \to \mathbb{R}$ be a Hölder continuous function. Let*

$$\mathscr{P} = \left\{ \int_\gamma \varphi \colon \gamma \in Per(X) \right\} \subset \mathbb{R}$$

be the set of periods of φ. If the rank of the additive group generated by $\mathscr{P}$ is one, then there are a smooth closed 1*-form ω and a Hölder continuous function u such that φ is an abelian coboundary, namely,*

$$\varphi = \omega(X) + L_X u.$$

Recall that integral cohomology $H^1(M, \mathbb{Z})$ is torsion-free and can be regarded as a Bruschlinsky group of homotopy classes of maps $\{[M \to S^1]\}$. Indeed, a smooth function $w\colon M \to S^1$ defines a closed integral form $\omega = dw$. The correspondence $[w] \mapsto [\omega] \in H^1(M, \mathbb{Z})$ is in fact an isomorphism.

Note that, if $w\colon M \to S^1$ is not smooth, then, by the Whitney Approximation Theorem it can be C^0 approximated by a smooth map $v\colon M \to S^1$. The function $\bar{u} = w - v$ has its image in a small interval and hence lifts to a function $u\colon M \to \mathbb{R}$.

Proof. Assume that $\mathscr{P} \subset c\mathbb{Z}, c \neq 0$. By a constant reparametrization we can assume that $c = 1$. Consider the cocycle

$$\hat{\varphi}(x, T) = \int_0^T \varphi(X^s(x))ds \in \mathbb{R}$$

and consider $\Phi(x, T) = [\hat{\varphi}(x, T)] \in S^1 = \mathbb{R}/\mathbb{Z}$. By assumption, for every periodic orbit p of period T_p we have $\Phi(p, T_p) = 0$. Hence, by applying the S^1-valued Livshits Theorem, there exists a function $w\colon M \to S^1$, which is differentiable along the flow direction, such that

$$\Phi(x, T) = w(X^T(x)) - w(x)$$

for every $x \in M$ and $T \in \mathbb{R}$. By applying the preceding discussion to w we have the decomposition $w = \bar{u} + v$, where $\omega = dv$ is smooth and $\bar{u}\colon M \to S^1$ lifts to a function $u\colon M \to \mathbb{R}$. Since v is smooth, $\bar{u}$ and u are also differentiable along the flow direction. So, taking the limit of

$$\frac{1}{T}\Phi(x, T) = \frac{1}{T}(w(X^T(x)) - w(x)) = \frac{1}{T}(\bar{u}(X^T(x)) - \bar{u}(x)) + \frac{1}{T}(v(X^T(x)) - v(x))$$

as $T \to 0$, we obtain that φ is an abelian coboundary:

$$\varphi = L_X\bar{u} + \omega(X) = L_X u + \omega(X).$$

□

Remark 3.8. If $\varphi = 1$, that is, the flow X^t only has integer length periodic orbits, then we have $\omega(X) = 1$. Therefore, by Schwartzman's theorem [Sch57] the flow is a suspension. Further, if we denote the section by $S \subset M$, then for any periodic orbit γ we have

$$\ell(\gamma) = \int_\gamma \omega(X)dt = \langle \omega, \gamma \rangle = c \cdot \#\{S \cap \gamma\}.$$

Hence, we can apply the Livshits Theorem to the roof function and obtain that the roof function is smoothly cohomologous to a constant c, that is, the flow X^t is a constant roof suspension over an Anosov diffeomorphism.

Question 3.9. In the setting of Proposition 3.7, assume that $\mathscr{P}$ has finite rank instead of rank one. Do there exist $\omega \in H^1(M, \mathbb{R})$ and a Hölder continuous function u such that

$$\varphi = \omega(X) + L_X u?$$

6.4 Reparametrizations of Flows

In this section we introduce some preliminaries on reparametrized flows (see Parry [Par86] for a more detailed introduction). We study how equilibrium states change under reparametrization. We also examine the behavior of Sharp's minimizer under reparametrization.

Given a X^t-invariant measure μ, we will denote by $h_\mu(X)$ the *metric entropy* of X^t. Also recall that, given a flow $X^t\colon M \to M$ and a Hölder continuous function $\varphi\colon M \to \mathbb{R}$, the *pressure* $P_X(\varphi)$ is defined by

$$P_X(\varphi) = \sup\left\{h_\nu(X) + \int \varphi d\nu\colon \nu \text{ invariant probability measure}\right\}.$$

When X^t is a transitive Anosov flow, the unique measure $\nu_\varphi = \nu_{\varphi,X}$ realizing the supremum is called the *equilibrium state* of φ with respect to X^t.

Let $X^t\colon M \to M$ be a flow generated by the vector field X, and let $\ell\colon M \to \mathbb{R}$ be a positive smooth function. Define $Z = \ell X$ to be the generator of the *reparametrized flow* Z^t. Then $Z^t = X^{\tau_t}$, where $\tau_t\colon M \to \mathbb{R}$ is the Z^t-cocycle with infinitesimal generator ℓ and is given by

$$\tau_t(x) = \int_0^t \ell(Z^s(x))ds.$$

Similarly, if $k = 1/\ell$, then $X^t = Z^{\kappa_t}$, where $\kappa_t\colon M \to \mathbb{R}$ is a X^t-cocycle given by

$$\kappa_t(x) = \int_0^t k(X^s(x))ds.$$

Lemma 4.1. *A reparametrization Z^t similar to the one given earlier is conjugate to X^t via a time-u map $X^u\colon M \to M$, $u\colon M \to \mathbb{R}$, if and only if*

$$L_X u = \frac{1}{\ell} - 1.$$

Further, two reparametrizations, $Z_i = \ell_i X$, $i = 1, 2$, are mutually conjugate via X^u if and only if

$$L_X u = \frac{1}{\ell_1} - \frac{1}{\ell_2}.$$

Proof. The conjugacy relation $X^u \circ Z^t = X^t \circ X^u$ yields

$$u(Z^t x) + \tau_t(x) = t + u(x)$$

or

$$u(Z^t x) - u(x) = t - \tau_t(x).$$

Dividing by t and taking the limit as $t \to 0$ gives $L_Z u = 1 - \ell$. It remains to notice that $L_Z u = \ell L_X u$. To show that X^u is a conjugacy when $L_X u = 1/\ell - 1$, one works backwards to obtain $X^u \circ Z^t = X^t \circ X^u$ by integrating.

For the last statement, notice that $Z_1 = \frac{\ell_1}{\ell_2} Z_2$ and apply the criterion. □

Also recall the following result of Anosov and Sinai.

Proposition 4.2. *If $X^t \colon M \to M$ is an Anosov flow and Z^t is a smooth reparametrization of X^t, that is, $Z = \ell X$, where ℓ is positive and smooth, then Z^t is also Anosov.*

If μ is a X-invariant measure, then

$$\mu_k = \frac{k}{\int k d\mu} \mu$$

is Z^t-invariant. Recall that by definition, entropy of a flow is the entropy of its time-1 map. Then the Abramov entropy formula gives

$$h_{\mu_k}(Z) = \frac{h_\mu(X)}{\int k d\mu}.$$

Proposition 4.3. *Let $X^t \colon M \to M$ be a transitive Anosov flow and let $Z = \ell X$, $\ell > 0$, be a smooth reparametrization. Let $\varphi \colon M \to \mathbb{R}$ be a Hölder continuous function and let $k = 1/\ell$. Then*

$$P_Z(\ell(\varphi - P_X(\varphi))) = 0.$$

Moreover, if $\nu_{\varphi,X}$ is the equilibrium measure for φ with respect to X^t, then the equilibrium measure $\nu_{\ell(\varphi - P_X(\varphi)),Z}$ for $\ell(\varphi - P_X(\varphi))$ with respect to Z^t is given by

$$\nu_{\ell(\varphi - P_X(\varphi)),Z} = \frac{k}{\int k \nu_{\varphi,X}} \nu_{\varphi,X}.$$

Proof. Take any X-invariant measure ν and let $\hat{\nu} = \frac{k}{\int k d\nu} \nu$. Then

$$\begin{aligned} h_{\hat{\nu}}(Z) + \int \ell(\varphi - P_X(\varphi)) d\hat{\nu} &= \frac{1}{\int k d\nu} \left(h_\nu(X) + \int (\varphi - P_X(\varphi)) d\nu \right) \\ &= \frac{1}{\int k d\nu} \left(\left(h_\nu(X) + \int \varphi d\nu \right) - P_X(\varphi) \right) \le 0. \end{aligned}$$

Further, the equality in the preceding inequality holds if and only if $\nu = \nu_{\varphi,X}$. Since the correspondence $\nu \mapsto \hat{\nu}$ is a one-to-one and onto correspondence between invariant measures for X and Z, uniqueness of equilibrium measures yields the posited result. □

Now let $X^t : M \to M$ be a homologically full Anosov flow. Recall the definition of β-functional, $\beta([\theta]) = P_X(\theta(X))$, and Sharp's minimizer ξ_X from Section 6.2. Denote by μ_ξ the equilibrium state of $\xi_X(X)$. Also recall that by Step 1 of the proof of [Sh93, Theorem 1]

$$\int_M \theta(X) d\mu_\xi = 0 \tag{6.4}$$

for any closed 1-form θ. It follows that

$$P_X(\xi_X(X)) = h_{\mu_\xi}(X). \tag{6.5}$$

Proposition 4.4. *Let X^t be a homologically full Anosov flow and let Z^t be a reparametrization given by $Z = \ell X$ where $\ell = 1/(a + \omega(X))$ for some positive constant a and some smooth closed 1-form ω with $a + \omega(X) > 0$. Then*

$$\xi_Z = \xi_X + \frac{P_X(\xi_X(X))}{a}[\omega].$$

Proof. By Proposition 4.3 we have $P_Z(\ell(\theta(X) - P_X(\theta(X))) = 0$ for every 1-form θ. Applying ω to

$$Z = \frac{X}{a + \omega(X)}$$

yields

$$\omega(Z) = \frac{\omega(X)}{a + \omega(X)}$$

and hence

$$\ell = \frac{1}{a + \omega(X)} = \frac{1}{a}(1 - \omega(Z)).$$

Then

$$\ell(\theta(X) - P_X(\theta(X))) = \theta(Z) + \frac{\omega(Z)}{a} P_X(\theta(X)) - \frac{1}{a} P_X(\theta(X))$$

and we obtain

$$P_Z\left(\theta(Z) + \frac{\omega(Z)}{a} P_X(\theta(X)) - \frac{1}{a} P_X(\theta(X))\right) = 0$$

or

$$P_Z\left(\theta(Z) + \frac{\omega(Z)}{a} P_X(\theta(X))\right) = \frac{1}{a} P_X(\theta(X))$$

for every θ.

We use the preceding formula to conclude that the map

$$\theta \mapsto \theta + \frac{P_X(\theta(X))}{a}\omega$$

is an invertible bijection on cohomology $H^1(M,\mathbb{R})$. Indeed the inverse is given by $\theta \mapsto \theta - P_Z(\theta(Z))\omega$. Hence, if θ minimizes

$$\theta \mapsto P_X(\theta(X)),$$

then

$$\hat{\theta} = \theta + \frac{P_X(\theta(X))}{a}\omega$$

minimizes $P_Z(\hat{\theta}(Z))$. Hence, by uniqueness of Sharp's minimizer,

$$\xi_Z = \xi_X + \frac{P_X(\xi_X(X))}{a}[\omega]. \qquad \square$$

Finally, we use Proposition 4.3 to show that any homologically full Anosov flow can be reparametrized so that Sharp's minimizer becomes zero. Sharp proved that such flows are special in the sense that periodic orbit growth is balanced in different homology classes [Sh93, Section 5]. We will need the following lemma.

Lemma 4.5. *Assume that, for a cohomology class μ and each X^t-invariant probability measure ν, we have $\int_M \mu(X)d\nu > -1$. Then μ can be represented by a 1-form ω such that $\omega(X) > -1$.*

The proof is very standard and is similar to the proof of uniform convergence of Birkhoff ergodic averages for uniquely ergodic systems. We just indicate the approach.

Fix a closed 1-form ω^0 which represents μ. For any $\lambda > 0$ let

$$\omega_x^\lambda = \frac{1}{\lambda}\int_x^{X^\lambda(x)} (X^t)^*\omega^0_{X^t(x)}dt.$$

We have $[\omega^\lambda] = \mu$ and $\omega_x^\lambda(X)$ is given by the ergodic average of $\omega^0(X)$

$$\omega_x^\lambda(X) = \frac{1}{\lambda}\int_x^{X^\lambda(x)} \omega^0_{X^t(x)}(X(X^t(x))dt.$$

Then the condition on the integrals of $\omega(X)$ implies that $\omega^\lambda(X) > -1$ for a sufficiently large λ.

Proposition 4.6. *Let $X^t\colon M \to M$ be a homologically full Anosov flow. Then there is a unique (up to conjugacy) reparametrization of the form*

$$Z = \frac{X}{1+\omega(X)},$$

which has zero Sharp's minimizer. Here ω is a closed 1-form.

Proof. By Proposition 4.3 we have $\xi_Z = \xi_X + P_X(\xi_X(X))[\omega]$ and hence the cohomology class $[\omega]$ is uniquely determined:

$$[\omega] = \frac{-1}{P_X(\xi_X(X))}\xi_X.$$

(Recall that by (6.5) $P_X(\xi_X(X)) = h_{\mu_\xi}(X) > 0$.)

Thus it remains to show that the class $[\omega]$ can be realized by a 1-form ω with $\omega(X) > -1$. According to Lemma 4.5, it is enough to show that

$$\int_M \omega(X) d\nu > -1$$

for every X^t-invariant probability measure ν. If $\nu = \mu_\xi$, then

$$\int_M \omega(X) d\nu = 0$$

by (6.4). Otherwise, if $\nu \neq \mu_\xi$, then

$$\int_M \omega(X) d\nu = \frac{-1}{P_X(\xi_X(X))} \int_M \xi_X(X) d\nu > \frac{-\int_M \xi_X(X) d\nu}{h_\nu(X) + \int_M \xi_X(X) d\nu} \geq -1. \qquad \square$$

6.5 Conjugacy for Homologically Full Flows

Recall that a transitive Anosov flow $X^t : M \to M$ is *homologically full* if every integral homology class contains a periodic orbit of X^t (cf. Theorem 2.4). Notice that being homologically full is a property which is invariant under any orbit equivalence.

Two flows $X_i^t : M \to M$, $i = 1, 2$, are *conjugate* if there exists a homeomorphism $H : M \to M$ such that

$$\forall t \quad H \circ X^t = Y^t \circ H.$$

We say that X_1^t and X_2^t are *orbit equivalent* if there exists a homeomorphism H which sends orbits of X_1^t to orbits of X_2^t preserving the time direction.

Let $X_i^t : M \to M$, $i = 1, 2$, be orbit-equivalent Anosov flows. Fix an orbit equivalence $H_0 : M \to M$ which sends orbits of X_1^t to orbits of X_2^t. We say that H_0 *matches period spectra* if, for every periodic point x, the X_1^t-period of x is the same as the X_2^t-period of $H_0(x)$. And we say that H_0 *matches homologically trivial period spectra* if only the periods of corresponding homologically trivial periodic orbits are assumed to be the same. Note that matching is not merely a property of X_1^t and X_2^t, but also depends on the choice of H_0 because flows can admit multiple non-equivalent orbit equivalences.

Recall the following classical application of the Livshits Theorem due to Katok.

Theorem 5.1. *Let X_1^t and X_2^t be transitive Anosov flows and let H_0 be an orbit equivalence which matches period spectra. Then X_1^t and X_2^t are conjugate,*

$$H \circ X_1^t = X_2^t \circ H,$$

where $H : M \to M$ is a bi-Hölder continuous homeomorphism.

Proof. The orbit equivalence H_0 is a bi-Hölder homeomorphism. By adjusting in the time direction we can also make H_0 continuously differentiable in the flow direction. Define

$$Z^t = H_0 \circ X_1^t \circ H_0^{-1}.$$

Then Z^t is a Hölder continuous reparametrization of X_2^t with the same periods.

We use the same notation as in Section 6.4: $Z = \ell X_2$, $Z^t = X_2^{\tau_t}$, $X_2^t = Z^{\kappa_t}$. For any periodic point x of period T we have

$$x = Z^T(x) = X_2^T(x) = Z^{\kappa_T}(x)$$

and, hence, $\kappa_T(x) = T$ or

$$\int_0^T \left(\frac{1}{\ell(X_2^s(x))} - 1 \right) ds = 0.$$

Now, by Livshits Theorem there exists a Hölder function $u : M \to \mathbb{R}$ which is continuously differentiable along X_2 such that $L_{X_2} u = \frac{1}{\ell} - 1$. We conclude from Lemma 4.1 that Z^t and X_2^t are bi-Hölder conjugate. □

Recall from Section 2 that $\xi_X \in H^1(M, \mathbb{R})$ denotes the Sharp's minimizer for a homologically full Anosov flow $X^t : M \to M$.

Theorem 5.2. *Let $X_1^t : M \to M$ be a homologically full Anosov flow. Assume that $X_2^t : M \to M$ is another Anosov flow which is orbit equivalent to X_1^t via H_0. Assume that H_0 matches homologically trivial period spectra. Then X_1^t is conjugate to the reparametrization of X_2^t generated by*

$$\frac{X_2}{1 + \omega(X_2)},$$

where ω is a smooth closed 1-form. If, moreover, $H_0^* \xi_{X_2} = \xi_{X_1}$, then ω can be chosen to be zero and, hence, X_1^t and X_2^t are conjugate.

Proof. The proof proceeds in exactly the same way as the proof of Theorem 5.1, but instead of applying the classical Livshits Theorem we apply Theorem 3.5 and obtain a function $u : M \to \mathbb{R}$ and a smooth closed 1-form ω such that

$$L_{X_2} u = \frac{1}{\ell} - 1 - \omega(X_2).$$

Note that we can approximate $L_{X_2}u$ with $L_{X_2}u'$, where u' is smooth. Then, after replacing ω with $\omega - du'$, we have (cf. Remark 3.1)

$$L_{X_2}(u-u') = \frac{1}{\ell} - 1 - \omega(X_2)$$

and for a sufficiently small $L_{X_2}(u-u')$ we have $1+\omega(X_2) > 0$. Then by Lemma 4.1 flow Z^t (and hence X_1^t) is conjugate to the reparametrization of X_2^t generated by $X_2/(1+\omega(X_2))$. This gives us the first part of the theorem.

Hence without loss of generality we can (and do) assume that

$$Z = \frac{1}{1+\omega(X_2)} X_2.$$

It is left to check that ω is exact if the cohomology classes ξ_{X_1} and ξ_{X_2} match.

Because Z^t is conjugate to X_1^t, we have $\xi_{X_1} = H_0^* \xi_Z$ and hence, by the assumption of the theorem, $\xi_Z = \xi_{X_2}$. Recall that by (6.5) we have $P_{X_2}(\xi_{X_2}(X_2)) = h_{\mu_{X_2}}(X_2) > 0$. Thus, applying Proposition 4.4 with $a = 1$ and $X = X_2$, we obtain that $[\omega] = 0$, that is, ω is exact. Hence, by Lemma 4.1, flows Z^t and X_2^t are conjugate. □

6.6 Sharpened Marked Length Spectrum Rigidity

Here we explain that our abelian Livshits theory can be used to improve marked length spectrum rigidity results on surfaces and higher-dimensional manifolds. Recall that Croke [Cr90] and Otal [Ot90] famously proved that marked lengths of closed geodesics determine the isometry class of a negatively curved surface. We offer the following enhancement.

Given a negatively curved surface (S, g), a free homotopy class of loops α on S admits a unique geodesic representative. Denote by $\ell_g(\alpha)$ the length of this geodesic.

Theorem 6.1. *Let g_1 and g_2 be two negatively curved metrics on a smooth compact surface S. Given a free homotopy class of loops α on S, denote by $\ell_{g_i}(\alpha)$ the length of the g_i-geodesic representative of α, $i = 1, 2$. Assume that their marked length spectra are the same for homologically trivial geodesics, that is, $\ell_{g_1}(\alpha) = \ell_{g_2}(\alpha)$ for every homologically trivial free homotopy class of loops α. Then g_1 and g_2 are isometric.*

In fact, the following more general result holds true.

Addendum 6.2. *Fix a homology class $c \in H_1(S, Z)$. If instead we assume that $\ell_{g_1}(\alpha) = \ell_{g_2}(\alpha)$ for every free homotopy class of loops in homology class c, then g_1 and g_2 are isometric.*

Remark 6.3. A precursor for the idea of considering a fixed homology class can be found in [K88, Theorem 3], where Katok proved that a marked length spectrum in a fixed homology class c determines the negatively curved metric on the surface in a fixed conformal class.

We proceed to prove Theorem 6.1 below. The addendum can be reduced to Theorem 6.1 in the following way. The length of a homologically trivial geodesic can be arbitrarily well approximated by the difference of lengths of two geodesics in homology class c. This approximation can be done in the same way as in the proof of Lemma 2.7. Hence the length of such a homologically trivial geodesic can be recovered from the lengths of geodesics in c. Further, the approximation procedure persists under orbit equivalence of geodesic flows and, hence, marked length spectrum in c recovers the homologically trivial length spectrum.

Lemma 6.4. *Let $X^t\colon T^1S \to T^1S$ be the geodesic flow on a negatively curved surface and let $I\colon T^1S \to T^1S$ be the involution given by $v \mapsto -v$. Assume that μ is an X^t-invariant measure such that $I_*\mu = \mu$. Then*

$$\int \omega(X)d\mu = 0$$

for every closed 1-form ω. In particular, this holds for the measure of maximal entropy.

Proof. Note that $I(x,v) = (x,-v)$ conjugates the geodesic flow and its inverse and interchanges the stable and unstable foliations. We have $X^{-t} = IX^tI$ or, infinitesimally, $DI(X) = -X$. Hence, if μ is the measure of maximal entropy for X^t, then $I_*\mu$ is the measure of maximal entropy for X^{-t} and hence, indeed, $I_*\mu = \mu$.

Recall that the bundle map $T^1S \to S$ induces an isomorphism on cohomology. Hence we can and do assume that ω is a pullback of a 1-form on the surface. It is easy to see that for such forms we have $\omega_{(x,-v)}(X(x,-v)) = -\omega_{(x,v)}(X(x,v))$. Now, for any μ such that $I_*\mu = \mu$, the claim of the lemma comes from the following calculation:

$$\begin{aligned}\int \omega(X)d\mu = \int \omega(X)dI_*\mu &= \int \omega_{(x,-v)}(X(x,-v))d\mu(x,v)\\ &= \int -\omega_{(x,v)}(X(x,v))d\mu(x,v) = -\int \omega(X)d\mu.\end{aligned}$$

□

Proof of Theorem 6.1. Denote by X_1 and X_2 the generating vector fields of geodesic flows on T^1S corresponding to g_1 and g_2, respectively. Then, it is well known that there exists $H\colon T^1S \to T^1S$, an orbit equivalence between X_1^t and X_2^t, which is homotopic to identity.

Recall the definition of the β-functional

$$\beta([\theta]) = P(\theta(X)) = \sup_{\mu} \left\{ h_\mu(X) + \int \theta(X) d\mu \right\},$$

where the supremum is taken among all X^t-invariant probability measures.

Denote by ξ_i, $i = 1, 2$, Sharp's minimizers of the β functional for X_i. Let μ_{ξ_i} be the equilibrium measures for $\xi_i(X)$ and also denote by μ_i, $i = 1, 2$, the measures of maximal entropy for X_i.

Then, using (6.4) which gives $\int \xi_i(X_i) d\mu_{\xi_i} = 0$ and Lemma 6.4, we have

$$\beta(\xi_i) = h_{\mu_{\xi_i}}(X_i) + \int \xi_i(X_i) d\mu_{\xi_i} = h_{\mu_{\xi_i}}(X_i) \leq h_{\mu_i}(X_i) = h_{\mu_i}(X_i) + \int \xi_i(X_i) d\mu_i.$$

Hence, by the definition of β, we have $\mu_{\xi_i} = \mu_i$ and, by uniqueness of Sharp's minimizer, $\xi_i = 0$, $i = 1, 2$. Then, obviously, $H^* \xi_2 = 0 = \xi_1$ and we can apply Theorem 5.2 to conclude that X_1^t and X_2^t are conjugate. Hence, we have complete matching of marked length spectra and, by the Croke–Otal rigidity theorem, g_1 is isometric to g_2. □

Remark 6.5. Recent results of Guillarmou and Lefeuvre on local marked length spectrum rigidity [GL18, Theorem 1] for higher-dimensional negatively curved manifolds can be enhanced in the same way – one only needs to assume that homologically trivial marked length spectra coincide.

6.7 Conjugacy for Contact Anosov Flows

Recall that a (positive) *contact form* on an oriented $(2k+1)$-dimensional manifold M is a smooth 1-form α such that $\alpha \wedge (d\alpha)^k > 0$. Associated to the contact form is its *Reeb vector field* X_α, which is uniquely determined by $\alpha(X_\alpha) = 1$ and $L_{X_\alpha} \alpha = 0$ (the latter is equivalent to $\iota_{X_\alpha} d\alpha = 0$). Call an Anosov flow X^t a *contact Anosov flow* if X is the Reeb vector field for a contact form α.

Theorem 7.1. *Let X_i, $i = 1, 2$, be contact Anosov flows. Assume one of the following:*

1. *X_i are flows on a 3-dimensional manifold, which are orbit equivalent via a C^1 orbit equivalence.*
2. *X_i are Anosov geodesic flows with C^1 Anosov splittings, which are orbit equivalent via a C^2 orbit equivalence.*

Then there exist a closed 1-form ω and a constant $c > 0$ such that X_1 is conjugated to $\frac{X_2}{c+\omega(X_2)}$. If, moreover, the orbit equivalence matches Sharp's minimizers $\xi_{X_1} \in H^1(M, \mathbb{R})$ to $\xi_{X_2} \in H^1(M, \mathbb{R})$, then ω can be taken to be 0, that is, X_1 is conjugate to a constant rescaling of X_2.

Recall that geodesic flows on perturbations of hyperbolic manifolds are $\frac{1}{2}$-pinched and hence have C^1 Anosov splittings.

Addendum 7.2. *In the first case, when X_i are 3-dimensional flows, the C^0 conjugacy is, in fact, smooth.*

The addendum follows from work of Fledman–Ornstein [FO87] who proved that a C^0 conjugacy must be C^1 and the bootstrap argument of de la Llave–Moriyón [LM88].

Proof of Theorem 7.1. Let α be the contact form for X_1 and let β be the contact form for X_2. Denote by H the orbit equivalence so that $H_*X_1 = \varphi X_2$ for some positive $\varphi \in C^0$.

Lemma 7.3. *There exists a constant $c > 0$ such that the 1-form $\mu = c\beta - H_*\alpha$ is closed.*

Proof. First, we prove the lemma when X_i are Anosov geodesic flows with C^1 Anosov splitting and H is C^2. Because of the C^2 hypothesis we have that $\varphi \in C^1$ and $H_*d\alpha = dH_*\alpha$ is exact. We claim that it is also X_2-invariant. Using functoriality, we have

$$0 = L_{X_1}d\alpha = L_{\varphi X_2}H_*d\alpha = \varphi L_{X_2}(H_*d\alpha) + d\varphi \wedge \iota_{X_2}H_*d\alpha.$$

Notice that $0 = \iota_{X_1}d\alpha = \iota_{\varphi X_2}H_*d\alpha = \varphi\iota_{X_2}H_*d\alpha$. Hence, indeed, we have $\varphi L_{X_2}(H_*d\alpha) = 0$.

Therefore, both $d\beta$ and $H_*d\alpha$ are exact X_2-invariant 2-forms. Then by [Ham95, Theorem A3] there is a constant $c > 0$ such that $H_*d\alpha = cd\beta$ and the lemma follows. (The constant c is positive because both α and β are positive contact forms and $\varphi > 0$.)

In the 3-dimensional case when H is merely C^1 (and, hence, we do not know that $H_*d\alpha$ is exact anymore) we can actually make a direct argument. We have that $H_*(\alpha \wedge d\alpha)$ is a φX_2-invariant C^0 volume form. Hence, both $\beta \wedge d\beta$ and $\varphi H_*(\alpha \wedge d\alpha)$ are X_2-invariant volume forms. Then, by ergodicity,

$$\varphi H_*(\alpha \wedge d\alpha) = c\beta \wedge d\beta,$$

where $c > 0$, again, because both α and β are positive.

Now note that $\iota_{X_2}(\beta \wedge d\beta) = d\beta$ and $\iota_{X_1}(\alpha \wedge d\alpha) = d\alpha$. We calculate

$$\begin{aligned} cd\beta &= c\iota_{X_2}(\beta \wedge d\beta) = \frac{c}{\varphi}\iota_{H_*X_1}(\beta \wedge d\beta) = \iota_{H_*X_1}\left(\frac{c}{\varphi}\beta \wedge d\beta\right) \\ &= \iota_{H_*X_1}\left(H_*(\alpha \wedge d\alpha)\right) = H_*\iota_{X_1}\left(\alpha \wedge d\alpha\right) \\ &= H_*d\alpha. \end{aligned}$$

□

Consider any homologically trivial X_1-periodic orbit γ. Then $H_*\gamma$ bounds a surface S and we can derive a relation between the periods of γ and $H_*\gamma$ as follows:

$$c\,\mathrm{per}_{X_2}(H_*\gamma) - \mathrm{per}_{X_1}(\gamma) = c\int_{H_*\gamma}\beta - \int_{\gamma}\alpha = \int_{H_*\gamma} c\beta - \int_{H_*\gamma} H_*\alpha$$
$$= \int_{H_*\gamma}\mu = \int_{S} d\mu = 0.$$

Hence, after rescaling by c, the marked length spectra for homologically trivial orbits match. Hence the result follows immediately from Theorem 2.9 and Theorem 5.2. □

Remark 7.4. For non-homotopically trivial periodic orbits γ, the above calculation gives that the periods $\mathrm{per}_{X_1}(\gamma)$ and $\mathrm{per}_{X_2}(H_*\gamma)$ are related as follows:

$$\mathrm{per}_{X_1}(\gamma) = c\,\mathrm{per}_{X_2}(H_*\gamma) + [\mu]([H_*\gamma]).$$

Note that $[H_*\gamma] = [\gamma]$ if H is homotopic to identity.

6.8 An Example

We begin by pointing out that the scenario of Theorem 7.1 actually occurs. Indeed, given a contact Anosov flow X with a contact form β and a closed 1-form ω with $\omega(X) > -1$, then the reparametrization

$$X_\omega = \frac{X}{1+\omega(X)}$$

is Anosov by Proposition 4.2 and contact with contact form $\beta + \omega$. Note that Remark 7.4 implies X^t is not conjugate to X^t_ω if X^t admits a periodic orbit on which $[\omega]$ does not vanish. In particular, if X^t admits at least one homologically nontrivial (in $H_1(M,\mathbb{R})$ namely non-torsion) periodic orbit, then there exists a small cohomology class $[\omega]$ such that X^t is not conjugate "to X^t_ω."

Hence we see that indeed, unlike in Theorem 6.1, matching of homologically trivial length spectra does not imply conjugacy for contact Anosov flows. This observation shows that that the conclusion of Theorem 7.1 is optimal. Further, by Proposition 4.6 we can find the "best" contact reparametrization with Sharp's minimizer equal to zero.

In this section we would like to present the same example from the point of view of deforming the Deck group rather than reparametrizing. At the end we will find out that this is exactly the same example. While there is some redundancy with what was already discussed, we consider this alternative description quite instructive and thus give a rather detailed and self-contained presentation.

We will describe an explicit deformation $X^t_\mu\colon M_\mu \to M_\mu$ of a flow $X^t = X^t_0$ for small $\mu \in H^1(M,\mathbb{R})$, such that the length spectrum deforms according to μ:

$$\mathrm{per}_{X_\mu}(h(p)) = \mathrm{per}_X(p) + \mu(\gamma),$$

where h is an orbit equivalence and γ is the homology class of closed orbit of x. The result of the construction will be summarized in what follows as Proposition 8.3. In the context of geodesic flows on surfaces of constant negative curvature the same example was given by Ghys [G87, Theorem 2.2]. He was interested in examples of Anosov flows which are not conjugate to algebraic flows and have analytic stable and unstable distributions. We give a different, more general construction.

We proceed with the description of the example. Let $X^t\colon M \to M$ be a flow. We lift the flow to the universal abelian cover $\tilde{X}^t\colon \tilde{M} \to \tilde{M}$. Then $\Gamma_0 \simeq H_1(M,\mathbb{Z})$ is the group of Deck transformations acting by isometries on $\tilde{M}$. Note that $\gamma \in \Gamma_0$ commutes with $\tilde{X}^t$.

Take a $\mu \in H^1(M,\mathbb{R}) \simeq \mathrm{Hom}(H_1(M,\mathbb{Z}),\mathbb{R})$. Let

$$\Gamma_\mu = \{\gamma \circ \tilde{X}^{\mu(\gamma)} \colon \gamma \in \Gamma_0\}$$

It is easy to see now that $I_\mu\colon \Gamma_0 \to \Gamma_\mu = Im(I_\mu)$, given by

$$I_\mu(\gamma) = \gamma \circ \tilde{X}^{\mu(\gamma)},$$

is a group homomorphism (which is one-to-one when X^t is not a periodic flow).

Then Γ_μ acts on $\tilde{M}$ by $\gamma\colon x \mapsto \gamma(\tilde{X}^{\mu(\gamma)}(x))$. First, we will see that, for all sufficiently small μ, the orbit space $M_\mu = \tilde{M}/\Gamma_\mu$ is a smooth manifold diffeomorphic to M.

Recall that the isomorphism $H^1(M,\mathbb{R}) \simeq \mathrm{Hom}(H_1(M,\mathbb{Z}),\mathbb{R})$ arises as follows. Let ω be a closed 1-form with cohomology class $[\omega] = \mu$. Then the lift $\tilde{\omega}$ of ω to $\tilde{M}$ is exact, that is, there exists a function $\alpha\colon \tilde{M} \to \mathbb{R}$ such that $\tilde{\omega} = d\alpha$. Then the homomorphism $\mu\colon H_1(M,\mathbb{Z}) \to \mathbb{R}$ is given by

$$\mu(\gamma) = \int_x^{\gamma(x)} \tilde{\omega} = \alpha(\gamma(x)) - \alpha(x)$$

for any $x \in \tilde{M}$.

Now define $H\colon \tilde{M} \to \tilde{M}$ by $H(x) = \tilde{X}^{\alpha(x)}(x)$. Clearly H sends orbits of $\tilde{X}^t$ to themselves. If

$$L_{\tilde{X}}\alpha > -1, \tag{6.6}$$

then H is invertible on every orbit and, hence, is a smooth diffeomorphism. Further H intertwines actions of Γ_0 and Γ_μ:

$$\forall \gamma \in \Gamma_0 \quad H \circ \gamma = I_\mu(\gamma) \circ H.$$

Indeed,

$$\begin{aligned} X^{\alpha(\gamma(x))}(\gamma(x)) &= X^{\alpha(x)+\mu(\gamma)}(\gamma(x)) \\ &= X^{\alpha(x)}(X^{\mu(\gamma)}(\gamma(x))) = \gamma \circ X^{\mu(\gamma)}(X^{\alpha(x)}(x)) = I_\mu(\gamma) \circ H. \end{aligned}$$

Hence, under condition (6.6), H induces a diffeomorphism $h\colon M \to M_\mu$. Moreover, if we denote by $X^t_\mu\colon M_\mu \to M_\mu$ the flow induced by $\tilde{X}^t$, then h is an orbit equivalence between X^t and X^t_μ.

Remark 8.1. Note that $X_\mu = (L_X\alpha + 1)Dh(X)$. Hence by Proposition 4.2, under the condition (6.6), if X is Anosov, then so is X_μ. Note also that if X is contact, then so is X_μ because the action of Γ_μ preserves the contact form. In fact, if β is the contact form for X^t, then $\beta + \mu$ is the contact form for $h^{-1} \circ X^t_\mu \circ h$.

Let $x \in M$ be a periodic point of period $\mathrm{per}_X(x)$ and let $\tilde{x} \in \tilde{M}$ be a lift of x. Then $\tilde{X}^{\mathrm{per}_X(x)} = \gamma(\tilde{x})$, where γ is the homology class of the orbit of x. Then

$$\tilde{X}^{\mathrm{per}_X(x)+\mu(\gamma)} = \tilde{X}^{\mu(\gamma)}(\gamma(\tilde{x})) = I_\mu(\gamma)(\tilde{x})$$

and hence

$$\mathrm{per}_{X_\mu}(h(p)) = \mathrm{per}_X(p) + \mu(\gamma). \tag{6.7}$$

Finally, we have the following lemma.

Lemma 8.2. *The set*

$$\mathcal{U} = \{\mu \in H^1(M,\mathbb{R})\colon \exists\omega\colon [\omega] = \mu,\ \|\omega\|_{C^0} < 1\}$$

is an open neighborhood of 0 in $H^1(M,\mathbb{R})$. Further, if $\mu = [\omega] \in \mathcal{U}$, then condition (6.6) holds.

We summarize all of the preceding discussion as follows.

Proposition 8.3. *Let $X^t\colon M \to M$ be a smooth flow on a compact manifold M. Then there exists a open neighborhood $\mathcal{U} \subset H^1(M,\mathbb{R})$ of zero and a deformation $X^t_\mu\colon M_\mu \to M_\mu$ such that*

1. *$X^t_0 = X^t$;*
2. *There exists a family of diffeomorphisms $h_\mu\colon M \to M_\mu$ which give orbit equivalences between X^t and X^t_μ;*
3. *The periods of periodic orbits deform according to (6.7);*
4. *If X^t is Anosov, then all X^t_μ, $\mu \in \mathcal{U}$ are Anosov;*
5. *If X^t is contact, then all X^t_μ, $\mu \in \mathcal{U}$ are contact.*

Remark 8.4. While Lemma 8.2 is very simple and elementary, the actual description of the set of admissible cohomology classes μ appears in Lemma 4.5.

It remains to prove the lemma.

Proof of Lemma 8.2. If $\mu = [\omega]$ with $\| \omega \|_{C^0} < 1$, then we have $L_{\tilde{X}}\alpha = d\alpha(\tilde{X}) = \tilde{\omega}(\tilde{X}) = \omega(X)$. Hence if $\mu \in \mathcal{U}$, then condition (6.6) is verified.

Now we check that $\mathcal{U}$ is open. Let $[\omega_1], [\omega_2], \dots [\omega_N]$ be a basis of $H^1(M, \mathbb{R})$. By rescaling if necessary, we can assume that $\| \omega_i \|_{C^0} = 1$, $i = 1, \dots N$. Then, obviously, the set

$$\mathcal{B}_\varepsilon = \left\{ \frac{1}{N} \left[\sum_{i=1}^{N} t_i \omega_i \right] : \sum_{i=1}^{N} t_i \leq \varepsilon \right\}$$

contains an open neighborhood of 0 in $H^1(M, \mathbb{R})$ and any $\mu \in \mathcal{B}_\varepsilon$ can be represented by a closed 1-form of norm $\leq \varepsilon$.

For any $\mu \in \mathcal{U}$ we have $\mu = [\omega]$ with $\| \omega \|_{C^0} < 1$. Let $\varepsilon = \frac{1}{2}(1 - \| \omega \|_{C^0})$. Then it is easy to see that $\mu + \mathcal{B}_\varepsilon \subset \mathcal{U}$, proving that $\mathcal{U}$ is open. □

References

[A08] M. Asaoka, *On invariant volumes of codimension-one Anosov flows and the Verjovsky conjecture.* Invent. Math. 174 (2008), no. 2, 435–462.

[Cr90] C. Croke, *Rigidity for surfaces of nonpositive curvature.* Comment. Math. Helv. 65 (1990), no. 1, 150–169.

[DS19] R. Dougall, R. Sharp, Anosov flows, growth rates on covers and group extensions of subshifts, Invent. Math. (2020).

[FO87] J. Feldman, D. Ornstein, *Semirigidity of horocycle flows over compact surfaces of variable negative curvature.* Ergod. Theory Dyn. Syst. 7 (1987), no. 1, 49–72.

[FH13] P. Foulon, B. Hasselblatt, *Contact Anosov flows on hyperbolic 3-manifolds.* Geom. Topol. 17 (2013), no. 2, 1225–1252.

[Fri83] D. Fried, *Transitive Anosov flows and pseudo-Anosov maps.* Topology 22 (1983), no. 3, 299–303.

[G87] E. Ghys, *Flots d'Anosov dont les feuilletages stables sont différentiables*, Annales scientifiques de la É.N.S. 4e série, tome 20, no. 2 (1987), pp. 251–270.

[GL18] C. Guillarmou, T. Lefeuvre, *The marked length spectrum of Anosov manifolds.* Ann. Math. (2) 190 (2019), no. 1, 321–344.

[Ham95] U. Hamenstädt, *Invariant two-forms for geodesic flows.* Math. Ann. 301 (1995), no. 4, 677–698.

[K88] A. Katok, *Four applications of conformal equivalence to geometry and dynamics.* Ergod. Theory Dyn. Syst. 8 (1988), Charles Conley Memorial Issue, 139–152.

[K03] A. Katok, *Combinatorial constructions in ergodic theory and dynamics.* University Lecture Series, 30. American Mathematical Society, Providence, RI, 2003. iv+121 pp.

[KK96] A. Katok, A. Kononenko, *Cocycles' stability for partially hyperbolic systems.* Math. Res. Lett. 3 (1996), no. 2, 191–210.

[KB37] N. Kryloff, N. Bogolioubofff, *La théorie générale de la mesure dans son application à l'étude des systèmes dynamiques de la mécanique non linéaire.* Ann. Math. (2) 38 (1937), no. 1, 65–113.

[Jou88] J.-L. Journé, *A regularity lemma for functions of several variables.* Rev. Mat. Iberoamericana 4 (1988), no. 2, 187–193.

[Liv72] A. N. Livsic, *Cohomology of dynamical systems.* Izv. Akad. Nauk SSSR Ser. Mat. 36 (1972), 1296–1320.

[LM88] R. de la Llave, R. Moriyón, *Invariants for smooth conjugacy of hyperbolic dynamical systems. IV.* Comm. Math. Phys. 116 (1988), no. 2, 185–192.

[LMM86] R. de la Llave, R, J. Marco, R. Moriyón, *Canonical perturbation theory of Anosov systems and regularity results for the Livsic cohomology equation.* Ann. Math. (2) 123 (1986), no. 3, 537–611.

[M96] D. Massart, *Normes stables des surfaces.* Mathématiques. École normale supérieure de Lyon, 1996.

[Ot90] J.-P. Otal, *Le spectre marque des longueurs des surfaces a courbure negative.* Ann. Math. (2) 131 (1990), no. 1, 151–162.

[Par86] W. Parry, *Synchronisation of canonical measures for hyperbolic attractors.* Comm. Math. Phys. 106 (1986), no. 2, 267–275.

[PP86] W. Parry, M. Pollicott, *The Chebotarov theorem for Galois coverings of Axiom A flows.* Ergod. Theory Dyn. Syst. 6 (1986), no. 1, 133–148.

[Pl72] J. Plante, *Anosov flows.* Am. J. Math. 94 (1972), 729–754.

[Pl73] J. Plante, *Homology of closed orbits of Anosov flows.* Proc. Amer. Math. Soc. 37 (1973), 297–300.

[S72] K. Sigmund, *On the space of invariant measures for hyperbolic flows.* Am. J. Math. 94 (1972), 31–37.

[Sh93] R. Sharp, *Closed orbits in homology classes for Anosov flows.* Ergod. Theory Dyn. Syst. 13 (1993), no. 2, 387–408.

[Sm67] S. Smale, *Differentiable dynamical systems.* Bull. Amer. Math. Soc. 73 (1967), 747–817.

[Sch57] S. Schwartzman. *Asymptotic cycles.* Ann. Math. 66 (1957), 270–284.

[W13] A. Wilkinson, *The cohomological equation for partially hyperbolic diffeomorphisms.* Astérisque No. 358 (2013), 75–165.

7

A Dynamical Approach to Validated Numerics

OLIVER JENKINSON AND MARK POLLICOTT

*In memoriam, Anatole Borisovich Katok**

We describe a method, using periodic points and determinants, for giving alternative expressions for dynamical quantities (including Lyapunov exponents and Hausdorff dimension of invariant sets) associated to analytic hyperbolic systems. This leads to validated numerical estimates on their values.

7.1 Introduction

7.1.1 An Overview

There are many well-known and important numerical invariants in the context of dynamical systems, including, for example, entropy, Lyapunov exponents, and the Hausdorff dimension of invariant sets. Even in the context of uniformly hyperbolic systems, many of these invariants do not have simple explicit expressions, nor are they easy to estimate. This leads naturally to the search for useful alternative expressions for each of these quantities which, in particular, lend themselves to accurate numerical evaluation.

Our aim in this account is to describe an approach developed over some years, based on periodic points for (hyperbolic) dynamical systems. In particular, in addition to presenting a general overview we will also give a general recipe for converting this information into formulae and useful numerical estimates for a variety of characteristic values, such as those just mentioned. Perhaps the most important aspect of this side of the work is that we obtain rigorous bounds on the errors. Given the data available, the idea is to minimise the error estimate, subject to the practical constraints imposed by the computational power available. There is

* The second author collaborated with A. Katok, G. Knieper, and H. Weiss over 30 years ago on the regularity of topological entropy for Anosov flows. One of the methods used there was based on the characterisation of the entropy as a pole for the dynamical zeta function, which is a complex function closely related to the determinant central to ideas we pursue here.

always some flexibility in the choices, which we can exploit in order to obtain the best error estimate, namely the smallest bound on the error.

Typically, the quantities that we can expect to study in this approach are those that can be expressed in terms of the thermodynamic pressure. The main ingredient in our approach is a complex function (called the determinant) which packages together the data on periodic orbits and from which can be derived estimates on the pressure, and consequently the quantities of interest.

In our presentation, no specialist knowledge is required to implement the general result. However, for completeness we give a fairly full outline of the proof (with some details on operator theory postponed to Appendix A).

7.1.2 A Historical Perspective

The starting point for our approach is the important work of Grothendieck in the 1950s on nuclear operators [15, 16]. This extended the classical theory of trace class operators and Fredholm determinants [12]. Although the impact of this theory in functional analysis and operator theory was well understood, it was not until 20 years later that Ruelle employed it with great effect in ergodic theory and dynamical systems, in his application to dynamical zeta functions [33]. This viewpoint was highly influential in the work of mathematical physicists (e.g., the well-known study by Cvitanović in his monumental online tome [5]).

With the advent of modern computers, an important component in ergodic theory and dynamical systems has been the focus on *explicit computation* of quantities arising in the context of dynamical systems. Typically, these approaches are based on the quantity in question being expressed in terms of an associated Ruelle transfer operator, implicitly assumed to act on some appropriate space of functions, and then a finite dimensional approximation to the operator is used to reduce this to a finite dimensional matrix problem, to be solved numerically. In contrast, the approach we will describe is to exploit *real-analytic* properties of the underlying dynamical system by introducing Ruelle transfer operators with strong spectral properties (in particular nuclearity), which allows us to exploit the earlier circle of ideas initiated by Grothendieck.

7.1.3 A Selection of Applications

By way of an appetiser to this approach, we now list a cross section of actual and potential applications to a number of different areas in mathematics. We will elaborate these later, but for purposes of motivation we list them here.

1. In number theory, a recent breakthrough in the understanding of the Markov and Lagrange spectra $M, L \subset (0, +\infty)$ from Diophantine approximation has

been brought about by the work of Matheus and Moreira, who were able to estimate the dimension of the difference $M \setminus L$ of these spectra, obtaining first a lower bound [27] and then an upper bound [28]. The methods, in both cases, involved the approximation of the dimension of certain fractal sets which would be amenable to the techniques developed in [21] for proving rigorous high-quality bounds on the dimension (cf. Example 4.12 (b)).

2. In the field of spectral geometry, there is a strong tradition of computing eigenvalues of the Laplacian, dating back to pioneering work of Hejhal [18] using classical methods. However, McMullen's approximations for the lowest eigenvalue for certain infinite-volume hyperbolic manifolds were based on the dimension of the limit set and, using this viewpoint, could be accurately estimated [26].
3. Within dynamical systems, among the most widely studied numerical quantities are Lyapunov exponents, measuring the exponential instability of solutions to various problems. There is also interest in estimating Lyapunov exponents in the theory of random matrix products; in particular, in information theory this leads to the computation of entropy rates for binary symmetric channels, related to examples of hidden Markov chains [17].

We will develop this last setting in the following subsection.

7.1.4 Illustrative Example: Lyapunov Exponents for Bernoulli Interval Maps

We begin by specifying a suitable class of hyperbolic transformations. In particular, we want to assume that the systems we are studying are both real analytic and uniformly hyperbolic, for example, either a real analytic expanding map, or a real analytic hyperbolic diffeomophism or flow on a locally hyperbolic set.

Before formulating statements in greater generality, let us consider a very specific example of a one-dimensional map (Figure 7.1). Let $X = \mathbb{R}/\mathbb{Z}$ be the unit circle and let $T\colon X \to X$ be a piecewise C^ω Bernoulli map of the interval which is expanding, that is, there exists $\lambda > 1$ such that $|T'(x)| \geq \lambda$ for all x. In this setting it is well known that there is a unique T-invariant probability measure μ which is absolutely continuous with respect to Lebesgue measure (i.e., $\frac{d\mu}{dx} \in L^1(X)$) by the famous Lasota–Yorke theorem (see [24, §5.1]).

Example 1.1. *More concretely, suppose* $T\colon \mathbb{R}/\mathbb{Z} \to \mathbb{R}/\mathbb{Z}$ *is defined by* $T(x) = 2x + \epsilon \sin(2\pi x) \pmod 1$ *where* $|\epsilon| < \frac{1}{4\pi}$, *so in particular* $|T'(x)| > (2 - 2\pi\epsilon) > \frac{3}{2} > 1$ *for all* $x \in \mathbb{R}/\mathbb{Z}$.

The Lyapunov exponent of the unique absolutely continuous T-invariant probability measure μ is defined by

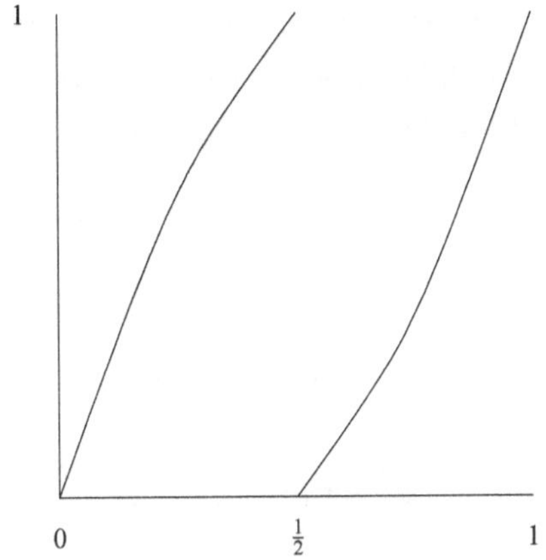

Figure 7.1 The graph of an expanding map of $\mathbb{R}/\mathbb{Z}$ represented on the unit interval.

$$L(\mu) = \int \log |T'(x))| \, d\mu(x) , \tag{7.1}$$

and, by the well-known Rohlin identity, this also equals the entropy $h(\mu)$.

We briefly summarise the method for estimating $L(\mu)$ in three steps.

Step 1 (Complex functions and coefficients). We wish to consider period-n points $T^n x = x$ and then define for $t \in \mathbb{R}$ the coefficients $a_1(t), a_2(t), \ldots$ using the Taylor expansion

$$\exp\left(- \sum_{n=1}^{\infty} \frac{z^n}{n} \sum_{T^n x = x} \frac{|(T^n)'(x)|^{-t}}{1 - ((T^n)'(x))^{-1}} \right) = 1 + \sum_{n=1}^{\infty} a_n(t) z^n + \cdots .$$

Step 2 (Coefficients and Lyapunov exponents). It can be shown that the Lyapunov exponent $L(\mu)$ given by (7.1) admits the alternative formulation

$$L(\mu) = \frac{-\sum_{n=1}^{\infty} a_n'(0)}{\sum_{n=1}^{\infty} n a_n(0)} ,$$

where both the numerator and denominator are absolutely convergent series. Truncating these series to give the computable quantity

$$L_N = \frac{-\sum_{n=1}^{N} a_n'(0)}{\sum_{n=1}^{N} n a_n(0)} ,$$

we note that

$$L_N \to L \text{ as } N \to +\infty ,$$

and so for a large natural number N, the quantity L_N is an approximation to L.

Step 3 (Error bounds). The quality of the approximation to L given by L_N may be indicated heuristically by comparing how closely L_N and L_{N-1} agree, though more accurate errors in the approximation can be obtained using the following:

1. A certain value $\theta \in (0, 1)$, which we refer to as the *contraction ratio*, measuring the extent to which a complex disc D is mapped inside itself by the inverse branches T_j of T;
2. The integrals $\beta_k = \frac{1}{r^{2k}} \int_0^1 \left| \sum_j T_j'(x + re^{2\pi it})(T_j(x + re^{2\pi it}))^k \right|^2 dt$, where $1 \leq k \leq L$; and
3. The weights $\alpha_k = \sqrt{\sum_{l=k+1}^{L} \beta_l^2}$ for $N \leq k \leq L$,

where $N < L$ are suitably chosen. In particular, we can then use these values to bound the coefficients $a_n(t)$, with $n > N$, for which it is impractical to explicitly compute them with effective error estimates. This will be explained in greater detail in Section 7.5.

Now that we have illustrated the general theme using the specific example of Lyapunov exponents for one-dimensional expanding maps, we can turn to more general dynamical settings, and more general characteristic values. In the next section we describe the broad context of the results and consider the pressure function, from which many of the quantities we want to consider can be derived.

7.2 Hyperbolic Maps and the Pressure Function

To set the scene, we begin by introducing two natural classes of discrete dynamical system, then consider the associated pressure function, which will be useful in providing the bridge between the dynamics and the various quantities we wish to describe.

7.2.1 Hyperbolic Maps

Let us begin with the discrete setting. Assume that $T: M \to M$ is either a smooth expanding map (perhaps on an invariant repeller $Y \subset M$) or Anosov diffeomorphism (see [24, §6.4]). Later we will consider the more general settings of repellers and the natural generalisation to flows. However, for the present we will restrict to the discrete cases just presented and, whenever more convenient, to the case of expanding maps.

Definition 2.1. *A partition* $X = \cup_{i=1}^{k} X_i$ *is called a Markov partition for the expanding map* $T: X \to X$ *if*

1. $\overline{\mathrm{int}(X_i)} \cap \overline{\mathrm{int}(X_j)}$ *for* $i \neq j$;
2. $X_i = \overline{\mathrm{int}(X_i)}$ *for* $i = 1, \cdots, k$; *and*
3. *each* $T(X_i)$ *for* $i = 1, \cdots, k$ *is a union of other elements of the partition.*

Eventually, we will want to assume that each of the restrictions $T|X_i$ ($i = 1, \cdots, k$) is real analytic, in the sense of having an analytic extension (via charts) to a complex neighbourhood U_i. However, to set up the definitions we only require that it be C^1.

In the case of Anosov diffeomorphisms, the approach is somewhat similar, except that one uses Markov partitions for invertible maps. In the case of Anosov flows, one can expect to use Markov Poincaré sections to reduce the analysis to the discrete case.

7.2.2 Pressure

The pressure function was introduced into the study of hyperbolic dynamical systems by Ruelle (see, e.g., [34]). The importance of pressure stems from the fact that it yields a unifying concept to describe dynamical and geometric invariants. For example, it is well known that various dynamically and geometrically defined fractals (e.g., limit sets and Julia sets) have the property that their Hausdorff dimension is given by solving an associated *pressure equation* (usually known as the Bowen formula). More generally, there are a host of other dynamical quantities that can be expressed in terms of the pressure function, some of which are listed in Subsection 7.2.3.

To define pressure P, since we are considering hyperbolic maps we have the luxury of expressing this in terms of periodic orbits $T^n x = x$, for $n \geq 1$, and define $P\colon C^0\left(\coprod_i X_i\right) \to \mathbb{R}$ ($i = 1, \cdots, k$) on the disjoint union of the elements of the Markov partition.

Definition 2.2. *The pressure of the continuous function g is given by*

$$P(g) := \limsup_{n\to+\infty} \frac{1}{n} \log\left(\sum_{T^n x = x} \exp\left(\sum_{j=0}^{n-1} g(T^j x)\right)\right)$$

and admits the alternative variational definition

$$P(g) = \sup\left\{h(\mu) + \int g d\mu \colon \mu \text{ is a } T\text{-invariant probability measure}\right\}.$$

When g is Hölder continuous, there is a unique probability measure μ_g realising the preceding supremum [2].

Definition 2.3. *The measure μ_g is called the equilibrium measure (or Gibbs measure) for g.*

If $g = 0$, then $P(0) = h_{top}(f)$ is the topological entropy, and the corresponding equilibrium measure μ_g is called the measure of maximal entropy (and the Bowen–Margulis measure in the case of Anosov systems).

Example 2.4. *For expanding maps $T: M \to M$, in the special case that $g(x) = -\log|\mathrm{Jac}(D_xT)|$ then $P(g) = 0$, and the corresponding equilibrium measure μ_g is a T-invariant probability measure equivalent to the volume on M. For Anosov maps $T: M \to M$, in the special case that $g(x) = -\log|\mathrm{Jac}(D_xT|E^u)|$, where E^u is the unstable bundle, then $P(g) = 0$ and the corresponding equilibrium measure μ_g is called the Sinai–Ruelle–Bowen measure (or SRB-measure) (see [24], §20.4).*

It is this pressure function that often helps to relate periodic orbits to the quantities in which we are interested, and which we ultimately want to numerically estimate. A simple, but important, application is the following result due to Ruelle (see [34, p. 99]):

Lemma 2.5. *For any Hölder continuous functions $g_0, g: X \to \mathbb{R}$ the function*

$$t \mapsto P(g_0 + tg) \in \mathbb{R}, \ \text{for } t \in \mathbb{R},$$

is analytic. Moreover,

1. $\frac{dP(g_0+tg)}{dt}|_{t=0} = \int g d\mu_{g_0}$, *and*
2. $\frac{d^2P(g_0+tg)}{dt^2}|_{t=0} = \lim_{n\to+\infty} \frac{1}{n} \int \left(\sum_{i=0}^{n-1} g(T^i x)\right)^2 d\mu_{g_0}(x)$ *provided* $\int g d\mu_{g_0} = 0$.

The quantity in part 2 of Lemma 2.5 is often called the *variance*.

It is a feature of the method we use that we can obtain fairly explicit expressions for derivatives of pressure. In particular, those quantities that can be written in terms of the derivative expressions can therefore, in turn, be written in terms of periodic points.

7.2.3 Relating Pressure to Characteristic Values

We can now consider a number of familiar quantities that we can write in terms of the pressure function. Here we list a few simple examples. Later we shall consider other applications, but for the present these three examples illustrate well our theme.

(I) Lyapunov exponents. For an expanding map $T\colon M \to M$ we can write the Lyapunov exponent for the absolutely continuous invariant measure μ by

$$L(\mu) = \int \log \|D_x T\| d\mu(x).$$

In the particular case of one dimension, this reduces to the situation described in Subsection 7.1.4.

The following follows immediately from part 1 of Lemma 2.5.

Lemma 2.6. *If we let* $g_0(x) = -\log|\mathrm{Jac}(D_x T)|$ *and* $g = -\log\|D_x T\|$*, then* $L(\mu) = \frac{d}{dt} e^{P(g_0+tg)}|_{t=0}$.

(II) Variance. For an expanding map $T\colon M \to M$ we can write the variance for the absolutely continuous T-invariant measure μ and a Hölder continuous function $g\colon X \to \mathbb{R}$ with $\int g d\mu = 0$ defined by

$$\Sigma(g,\mu) := \lim_{n\to+\infty} \frac{1}{n} \int \left(\sum_{i=0}^{n-1} g(T^i x) \right)^2 d\mu(x).$$

The next lemma follows immediately from part 2 of Lemma 2.5.

Lemma 2.7. *We can write*

$$\frac{d^2 P(g_0 + tg)}{dt^2}|_{t=0} = \Sigma(g,\mu).$$

This plays an important role in the Central Limit Theorem [2], that is, for any real numbers $a < b$ we have

$$\lim_{n\to+\infty} \mu\left(\left\{ x \in X : \frac{1}{\sqrt{n}} \sum_{i=0}^{n-1} g(T^i x) \in [a,b] \right\} \right) = \frac{1}{\sqrt{2\pi}} \int_a^b e^{-u^2/2\sigma} du.$$

(III) Linear response. Let $T_\lambda\colon M \to M$ be a smooth family of expanding maps $(-\epsilon < \lambda < \epsilon)$ and let μ_{g_λ} be the associated absolutely continuous measure, arising as the Gibbs measure for $g_\lambda(x) = -\log|\mathrm{Jac}(D_x T_\lambda)|$. Then by part 2 of Lemma 2.5 we have

$$\frac{\partial P(g_\lambda + tg)}{\partial t}|_{t=0} = \int g d\mu_{g_\lambda}.$$

Thus differentiating in λ formally gives

$$\frac{d^2 P(g_\lambda + tg)}{dt\, d\lambda}|_{\lambda=0} = \frac{d}{d\lambda}\left(\int g d\mu_{g_\lambda} \right)|_{\lambda=0}.$$

(A slight subtlety here is that the differentiation of the pressure is easier with respect to the fixed expanding map $T_0\colon M \to M$ and thus it is appropriate to introduce a family of conjugacies $\mu_\lambda\colon M \to M$ between T_0 and T_λ and to consider $\frac{d^2 P(g_0\circ\pi_\lambda + tg\circ\pi_\lambda)}{dtd\lambda}|_{\lambda=0}$.)

The preceding list does not exhaust the possible quantities that can be derived from the pressure, but it gives a selection that, we hope, illustrates our general approach.

In the next section, we will introduce a standard tool, the *transfer operator*, which allows us to analyse the pressure, and thus its many derivative properties, using basic ideas from linear operator theory. We will also describe the connection with a family of complex functions called determinants.

7.3 Transfer Operators and Determinants

A central object in thermodynamic formalism is the *transfer operator*. From this operator important dynamical and geometric invariants such as entropy, Lyapunov exponents, invariant measures, and Hausdorff dimension can be obtained.

Let us now restrict (for the present) to the case of expanding maps $T\colon X \to X$. The analyticity of the pressure, as well as other properties including the proof of Lemma 2.5, depend on the use of transfer operators. Eventually, we will want to consider operators acting on spaces of analytic functions, but for purposes of defining them it suffices for the present to consider the Banach space of C^1 function $C^1(X,\mathbb{C})$ with the norm $\|f\| = \|f\|_\infty + \|Df\|_\infty$. The operators are then defined as follows:

Definition 3.1. *If $T\colon X \to X$ is a C^1 expanding map, and $g\colon X \to \mathbb{R}$ is C^1, then we define the transfer operator $\mathcal{L}_g\colon C^1(X,\mathbb{C}) \to C^1(X,\mathbb{C})$ by*

$$\mathcal{L}_g w(x) = \sum_{Ty=x} e^{g(y)} w(y),$$

the summation being over the inverse images y of the point $x \in X$.

This operator preserves various function spaces and exhibits certain positivity properties which ensure that a Perron–Frobenius-type theorem holds: when acting on $C^1(X,\mathbb{C})$, $\mathcal{L}_g$ has a leading eigenvalue which is simple, positive, and isolated. Moreover, the connection with the pressure comes from the following basic result due to Ruelle [34] (see also [2]).

Lemma 3.2. *The spectral radius of $\mathcal{L}_g$ is $e^{P(g)}$. In particular, $e^{P(g)}$ is a maximal isolated eigenvalue for $\mathcal{L}_g$.*

In particular, the differentiability (indeed analyticity) of the pressure follows by standard perturbation theory, and the expressions for the derivatives in the lemma follow by explicit manipulations.

Example 3.3. *In the special case that* $g(x) = -\log|\mathrm{Jac}(D_xT)|$, *then* $P(g) = 0$ *and* $\mathcal{L}_g$ *is known as the Ruelle–Perron–Frobenius operator. The eigenmeasure* $m_g = \mathcal{L}_g^* m_g$ *is the normalised Lebesgue measure, and the equilibrium measure* $\mu_g = h_g m_g$ *is the unique T-invariant measure absolutely continuous with respect to Lebesgue measure, where* $h_g = \mathcal{L}_g h_g$ *is the maximal eigenfunction [2].*

Thus far, we have been following a very traditional approach. However, now we introduce an extra ingredient.

7.3.1 Determinants and Their Coefficients

Let $T: X \to X$ be a C^1 expanding map. For any continuous function $G: X \to \mathbb{R}$ and each period-n point $T^n x = x$, $n \geq 1$, we can associate the weight

$$G^n(x) := \sum_{i=0}^{n-1} G(T^i x) \in \mathbb{R}.$$

Later we will want to assume that T and G are real analytic, but for purposes of introducing the determinant we need only assume these weaker hypotheses. It is convenient to package up the information from individual periodic points into a single (generating) complex function.

Definition 3.4. *Given a continuous function* $G: X \to \mathbb{R}$, *we can formally define a function of the single complex variable by*

$$D(z) = D_{G,T}(z) = \exp\left(-\sum_{n=1}^{\infty} \frac{z^n}{n} \sum_{T^n x = x} \frac{\exp\left(\sum_{i=0}^{n-1} G(T^i x)\right)}{\det(I - [D(T^n)(x)]^{-1})}\right), \quad z \in \mathbb{C}.$$

The radius of convergence of the infinite series $D(z)$ is related to the pressure. More precisely, we can see that this converges to an analytic function, provided the series converges, namely $|z|e^{P(G)} < 1$ where

$$e^{P(G)} = \lim_{n\to+\infty} \left| \sum_{T^n x = x} \frac{\exp\left(\sum_{i=0}^{n-1} G(f^i x)\right)}{\det(I - [D(T^n)(x)]^{-1})} \right|^{1/n}$$

$$\left(= \lim_{n\to+\infty} \left| \sum_{T^n x = x} \exp\left(\sum_{i=0}^{n-1} G(f^i x)\right) \right|^{1/n} \right).$$

In particular, writing $D(z)$ as a power series

$$D(z) = 1 + \sum_{n=1}^{\infty} a_n z^n, \tag{3.1}$$

with coefficients $a_n = a_n(T, G)$ depending on T and G, we see it has radius of convergence of at least $e^{-P(G)}$.

Example 3.5 (Expanding maps of the interval). *In the particular case of an expanding map $T\colon X \to X$ of the interval X, given a continuous function $G\colon X \to \mathbb{R}$, the function $D(z)$ takes the simpler form:*

$$D(z) = \exp\left(-\sum_{n=1}^{\infty} \frac{z^n}{n} \sum_{T^n x = x} \frac{\exp\left(\sum_{i=0}^{n-1} G(T^i x)\right)}{1 - 1/(T^n)'(x)} \right), \quad z \in \mathbb{C}.$$

This naturally leads to asking about the meromorphic extension of $D(z)$. To proceed further, we need to assume more regularity on the function G. This brings us to the following important result of Ruelle [33].

Lemma 3.6. *If T is real analytic, then*

1. *$D_{G,T}(z)$ is analytic in all of $\mathbb{C}$*
2. *The value $z = e^{-P(G)}$ is a simple zero for $D_{G,T}(z)$ in this extension.*

In particular, we see from part 1 of Lemma 3.6 that we can improve the result on the radius of convergence of the power series to $\lim_{n\to+\infty} |a_n|^{1/n} = 0$, namely for any $0 < \theta < 1$ there exists $C > 0$ such that $|a_n| \leq C\theta^n$. In fact, the original proof of Lemma 3.6 due to Ruelle, and inspired by work of Grothendieck, works in this way by giving estimates on the coefficients a_n. We will later describe quite precise bounds on the coefficients a_n which establishes part 1. We will return to this point in the next subsection.

Remark 3.7. If we assume that T and G are C^∞, then we would still have that $\lim_{n\to+\infty} |a_n|^{1/n} = 0$. However, as we shall see, in the analytic case we have more effective estimates on $|a_n|$.

7.3.2 Pressure and the Characteristic Quantities

In order to relate $D(z)$ back to the pressure, and thus the various dynamical quantities, we need to make different choices for G. More precisely, we can consider the special case of the function $G = g_0 + tg$ where $g_0, g\colon X \to \mathbb{C}$ and $t \in \mathbb{R}$. This leads to the following particular case of the previous definition.

Definition 3.8. *We formally define the determinant for* g_0, g_1 *to be the bi-complex function*

$$d_{g_0,g}(z,t) := \exp\left(-\sum_{n=1}^{\infty}\frac{z^n}{n}\sum_{T^n x = x}\frac{\exp\left(\sum_{j=0}^{n-1}(g_0+tg)(T^i x)\right)}{1-1/(T^n)'(x)}\right),$$

where $z, t \in \mathbb{C}$.

The relationship between the determinant and the pressure in Lemma 3.6 now implies the following.

Corollary 3.9. *Assume that* g_0, g_1 *are* C^ω.

1. *The function* $d_{g_0,g}(z,t)$ *has a bi-analytic extension to all of* $\mathbb{C}^2$.
2. *The value* $z = e^{-P(g_0+tg)}$ *occurs as a simple zero for* $z \mapsto d_{g_0,g}(z,t)$.

We can make the further simplifying assumption that $P(g_0) = 0$, where we can replace g_0 by $g_0 - P(g_0)$, if necessary. In particular, the first zero for $z \mapsto d_{g_0,g}(z,0)$ (where $t = 0$) occurs at $z = e^{-P(g_0)}$. We can use part 2 of Corollary 3.9, and the implicit function theorem, to write the derivative of the pressure as

$$\frac{dP(g_0+tg)}{dt}\Big|_{t=0} = \frac{\partial d_{g_0,g}(1,t)}{\partial t}\Big|_{t=0} \Big/ \frac{\partial d_{g_0,g}(z,0)}{\partial z}\Big|_{z=1}$$

in terms of partial derivatives of the determinant. Furthermore, in light of part 1 of Corollary 3.9, for each $t \in \mathbb{R}$ we can formally expand

$$d_{g_0,g}(z,t) = 1 + \sum_{n=1}^{\infty} a_n(t) z^n.$$

It is these values which we need to relate to the quantities in which we are interested, and which we ultimately want to numerically estimate.

This gives us the simplest definition of the numbers $(a_n(t))_{n=1}^{\infty}$, although we can explicitly expand these in terms of periodic points too. This reveals the following simple, but crucial, fact.

Lemma 3.10. *For each* $n \in \mathbb{N}$, *we can express the value* $a_n(t)$ *in terms of the periodic points of period at most* n.

In particular, this ensures that the coefficients $a_n(t)$ are relatively easy to estimate. Moreover, we can rapidly approximate $d_{g_0,g}(z,t) = 1 + \sum_{n=1}^{\infty} a_n(t) z^n$ by the truncated series

$$1 + \sum_{n=1}^{N} a_n(t) z^n$$

for N moderately large. The rapidity of the approximation is justified by the following result.

Corollary 3.11. *The coefficients $a_n = a_n(t)$ tend to zero at a superexponential rate.*

When X is one-dimensional, then there exists $\theta \in (0, 1)$ such that $|a_n| = O(\theta^{n^2})$ as $n \to \infty$. We will give very explicit estimates for the implied constants in the $O(\cdot)$ term.

Remark 3.12. When X is d-dimensional (with $d \geq 2$), then there exists $0 < \theta < 1$ such that $|a_n| = O(\theta^{n^{1+1/d}})$ as $n \to \infty$.

Remark 3.13. More generally, we can assume we have a family of real analytic functions $g_1, \cdots, g_m : X \to \mathbb{C}$ and

$$d(z, t) := \exp\left(-\sum_{n=1}^{\infty} z^n \sum_{T^n x = x} \exp\left(\sum_{j=0}^{n-1} (g_0 + t_1 g_1 + \cdots + t_m g_m)(T^j x)\right)\right)$$

for $z \in \mathbb{C}$ and $t_1, \ldots, t_m \in \mathbb{R}$.

To summarise, we now have a method of approaching the pressure function which might be considered to have a simple analogy to that of estimating the largest eigenvalue of a matrix by computing the characteristic polynomial, which is a complex function whose zeros give the eigenvalues. This simple viewpoint relating the determinant and transfer operators ultimately leads to a surprisingly efficient method of computing pressure. Furthermore, we can accurately estimate the error terms (see Section 7.5).

Having related the various quantities which we want to estimate to the zero(s) of the determinant $d(z, t)$ (often via the pressure function; cf. Subsection 7.2.3) we need to answer three key questions: *How can we use this formulation to get numerical estimates? Why does this approach lead to superior approximation estimates? How can we estimate the quantities with validated rigour?*

We can now turn to the practical problem of writing explicit expressions for the approximations to quantities we described in Subsection 7.2.3, in terms of the first N coefficients in the expansion of the determinant.

7.3.3 Dynamical Quantities and Coefficients

We have established (in Subsection 7.2.3) that many of the quantities that we want to estimate can be expressed in terms of pressure, and its derivatives, which in turn can be written in terms of the determinant and its derivatives. Since the determinant has a power series expansion, it is a straightforward, but useful, exercise to write these expressions explicitly in terms of the derivatives of the coefficients. More precisely, let us write

$$A = \sum_{n=1}^{\infty} n a_n(0), \quad B = \sum_{n=1}^{\infty} n(n-1) a_n(0), \quad C = \sum_{n=1}^{\infty} a_n'(0),$$

$$D = \sum_{n=1}^{\infty} n a_n'(0), \quad E = \sum_{n=1}^{\infty} a_n''(0)$$

and the associated finite sums

$$A_N = \sum_{n=1}^{N} n a_n(0), \quad B_N = \sum_{n=1}^{N} n(n-1) a_n(0), \quad C_N = \sum_{n=1}^{N} a_n'(0),$$

$$D_N = \sum_{n=1}^{N} n a_n'(0), \quad E_N = \sum_{n=1}^{N} a_n''(0)$$

for $N \geq 1$.

A recurrent theme in our discussions is that we want to express the dynamical quantities in terms of A, B, C, D, E, and so on, and then approximate these expressions by using instead the more computationally tractable quantities A_N, B_N, C_N, D_N, E_N.

Remark 3.14. In practical applications, even for quite simple examples, we might currently only expect to compute these values up to $N = 25$, say, in a reasonable time frame. However, with further technological advances, one might expect that this value can be improved.

To illustrate this principle, we can now reformulate the three key quantities described in Subsection 7.2.3, and their approximations, in terms of these series and summations. We list these here.

(I) Lyapunov exponents. We can write the Lyapunov exponent for μ as

$$L(\mu) = -\frac{\sum_{n=1}^{\infty} a_n'(0)}{\sum_{n=1}^{\infty} n a_n(0)} = -\frac{C}{A},$$

and in particular, the Lyapunov exponent can be approximated by the computable quantities

$$-\frac{C_N}{A_N}, \quad N \geq 1.$$

(II) Variance. The variance is given by

$$\Sigma^2 = \left(\frac{C}{A}\right)^2 + \frac{1}{A}\left(B\left(\frac{C}{A}\right)^2 - 2DB\left(\frac{C}{A}\right) + E\right),$$

and in particular, we can approximate the variance by

$$\left(\frac{C_N}{A_N}\right)^2 + \frac{1}{A_N}\left(B_N\left(\frac{C_N}{A_N}\right)^2 - 2D_N B_N\left(\frac{C_N}{A_N}\right) + E_N\right), \quad N \geq 1.$$

(III) Linear response. We can write

$$\int g d\mu_T = -\frac{C}{A},$$

and in particular, we can approximate the integral by

$$-\frac{C_N}{A_N}, \quad N \geq 1.$$

Finally, by replacing T by T_λ and differentiating both sides in λ, we get an expression for the linear response in terms of the (derivatives of the) coefficients a_n.

7.4 Rates of Mixing and Dimension

In this section we want to make a slight detour to introduce another two important quantities which, although not quite fitting into the same framework described in the previous section, can also be approximated using the determinant.

First we consider the rate(s) of mixing, which can be studied via the zeros of the determinant.

7.4.1 Rates of Mixing

Let X be d-dimensional and let $T: X \to X$ be a C^ω conformal expanding map. More precisely, we can write the derivative $DT(x) = w(x)\Theta(x)$ where $w: X \to \mathbb{R}$ and $\Theta: X \to SO(d)$. In the particular case that $d = 1$, then the one-dimensional map T is automatically conformal.

Let $g_0: X \to \mathbb{R}$ be real analytic and let $\mu = \mu_{g_0}$ be the equilibrium-Gibbs measure associated to g_0.

Example 4.1. *As we observed in Example 2.4, when* $g_0(x) = -\log|\mathrm{Jac}(T)(x)|$, *the associated measure* μ_{g_0} *is the unique absolutely continuous invariant probability measure.*

The following is an important object in ergodic theory.

Definition 4.2. *Given a real analytic function g, we can consider the correlation function defined by*

$$c(n) = \int g \circ T^n g d\mu - \left(\int g d\mu\right)^2, \quad n \geq 1.$$

Since $T\colon (X, \mu) \to (X, \mu)$ is mixing, we know that $c(n) \to 0$, as $n \to +\infty$. The (exponential) rate of mixing is given by the smallest value $0 < \lambda_1 < 1$ such that $c(n) = O(\lambda_1^n)$ for all such g and $n \geq 1$. Since λ_1 corresponds to the modulus of the second eigenvalue of the transfer operator, the connection to the determinant comes through the following simple lemma.

Lemma 4.3. *The rate of mixing $0 < \lambda_1 < 1$ is the reciprocal of the modulus $\rho > 1$ of the second smallest (in modulus) zero of $d_{g_0,g}(z, 0)$, namely $\lambda_1 = 1/\rho$.*

In particular, the value ρ in Lemma 4.3 is a zero of the (real-valued) series

$$d_{g_0,g}(z, 0) = 1 + \sum_{n=1}^{\infty} z^n a_n(0)$$

and so in order to get rigorous bounds on ρ we can use the intermediate value theorem. More precisely, given $\epsilon_1, \epsilon_2 > 0$ and $N \in \mathbb{N}$, we look for bounds $\alpha_N < \rho < \beta_N$ by choosing α_N, β_N such that

$$1 + \sum_{n=1}^{N} \alpha_N^n a_n(0) \leq -\epsilon_1 \text{ and } 1 + \sum_{n=1}^{N} \beta_N^n a_n(0) \geq \epsilon_2$$

and

$$\epsilon_1 > \left| \sum_{n=N+1}^{\infty} \alpha_N^n a_n(0) \right| \text{ and } \epsilon_2 > \left| \sum_{n=N+1}^{\infty} \beta_N^n a_n(0) \right|.$$

Thus, finding good bounds $\alpha_N < \rho < \beta_N$ reduces to the following:

1. Getting good estimates on the coefficients $a_i(0)$ ($i = 1, \cdots, N$); and
2. Finding effective bounds on the tails $\sum_{n=N+1}^{\infty} a_n(0)$.

The first is a basic problem in practical computing. The second is a more challenging mathematical problem.

We can illustrate this with a simple example.

Example 4.4 (Lanford map, cf. [23, 25]). *Let $T\colon [0, 1] \to [0, 1]$ be defined by*

$$T(x) = 2x + \frac{1}{2}x(1 - x) \pmod{1}.$$

Let μ denote the unique absolutely continuous T-invariant probability measure. We can now evaluate dynamical quantities such as the rate of mixing by approximating the infinite series for the determinant by finite truncations. For example, using periodic points up to period $N = 16$, we might estimate the first ten zeros (i.e., the ten zeros of smallest modulus) of the determinant by locating the first ten zeros of the degree-16 polynomial truncation (see Table 7.1).

Table 7.1 *Estimates on the first* 10 *zeros of the determinant.*

n	zero z
1	1.0000000000000000000000033711203720152
2	1.7298653106643168192789406951918162 9
3	2.6922698183465737455729975178528581
4	4.1132466756759777783645672672979956
5	6.2454853205721291176033177124804291
6	9.4538916717397326544473431332123370
7	14.282734336458524434000313080802510
8	21.549229994532327179757991408669084
9	32.516266102701803490675636630907193
10	47.829104847022187584467732898137 53

More accurate estimates on the second eigenvalue λ_1 of the transfer operator, namely the exponential rate of mixing, are obtained by approximating the reciprocal of the second zero of the determinant, given in Table 7.2, using degree-N truncations with N ranging up to $N = 25$.

We will address the rigorous bounds on the error later.

Remark 4.5. The rate of mixing has a more subtle generalisation of the following form. The speed of mixing is controlled by a sequence of complex numbers $\lambda_j \to 0$ ordered to be decreasing in modulus, and polynomials P_j. More precisely, for any $\delta > 0$ there exist M and $c_j = c_j(g)$ $(j = 1, \cdots, M)$ such that

$$c(n) = \sum_{j=1}^{M} c_j P_j(n) \lambda_j^n,$$

where $P(n) = 1$ in the case of a single zero of the determinant. Since we can rewrite

$$c(n) = \int g \left(\mathcal{L}_{g0}^n g \right) d\mu - \left(\int g d\mu \right)^2,$$

the values λ_j are actually the *other* nonzero eigenvalues of the transfer operator. Furthermore,

$$P_j(n) = \begin{cases} 1 & \text{if } \lambda_j \text{ has multiplicity } m = 1 \\ \text{a polynomial of degree at most } n & \text{if } \lambda_j \text{ has multiplicity } > 1. \end{cases}$$

Furthermore, one can identify the eigenvalues in terms of the *other* zeros $\rho_j = 1/\lambda_j$ for $z \mapsto d_{f,z}(z, 1)$.

All this said, in the case of the Lanford map the first 10 zeros of the determinant are simple.

Table 7.2 *Estimates on the second eigenvalue* λ_1 *coming from the reciprocals of zeros for* $1 + \sum_{n=1}^{N} z^n a_n(0)$, *for* $12 \le N \le 25$.

N	second eigenvalue λ_1 estimate
12	0.5780796887515271422742765368788953299348846128812023109203951947004787498004165
13	0.5780796885356306834127405345836355663641109763750019611087170244976104563627485
14	0.5780796885371288506764371131157188309769151998850254045247866596035386808066373
15	0.5780796885371219470570630291328371225537224787114789966418506438634692131905786
16	0.5780796885371219681960432055118626393344991913606205477442507113706445878179303
17	0.5780796885371219681530107872274433995896003010891980049121721575572541941602200
18	0.5780796885371219681530690475964044549434630264578745046610737538545621059499499
19	0.5780796885371219681530689951228630661230046404665596121577924787108887084624467
20	0.5780796885371219681530689951543111607290086789469044407358342311002102577614591
20	0.5780796885371219681530689951543111607290086789469044407358342311002102577614591
21	0.5780796885371219681530689951542986173994750355886730998629996708085266331162364
22	0.5780796885371219681530689951542986207295902353572376170142239221705813115693187
23	0.5780796885371219681530689951542986207290016813780055122529428636713090405312973
24	0.5780796885371219681530689951542986207290017506309910801987018396260494147990458
25	0.5780796885371219681530689951542986207290017506255654011865278736341388970140860

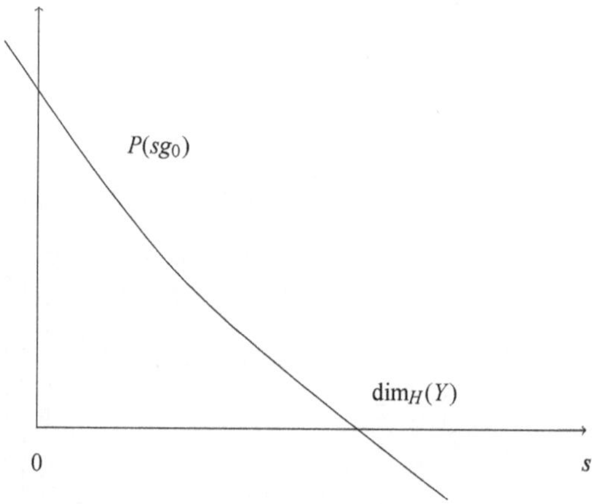

Figure 7.2 Graph of a pressure curve $s \mapsto P(sg_0)$

7.4.2 Dimension of Repellers

Let $T: X \to X$ be a C^ω conformal expanding map (Figure 7.2).

Definition 4.6. *A closed invariant set $Y \subset X$ is called a repeller if there is an open set U satisfying $Y \subset U \subset X$ such that $Y = \cap_{n=1}^{\infty} T^{-n}U$.*

We refer the reader to the book of Falconer for the definition and basic properties of Hausdorff dimension [9]. In the present context we have a convenient dynamical formulation. Let $g_0(x) = -\log|\mathrm{Jac}(T)(x)|$ and then consider the restriction $T: Y \to Y$. The following standard result relates the dimension $\dim_H(X)$ to the pressure function for the transformation $T: Y \to Y$ and function $sg_0: Y \to \mathbb{R}$, for $s > 0$ (see [35]).

Lemma 4.7 (Bowen–Ruelle). *We can characterise the Hausdorff dimension of the limit set by $s = \dim_H(Y)$ such that $P(sg_0) = 0$.*

The following examples fit into the setting of Lemma 4.7.

Example 4.8. *If $T: J \to J$ is a hyperbolic rational map acting on its Julia set J, and $f = -s\log|T'|$, then the Hausdorff dimension $\dim_H(J)$ is given by the unique zero s of $P(-s\log|T'|)$ by Lemma 4.7.*

It is a result of Ruelle [35] that the dimension of a hyperbolic Julia set J varies analytically with f. In the interests of clarity, for the following discussion we shall restrict attention to the specific example of the quadratic map $f_c(z) = z^2 + c$, where c is in the main cardioid of the Mandelbrot set.

Theorem 4.9 (Ruelle). *Let c be in the main cardioid of the Mandelbrot set. Then the mapping $c \mapsto \dim_H(J_c)$ is real analytic. Moreover, if $|c|$ is sufficiently small, then there is an asymptotic expansion*

$$\dim_H(J_c) = 1 + \frac{|c|^2}{2\log 2} + O(|c|^3).$$

We briefly sketch the argument for the analyticity in what follows. Although not central to our survey, for completeness we also include a brief account of the asymptotic expansion in Appendix B.

To make use of Lemma 4.7, we first need that there exists a holomorphic function Φ_c which conjugates f_c to f_0, that is, $f_0 \circ \Phi_c = \Phi_c \circ f_c$ (see [36]), and if z has $|z| > 1$, then $\Psi_0(z) = z$. Moreover, this function is holomorphic in c for c in the main cardioid $\mathcal{C} = \{w(1-w) \colon |w| < \frac{1}{2}\}$ of the Mandelbrot set, as is $\Psi_c = \Phi_c^{-1}$. Moreover, for c inside this main cardioid:

1. For any z with $|z| > 1$, the map $c \mapsto \Psi_c(z)$ is holomorphic in c.
2. For z in the unit circle J_0, the map $z \mapsto \log 2|\Psi_c(z)|$ is Hölder continuous.
3. For any $c \in \mathcal{C}$, the map $z \mapsto \Psi_c(z)$ is injective on $\{z \colon |z| > 1\}$.

The function Ψ_c extends to a holomorphic map on $\{z \colon |z| \geq 1\}$, with $\Psi_c|_{J_0} \colon J_0 \to J_c$ and for z on the unit circle, $z \mapsto \log 2|\Psi_c(z)|$ is Hölder continuous. By Lemma 4.7,

$$P(-t\log(2|\Psi_c(z)|)|_{J_0}) = 0.$$

So $(c, t) \mapsto P(-t\log 2|\Psi_c|)$ is analytic in a neighbourhood of $(0, 1)$. We then have

$$\frac{\partial}{\partial t}(P(-t\log 2|\Psi_{c=0}(z)|))|_{t=0} = \int_{J_0} -\log 2|\Psi_{c=0}(z)|\,dz \neq 0.$$

Remark 4.10. Theorem 4.9 can be generalised to the context of hyperbolic Julia sets of rational maps (see [35]).

Remark 4.11. The real analyticity of Theorem 4.9 may fail outside the main cardioid of the Mandelbrot set. For example, when $c = \frac{1}{4}$, then $f_c(\frac{1}{2}) = \frac{1}{2}$ and $|f'(1/2)| = 1$, so J_c is no longer hyperbolic, and in fact $c \mapsto \dim_H(J_c)$ is not even continuous at $\frac{1}{4}$ (see [7]).

The same ideas apply to the Hausdorff dimension of limit sets of certain Fuchsian groups (see, for example, [3]).

Lemma 4.7 allows us to define the dimension $\dim_H(Y)$ implicitly in terms of the determinant (defined in terms of periodic points for $T: Y \to Y$, i.e., those contained in Y). In particular, setting $g_0 = 0$, $g = -\log|T'|$, and $z = 1$ in Definition 3.8, we have $s = \dim_H(X)$ satisfies $d_{0,g}(1,s) = 0$. Equivalently, the Hausdorff dimension of the limit set is a solution $s = \dim_H(X)$ to the absolutely convergent series

$$1 + \sum_{n=0}^{\infty} a_n(s) = 0.$$

As before, when studying the rate of mixing in the previous subsection, in practice we can use the intermediate value theorem to get effective bounds $\alpha_N < \rho < \beta_N$ by choosing α_N, β_N such that

$$1 + \sum_{n=1}^{N} a_n(\alpha_N) \geq \epsilon_1 \text{ and } 1 + \sum_{n=1}^{N} a_n(\beta_N) \leq -\epsilon_2$$

where

$$\epsilon_1 > \sum_{n=N+1}^{\infty} a_n(\alpha_1) \text{ and } \epsilon_2 > \sum_{n=N+1}^{\infty} a_n(\alpha_2).$$

Thus, as before, finding good bounds comes down to the following:

1. Getting good estimates on the coefficients $a_i(s)$ $(i = 1, \cdots, N)$; and
2. Finding effective bounds on the tail $\sum_{n=N+1}^{\infty} a_n(s)$.

As we mentioned before, the first is a basic problem in computing, and the second is a more challenging mathematical problem which we will address in Section 7.5.

We now turn to a class of deceptively simple examples.

Example 4.12 (Continued fractions and deleted digits). *Consider a finite set $F \subset \mathbb{N}$ and the set $E_F \subset [0,1]$ given by*

$$E_F = \{x = [x_1, x_2, \ldots] : x_1, x_2, \ldots \in F\},$$

namely, the Cantor set of points whose continued fraction expansion has all coefficients lying in F. This can be viewed as a repeller for the map $T: E_F \to E_F$ defined by $T(x) = \frac{1}{x}$ (mod 1).

(a) In the case $F = \{1,2\}$ the set E_F is usually denoted E_2 and is of historical interest, with its Hausdorff dimension studied by Good [14] as far back as 1941, after the even earlier work of Jarnik [20]. Using our algorithm, we were able to compute the dimension accurately, and rigorously, to over 100 decimal places (see [21]), improving on the previous

best rigorous estimate due to Falk and Nussbaum [10], using a subtler variant of Ulam's method.

(b) In the case $F = \{1, 2, 3, 4, 5\}$*, the dimension* $\dim_H(E_{\{1,2,3,4,5\}})$ *appears as a crucial ingredient in the work of Huang [19], refining the work of Bourgain and Kontorovich [1] on a density-one solution to the Zaremba Conjecture. Here we were able to use the algorithm to compute the dimension accurately, and rigorously, to eight decimal places [22].*

Remark 4.13. The Cantor sets just presented are also special cases of limit sets of one-dimensional iterated function schemes. More precisely, we can consider a (finite) set of C^2 contractions $\psi_i : I \to I$ $(i = 1, \cdots, N)$ on the unit interval I (with $\max_{1 \le i \le N} \sup_{x \in I} |\psi_i'(z)| < 1$). The associated limit set $X \subset I$ is the smallest non-empty closed set satisfying $X = \cup_{i=1}^{N} \psi_i(X)$. The Hausdorff dimension $\dim_H(X)$ can then be expressed in terms of the pressure function and the corresponding version of Lemma 4.7. In the particular case that $\psi_i(x) = \frac{1}{x+i}$, for $i \in F$, the limit set recovers the Cantor sets E_F.

Remark 4.14. There are natural analogues of continued fractions for which the contractions are $\psi_b(z) = \frac{1}{z+b}$, where $b \in \mathcal{B} \subset \{m + in : m \in \mathbb{N}, n \in \mathbb{Z}\}$ [29]. These map the domain $D = \{z \in \mathbb{C} : |z - \frac{1}{2}| \le \frac{1}{2}\}$ inside itself. We denote by $X_{\mathcal{B}}$ the associated limit set, namely the smallest non-empty closed subset of D such that $\cup_{b \in \mathcal{B}} \psi_b(X_{\mathcal{B}}) = X_{\mathcal{B}}$. The dimension of this set was studied in [35], and in [11] the bounds $1.85574 \le \dim_H(X_{\mathcal{B}}) \le 1.85590$ were established.

Remark 4.15. These examples also help to highlight some of the limitations of the periodic point method for estimating values. Most of the examples we have studied have been one-dimensional and involved uniformly expanding maps (or uniformly contracting iterated function schemes) with only finitely many branches. If we consider analytic maps in higher dimensions, then the method still applies, although it is less efficient as the dimension grows. However, if we consider the case of infinitely many branches, then we have the complication that there can be infinitely many periodic points of any given period, and this genuinely makes the method less applicable.

7.5 Rigorous Error Bounds

We now come to one of the most interesting and challenging aspects of the estimation problem: *finding rigorous upper bounds for the errors in the approximations.*

In the interests of clarity, and notational simplicity, we shall explain the ideas in the particular case of estimating the Hausdorff dimension of limit sets (corresponding to conformal iterated function schemes). The more general settings require variants of this basic approach.

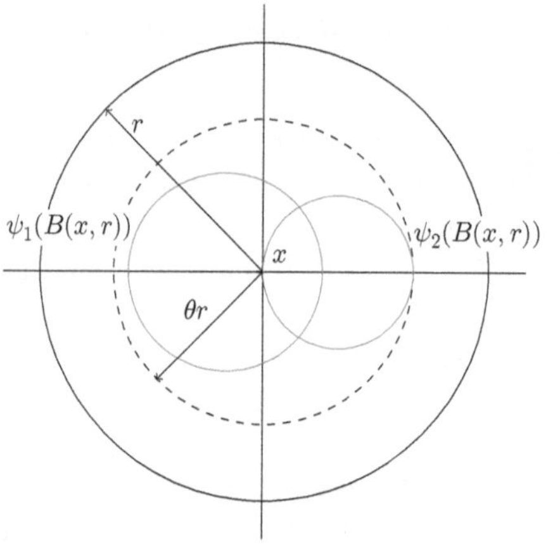

Figure 7.3 The choice of $0 < \theta < 1$ for two contractions $\psi_1, \psi_2 \colon B(c,r) \to B(c,r)$ with $B(c,\theta r) \supset \psi_1(B(c,r)) \cup \psi_2(B(c,r))$.

In particular, we need to provide an estimate on the error when we truncate the series which comes from bounds on the terms $|a_n(t)|$ for large values of n. Our bounds will involve a number of variables in whose choice we have some limited flexibility. In particular, we can select these so as to optimise the error terms.

7.5.1 The Contraction Ratio θ for Expanding Maps

Let us assume that we can associate a Markov partition $\mathcal{P} = \{P_1, \ldots, P_K\}$. In the particular case that $T \colon X \to X$ is an expanding map (see Figure 7.3) we can consider the following:

1. Charts and the complexification of the maps T to neighbourhoods $\mathbb{C}^d \supset U_i \supset P_i$ $(i = 1, \cdots, K)$; and
2. Associated contractions $\psi_{ij} \colon U_i \to U_j$ wherever $T(\mathrm{int}(U_j)) \supset \mathrm{int}(U_i)$.

In more fortunate situations we can assume that we have *Bernoulli* contractions where $U = U_i$ $(i = 1, \ldots, K)$ are identical and $\psi_{ij} = \psi_i$ $(i = 1, \ldots, K)$ are independent of j. (This applies in the case of Example 1.1 and the Lanford map in Example 4.4.)

Definition 5.1. *Choose* $0 < \Theta_i < 1$ *such that we can choose complex polydiscs* $P_i \subset B(c_i, r_i) \subset U_i$ *where*

$$B(c_i, r_i) = \{\underline{z} \in \mathbb{C} \colon |z_i - c_i| < r_i, \quad i = 1, \ldots, K\}$$

such that

$$\overline{\psi_{ij}(B(c_i, r_i))} \subset B(c_j, \Theta_j r_j).$$

Let $\theta := \max_i\{\Theta_i^{1/K}\}$. *In the particular case of Bernoulli contractions we can take* $\theta := \max_i\{\Theta_i\}$.

Remark 5.2. It may not always be possible to extend the contractions analytically to discs about elements of the partitions. However, in that case we can consider instead the more refined partition by elements $P_{i_0} \cap T^{-1}P_{i_1} \cap \cdots \cap T^{-n}P_{i_n}$, for suitable n.

Example 5.3. *For an expanding map of the interval then we can take* $d = 1$. *For a Bernoulli expanding map* T *we would like to take* $k = 1$ *(i.e., a single disc* $B(c,r)$ *and contractions* $\psi_i \colon B(c,r) \to B(c,r)$ *for* $i = 1,\ldots,k$*) arising from the inverse branches of* T*. The* P_i *will be a partition of the interval into subintervals and the discs* $P_i \subset U_i \subset \mathbb{C}$ *extend into the complex plane.*

First parameter choice: Choose a real number $0 < \theta < 1$.

There is no canonical choice of θ and one can try to arrange the partition and the polydiscs so as to minimise the possible choices. This can be achieved either by trial and error, or by simple calculus.

In the next subsection we begin to elaborate on the description of the error bounds mentioned in Subsection 7.1.4.

7.5.2 Bounds on the Determinant Coefficients

The key to obtaining validated estimates on characteristic values is to get accurate bounds on the coefficients a_n, especially for large $n \geq 1$.

Second parameter choice: Choose a natural number $N > 0$.

This should be chosen as large as is practicable. Typically this will depend on the time, computer memory, and computing power available to compute periodic points. For ℓ contractions, to compute a_n for $1 \leq n \leq N$ would require making estimates on up to ℓ^N periodic points. This involves quantities β_k, t_m, and B_k, which we define below.

i) Let $\epsilon_1 = \epsilon_1(N) > 0$ be the corresponding error bound, by which we mean that, for $1 \leq n \leq N$, the coefficient a_n can be computed to a guaranteed accuracy of no more than ϵ_1.

Third parameter choice: Choose a natural number $L > N$.

Typically this choice will depend on the time, computer memory, and computing power available to compute integrals. This will involve us in numerically integrating (to appropriate precision) approximately $L - N$ integrals.

ii) Let $\epsilon_2 = \epsilon_2(L) > 0$ be the bound on the tail

$$\sum_{k=L}^{\infty} \|\mathcal{L}q_k\|^2 \leq \frac{C\theta^L}{1-\theta} < \epsilon_2,$$

where

$$\|\mathcal{L}q_k\|^2 = r_i^{-2k} \int_0^1 |\sum_j \psi_j'(c_j + r_i e^{2\pi it})(c_i + r_i\psi_j(e^{2\pi it}))^k|^2 dt \leq C\theta^n$$

and where:
(a) $C = K^2 \max_j \|\psi_j'\|_\infty^s$; and
(b) $q_k(z) = (z - c_i)^k r_i^{-k}$, with $i = i(k) \pmod K$),
where K is the number of discs needed. (In the Bernoulli case we have the integral around the same curve.)

iii) Let $\epsilon_3 = \epsilon_3(L, N) > 0$ be the bound on rigorous computational estimates $(\beta_k)_{k=N}^L$ such that

$$\beta_k := \|\mathcal{L}q_k\|^2 = r_i^{-2k} \int_0^1 \left| \sum_j \psi_j'(c_j + r_i e^{2\pi it})(c_i + r_i\psi_j(e^{2\pi it}))^k \right|^2 dt \text{ for } N < k < L$$

up to an error of ϵ_3.

We can now use these choices to define a sequence (t_m) for the next set of bounds.

Definition 5.4. *We can define a sequence of positive real numbers*

$$t_m := \begin{cases} \left(\sum_{k=m+1}^L \beta_k + (L-m)\epsilon_3 + \epsilon_2\right)^{1/2} & \text{for } m \leq L \\ C\theta^m & \text{for } m > L \end{cases}$$

for $C > 0$ as earlier.

In particular, these numbers will tend to zero as $m \to \infty$. We can combine the values of t_n by introducing the following definition of B_k.

Definition 5.5. *For $1 \leq k \leq L$, define positive real numbers*

$$B_k := \sum_{m_1 < \cdots < m_k \leq L} t_{m_1} \cdots t_{m_k}.$$

Typically, the coefficients B_k will tend to zero quite quickly. Moreover, B_k is defined up to an error

$$\epsilon_4 := \epsilon_3 \left(\max\{t_i\}\right)^{L-1} L^k.$$

The quantities β_k, t_m, and B_k are used in giving the following bounds on the coefficients a_n for the determinant.

Theorem 5.6 (Coefficient bounds). *Let* $c = \frac{1}{\prod_{k=1}^{\infty}(1-\theta^j)}$. *Then:*

1. *For* $N < n \leq L$ *we can bound*

$$|a_n| \leq \gamma := c \sum_{k=1}^{n} (B_k + \epsilon_4)(\theta^L C)^{n-k}; \textit{and}$$

2. *For* $n > L$ *we can bound*

$$|a_n| \leq \xi(\theta^L C)^n \textit{ where } \xi := c\left(\sum_{k=1}^{L} (B_k + \epsilon_4)(\theta^L C)^{-k}\right).$$

In particular, we get effective bounds for $|a_n|$ when $n > N$ (combining these different bounds for $N < n \leq L$ and $n > L$).

Remark 5.7. In part 1 of Proposition 5.6 we have used the simply proven bound

$$\sum_{r_1,\cdots,r_{n-k}=L}^{\infty} C^{n-k}\theta^{r_1+\cdots+r_{n-k}} \leq c(\theta^L C)^{n-k}.$$

The proof of Theorem 5.6 follows the same lines as the arguments in [21, 22, 23]. For the convenience of the reader we give a brief account of the underlying operator theory ideas in the proof in Appendix A.

7.5.3 Applying the Bounds

There are two different ways that the bounds in Theorem 5.6 might be applied to estimate the accuracy of our approximations to the relevant characteristic values, depending on the quantity in question.

(a) Explicit values. Assume that we have an expression that can be written in terms of $d(z,t)$ and its derivatives $\frac{\partial^{i+j}}{\partial z^i \partial s^j} d(z,s)$ $(i,j \geq 0)$. The value of $d(z,t)$ can be approximated using the preceding estimates:

$$\left| d(z,t) - \left(1 + \sum_{n=1}^{N} a_n(t) z^n\right)\right| \leq N|z|^N \epsilon_1 + \sum_{n=N+1}^{M} \gamma_n |z|^n + \xi \frac{|z|(\theta^L C)^L}{1 - |z|(\theta^L C)}.$$

More generally, we can bound

$$\left| \frac{\partial^{i+j}}{\partial z^i \partial s^j} d(z,s) - \sum_{n=1}^{N} z^n \frac{n!}{(n-i)!} \frac{\partial^i}{\partial s^i} a_n(s) \right|$$

where we bound $\frac{\partial^i}{\partial s^i} d(z,s)$ using Cauchy's theorem.

These estimates can be applied, for example, to computing the error terms in the estimates on Lyapunov exponents, variance, and linear response as in Subsection 7.3.3.

(b) Implicit values. If we are seeking a zero of $d(z, s)$, then provided we can choose $z_1 < z_2$ close to each other and with validated bounds

$$d(z_1, s) < 0 < d(z_2, s) \text{ or } d(z_2, s) < 0 < d(z_1, s)$$

using $z = z_1$ or $z = z_2$, then there must be a zero in the interval $[z_0, z_1]$.

This approach can be used to estimate errors in computing the rates of mixing and Hausdorff dimension as described in Section 7.4. We will illustrate this in a specific instance in the next section.

7.6 A Worked Example: Hausdorff Dimension of the Set E_2

To illustrate how Theorem 5.6 can be applied in practice, we want to consider a particular concrete problem. In particular, we will describe its use in estimating the Hausdorff dimension of a specific Cantor set (see [21]).

Recall from Example 4.12 that E_2 is the subset of [0, 1] consisting of those reals whose continued fraction expansion contains only the numbers 1 and 2. In other words, if

$$T_1(x) := \frac{1}{1+x} \text{ and } T_2(x) := \frac{1}{2+x},$$

then E_2 is the corresponding limit set (i.e., the smallest non-empty closed set X such that $T_1(X) \cup T_2(X) = X$). Let us consider the estimation of the Hausdorff dimension of E_2, referring to [21] for full details.

Defining $\underline{i} = (i_1, \cdots, i_n) \in \{1, 2\}^n$ and $|\underline{i}| = n$, and letting $x_{\underline{i}} = T_{\underline{i}}(x_{\underline{i}})$ be the fixed point for

$$T_{\underline{i}} = T_{i_1} \circ \cdots \circ T_{i_n} : [0, 1] \to [0, 1],$$

we have the determinant

$$d(z, t) := \exp\left(-\sum_{n=1}^{\infty} \frac{z^n}{n} \sum_{|\underline{i}|=n} \frac{|(T_{\underline{i}})'(x_{\underline{i}})|^t}{1 - (T_{\underline{i}})'(x_{\underline{i}})}\right),$$

and $\dim_H(E_2)$ is the value t such that $d(1, t) = 0$. This value is then approximated using the following four steps:

1. For each t, approximate $z \mapsto d(z,t)$ by a polynomial $z \mapsto d_N(z,t)$;
2. Set $z = 1$ and consider $t \mapsto d_N(z,1)$;
3. Solve for $t_N = t$: $d_N(1,t) = 0$;
4. Then $t_N \to \dim_H(E_2)$ as $N \to +\infty$.

In particular, for each $N \in \mathbb{N}$ we obtain an approximation t_N to $\dim_H(E_2)$. The sequence t_N gives an intuitive estimate on the quality of the approximation in terms of the difference $|t_N - t_{N-1}|$.

We can write the series expansion

$$d(z,t) = 1 + \sum_{n=1}^{\infty} a_n(t)z^n = 1 + \underbrace{\sum_{n=1}^{N} a_n(t)z^n}_{=:d_N(z,t)} + \underbrace{\sum_{n=N+1}^{\infty} a_n(t)z^n}_{=:\epsilon_N(z,t)}$$

for some $N \geq 1$, and take for the approximating polynomial

$$d_N(z,t) = 1 + \sum_{n=1}^{N} a_n(t)z^n,$$

where N is chosen to be sufficiently large that (with $z = 1$ and $0 \leq t \leq 1$) the error ϵ_N is small, but sufficiently small that the terms $a_n(t)$, $n = 1, 2, \cdots, N$ can be calculated in a reasonable time. In the present setting one might choose $N = 25$, say.

We can explain part of the mechanism for effective estimates on the coefficients a_n $(n \geq 1)$ as follows.

Step 1. Choose $z_0 \in \mathbb{R}$ and $r > 0$ such that

$$D = \{z \in \mathbb{C}\colon |z - z_0| < r\} \supset [0,1] \text{ and } T_1 D, T_2 D \subset D.$$

For example, we could let $z_0 = 1$ and $r = \frac{3}{2}$. Consider the family of transfer operators defined on analytic functions $w\colon D \to \mathbb{C}$ by

$$\mathcal{L}_t w(z) = \frac{1}{(z+1)^{2t}} w\left(\frac{1}{z+1}\right) + \frac{1}{(z+2)^{2t}} w\left(\frac{1}{z+2}\right), \quad t \in \mathbb{R}.$$

Step 2. Let $q_k(z) = \frac{(z-z_0)^k}{r^k}$, for $z \in D$ and $k \geq 1$. We can then define

$$t_m = \left(\sum_{k=m-1}^{\infty} \|\mathcal{L}(q_k)\|^2\right)^{\frac{1}{2}} \ (m \geq 1) \text{ where } \|q_k\|^2 = \int_0^1 |q_k(z_0 + re^{2\pi it})|^2 dt \quad (k \geq 1).$$

Step 3. We can bound the coefficients a_n $(n > N)$ by

$$|a_n| \leq \sum_{m_1 < \cdots < m_n} t_{m_1} t_{m_2} \cdots t_{m_n}.$$

We will give more details of the underlying operator theory in Appendix A.

Given $M > 0$ (in the present setting one can choose $M = 600$), we can numerically estimate $\|\mathcal{L}(q_k)\|$ for $k \leq M$, and we can trivially bound $\|\mathcal{L}(q_k)\|$ for $k > M$. Combining these various bounds gives the estimate.

7.7 Future Directions

In this note we have discussed how the periodic point algorithm works and how the error terms can be efficiently estimated. However, there may be further scope to fine-tune the underlying analysis and estimates, perhaps by using different Hilbert spaces or other operators. This leads to the first very general question.

Objective 7.1. *Can we improve the approach to computing the dynamical determinant and, in particular, the error estimates?*

We have illustrated the general approach with a number of applications and examples. However, we now want to propose some further potential applications.

7.7.1 Dynamical Invariants and Lyapunov Exponents

We have already discussed the theoretical use of our method to rigorously estimate certain dynamical quantities. However, this has been done practically in only a small number of cases (e.g., Lyapunov exponents, variance). It remains a task to be completed to explore how to estimate rigorously other quantities (e.g., resonances, linear response) for simple test examples (such as the Lanford map).

Objective 7.2. *Apply this approach to compute more dynamical invariants in simple examples.*

Lyapunov exponents also occur naturally in the theory of random matrix products, which has been an active area of research since the pioneering work of Kesten and Furstenberg in the 1960s. For example, given $k \times k$ square matrices $A_1, \ldots, A_k$ *with positive entries*, we define the *Lyapunov exponent* by

$$\lambda = \lim_{n \to +\infty} \frac{1}{d^n} \sum_{i_1, \cdots, i_n \in \{1, \cdots, d\}} \frac{\log \|A_{i_1} \cdots A_{i_n}\|}{n}.$$

There was an implementation of the basic algorithm in [31].

Objective 7.3. *Can the error bounds in the computation of λ be made more effective?*

An interesting application is to the case of binary symmetric channels in information theory. In this context there are positive matrices, and λ is related to a useful value called the entropy rate.

7.7.2 Connections to Number Theory

We have already mentioned the application to the density-one Zaremba conjecture (see [22]). However, we now want to describe a different application to number theory.

Given an irrational number α, we can associate the number

$$\mu(\alpha) = \limsup_{p,q\to+\infty} |q^2| \left|\alpha - \frac{p}{q}\right|$$

(i.e., the best constant in Diophantine approximation for α). The *Lagrange spectrum* is defined to be the set $\mathcal{L} = \{1/\mu(\alpha) : \alpha \in \mathbb{R} - \mathbb{Q}\}$. On the other hand, we can consider those binary quadratic forms $f(x,y) = ax^2 + bxy + cy^2$ $(a, b, c \in \mathbb{R})$ with discriminant $D(f) = b^2 - 4ac > 0$ and denote

$$\lambda(f) = \inf\left\{|f(x,y)| : (x,y \in \mathbb{Z}^2 - \{(0,0)\})\right\} / \sqrt{D(f)}.$$

The *Markov spectrum* is defined to be the set $\mathcal{M} = \{1/\lambda(f) : \alpha \in \mathbb{R} - \mathbb{Q}\}$. It is known that $\mathcal{L} \subset \mathcal{M}$ and Matheus and Moreira showed that $0.513\ldots < \dim_H(\mathcal{M} - \mathcal{L}) < 0.98\ldots$ [28]. Moreover, in their article they conjecture that the bounds can be improved to $\dim_H(\mathcal{M} - \mathcal{L}) < 0.88$, based on empirical estimates using our algorithm.

Objective 7.4. *Obtain improved rigorous bounds on $\dim_H(\mathcal{M} - \mathcal{L})$.*

This involves rigorously computing the Hausdorff dimension of limit sets associated to iterated function systems $\{\phi_i : I \to I\}$, but with a Markov condition, namely, there is a 0-1 matrix A and compositions $\phi_i \circ \phi_j$ are only allowed if $A(i,j) = 1$. Whereas the basic algorithm still applies in this setting, the major complication is to get effective error estimates.

7.7.3 Spectral Geometry and the Selberg Zeta Function

Given a compact surface V, with a metric ρ of constant negative curvature, we can associate the Selberg zeta function defined by

$$Z_\rho(s) = \prod_{n=0}^{\infty} \prod_{\gamma} \left(1 - e^{-s(n+s)l(\gamma)}\right), \quad s \in \mathbb{C},$$

where γ denotes a closed geodesic of length $l(\gamma)$. To relate this to our analysis, we recall that we can associate a piecewise C^ω expanding map of the circle (using the Bowen–Series approach) and then the zeta function can be written in terms of the determinants $\det(I - \mathcal{L}_s)$ of the associated transfer operators $\mathcal{L}_s$ (see, e.g., [30, 33]).

The zeros of $Z_\rho(s)$ have a spectral interpretation, in terms of eigenvalues of the Laplacian on (V, ρ), but these can be well estimated using other techniques. Other special values such as $Z_\rho'(0)$ can be written in terms $\frac{d}{ds}\det(I - \mathcal{L}_s)|_{s=0}$, and this is proportional to the much studied *determinant of the Laplacian*, originally defined in terms of the spectrum of the Laplacian (see, e.g., [6, 13]).

Objective 7.5. *Can we get useful bounds on* $Z_\rho'(0)$*?*

There is a well-known problem of Sarnak to show that there is a (local) minimum for the determinant that occurs at very symmetric hypergeometric surfaces which could be addressed using this approach.

In a different direction, the Weil–Petersson metric is a classical distance on the space of such Riemann metrics ρ. There is a particularly useful thermodynamic interpretation for the Weil–Petersson metric due to McMullen in terms of the second derivative of the associated thermodynamic pressure in the preceding context of the piecewise C^ω expanding maps.

Objective 7.6. *Can one get effective estimates on the Weil–Petersson metric?*

This could then be used to explore empirically the Weil–Petersson metric on the space of metrics. Moreover, there are higher-dimensional analogues of Weil–Petersson metrics in [4] which could be similarly analysed.

Appendices

Appendix A Operator Theory and Coefficient Bounds

We now give a little more detail on how the bounds on the coefficients a_n arise, using operator theory.

A.1 Operator Theory

Assume for notational convenience that $T: X \to X$ is a Bernoulli expanding map of the unit interval X. Assume that we can choose an open disc $X \subset U \subset \mathbb{C}$ so that T extends analytically to U. In the particular case that $D = \{z \in \mathbb{C}: |z - z_0| < r\}$ we can let $\mathcal{H}$ be the Hardy Hilbert space of analytic functions on U for which the norm $\|\cdot\|$ is given by

$$\|f\|^2 = \sup_{\rho<r} \int_0^1 |f(z_0 + \rho e^{2\pi it})|^2 dt < \infty, \quad f \in \mathcal{H}.$$

Example A.1.1 If $D = \left\{z \in \mathbb{C}: |z - \frac{1}{2}| < 1\right\}$, then $f \in \mathcal{H}$ presented in the form

$$f(z) = \sum_{n=0}^{\infty} b_n \left(z - \frac{1}{2}\right)^n$$

has norm $\|f\|^2 = \sum_{n=0}^{\infty} |b_n|^2$.

Let $\mathcal{L} = \mathcal{L}_{G,T}: \mathcal{H} \to \mathcal{H}$ be a transfer operator. We can then assume that $\mathcal{L}$ is a trace class operator and a simple Lidskii-type identity relates periodic orbits to the spectrum

$$\text{trace}(\mathcal{L}^n) = \sum_{T^n x = x} \frac{\exp\left(\sum_{i=0}^{n-1} G(T^i x)\right)}{1 - 1/(T^n)'(x)}$$

and then

$$D(z) = \det(I - z\mathcal{L}) = \exp\left(-\sum_{n=1}^{\infty} \frac{z^n}{n} \text{trace}(\mathcal{L}^n)\right),$$

where both sides depend on G and T [32]. The basic idea is to use the operator description of $D(z)$ to get estimates on the a_n via the approximation numbers s_l ($l \geq 1$) defined by

$$s_l := \inf\{\|\mathcal{L} - K\| : K \text{ bounded linear operator of finite rank } l\}, \quad l \geq 1.$$

For definiteness, let $(e_n)_{n=0}^{\infty}$ be a complete orthonormal family for $\mathcal{H}$. For any

$$f = \sum_{n=0}^{\infty} c_n e_n \in \mathcal{H} \text{ with } \|f\| = \sqrt{\sum_{n=0}^{\infty} |c_n|^2} = 1$$

and each $l \geq 1$ we can define $K_l f = \sum_{n=0}^{l-1} c_n \mathcal{L}(e_n) \in \mathcal{H}$. In particular, we can bound

$$\begin{aligned} s_l &\leq \|(\mathcal{L} - K_l)f\| \leq \sum_{n=l}^{\infty} |c_n| \|\mathcal{L}(e_n)\| \\ &\leq \sqrt{\sum_{l=n}^{\infty} |c_n|^2} \sqrt{\sum_{n=l}^{\infty} \|\mathcal{L}(e_n)\|^2} \leq \sqrt{\sum_{n=l}^{\infty} \|\mathcal{L}(e_n)\|^2}, \end{aligned} \tag{A.1}$$

using the Cauchy–Schwarz inequality.

Example A.1.2 (Example A.1.1 revisited). If $D = \{z \in \mathbb{C} \colon |z - \frac{1}{2}| < 1\}$, then we can take $e_n(z) = (z - \frac{1}{2})^n / \sqrt{n}$, for $n \geq 0$.

We can now bound the coefficients a_n $(n > N)$ by

$$|a_n| \leq \sum_{m_1 < \cdots < m_n} s_{m_1} s_{m_2} \cdots s_{m_n}, \tag{A.2}$$

where the summation is over distinct natural numbers $m_1, \ldots, m_n \in \mathbb{N}$ such that $m_1 < \cdots < m_n$, using [13, Corollary VI.2.6].

Given $M > 0$ (in the present setting one can choose $M = 600$), we can numerically estimate $\|\mathcal{L}(e_k)\|$ for $k \leq M$, and we can trivially bound $\|\mathcal{L}(e_k)\|$ for $k > M$.

A.2 Basic Convergence

The basic convergence (which follows from work of Ruelle [33], after Grothendieck) only requires the following bound (referred to in [21, 22, 23] as an *Euler bound*, after [8]):

If there exists $C > 0$ and $0 < r < 1$ with $s_l \leq Cr^l$ for $l \geq 1$, then

$$|a_n| \leq \frac{C^n r^{n(n+1)/2}}{(1-r)(1-r^2)\cdots(1-r^n)} = O(r^{n(n+1)/2}) \tag{A.3}$$

for $n \geq 1$.

In particular, the coefficients a_n tend to zero fast enough to ensure absolute convergence of the series describing the various quantities of interest (cf. Part 1 of Corollary 3.9 and Corollary 3.11).

Remark A.2.1. We can now see the technical limitations of this method. In (A.3) we have a bound $|a_n| = O(r^{n^2})$. However, if we look at d-dimensional analytic maps, then we have a slightly weaker bound $|a_n| = O(r^{n^{1+\frac{1}{d}}})$, which will make the approach marginally less efficient. Moreover, if there are infinitely many branches, then typically we lose the advantages of this method because the terms a_n involve infinite summations which need to be approximated individually.

A.3 Better Estimates

Assume that we can compute numerically $a_1, \ldots, a_N$ for some N. We could then use the preceding bound (A3) to bound the terms a_n for $n > N$ (although this was not the original purpose of this approach nor is it necessarily a very efficient approach). The basis of the improved approach involves the following:

1. Again computing $a_1, a_2, \ldots, a_N$ using the periodic points of period at most N (where N is chosen depending on the limitations of our computer).
2. Bounding $|a_n|$, for $n \geq N$ using (A.2) and different bounds on the s_l:
   (a) For $N < l \leq M$, we can bound s_l using (A1) (where M is chosen depending on the limitations of our computer); and
   (b) For $M < l$, we can use the bound $s_l \leq Cr^l$.

There are further refinements possible, but this explains the gist of the approach (see [21]).

A.4 Generalisations

Finally, we can describe a plan for how this method can be extended to Anosov diffeomorphisms and flows.

If we have a C^ω Anosov diffeomorphism T, then it is more appropriate to write

$$D(z) = D_{G,T}(z) = \exp\left(-\sum_{n=1}^{\infty} \frac{z^n}{n} \sum_{T^n x = x} \frac{\exp\left(\sum_{i=0}^{n-1} G(T^i x)\right)}{\det(I - (DT)^{-1}(x))} \right), \qquad z \in \mathbb{C}.$$

An extra feature here is that we only need to take $G_t(x) = tg(x)$ because of the contributions from $\det(I - (DT)^{-1}(x))$.

One can adapt this approach using either the less fashionable device of Markov partitions and the approach of Rugh, or using anisotropic spaces of functions when applicable (for example, on the torus in the context of Faure–Roy). The key point is simply to find a setting to which the functional analysis applies.

For C^ω Anosov flows it is perhaps simpler to use Markov Poincaré sections to accommodate the operator theory being used.

Appendix B Asymptotic Expansion for the Dimension of Quadratic Julia Sets

As mentioned in Section 7.4.2, we detail here a derivation of the asymptotic expansion $\dim_H(J_c) = 1+\frac{|c|^2}{2\log 2}+O(|c|^3)$ from Theorem 4.9. We will follow the approach of Zinsmeister in [36], taking the opportunity to correct a few minor errors.[1] We can write $c = x + iy$ and define

$$F(d,x,y) := P\left(-d(c)\log(2|\Psi_{x+iy}(z)|)|_{J_0}\right).$$

We can then solve for $d(x,y) = \dim_H(J_c)$ satisfying $F(d(c),x,y) = 0$. Expanding $d(c)$ as a Taylor series around $c = 0$, we obtain

$$d(x,y) = 1 + \alpha x + \beta y + \gamma x^2 + \delta y^2 + \epsilon xy + o(x^2 + y^2).$$

For c close to 0, we have that $\dim_H(J_c) \geq \dim(J_0) = 1$ and it follows that α and β must be zero. By expanding F as a series around the point $(1,0,0)$, we get

$$\begin{aligned} 0 = F(d,x,y) &= xF_x + yF_y + x^2\left(\delta F_d + \frac{1}{2}F_{xx}\right) + y^2\left(\gamma F_d + \frac{1}{2}F_{yy}\right) \\ &\quad + xy\left(\epsilon F_d + F_{xy}\right) + o(x^2 + y^2), \end{aligned}$$

where F_x, F_y, F_{xx}, F_{yy}, and F_{xy} are the partial derivatives as $x = d$ and $y = 0$, and thus

$$F_x = F_y = 0, \ \gamma = -\frac{F_{xx}}{2F_d}, \ \delta = -\frac{F_{yy}}{2F_d}, \ \epsilon = -\frac{F_{xy}}{F_d}.$$

To begin, since $F(d,0,0) = P(-d\log 2) = \log 2 - d\log 2$ we have $F_d = -\log 2$. Next, we can expand Ψ_c in terms of c:

$$\Psi_c(z) = z + a_1(z)c + a_2(z)c^2 + o(c^2). \qquad (B.1)$$

Since Ψ_c satisfies $\Psi_c \circ f_0 = f_c \circ \Psi_c$, we can write $\Psi_c(z^2) = (\Psi_c(z))^2 + c$. Substituting the expansion (B.1) and comparing the coefficients for c and c^2, we obtain[2]

$$a_1(z^2) = 2za_1(z) + 1 \ \text{ and } \ (a_1(z))^2 + 2za_2(z) = a_2(z^2). \qquad (B.2)$$

The first equation in (B.2) has a series solution:[3]

$$a_1(z) = -z\sum_{n=0}^{\infty} \frac{1}{2^{n+1}} z^{-2^{n+1}}.$$

[1] We are grateful to Michel Zinsmeister for useful discussions on these points.

[2] This is the first of the typographical inaccuracies in [36], replacing $(a_1(z))^2$ with $a_1(z^2)$.

[3] This is the second of the minor changes to [36], replacing z^{-2^n} with $z^{-2^{n+1}}$.

We next write $a_2(z) = \sum_{n=-\infty}^{\infty} d_n z^n$ and compare coefficients in the second equation in (B.2). In particular, we obtain $d_n = 0$ for all $n \geq 0$, and $d_{-1} = d_{-2} = -1/8$. For $|z| = 1$ we have

$$|\Psi_c(z)| = |1 + b(z)c + \bar{z}a_2(z)c^2 + \dots|,$$

where we write $b(z) := \bar{z}a_1(z)$ to make notation easier. Writing $|\Psi_c(z)|^2 = (\mathrm{Re}(\Psi_c(z)))^2 + \mathrm{Im}(\Psi_c(z))^2$, we have that

$$\begin{aligned}
\log|\Psi_c(z)| &= \log\left((1 + \mathrm{Re}(b(z)c) + \mathrm{Re}(\bar{z}a_2(z))^2 + (\mathrm{Im}(b(z)c)))^2 + o(|c|^2)\right)^{1/2} \\
&= \frac{1}{2}\log\left(1 + 2\mathrm{Re}(b(z)c) + |b(z)c|^2 + 2\mathrm{Re}(\bar{z}a_2(z)c^2) + o(|c|^2)\right).
\end{aligned}$$

Using the series $\log(1 + w) = w - w^2/2 + w^3/3 + \dots$ yields

$$\log|\Psi_c(z)| = \mathrm{Re}(b(z)c) + \frac{1}{2}|b(z)c|^2 + \mathrm{Re}(\bar{z}a_2c^2) - (\mathrm{Re}(b(z)c))^2 + o(|c|^2)$$

and then writing $d = 1 + h$:

$$\begin{aligned}
F(1 + h, x, y) &= P(-(1 + h)\log 2 - (1 + h)\log|\Psi_c|) \\
&= P\Big(- \log 2 - \mathrm{Re}(b(z)c) - \frac{1}{2}|b(z)c|^2 - \mathrm{Re}(\bar{z}a_2c^2) \\
&\qquad + \mathrm{Re}(b(z)c^2) - h\log 2 - \mathrm{Re}(b(z)c) + \dots\Big) \\
&= 0 + P'_g(-\log 2) + \frac{1}{2}P''_{g,g}(-\log 2) + o(|c|^2 + h^2),
\end{aligned}$$

where $g = -\mathrm{Re}(b(z)c) - \frac{1}{2}|b(z)c|^2 - \mathrm{Re}(\bar{z}a_2c^2) + (\mathrm{Re}(b(z)c))^2 - h\log 2 - h\mathrm{Re}(b(z)c)$. We now compute $\int_{J_0} g\, d\lambda$, where λ is the normalised Haar measure on J_0. We begin by calculating the first term

$$\int_{J_0} \mathrm{Re}(bc)\, d\lambda = \frac{x}{2\pi}\int_0^{2\pi}\sum_{n=0}^{\infty}\frac{\cos(2^{n+1}2\pi t)}{2^{n+1}}\, dt - \frac{y}{2\pi}\int_0^{2\pi}\sum_{n=0}^{\infty}\frac{\sin(2^{n+1}2\pi t)}{2^{n+1}}\, dt = 0.$$

Similarly we can show that:

$$\int_{J_0} \mathrm{Re}(\bar{z}a_2(z)c)\, d\lambda = 0.$$

We briefly describe how to compute the remaining values. We have:[4]:

$$\int_{J_0} (\mathrm{Re}(b(z)c))^2\, d\lambda = \int_{J_0} \mathrm{Re}\left((x + iy)\sum_{n=0}^{\infty}\frac{1}{2^{n+1}}\left(z^{-2^{n+1}}\right)\right)^2 d\lambda$$

[4] This is another slight deviation from [36], where the integral is computed to be $2(x^2 + y^2)/3$ rather than $(x^2 + y^2)/6$.

$$= \frac{x^2}{2\pi}\int_0^{2\pi}\left(\sum_{n=0}^{\infty}\frac{\cos(2^{n+1}2\pi t)}{2^{2(n+1)}}\right)^2 dt$$
$$+\frac{y^2}{2\pi}\int_0^{2\pi}\left(\sum_{n=0}^{\infty}\frac{\sin(2^{n+1}2\pi t)}{2^{2(n+1)}}\right)^2 dt$$
$$= (x^2+y^2)\sum_{n=0}^{\infty}\frac{1}{2^{2n+3}} = \frac{x^2+y^2}{6}.$$

We also compute

$$\int_{J_0}|b(z)|^2\,d\lambda = \int_{J_0}\left(\sum_{n=0}^{\infty}\left(\frac{1}{2^{n+1}}z^{-2^{n+1}}\right)\sum_{n=0}^{\infty}\left(\frac{1}{2^{n+1}}\overline{z}^{-2^{n+1}}\right)\right)d\lambda$$
$$= \int\sum_{n=1}^{\infty}\frac{1}{4^n}\,d\lambda = \frac{1}{3}.$$

Therefore, $\frac{1}{2}\int|b(z)c|^2\,d\lambda = |c|^2/6$ and we conclude: $\int_{J_0} g\,d\lambda = -h\log 2$. Writing $b(z) = b_1 + ib_2$ gives

$$F(1+h,x,y) = -h\log 2 + x^2\left(\lim_{m\to\infty}\frac{1}{N}\int_{J_0}(b_1^N)^2\,d\lambda\right) + y^2\left(\lim_{N\to\infty}\frac{1}{N}\int_{J_0}(b_2^N)^2\,d\lambda\right) + \ldots,$$

where $b_i^N(z) = \sum_{k=0}^{N-1} b_i\circ f^k(z)$. Calculations similar to that for $\int(\mathrm{Re}(b(z)c))^2 d\lambda$ give that

$$\int_{J_0}(b_1^N(z))^2\,d\lambda = \frac{1}{2\pi}\int_0^{2\pi}\left(\sum_{k=0}^{N-1}\sum_{n=0}^{\infty}\frac{1}{2^{n+1}}\cos\left(2\pi t(2^k-2^{n+1})\right)\right)^2 dt = \sum_{k=0}^{N-1}\frac{1}{4} = \frac{N}{4}$$

and similarly $\int_{J_0}(b_2^N(z))^2\,d\lambda = N/2$. Therefore, since $F_{xx} = F_{yy} = \frac{1}{2}$ and $F_{xy} = 0$, we can write

$$F(1+h,x,y) = -h\log 2 + \frac{1}{4}x^2 + \frac{1}{4}y^2 + o(x^2+y^2+h^2).$$

So we obtain that $\gamma = \delta = \frac{1}{4\log 2}$ and $\epsilon = 0$. Therefore:

$$d(c) = 1 + \frac{|c|^2}{4\log 2} + o(|c|^2),$$

as claimed.

References

[1] J. Bourgain and A. Kontorovich, On Zaremba's conjecture, *Ann. Math.*, **180** (2014), 137–96.

[2] R. Bowen, *Equilibrium States and the Ergodic Theory of Anosov Diffeomorphisms*, Springer Lecture Notes in Mathematics **470** Berlin, Springer (1975).
[3] R. Bowen, Hausdorff dimension of quasi-circles, *Publ. Math. (IHES)* **50** (1979) 11–25.
[4] M. Bridgeman, R. D. Canary, F. Labourie, and A. Sambarino, The pressure metric for Anosov representations, *Geom. Funct. Anal.*, **25** (2015), 1089–179.
[5] P. Cvitanovic, R. Artuso, R. Mainieri, G. Tanner, and G. Vattay, *Chaos: Quantum and Classical*, chaosbook.org/version12/chapters/ChaosBook.pdf.
[6] E. D'Hoker and D. Phong, On determinants of Laplacians on Riemann surfaces, *Comm. Math. Phys.*, **104** (1986), 537–45.
[7] A. Douady, P. Sentenac, and M. Zinsmeister, Implosion parabolique et dimension de Hausdorff, *C. R. Acad. Sci. Paris Sér. I Math.*, **325** (1997), 765–72.
[8] L. Euler, *Introductio in Analysin Infinitorum*, Marcum-Michaelem Bousquet, Lausannae, 1748.
[9] K. Falconer *Fractal Geometry: Mathematical Foundations and Applications*, Wiley, 1990.
[10] R. S. Falk and R. D. Nussbaum, C^m eigenfunctions of Perron–Frobenius operators and a new approach to numerical computation of Hausdorff dimension: applications in $\mathbb{R}^1$, *J. Frac. Geom.*, **5** (2018), 279–337.
[11] R. S. Falk and R. D. Nussbaum, A new approach to numerical computation of Hausdorff dimension of iterated function systems: applications to complex continued fractions, *Integral Equ. Oper. Theory*, 90 (2018).
[12] I. Fredholm, Sur une classe d'équations fonctionnelles, *Acta Math.*, **27** (1903), 365–90.
[13] I. Gohberg, S. Goldberg, and M. A. Kaashoek, *Classes of Linear Operators Vol. 1*, 1990, Birkhäuser, Berlin.
[14] I. J. Good, The fractional dimensional theory of continued fractions, *Proc. Camb. Phil. Soc.*, **37** (1941), 199–228.
[15] A. Grothendieck, Produits tensoriels topologiques et espaces nucléaires, *Mem. Amer. Math. Soc.*, **16**, 1955.
[16] A. Grothendieck, La théorie de Fredholm, *Bull. Soc. Math. France*, **84** (1956), 319–84.
[17] T. Holliday, P. Glynn, and A. Goldsmith, Shannon meets Lyapunov: connections between information theory and dynamical systems, Proceedings of the 44th IEEE Conference on Decision and Control, and European Control Conference, pp. 1756–1763, Elsevier, 2005.
[18] D. A. Hejhal, Eigenvalues of the Laplacian for$PSL(2,\mathbb{Z})$: some new results and computational techniques, in S. Gong, Q. K. Lu, Y. Wang, and L. Yang, (eds.), International Symposium in Memory of Hua Loo Keng, Springer, Berlin (1991), 59–102
[19] S. Huang, An improvement to Zaremba's conjecture, *Geom. Funct. Anal.*, **25** (2015), 860–914.
[20] I. Jarnik, Zur metrischen Theorie der diophantischen Approximationen, *Prace Mat.-Fiz.*, **36** (1928), 91–106.
[21] O. Jenkinson and M. Pollicott, Rigorous effective bounds on the Hausdorff dimension of continued fraction Cantor sets: a hundred decimal digits for the dimension of E_2, *Adv. Math.*, **325** (2018), 87–115.
[22] O. Jenkinson and M. Pollicott, Rigorous dimension estimates for Cantor sets arising in Zaremba theory, *Contemp. Math.*, **744** (2020), 83–108.

[23] O. Jenkinson, M. Pollicott, and P. Vytnova, Rigorous computation of diffusion coefficients for expanding maps, *J. Stat. Phys.*, **170** (2018), 221–53.
[24] A. Katok and B. Hasselblatt, Introduction to the modern theory of dynamical systems, Cambridge University Press, Cambridge, 1995.
[25] O. E. Lanford III, Informal remarks on the orbit structure of discrete approximations to chaotic maps, *Exp. Math.*, **7** (1998), 317–24.
[26] C. McMullen, Hausdorff dimension and conformal dynamics. III. Computation of dimension *Amer. J. Math.*, **120** (1998), 691–721.
[27] C. Matheus and C. G. Moreira, $HD(M \setminus L) > 0.353$, *arXiv:1703.04302*.
[28] C. Matheus and C. G. Moreira, $HD(M \setminus L) < 0.986927$, *arXiv:1708.06258*.
[29] R. Daniel Mauldin and Mariusz Urbański, Dimensions and measures in infinite iterated function systems, Proc. London Math. Soc. (3) 73 (1996), 105–54.
[30] D. H. Mayer, On the thermodynamic formalism for the Gauss map, *Comm. Math. Phys.*, **130** (1990), 311–33.
[31] M. Pollicott, Maximal Lyapunov exponents for random matrix products, *Invent. Math.*, **181** (2010), 209–26.
[32] D. Robert, Sur les traces d'opérateurs (de Grothendieck à Lidskii), *Gaz. Math.*, **141** (2014), 76–91.
[33] D. Ruelle, Zeta-functions for expanding maps and Anosov flows, *Invent. Math.*, **34** (1976), 231–42
[34] D. Ruelle, *Thermodynamic Formalism*, Reading, MA, Addison-Wesley, 1978.
[35] D. Ruelle, Repellers for real analytic maps, *Ergod. Th. Dynam. Sys.*, **2** (1982), 99–107.
[36] M. Zinsmeister, *Thermodynamic Formalism and Holomorphic Dynamical Systems*, Translated from the 1996 French original by C. Greg Anderson. SMF/AMS Texts and Monographs, 2. American Mathematical Society, Providence, RI; Société Mathématique de France, Paris, 2000.

8

Non-stationary Normal Forms for Contracting Extensions

BORIS KALININ*

Dedicated to the memory of Anatole Katok.

We present the theory of non-stationary normal forms for uniformly contracting smooth extensions with sufficiently narrow Mather spectrum. We give coherent proofs of existence, (non)uniqueness, and a description of the centralizer results. As a corollary, we obtain corresponding results for normal forms along an invariant contracting foliation. The main improvements over the previous results in the narrow spectrum setting include explicit description of non-uniqueness and obtaining results in any regularity above the precise critical level, which is especially useful for the centralizer. In addition to sub-resonance normal form, we also prove corresponding results for resonance normal form, which is new in the narrow spectrum setting.

8.1 Introduction

The theory of normal forms for smooth maps goes back to Poincare and Sternberg [St57] and plays an important role in dynamics. It has been extensively studied in the classical setting of normal forms at fixed points and invariant manifolds [BrKo]. The theory of non-stationary normal forms was developed more recently in the context of extensions and invariant foliations. The primary motivations and applications were various rigidity results for systems and actions exhibiting some form of hyperbolicity.

In the setting of an invariant contracting foliation $\mathcal{W}$ for a diffeomorphism f of a compact manifold X, the goal is to obtain a family of diffeomorphisms $\mathcal{H}_x\colon \mathcal{W}_x \to T_x\mathcal{W}$ such that the maps

$$\mathcal{P}_x = \mathcal{H}_{fx} \circ f \circ \mathcal{H}_x^{-1}\colon\ T_x\mathcal{W} \to T_{fx}\mathcal{W} \tag{8.1}$$

* Supported in part by Simons Foundation grant 426243.

are as simple as possible. The maps $\mathcal{P}_x$, called the normal form of f on $\mathcal{W}$, will be polynomial in our setting. The case of linear $\mathcal{P}_x$ is called non-stationary linearization. Some of the theory can be developed in a more general context of a smooth extension $\mathcal{F}$ of f to a vector bundle $\mathcal{E}$ over X. The foliation setting produces such a smooth extension on the tangent bundle to the foliation $\mathcal{E} = T\mathcal{W}$ as follows. We take $\mathcal{F}_x = f|_{W_x} : \mathcal{E}_x \to \mathcal{E}_{fx}$ after local identification of the leaf $\mathcal{W}_x$ with its tangent space $\mathcal{E}_x = T_x\mathcal{W}$. In this setting, the map $\mathcal{H}_x$ is a coordinate change on $\mathcal{E}_x$.

The non-stationary linearization for one-dimensional extensions was obtained by Katok and Lewis [KtL91] and applied to the study of rigidity for $SL(n,\mathbb{Z})$ actions on $\mathbb{T}^n$. For higher-dimensional foliations under the assumption of constant 1/2 pinched bounds for contraction rates, non-stationary linearization follows from results of Guysinsky and Katok [GuKt98] or from results of Feres in [Fe95]. Under a weaker assumption of pointwise 1/2 pinching, it was obtained by Sadovskaya [S05] and some further properties were established by Kalinin and Sadovskaya [KS06]. This is a less technical but very important case of non-stationary normal forms and we give a brief summary of these results in Section 8.2. They were used extensively in the study of rigidity of Anosov systems and higher rank actions; see for example [S05, KS03, KS06, F07, FFH10, GKS11, Bu18].

In higher-dimensional setting without 1/2 pinching, there may be no smooth non-stationary linearization, and so a polynomial normal form is sought. Under the narrow band spectrum assumption, such forms were developed by Guysinsky and Katok [GuKt98, Gu02] and used by Katok and Spatzier to obtain local rigidity of algebraic higher-rank Anosov abelian actions [KtSp97]. A geometric point of view on normal forms was developed by Feres in [Fe04]. The narrow band assumption ensures that the polynomial maps involved belong to a finite-dimensional Lie group of so-called sub-resonance generated polynomials. In [KS16] Kalinin and Sadovskaya obtained stronger results, constructing $\mathcal{H}_x$ which depend smoothly on x along the leaves and proving that they define an atlas with transition maps in a similar finite-dimensional Lie group.

A parallel theory of non-stationary normal forms was also developed for non-uniform contractions. Basic results were formulated by Kalinin and Katok [KKt01] along with a program of applications to measure rigidity for non-uniformly hyperbolic systems and actions. The existence of $\mathcal{H}_x$ for a general contracting C^∞ extension was proved by Li and Lu [LL05] in the setting of random dynamical systems. Some results, such as existence of Taylor polynomial or formal series for $\mathcal{H}_x$, can be obtained for extensions more general than contractions; see [AK92, A, LL05]. Non-stationary linearization of a $C^{1+\alpha}$ diffeomorphism along a non-uniformly contracting foliation was obtained by Kalinin and Katok [KKt07] for one-dimensional leaves and by Katok and Rodriguez Hertz [KtR15] under pinching assumption on Lyapunov exponents. The former result was used in the study of measure rigidity by Kalinin, Katok, and Rodriguez Hertz [KKt07, KKtR11]. Following [KS16], the advanced theory of normal forms on non-uniformly contracting foliations, including the consistency of normal form coordinates along the leaves, were obtained independently by Kalinin and Sadovskaya [KS17] and, in differential geometric formulations, by Melnick [M19]. In addition to measurable properties of non-uniformly hyperbolic systems, this theory is useful in global, rather than local, smooth rigidity of uniformly hyperbolic systems, where the spectrum may not be narrow [FKSp11]. The main disadvantage of this setting is that dependence of $\mathcal{H}_x$ on x is only measurable.

Our main goal is to present the theory of non-stationary normal forms, both in extension and foliation settings, assuming that the spectrum is sufficiently narrow. In particular, the results apply to perturbations of algebraic and point spectrum systems. We follow the approach developed in [KS16, KS17] and give a coherent treatment of existence, (non)uniqueness, and centralizer results. The main improvements over the previous results in the narrow band spectrum setting are the following. Our construction allows us to describe the exact extent of non-uniqueness in $\mathcal{H}_x$ and $\mathcal{P}_x$ and hence gives the description of the centralizer. It works in any regularity of $\mathcal{F}$ above the precise critical level in Hölder classes. This is especially important for the centralizer results as they yield an automatic bootstrap of regularity for a commuting system, from critical one to that of $\mathcal{H}_x$. These improvements proved useful in rigidity results [DWX19, GKS19]. In addition to sub-resonance normal form, we also prove existence, (non)uniqueness, and centralizer results for resonance normal form, which has not been done in the narrow band setting.

8.2 Non-stationary Linearization

In this section we summarize the results on non-stationary linearization, that is, existence of $\mathcal{H}_x$ so that $\mathcal{P}_x$ in (8.1) are linear. This important particular case of non-stationary normal form theory is easier to formulate and provides a point of comparison for the more technical general case. Also, it is the only result obtained under pointwise assumption on contraction rates. We formulate it in a foliation setting, where it is most interesting. We state the results for C^∞ case, as established in [S05, Proposition 4.1, Lemma 4.1] and [KS06, Proposition 4.6], and then make remarks on lower regularity.

Theorem 2.1 (Non-stationary linearization). [S05, KS06]
Let f be a diffeomorphism of a compact manifold X and let $\mathcal{W}$ be an f-invariant continuous foliation of X with uniformly C^∞ leaves. Suppose that $\|Df|_{T\mathcal{W}}\| < 1$, and there exist $C > 0$ and $\gamma < 1$ such that

$$\| (Df^n|_{T_x\mathcal{W}})^{-1} \| \cdot \| Df^n|_{T_x\mathcal{W}} \|^2 \leq C\gamma^n \quad \text{for all } x \in X,\, n \in \mathbb{N}. \tag{8.2}$$

Then for every $x \in X$ there exists a C^∞ diffeomorphism $\mathcal{H}_x\colon \mathcal{W}_x \to T_x\mathcal{W}$ such that

(i) $\mathcal{H}_{fx} \circ f \circ \mathcal{H}_x^{-1} = Df|_{T_x\mathcal{W}}$,
(ii) $\mathcal{H}_x(x) = 0$ *and* $D_x\mathcal{H}_x$ *is the identity map,*
(iii) $\mathcal{H}_x$ *depends continuously on* $x \in X$ *in* C^∞ *topology.*
(iv) *Such a family* $\mathcal{H}_x$ *is unique and depends smoothly on* x *along the leaves of* $\mathcal{W}$.
(v) *The map* $\mathcal{H}_y \circ \mathcal{H}_x^{-1}\colon T_x\mathcal{W} \to T_y\mathcal{W}$ *is affine for any* $x \in X$ *and* $y \in \mathcal{W}_x$. *Hence the non-stationary linearization* $\mathcal{H}$ *defines affine structures on the leaves of* $\mathcal{W}$.

We note that non-stationary linearization is the only case when the $\mathcal{H}_x$, and hence $\mathcal{P}_x$, are unique under assumptions (i) and (ii) and sufficient regularity. This uniqueness immediately implies the description of centralizer as in part (3) of Theorem 4.6. Together with (v), it also easily gives smooth dependence on x along the leaves of $\mathcal{W}$.

Under the stronger 1/2 pinching assumption on rates in place of (8.2), the finite regularity version follows from our general results; see Corollary 4.8. However, finite regularity results can also be obtained under assumption (8.2), as we explain in the next remark.

Remark 2.2. *The proof of existence of* $\mathcal{H}$ *in* [S05, Proposition 4.1] *is for* C^N *with* N *sufficiently large and can be seen to work for* $N = 2$.

Parts (iv) *and* (v) *hold under the assumption that* f *and* $\mathcal{H}_x$ *are* C^2. *This is clear from the proof of* [S05, Lemma 4.1] *and* [KS06, Proposition 4.6]. *More precisely, it suffices to assume that* $Df|_{T_x\mathcal{W}}$ *and* $D\mathcal{H}_x$ *are Lipschitz along the leaves of* $\mathcal{W}$ *with uniform constant for all local leaves. More generally, uniqueness holds if they are* α*-Hölder under the stronger pinching assumption with the term* $\| Df^n|_{T_x\mathcal{W}} \|^2$ *in* (8.2) *replaced by* $\| Df^n|_{T\mathcal{W}} \|^{1+\alpha}$ *(cf. Corollary 4.8). In the particular case when* $\mathcal{W}$ *is a one-dimensional foliation, the uniqueness holds if* $\mathcal{H}$ *is* C^1 [KtL91].

We note that equation (8.2) with $1+\beta$ in place of 2 is precisely the *fiber bunching* assumption on the linear cocycle $Df|_{T_x\mathcal{W}}$ relative to the contraction along $\mathcal{W}$, which in particular ensures existence of cocycle holonomies along $\mathcal{W}$. In fact, the holonomies are given by the derivatives of the transition maps in (v).

8.3 Preliminaries and Notations

8.3.1 Smooth Extensions

Let $\mathcal{E}$ be a continuous vector bundle over a compact metric space X, let $\mathcal{V}$ be a neighborhood of the zero section in $\mathcal{E}$, and let f be a homeomorphism of X. We consider an extension $\mathcal{F}\colon \mathcal{V} \to \mathcal{E}$ that projects to f and preserves the zero section. We assume that the corresponding fiber maps $\mathcal{F}_x\colon \mathcal{V}_x \to \mathcal{E}_{f(x)}$ are C^r diffeomorphisms.

If $r = N \in \mathbb{N}$, we will assume that $\mathcal{F}_x$ depend continuously on x in C^N topology. To obtain sharper results, we will also use *Hölder condition at the zero section.* We assume that the fibers $\mathcal{E}_x$ are equipped with a continuous family of Riemannian norms. We denote by $B_{x,\sigma}$ the closed ball of radius $\sigma > 0$ centered at $0 \in \mathcal{E}_x$. For $N \in \mathbb{N}$ and $0 \le \alpha \le 1$ we denote by $C^{N,\alpha}(B_{x,\sigma}) = C^{N,\alpha}(B_{x,\sigma}, \mathcal{E}_{fx})$ the space of functions $R\colon B_{x,\sigma} \to \mathcal{E}_{fx}$ with continuous derivatives up to order N on $B_{x,\sigma}$ and, if $\alpha > 0$, with N-th derivative satisfying α-Hölder condition at 0:

$$\|D^{(N)}R\|_\alpha = \sup\{\, \|D_t^{(N)}R - D_0^{(N)}R\| \cdot \|t\|^{-\alpha} :\ 0 \ne t \in B_{x,\sigma}\} < \infty. \tag{8.3}$$

We call $\|D^{(N)}R\|_\alpha$ the α-Hölder constant of $D^{(N)}R$ at 0. We equip the space $C^{N,\alpha}(B_{x,\sigma})$ with the norm

$$\|R\|_{C^{N,\alpha}(B_{x,\sigma})} = \max\{\, \|R\|_0,\ \|D^{(1)}R\|_0,\ \ldots,\ \|D^{(N)}R\|_0,\ \|D^{(N)}R\|_\alpha \,\}, \tag{8.4}$$

where $\|D^{(k)}R\|_0 = \sup\{\|D_t^{(k)}R\| :\ t \in B_{x,\sigma}\}$ and the last term is omitted if $\alpha = 0$.

Definition 3.1. *We say that $\mathcal{F}$ is a $C^{N,\alpha}$ extension of f, $N \in \mathbb{N}$ and $0 \le \alpha \le 1$, if, for some $\sigma > 0$, the fiber maps $\mathcal{F}_x\colon B_{x,\sigma} \to \mathcal{E}_{f(x)}$ are $C^{N,\alpha}$ diffeomorphisms which depend continuously on x in C^N topology and the norms $\|\mathcal{F}_x\|_{C^{N,\alpha}(B_{x,\sigma})}$ are uniformly bounded.*

Similarly, we say that $\mathcal{H} = \{\mathcal{H}_x\}_{x\in X}$, where $\mathcal{H}_x\colon B_{x,\sigma} \to \mathcal{E}_{f(x)}$, is a $C^{N,\alpha}$ coordinate change if it is a $C^{N,\alpha}$ extension of $f = Id$ which preserves the zero section.

8.3.2 Mather Spectrum of the Derivative

For a smooth extension $\mathcal{F}$ we will denote by F its derivative at the zero section, that is $F\colon \mathcal{E} \to \mathcal{E}$ is a continuous linear extension of f whose fiber maps are linear isomorphisms $F_x = D_0\mathcal{F}_x\colon \mathcal{E}_x \to \mathcal{E}_{fx}$. Such a linear extension F induces a bounded linear operator F^* on the space of continuous sections of $\mathcal{E}$ by $F^*v(x) = F(v(f^{-1}x))$. The spectrum $Sp\,F^*$ of complexification of F^* is called *Mather spectrum* of F. Under a mild assumption that non-periodic points of f are dense in X, the Mather spectrum consists of finitely many closed annuli centered at 0, see for example [P], and its *characteristic set* $\Lambda(F) = \{\lambda \in \mathbb{R} :\ \exp\lambda \in Sp\,F^*\}$ consists of finitely many closed intervals. We will assume that F is a contraction and that the Mather spectrum of F is sufficiently narrow.

Definition 3.2. *Let $\varepsilon > 0$ and $\chi = (\chi_1, \ldots, \chi_\ell)$, where $\chi_1 < \cdots < \chi_\ell < 0$. We say that a linear extension F has (χ, ε)-spectrum if*

$$\Lambda(F) = \{\lambda \in \mathbb{R} :\ \exp\lambda \in Sp\,F^*\} \subset \bigcup_{i=1}^{\ell}(\chi_i - \varepsilon, \chi_i + \varepsilon). \tag{8.5}$$

If F has (χ, ε)-spectrum with disjoint intervals, then the bundle $\mathcal{E}$ splits into direct sum

$$\mathcal{E} = \mathcal{E}^1 \oplus \cdots \oplus \mathcal{E}^\ell \tag{8.6}$$

of continuous F-invariant sub-bundles so that $\Lambda(F|_{\mathcal{E}^i})$ is contained in $(\chi_i - \varepsilon, \chi_i + \varepsilon)$. This can be expressed using a convenient metric [GuKt98]: for each $i = 1, \ldots, \ell$ there exists a continuous family of Riemannian norms $\|.\|_x$ on $\mathcal{E}_x$ such that the splitting (8.6) is orthogonal and

$$e^{\chi_i - \varepsilon} \|t\|_x \le \|F_x(t)\|_{fx} \le e^{\chi_i + \varepsilon} \|t\|_x \quad \text{for every } t \in \mathcal{E}^i_x. \tag{8.7}$$

We will equip $\mathcal{E}$ with such a norm and will suppress the dependence on x. We can also summarize (8.7) using operator norms

$$\|F|_{\mathcal{E}^i_x}\| \le e^{\chi_i + \varepsilon}, \quad \|(F|_{\mathcal{E}^i_x})^{-1}\| \le e^{-\chi_i + \varepsilon}, \quad \|F_x\| \le e^{\chi_\ell + \varepsilon}, \quad \|(F_x)^{-1}\| \le e^{-\chi_1 + \varepsilon}. \tag{8.8}$$

8.3.3 Sub-resonance and Resonance Polynomials

We say that a map between vector spaces is *polynomial* if each component is given by a polynomial in some, and hence every, basis. We will consider a polynomial map $P: \mathcal{E}_x \to \mathcal{E}_y$ with $P(0_x) = 0_y$ and split it into components $(P_1(t), \ldots, P_\ell(t))$, where $P_i: \mathcal{E}_x \to \mathcal{E}^i_y$. Each P_i can be written uniquely as a linear combination of polynomials of specific homogeneous types. We say that $Q: \mathcal{E}_x \to \mathcal{E}^i_y$ has *homogeneous type* $s = (s_1, \ldots, s_\ell)$, where $s_1, \ldots, s_\ell$ are non-negative integers, if, for any real numbers $a_1, \ldots, a_\ell$ and vectors $t_j \in \mathcal{E}^j_x$, $j = 1, \ldots, \ell$, we have

$$Q(a_1 t_1 + \cdots + a_\ell t_\ell) = a_1^{s_1} \cdots a_\ell^{s_\ell} \cdot Q(t_1 + \cdots + t_\ell). \tag{8.9}$$

Definition 3.3. *We say that a homogeneous type $s = (s_1, \ldots, s_\ell)$ for $P_i: \mathcal{E}_x \to \mathcal{E}^i_y$ is*

$$\textbf{\textit{sub-resonance}}\ \textit{if}\ \ \chi_i \le \sum_{j=1}^{\ell} s_j \chi_j, \quad \textit{and} \quad \textbf{\textit{resonance}}\ \textit{if}\ \ \chi_i = \sum_{j=1}^{\ell} s_j \chi_j. \tag{8.10}$$

We say that a polynomial map $P: \mathcal{E}_x \to \mathcal{E}_y$ is sub-resonance (resp. resonance) *if each component P_i has only terms of sub-resonance (resp. resonance) homogeneous types. We denote by $\mathcal{S}_{x,y}$ (resp. $\mathcal{R}_{x,y}$) the set of all sub-resonance (resp. resonance) polynomials $P: \mathcal{E}_x \to \mathcal{E}_y$ with $P(0) = 0$ and invertible derivative at* 0.

Clearly, for any sub-resonance relation we have $s_j = 0$ for $j < i$ and $\sum s_j \le \chi_1 / \chi_\ell$. It follows that sub-resonance polynomials have degree at most

$$d = d(\chi) = \lfloor \chi_1 / \chi_\ell \rfloor. \tag{8.11}$$

We will denote $\mathcal{S}_{x,x}$ by $\mathcal{S}_x$, which is a finite-dimensional Lie group group with respect to the composition [GuKt98]. All groups $\mathcal{S}_x$ are isomorphic; moreover, any map $P \in \mathcal{S}_{x,y}$ induces an isomorphism between $\mathcal{S}_x$ and $\mathcal{S}_y$ by conjugation. Any invertible linear map $A: \mathcal{E}_y \to \mathcal{E}_x$ which respects the splitting induces an isomorphism between the groups $\mathcal{S}_x$ and $\mathcal{S}_y$. Similar statements hold for the resonance groups $\mathcal{R}_x = \mathcal{R}_{x,x}$.

Note that a linear map is resonance, resp. sub-resonance, if and only if it preserves the splitting (8.6), resp. the associated flag of fast sub-bundles:

$$\mathcal{E}_x^1 = \mathcal{V}_x^1 \subset \mathcal{V}_x^2 \subset \ldots \subset \mathcal{V}_x^l = \mathcal{E}_x, \quad \text{where } \mathcal{V}_x^i = \mathcal{E}_x^1 \oplus \cdots \oplus \mathcal{E}_x^i. \tag{8.12}$$

While the notion of resonance polynomials depends on the splitting, the notion of sub-resonance polynomials depends only on the flag (8.12) and sub-resonance polynomials preserve the flag; see [KS17, Proposition 3.2].

8.3.4 Narrow Spectrum

Now for a given $\chi = (\chi_1, \ldots, \chi_\ell)$, where $\chi_1 < \cdots < \chi_\ell < 0$, we will define $\varepsilon_0 = \varepsilon_0(\chi) > 0$ which ensures that the spectrum is sufficiently narrow. Informally, we choose it so that if F has (χ, ε)-spectrum for some $\varepsilon < \varepsilon_0(\chi)$, then all Mather spectrum sub-resonances and resonances of F come from the point spectrum χ and also any non-resonance homogeneous type is contracted by forward or backward iterates of F. This condition is stronger than the narrow band spectrum in [GuKt98].

We define $\tilde{\lambda} < 0$ as the largest value of $-\chi_i + \sum_{j=1}^{\ell} s_j \chi_j$ over all $i \in \{1, \ldots, \ell\}$ and nonnegative integers $s_1, \ldots, s_\ell$ such that this value is negative, that is, they do not satisfy any sub-resonance relation (8.10):

$$\tilde{\lambda} = \max\{-\chi_i + \sum s_j \chi_j < 0\} \text{ and let } \lambda = \max\{\tilde{\lambda}, -\chi_1 + (d+1)\chi_\ell\} < 0. \tag{8.13}$$

The maximum exists since there are at most finitely many values of $-\chi_i + \sum s_j \chi_j$ greater than any given number.

Similarly, we define $\mu < 0$ as the largest value of $\chi_i - \sum_{j=1}^{\ell} s_j \chi_j$ over all $i \in \{1, \ldots, \ell\}$ and non-negative integers $s_1, \ldots, s_\ell$ such that this value is negative, that is, they satisfy some sub-resonance relation which is not a resonance one (8.10) (we will refer to such homogeneous types as *strict sub-resonance*):

$$\mu = \max\{\chi_i - \sum s_j \chi_j < 0\}. \tag{8.14}$$

The maximum exists since there are at most finitely many sub-resonance relations. Finally, we define

$$\varepsilon_0 = \varepsilon_0(\chi) = \min\{-\chi_\ell, -\lambda/(d+2), -\mu/(d+1)\} > 0. \tag{8.15}$$

8.4 Statements of Results

First we summarize our basic notations and assumptions.

Assumptions 4.1. *In this section,*
$f: X \to X$ is a homeomorphism of a compact metric space X,
$\mathcal{E}$ is a continuous vector bundle over X, equipped with a continuous Riemannian metric,
$\mathcal{V}$ is a neighborhood of zero section in $\mathcal{E}$ and $B_{x,\sigma} \subset \mathcal{V}_x$ for some $\sigma > 0$ and all $x \in X$,
$\mathcal{F}: \mathcal{V} \to \mathcal{E}$ is a $C^{N,\alpha}$ extension of f (see Def. 3.1) that preserves the zero section,
$\mathcal{F}$ is contracts in the sense $\|\mathcal{F}_x(t)\| \le \xi \|t\|$ for some $\xi < 1$, and all $x \in X$ and $t \in B_{x,\sigma}$,
$F: \mathcal{E} \to \mathcal{E}$ is the derivative of $\mathcal{F}$ at the zero section, $F_x = D_0\mathcal{F}_x: \mathcal{E}_x \to \mathcal{E}_{fx}$,
F has (χ, ε)-spectrum (see Def. 3.2) for some $\chi = (\chi_1, \dots, \chi_\ell)$, with $\chi_1 < \dots < \chi_\ell < 0$, and some $\varepsilon < \varepsilon_0 = \varepsilon_0(\chi)$ given by (8.15).

Remark 4.2. *As we outlined in Section 8.3.2, if F has (χ, ε)-spectrum, then there is a continuous Riemannian metric on $\mathcal{E}$ and a continuous orthogonal F-invariant splitting $\mathcal{E} = \mathcal{E}^1 \oplus \dots \oplus \mathcal{E}^l$ such that for each $i = 1, \dots, \ell$ and all $x \in X$ we have*

$$e^{\chi_i - \varepsilon}\|t\| \le \|F_x(t)\| \le e^{\chi_i + \varepsilon}\|t\| \quad \textit{for every } t \in \mathcal{E}^i_x, \tag{8.16}$$

which also yields (8.8)*. This is the property that we use in the proof of Theorem 4.3.*

We recall that $C^{N,\alpha}$ is the space of C^N functions with N^{th} derivative satisfying α-Hölder condition at 0 (8.3). We will require that the smoothness $N+\alpha$ is higher than the "critical regularity" $\chi_1/\chi_\ell \ge 1$. If $N \ge 2$, we allow $\alpha = 0$.

Theorem 4.3 (Normal forms for contracting extensions).
Let $\mathcal{F}$ be an extension of f satisfying Assumptions 4.1. Suppose that $N \in \mathbb{N}$, $0 \le \alpha \le 1$,

$$\nu = \chi_1 - (N+\alpha)\chi_\ell > 0 \quad \textit{and} \quad \varepsilon < \nu/(N+\alpha+1). \tag{8.17}$$

Then
(1) There exists a $C^{N,\alpha}$ coordinate change $\mathcal{H} = \{\mathcal{H}_x\}_{x\in X}$ (see Def. 3.1) with diffeomorphisms $\mathcal{H}_x: B_{x,\sigma} \to \mathcal{E}_x$ satisfying $\mathcal{H}_x(0) = 0$ and $D_0\mathcal{H}_x = \text{Id}$ which conjugates $\mathcal{F}$ to a continuous polynomial extension $\mathcal{P}$ of sub-resonance type (see Def. 3.3):

$$\mathcal{H}_{fx} \circ \mathcal{F}_x = \mathcal{P}_x \circ \mathcal{H}_x, \quad \text{where } \mathcal{P}_x \in \mathcal{S}_{x,fx} \text{ for all } x \in X. \tag{8.18}$$

(1') There exists a $C^{N,\alpha}$ coordinate change $\mathcal{H}' = \{\mathcal{H}'_x\}_{x\in X}$ with diffeomorphisms $\mathcal{H}'_x: B_{x,\sigma} \to \mathcal{E}_x$ satisfying $\mathcal{H}'_x(0) = 0$ and $D_0\mathcal{H}'_x = \text{Id}$ which conjugates $\mathcal{F}$ to a continuous polynomial extension $\mathcal{P}'$ of resonance type:

$$\mathcal{H}'_{fx} \circ \mathcal{F}_x = \mathcal{P}'_x \circ \mathcal{H}'_x, \quad \text{where } \mathcal{P}_x \in \mathcal{R}_{x,fx} \text{ for all } x \in X. \tag{8.19}$$

(2) Suppose $\tilde{\mathcal{H}} = \{\tilde{\mathcal{H}}_x\}_{x\in X}$ is another $C^{N,\alpha}$ coordinate change as in (1) conjugating $\mathcal{F}$ to a sub-resonance polynomial extension $\tilde{\mathcal{P}}$. Then there exists a continuous family $\{G_x\}_{x\in X}$ with $G_x \in \mathcal{S}_x$ such that $\mathcal{H}_x = G_x \circ \tilde{\mathcal{H}}_x$. Moreover, if $D_0^{(n)}\tilde{\mathcal{H}}_x = D_0^{(n)}\mathcal{H}_x$ for all $n = 2, \dots, d = \lfloor \chi_1/\chi_\ell \rfloor$, then $\mathcal{H}_x = \tilde{\mathcal{H}}_x$ for all $x \in X$.

(2') Suppose $\tilde{\mathcal{H}}' = \{\tilde{\mathcal{H}}'_x\}_{x\in X}$ is another $C^{N,\alpha}$ coordinate change as in (1') conjugating $\mathcal{F}$ to a resonance polynomial extension $\tilde{\mathcal{P}}'$. Then there exists a continuous family $\{G'_x\}_{x\in X}$ with $G'_x \in \mathcal{R}_x$ such that $\mathcal{H}'_x = G'_x \circ \tilde{\mathcal{H}}'_x$. Moreover, if $D_0^{(n)}\tilde{\mathcal{H}}'_x = D_0^{(n)}\mathcal{H}'_x$ for all $n = 2, \dots, d = \lfloor \chi_1/\chi_\ell \rfloor$, then $\mathcal{H}'_x = \tilde{\mathcal{H}}'_x$ for all $x \in X$.

(3) Let $g\colon X \to X$ be a homeomorphism commuting with f and $\mathcal{G}\colon \mathcal{V} \to \mathcal{E}$ be a $C^{N',\alpha'}$ extension of g which preserves the zero section and commutes with $\mathcal{F}$. Suppose that $N' \in \mathbb{N}$ and $0 \le \alpha' \le 1$ satisfy $N' \le N$, $N' + \alpha' \le N + \alpha$, and

$$\nu' = \chi_1 - (N' + \alpha')\chi_\ell > 0 \quad \text{and} \quad \varepsilon < \nu'/(N' + \alpha' + 1). \tag{8.20}$$

Then the coordinate changes $\mathcal{H}$ and $\mathcal{H}'$ conjugate $\mathcal{G}$ to continuous sub-resonance and resonance polynomial extension respectively, that is

$$\mathcal{H}_{gx} \circ \mathcal{G}_x \circ \mathcal{H}_x^{-1} \in \mathcal{S}_{x,fx} \text{ and } \mathcal{H}'_{gx} \circ \mathcal{G}_x \circ (\mathcal{H}'_x)^{-1} \in \mathcal{R}_{x,fx}. \text{ for all } x \in X. \tag{8.21}$$

In particular, $\mathcal{G}$ is a $C^{N,\alpha}$ extension.

Corollary 4.4. *Suppose that $\mathcal{F}$ in the theorem is a C^∞ extension. Then the coordinate changes $\mathcal{H}$ in part (1) and $\mathcal{H}'$ in part (1') are also C^∞.*

Remark 4.5 (Global version). *Suppose that $\mathcal{F}\colon \mathcal{E} \to \mathcal{E}$ is a globally defined extension which satisfies the assumptions of Theorem 4.3 and either contracts fibers or, more generally, satisfies the property that for any compact set $K \subset \mathcal{V}$ and any neighborhood U of the zero section we have $\mathcal{F}^n(K) \subset U$ for all sufficiently large n. Then the coordinate changes $\mathcal{H}$ and $\mathcal{H}'$ can be uniquely extended "by invariance" $\mathcal{H}_x = (P_x^n)^{-1} \circ \mathcal{H}_{f^n(x)} \circ \mathcal{F}_x^n$ to the family of global $C^{N,\alpha}$ coordinate changes with diffeomorphisms $\mathcal{H}_x, \mathcal{H}'_x\colon \mathcal{E}_x \to \mathcal{E}_x$ satisfying* (8.18) *and* (8.19) *respectively. Moreover, if the extension $\mathcal{G}$ as in (3) is also globally defined, then it satisfies* (8.21) *globally.*

8.4.1 Normal Forms for Contracting Foliations

Now we apply the preceding results in the context of diffeomorphisms with invariant contracting foliation.

Let f be a C^r diffeomorphism of a compact manifold X. We will consider $r > 1$, and for $r \notin \mathbb{N}$ we will understand C^r in the usual sense that the derivative of order $N = \lfloor r \rfloor$ is Hölder with exponent $\alpha = r - \lfloor r \rfloor$. We will consider an f-invariant continuous foliation $\mathcal{W}$ of X with *uniformly C^r leaves*, by which we mean that, for some $R > 0$, the balls $B^{\mathcal{W}}(x, R)$ of radius R in the intrinsic Riemannian metric of the leaf can be given by C^r embeddings which depend continuously on x in C^N topology and, if $r \notin \mathbb{N}$, have α-Hölder derivative of order N with uniformly bounded Hölder constant. Similarly, for such a foliation we will say that a function g is *uniformly C^r along $\mathcal{W}$* if its restrictions to $B^{\mathcal{W}}(x,R)$ depend continuously on x in C^N topology and have α-Hölder derivative of order N with uniformly bounded Hölder constant. We also allow $r = \infty$, in which case uniformly C^∞ means uniformly C^N for each N.

Theorem 4.6 (Normal forms for contracting foliations). *Let f be a C^r, $r \in (1,\infty]$, diffeomorphism of a smooth compact manifold X, and let $\mathcal{W}$ be an f-invariant topological foliation of X with uniformly C^r leaves. Suppose that the linear extension $F = Df|_{T\mathcal{W}}$ has (χ,ε)-spectrum (or alternatively F satisfies the condition in Remark 4.2) for some $\chi = (\chi_1,\dots,\chi_\ell)$, where $\chi_1 < \cdots < \chi_\ell < 0$, and some $\varepsilon < \varepsilon_0 = \varepsilon_0(\chi)$ given by (8.15). Suppose also that $r > \chi_1/\chi_\ell$ and*

$$\varepsilon < \nu/(r+1), \quad \text{where} \quad \nu = \chi_1 - r\chi_\ell > 0. \tag{8.22}$$

Then **(1)** *There exists a family $\{\mathcal{H}_x\}_{x\in X}$ of C^r diffeomorphisms $\mathcal{H}_x\colon \mathcal{W}_x \to T_x\mathcal{W}$ satisfying $\mathcal{H}_x(0) = 0$ and $D_0\mathcal{H}_x = Id$ such that for each $x \in X$,*

$$\mathcal{P}_x = \mathcal{H}_{f(x)} \circ f \circ \mathcal{H}_x^{-1} : T_x\mathcal{W} \to T_{f(x)}\mathcal{W} \text{ is in } \mathcal{S}_{x,fx}.$$

The maps $\mathcal{H}_x|_{B^{\mathcal{W}}(x,R)}$ depend continuously on $x \in X$ in C^N topology with $N = \lfloor r \rfloor$ and, if $\alpha = r - N > 0$, they have α-Hölder derivative of order N with uniformly bounded Hölder constant.

(1') *There exists a family $\{\mathcal{H}'_x\}_{x\in X}$ of diffeomorphisms as in (1) so that for all $x \in X$,*

$$\mathcal{P}_x = \mathcal{H}'_{f(x)} \circ f \circ (\mathcal{H}'_x)^{-1} \colon T_x\mathcal{W} \to T_{f(x)}\mathcal{W} \text{ is in } \mathcal{R}_{x,fx}.$$

(2) *Suppose $\tilde{\mathcal{H}} = \{\tilde{\mathcal{H}}_x\}_{x\in X}$ is another family of diffeomorphisms as in (1) conjugating $f|_{\mathcal{W}}$ to sub-resonance polynomials $\tilde{\mathcal{P}}_x \in \mathcal{S}_{x,fx}$. Then there exists a continuous family $\{G_x\}_{x\in X}$ with $G_x \in \mathcal{S}_x$ such that $\mathcal{H}_x = G_x \circ \tilde{\mathcal{H}}_x$. Moreover, if $D_0^{(n)}\tilde{\mathcal{H}}_x = D_0^{(n)}\mathcal{H}_x$ for all $n = 2,\dots,$ $d = \lfloor \chi_1/\chi_\ell \rfloor$, then $\mathcal{H}_x = \tilde{\mathcal{H}}_x$ for all $x \in X$.*

(2') *Suppose $\tilde{\mathcal{H}}' = \{\tilde{\mathcal{H}}'_x\}_{x\in X}$ is another family of diffeomorphisms as in (1') conjugating $f|_{\mathcal{W}}$ to resonance polynomials $\tilde{\mathcal{P}}_x \in \mathcal{R}_{x,fx}$ Then there exists a continuous family $\{G'_x\}_{x\in X}$ with $G'_x \in \mathcal{R}_x$ such that $\mathcal{H}'_x = G'_x \circ \tilde{\mathcal{H}}'_x$. Moreover, if $D_0^{(n)}\tilde{\mathcal{H}}'_x = D_0^{(n)}\mathcal{H}'_x$ for all $n = 2,\dots,d = \lfloor \chi_1/\chi_\ell \rfloor$, then $\mathcal{H}'_x = \tilde{\mathcal{H}}'_x$ for all $x \in X$.*

(3) *Let g be a homeomorphism of X which commutes with f, preserves $\mathcal{W}$, and is uniformly $C^{r'}$ along the leaves of $\mathcal{W}$. Suppose that $1 < r' \le r$ satisfies*

$$\nu' = \chi_1 - r'\chi_\ell > 0 \quad \text{and} \quad \varepsilon < \nu'/(r'+1). \tag{8.23}$$

Then $\mathcal{Q}_x = \mathcal{H}_{g(x)} \circ g \circ \mathcal{H}_x^{-1} \in \mathcal{S}_{x,gx}$ and $\mathcal{Q}'_x = \mathcal{H}'_{g(x)} \circ g \circ (\mathcal{H}'_x)^{-1} \in \mathcal{R}_{x,gx}$ for all $x \in X$. In particular, $\mathcal{G}_x$ is uniformly C^r along the leaves of $\mathcal{W}$.

(4) *For any $x \in X$ and $y \in \mathcal{W}_x$, the maps $\mathcal{H}_y \circ \mathcal{H}_x^{-1}$ and $\mathcal{H}'_y \circ (\mathcal{H}'_x)^{-1}$ from $T_x\mathcal{W}$ to $T_y\mathcal{W}$ are compositions of a sub-resonance polynomial in $\mathcal{S}_{x,y}$ with a translation.*

(5) *The family* $\{\mathcal{H}_x\}_{x\in X}$ *as in (1) can be chosen so that* $\mathcal{H}_x$ *depends* $C^{\lfloor r\rfloor}$ *smoothly on* x *along the leaves of* $\mathcal{W}$.

We note that there is no analog of (5) for the resonance case. Also, the transition maps $\mathcal{H}'_y \circ (\mathcal{H}'_x)^{-1}$ in (4) are only sub-resonance, so this is just a particular case of the result for $\mathcal{H}_y \circ (\mathcal{H}_x)^{-1}$.

Another way to interpret (4) is to view $\mathcal{H}_x$ as a coordinate chart on $\mathcal{W}_x$, identifying it with $T_x\mathcal{W}$, and in particular identifying $T_y\mathcal{W}$ with $T_{\mathcal{H}_x(y)}(T_x\mathcal{W})$ by $D_y\mathcal{H}_x$. In this coordinate chart, (4) yields that all transition maps $\mathcal{H}_y \circ \mathcal{H}_x^{-1}$ for $y \in \mathcal{W}_x$ are in the group $\bar{\mathcal{S}}_x$ generated by $\mathcal{S}_x$ and the translations of $T_x\mathcal{W}$. Clearly, this is a finite-dimensional Lie group which acts transitively on $T_x\mathcal{W}$.

Remark 4.7. *The diffeomorphism* $\mathcal{H}_x$ *maps the sub-foliations of* $\mathcal{W}_x$ *by fast leaves tangent to the fast flag* (8.12) *to the linear sub-foliations of* $T_x\mathcal{W}$ *by subspaces parallel to the flag, and the compositions* $\mathcal{H}_y \circ \mathcal{H}_x^{-1}$ *map flag to flag; see* [KS16, Section 3.2]. *It follows that these fast sub-foliations of* $\mathcal{W}_x$ *are as smooth inside* $\mathcal{W}_x$ *as* $\mathcal{H}_x$. *While smoothness of fast sub-foliations is a well-known phenomenon, normal form results give an alternative proof.*

Corollary 4.8. *Under the assumptions of Theorem 4.6, if* $\ell = 1$, *that is,* $\chi_1 = \chi_\ell = \chi$, *then* $\mathcal{P}_x = Df|_{T_x\mathcal{W}}$ *in (1) and* $\mathcal{Q}_x = Dg|_{T_x\mathcal{W}}$ *in (3) are linear normal forms, the family* $\{\mathcal{H}_x\}_{x\in X}$ *as in (1) is unique, the maps* $\mathcal{H}_y \circ \mathcal{H}_x^{-1} \colon \mathcal{E}_x \to \mathcal{E}_y$ *are affine for all* $x \in X$ *and* $y \in W_x$, *and* $\mathcal{H}_y$ *depends* $C^{\lfloor r\rfloor}$*-smoothly on* y *along the leaves of* $\mathcal{W}$.

In this case we can take $\varepsilon < -\alpha\chi/(2+\alpha)$, *where* $\alpha = \min\{1, r-1\}$. *This means that* $Df|_{T\mathcal{W}}$ *is a contraction whose characteristic set is contained in an open interval with ratio of endpoints at most* $1+\alpha$. *This is the "constant rate" version of* $1/2$ *pinching* (8.2) *(cf. Remark 2.2).*

8.5 Proof of Theorem 4.3 and Corollary 4.4

We begin with the proof of part (1). First, we construct Taylor polynomials at zero of degree N for $\mathcal{H}_x$ and the corresponding terms of sub-resonance polynomials $\mathcal{P}_x$. Since (8.17) implies $N \geq d = \lfloor \chi_1/\chi_\ell \rfloor$, this will fully define $\mathcal{P}_x$, whose degree is at most d.

8.5.1 Construction of $\mathcal{P}$ *and the Taylor Polynomial for* $\mathcal{H}$

For each $x \in X$ and map $\mathcal{F}_x \colon \mathcal{E}_x \to \mathcal{E}_{fx}$ we consider its Taylor polynomial of degree N at $t = 0$:

$$\mathcal{F}_x(t) \sim \sum_{n=1}^{N} F_x^{(n)}(t). \tag{8.24}$$

As a function of t, $F_x^{(n)}(t) \colon \mathcal{E}_x \to \mathcal{E}_{fx}$ is a homogeneous polynomial map of degree n. We will view the family $F^{(n)} = \{F_x^{(n)}\}_{x\in X}$ as a section of the corresponding bundle of homogeneous polynomials. We will use similar notations for the Taylor polynomials at $t = 0$ of the desired coordinate change $\mathcal{H}_x(t)$ and the corresponding sub-resonance polynomial extension $\mathcal{P}_x(t)$:

$$\mathcal{H}_x(t) \sim \sum_{n=1}^{N} H_x^{(n)}(t) \quad \text{and} \quad \mathcal{P}_x(t) = \sum_{n=1}^{d} P_x^{(n)}(t).$$

We will inductively construct the Taylor terms $H^{(n)} = \{H_x^{(n)}\}_{x\in X}$ and $P^{(n)} = \{P_x^{(n)}\}_{x\in X}$ as continuous sections of the corresponding bundle of homogeneous polynomials.

For the first derivative we define

$$H_x^{(1)} = \mathrm{Id}\colon \mathcal{E}_x \to \mathcal{E}_x \quad \text{and} \quad P_x^{(1)} = F_x \quad \text{for all } x \in X.$$

Now we assume that the terms of order less than n are constructed. Using these linear terms in the conjugacy equation $\mathcal{H}_{fx} \circ \mathcal{F}_x = \mathcal{P}_x \circ \mathcal{H}_x$, we write

$$\left(\mathrm{Id} + \sum_{i=2}^{N} H_{fx}^{(i)}\right) \circ \left(F_x + \sum_{i=2}^{N} F_x^{(i)}\right) \sim \left(F_x + \sum_{i=2}^{d} P_x^{(i)}\right) \circ \left(\mathrm{Id} + \sum_{i=2}^{N} H_x^{(i)}\right)$$

and, considering the terms of degree n, $2 \le n \le N$, we obtain

$$F_x^{(n)} + H_{fx}^{(n)} \circ F(x) + \sum H_{fx}^{(i)} \circ F_x^{(j)} = F_x \circ H_x^{(n)} + P_x^{(n)} + \sum P_x^{(j)} \circ H_x^{(i)},$$

where the summations are over all i and j such that $ij = n$ and $1 < i,j < n$. We rewrite the equation as

$$F_x^{-1} \circ P_x^{(n)} = -H_x^{(n)} + F_x^{-1} \circ H_{fx}^{(n)} \circ F_x + Q_x, \tag{8.25}$$

where

$$Q_x = F_x^{-1}\left(F_x^{(n)} + \sum_{ij=n,\ 1<i,j<n} H_{fx}^{(i)} \circ F_x^{(j)} - P_x^{(j)} \circ H_x^{(i)}\right). \tag{8.26}$$

We note that Q_x is composed only of terms $H^{(i)}$ and $P^{(i)}$ with $1 < i < n$, which are already constructed, and terms $F^{(i)}$ with $1 < i \le n$, which are given. Thus by the inductive assumption Q_x is defined for all $x \in X$ and is continuous in x.

Let $\mathcal{Q}_x^{(n)}$ be the space of all homogeneous polynomial maps on $\mathcal{E}_x$ of degree n, and let $\mathcal{S}_x^{(n)}$ and $\mathcal{N}_x^{(n)}$ be the subspaces of sub-resonance and non–sub-resonance polynomials respectively. We seek $H_x^{(n)}$ so that the right side of (8.25) is in $\mathcal{S}_x^{(n)}$, and hence so is $P_x^{(n)}$ when defined by this equation.

Projecting (8.25) to the factor bundle $\mathcal{Q}^{(n)}/\mathcal{S}^{(n)}$, our goal is to solve the equation

$$0 = -\bar{H}_x^{(n)} + F_x^{-1} \circ \bar{H}_{fx}^{(n)} \circ F_x + \bar{Q}_x, \tag{8.27}$$

where $\bar{H}^{(n)}$ and $\bar{Q}$ are the projections of $H^{(n)}$ and Q respectively.

We consider the bundle automorphism $\Phi\colon \mathcal{Q}^{(n)} \to \mathcal{Q}^{(n)}$ covering $f^{-1}\colon X \to X$ given by the maps $\Phi_x\colon \mathcal{Q}^{(n)}_{fx} \to \mathcal{Q}^{(n)}_x$

$$\Phi_x(R) = F_x^{-1} \circ R \circ F_x. \tag{8.28}$$

Since F preserves the splitting $\mathcal{E} = \mathcal{E}^1 \oplus \cdots \oplus \mathcal{E}^\ell$, it follows from the definition that the sub-bundles $\mathcal{S}^{(n)}$ and $\mathcal{N}^{(n)}$ are Φ-invariant. We denote by $\bar{\Phi}$ the induced automorphism of $\mathcal{Q}^{(n)}/\mathcal{S}^{(n)}$ and conclude that (8.27) is equivalent to

$$\bar{H}_x^{(n)} = \tilde{\Phi}_x(\bar{H}_{fx}^{(n)}), \quad \text{where } \tilde{\Phi}_x(R) = \bar{\Phi}_x(R) + \bar{Q}_x. \tag{8.29}$$

Thus a solution of (8.27) is a $\tilde{\Phi}$-invariant section of $\mathcal{Q}^{(n)}/\mathcal{S}^{(n)}$. In Lemma 5.2 we will show that $\tilde{\Phi}$ is a contraction and hence has a unique continuous invariant section, which can be explicitly written as

$$\bar{H}_x = \sum_{k=0}^{\infty} (F_x^k)^{-1} \circ \bar{Q}_{f^k x} \circ F_x^k, \quad \text{where } F_x^k = F_{f^{k-1}x} \circ \cdots \circ F_{fx} \circ F_x. \tag{8.30}$$

We equipped $\mathcal{E}$ with a continuous Riemannian metric as in Remark 4.2 for which the splitting $\mathcal{E} = \mathcal{E}^1 \oplus \cdots \oplus \mathcal{E}^l$ is orthogonal and so that (8.16) and hence (8.8) hold.

The norm of a homogeneous polynomial map $R\colon \mathcal{E}_x \to \mathcal{E}_y$ of degree n is defined as

$$\|R\| = \sup\{\, \|R(v)\| : v \in E,\ \|v\| = 1\,\}. \tag{8.31}$$

It follows that for any other homogeneous polynomial map $P\colon \mathcal{E}_z \to \mathcal{E}_x$ we have

$$\|\, R \circ P \,\| \le \|R\| \cdot \|P\|^n. \tag{8.32}$$

First, we look at the action of Φ on polynomials of specific homogeneous type.

Lemma 5.1. *Let $Q \in \mathcal{Q}^{(n)}_x$ and $R \in \mathcal{Q}^{(n)}_{fx}$ be polynomials of homogeneous type $s = (s_1, \ldots, s_\ell)$ with $s_1 + \cdots + s_\ell = n$. Then*

$$\|\Phi_x(R)\| \le e^{-\chi_i + \sum s_j\chi_j + (n+1)\varepsilon}\, \|R\| \quad \text{and} \quad \|\Phi_x^{-1}(Q)\| \le e^{\chi_i - \sum s_j\chi_j + (n+1)\varepsilon}\, \|R\|. \tag{8.33}$$

Proof. We will prove the first inequality, the second one is obtained similarly. Suppose that $v = v_1 + \cdots + v_\ell$, where $v_j \in \mathcal{E}_x^j$, and $\|v\|_x = 1$. We denote $a_j = \|F|_{\mathcal{E}_x^j}\|$ and observe that $F_x(v_j) = a_j v_j' \in \mathcal{E}_{fx}^j$ with $\|v_j'\| \le \|v_j\|$. Since R has homogeneous type $s = (s_1, \ldots, s_\ell)$, we obtain by (8.9) that

$$(R \circ F_x)(v) = R(a_1 v_1' + \cdots + a_\ell v_\ell') = a_1^{s_1} \cdots a_\ell^{s_\ell} \cdot R(v_1' + \cdots + v_\ell'), \tag{8.34}$$

where $v' = v'_1 + \cdots + v'_\ell$ has $\|v'\|_{fx} \le \|v\|_x = 1$ by orthogonality of the splitting. Thus

$$\|(R \circ F_x)(v)\| = a_1^{s_1} \cdots a_\ell^{s_\ell} \cdot \|R(v')\| \le a_1^{s_1} \cdots a_\ell^{s_\ell} \cdot \|R\|$$

for any $v \in \mathcal{E}_x$ with $\|v\|_x = 1$, so by definition (8.31) we obtain $\|R \circ F_x\| \le a_1^{s_1} \cdots a_\ell^{s_\ell} \cdot \|R\|$. Now (8.32) yields

$$\begin{aligned}\|\Phi_x(R)\| &= \|F|_{\mathcal{E}_x^i}^{-1} \circ R \circ F_x\| \le \|F|_{\mathcal{E}_x^i}^{-1}\| \cdot \|R \circ F_x\| \le \\ &\le \|F|_{\mathcal{E}_x^i}^{-1}\| \cdot a_1^{s_1} \cdots a_\ell^{s_\ell} \cdot \|R\| \le e^{-\chi_i+\varepsilon} \cdot \prod_j (e^{\chi_j+\varepsilon})^{s_j} \cdot \|R\|.\end{aligned}$$

Since $a_j = \|F|_{\mathcal{E}_x^j}\| \le e^{\chi_j+\varepsilon}$ and $\|F|_{\mathcal{E}_x^i}^{-1}\| \le e^{-\chi_i+\varepsilon}$ by (8.8). □

Lemma 5.2. *The map $\Phi \colon \mathcal{N}^{(n)} \to \mathcal{N}^{(n)}$ given by* (8.28) *is a contraction over f^{-1}, and hence so is $\tilde{\Phi} \colon \mathcal{Q}^{(n)}/\mathcal{S}^{(n)} \to \mathcal{Q}^{(n)}/\mathcal{S}^{(n)}$ given by* (8.29). *More precisely, $\|\Phi_x(R)\| \le e^{\lambda+(d+2)\varepsilon} \cdot \|R\|$.*

Proof. The statement about $\tilde{\Phi}$ follows since the linear part $\bar{\Phi}$ of $\tilde{\Phi}$ is given by Φ when $\mathcal{Q}^{(n)}/\mathcal{S}^{(n)}$ is naturally identified with $\mathcal{N}^{(n)}$.

For all non sub-resonance homogeneous types we have $-\chi_i + \sum s_j\chi_j \le \tilde{\lambda}$ by the definition of $\tilde{\lambda}$ (8.13) and hence, for any $R \in \mathcal{N}_{fx}^{(n)}$, Lemma 5.1 yields the estimate $\|\Phi_x(R)\| \le e^{\tilde{\lambda}+(n+1)\varepsilon} \cdot \|R\|$. For all $n \le d$ the exponent satisfies $\tilde{\lambda} + (n+1)\varepsilon \le \lambda + (d+1)\varepsilon < 0$ since $\varepsilon < \varepsilon_0$ given by (8.15).

If $d + 1 \le n \le N$, then $\mathcal{S}^{(n)} = 0$ and $\mathcal{Q}^{(n)} = \mathcal{N}^{(n)}$. In this case for any $R \in \mathcal{Q}_{fx}^{(n)}$ we can estimate $\|\Phi_x(R)\| \le e^{-\chi_1+n\chi_\ell+(n+1)\varepsilon} \cdot \|R\|$ and by the definition of λ (8.13) we have $-\chi_1 + (d+1)\chi_\ell \le \lambda < 0$ and hence the exponent satisfies

$$-\chi_1 + n\chi_\ell + (n+1)\varepsilon \le \lambda + (d+2)\varepsilon + (n-(d+1))(\varepsilon + \chi_\ell) < 0$$

since $\varepsilon < \varepsilon_0$ given by (8.15). □

We conclude that the unique continuous invariant section (8.30) for $\tilde{\Phi}$ is the unique continuous solution $\bar{H}^{(n)}$ of (8.27). Now we can choose a continuous section $H^{(n)}$ of $\mathcal{Q}^{(n)}$ which projects to $\bar{H}^{(n)}$. Such $H^{(n)}$ is defined uniquely up to a continuous section $\mathcal{S}^{(n)}$.

Remark 5.3. *For example, one can take $H^{(n)}$ in $\mathcal{N}^{(n)}$. However, in the foliation setting the bundles $\mathcal{Q}^{(n)}$, $\mathcal{S}^{(n)}$, their quotient, and $\bar{H}^{(n)}$ are often more regular along the leaves than $\mathcal{N}^{(n)}$. In this case one can make a more regular choice for $H^{(n)}$, leading to better dependence of $\mathcal{H}_x$ on x along the leaves. See* [KS16] *for more details of this argument.*

Once a specific lift $H^{(n)}$ is chosen, $P_x^{(n)}$ is uniquely defined by Eq. (8.25) and is a continuous section $P^{(n)}$. This completes the inductive step and the construction of $H^{(n)}$ and $P^{(n)}$, $n = 1, \dots, N$.

Thus we have constructed the N-th Taylor polynomials for the coordinate changes

$$\mathcal{H}_x^N(t) = \sum_{n=1}^{N} H_x^{(n)}(t) \quad \text{of degree } N \geq d = \lfloor \chi_1/\chi_\ell \rfloor \tag{8.35}$$

and the polynomial maps $\mathcal{P}_x(t) = \sum_{n=1}^{d} P_x^{(n)}(t)$.

8.5.2 Construction of the Coordinate Change $\mathcal{H}$

In this section we complete the proof of part (1) by constructing the actual coordinate changes $\mathcal{H}_x$ with the Taylor polynomial $\mathcal{H}_x^N$ given by (8.35). To simplify the calculations, we note that $\mathcal{H}_x^N(t)$ is a diffeomorphism on some neighborhood $\tilde{\mathcal{V}}_x$ of $0 \in \mathcal{E}_x$ since its differential at 0 is Id; moreover, the size of $\tilde{\mathcal{V}}_x$ can be bounded away from 0 by compactness of X. Thus we can consider extension $\tilde{\mathcal{F}}_x(t) = \mathcal{H}_{fx}^N \circ \mathcal{F}_x \circ (\mathcal{H}_x^N)^{-1}$. By the construction of $\mathcal{H}^N$, the maps $\tilde{\mathcal{F}}_x$ and $\mathcal{P}_x$ have the same derivatives at $t = 0$ up to order N for each $x \in X$. Since the construction is done in a sufficiently small neighborhood of zero section, we can replace $\mathcal{F}$ by $\tilde{\mathcal{F}}_x$, and so henceforth we assume that $\mathcal{F}$ itself has this property.

We rewrite the conjugacy equation $\mathcal{H}_{fx} \circ \mathcal{F}_x = \mathcal{P}_x \circ \mathcal{H}_x$ in the form

$$\mathcal{H}_x = \tilde{T}(\mathcal{H})_x = \mathcal{P}_x^{-1} \circ \mathcal{H}_{fx} \circ \mathcal{F}_x, \tag{8.36}$$

so that solution $\mathcal{H} = \{\mathcal{H}_x\}$ is a fixed point of the operator $\tilde{T}$. Since $\mathcal{P}_x$ is a sub-resonance polynomial with invertible linear part $P_x^{(1)} = F_x$, by the group property the inverse $\mathcal{P}_x^{-1}$ is also a sub-resonance polynomial and thus has degree at most $d \leq N$.

Denoting $\bar{H}_x = \mathcal{H}_x - \text{Id}$ we rewrite (8.36) as

$$\bar{H}_x = T(\bar{H})_x = \mathcal{P}_x^{-1} \circ (\text{Id} + \bar{H}_{fx}) \circ \mathcal{F}_x - \text{Id}. \tag{8.37}$$

Thus the coordinate change $\mathcal{H}$ corresponds to the fixed point $\bar{H} = T(\bar{H})$ of the operator T on the space of continuous families of smooth functions. We will show that T is a contraction on an appropriate space and thus has a unique fixed point.

For any $x \in X$ we consider the ball $B_{x,r}$ in $\mathcal{E}_x$ centered at 0 of radius r and denote

$$\mathcal{C}_x = \mathcal{C}_{x,r} = \{R \in C^{N,\alpha}(B_{x,r}, \mathcal{E}_x) : \; D_0^{(k)} R = 0, \;\; k = 0, \dots, N\},$$

where $C^{N,\alpha}(B_{x,r},\mathcal{E}_x)$ and its norm are defined as in (8.4). We note that, for any $R \in \mathcal{C}_x$, the α-Hölder constant (8.3) of $D^{(N)}R$ at 0 is

$$\|D^{(N)}R\|_\alpha = \sup\{\|D_t^{(N)}R\| \cdot \|t\|^{-\alpha} : \ 0 \neq t \in B_{x,r}\}. \tag{8.38}$$

For any $R \in \mathcal{C}_x$ lower derivatives can be estimated by the mean value theorem as

$$\|D_t^{(n)}R\| \le \|t\|^{N-n} \cdot \sup\{\|D_s^{(N)}R\| : \ \|s\| \le \|t\|\}, \tag{8.39}$$

so using the above Hölder constant we obtain that for any $0 \le n < N$ and $t \in B_{x,r}$,

$$\|D_t^{(n)}R\| \le \|t\|^{1+\alpha} \cdot \|D^{(N)}R\|_\alpha. \tag{8.40}$$

Thus, for $r < 1$, the norms of all derivatives are dominated by the Hölder constant and hence

$$\|R\|_{C^{N,\alpha}(B_{x,r},\mathcal{E}_x)} = \|D^{(N)}R\|_\alpha. \tag{8.41}$$

It follows that $\mathcal{C}_x$ equipped with the norm $\|D^{(N)}R\|_\alpha$ is a Banach space. We denote by $\mathcal{C}$ the bundle over X with fibers $\mathcal{C}_x$ and by $\mathcal{B}$ the space of sections $\bar{R} = \{R_x\}_{x\in X}$ of $\mathcal{C}$ which are bounded in $C^{N,\alpha}$ norm and continuous in C^N norm. Then $\mathcal{B}$ is a Banach space with the norm $\|\bar{R}\|_{\mathcal{C}} = \sup_x \|R_x\|_\alpha$.

We consider T as an operator on $\mathcal{B}$. It follows from the definition of $\mathcal{C}_x$ and the coincidence of the derivatives of $\mathcal{P}_x$ and $\mathcal{F}_x$ at 0 that $T(\bar{R})$ is in $\mathcal{B}$. We will show that, for a sufficiently small r, T is a contraction on some ball B_γ in $\mathcal{B}$.

We will now define the parameters for this argument. First we note that $0 < \varepsilon < -\chi_\ell$ since $\varepsilon < \varepsilon_0$ given by (8.15), and that we have $\nu - (N+1+\alpha)\varepsilon > 0$ by assumption (8.17) so we can take $\varepsilon' > 0$ satisfying

$$\chi_\ell + \varepsilon + \varepsilon' < 0 \quad \text{and} \quad \delta = \nu - (N+1+\alpha)(\varepsilon+\varepsilon') > 0. \tag{8.42}$$

We also recall that by (8.8) we have

$$D_0\mathcal{P}_x = D_0\mathcal{F}_x = F_x, \quad \|F_x\| \le e^{\chi_\ell+\varepsilon}, \quad \text{and} \quad \|F_x^{-1}\| \le e^{-\chi_1+\varepsilon}. \tag{8.43}$$

Now we can choose $\rho < \min\{1,\sigma\}$ sufficiently small so that for all $x \in X$ we have

$$\|D_t\mathcal{F}_x\| \le e^{\chi_\ell+\varepsilon+\varepsilon'} \quad \text{and} \quad \|D_t(\mathcal{P}_x)^{-1}\| \le e^{-\chi_1+\varepsilon+\varepsilon'} \quad \text{for all } t \in B_{x,\rho}, \tag{8.44}$$

so in particular $\mathcal{F}_x : B_{x,\rho} \to B_{fx,\rho}$ is a contraction. We choose K so that

$$\|\mathcal{F}\|_{C^N(B_{x,\rho})} \le K \quad \text{and} \quad \|(\mathcal{P}_x)^{-1}\|_{C^N(B_{x,\rho})} \le K \quad \text{for all } x \in X. \tag{8.45}$$

Since δ given by (8.42) is positive, we can define $\theta > 0$ by

$$1 - 2\theta = e^{-\delta} < 1 \quad \text{and let} \quad \gamma = \max\{1, \|T(\bar{0})\|_{\mathcal{B}}/\theta\}. \tag{8.46}$$

Here, as r is not yet defined, we take $r = \rho$ in the definition of the norm $\|T(\bar{0})\|_{\mathcal{B}}$ (this does not create problems since the norm decreases with r). To show that T is a contraction on the ball B_γ in $\mathcal{B}$ centered at $\bar{0}$ of radius γ, we will estimate the norm of its differential by $1 - \theta$. For this we choose $r > 0$, satisfying

$$r < \rho < 1, \quad r < \rho/(1+\gamma), \quad r \le \theta/(c_3(K,N)\,\gamma^N), \tag{8.47}$$

where the constant $c_3(K,N)$ from (8.54) depends only on N and K.

Now we will calculate the differential of T on B_γ and estimate its norm. For any $\bar{R}, \bar{S} \in B_\gamma$, we can write

$$(T(\bar{R}+\bar{S}) - T(\bar{R}))_x = (\mathcal{P}_x)^{-1} \circ (\mathrm{Id} + R_{fx} + S_{fx}) \circ \mathcal{F}_x - (\mathcal{P}_x)^{-1} \circ (\mathrm{Id} + R_{fx}) \circ \mathcal{F}_x.$$

Differentiating $(\mathcal{P}_x)^{-1}$ and denoting

$$y(t) = (\mathrm{Id} + R_{fx})(\mathcal{F}_x(t)) = \mathcal{F}_x(t) + R_{fx}(\mathcal{F}_x(t)) \quad \text{and} \quad z(t) = S_{fx}(\mathcal{F}_x(t)),$$

we obtain

$$(T(\bar{R}+\bar{S}) - T(\bar{R}))_x(t) = D_{y(t)}(\mathcal{P}_x)^{-1}\, z(t) + E(z(t)),$$

where E is a polynomial with terms of degree at least two. It follows that $\|E(z(t))\|_{\mathcal{C}} = O(\|\bar{S}\|_{\mathcal{C}}^2)$ and so the differential of T is given by

$$([D_{\bar{R}}T]\bar{S})_x(t) = D_{y(t)}(\mathcal{P}_x)^{-1}\, S_{fx}(\mathcal{F}_x(t)) = A_x(y(t))z(t),$$

where $A_x(s) = D_s(\mathcal{P}_x)^{-1}$. To estimate the norm we consider the derivative of order N.

Since $A_x(y(t))$ is a linear operator on z, the product rule yields

$$D^{(N)}[A_x(y(t))z(t)] = A_x(y(t))D^{(N)}z(t) + \sum c_{m,l}\, D^{(m)}A_x(y(t))D^{(l)}z(t), \tag{8.48}$$

where $m + l = N$ and $l < N$ for all terms in the sum. Differentiating $z(t)$, we get

$$D^{(l)}z(t) = D^{(l)}S_{fx}(\mathcal{F}_x(t)) = \sum D_{t'}^{(i)} S_{fx} \circ D_t^{(j)} \mathcal{F}_{x_k},$$

where $ij = l$ and $t' = \mathcal{F}_x(t)$. Only the first term in (8.48) contains $D^{(N)}S_{fx}$ so

$$D_t^{(N)}([D_{\bar{R}}T]\bar{S})_x = D_{y(t)}^{(1)}(\mathcal{P}_{x_k})^{-1} \circ D_{t'}^{(N)} S_{fx} \circ D_t^{(1)} \mathcal{F}_x + J_x, \tag{8.49}$$

where J_x consists of a fixed number $c(N)$ of terms of the type

$$D_t^{(m)} A_x(y(t)) \left(D_{t'}^{(i)} S_{fx} \circ D_t^{(j)} \mathcal{F}_x \right), \quad i < N,\ m + ij = N,$$

whose norms can be estimated by

$$\|A_x(y(t))\|_{C^N} \cdot \|D^{(i)}_{t'} S_{fx}\| \cdot \|\mathcal{F}_x\|^{N-1}_{C^N}. \tag{8.50}$$

We start with the middle term and observe that

$$\|t'\| = \|\mathcal{F}_x(t))\| \leq e^{\chi_\ell+\varepsilon+\varepsilon'}\|t\| < \|t\| \leq r \tag{8.51}$$

by (8.44). Since $i < N$, we can estimate the middle term using (8.40):

$$\|D^{(i)}_{t'} S_{fx}\| \leq \|t'\|^{1+\alpha} \cdot \|D^{(N)} S_{fx}\|_\alpha < \|t\|^{1+\alpha} \cdot \|\bar{S}\|_C \leq r\|t\|^\alpha \cdot \|\bar{S}\|_C. \tag{8.52}$$

For the first term we note that $t'' = y(t) \in B_{x,\rho}$; indeed since $\|\bar{R}\|_C \leq \gamma$ we get

$$\|t''\| = \|y(t)\| = \|(\mathrm{Id} + R_{fx})(\mathcal{F}_x(t))\| = \|t' + R_{fx}(t')\| < (r + \gamma r) < \rho \tag{8.53}$$

by (8.51) and the choice of r (8.47). We also use an estimate for the norm of composition of smooth maps, see for example [dlLO98]:

Lemma 5.4. [dlLO98, Theorem 4.3(ii.3)] *For any $N \geq 1$ there exists a constant M_N such that $\|h \circ g\|_{C^N} \leq M_N \|h\|_{C^N}(1 + \|g\|_{C^N})^N$ for any $h, g \in C^N(B_{x,\rho})$.*

Using this together with (8.45), $\|\bar{R}\|_C \leq \gamma$, and $\gamma \geq 1$, we can write

$$\|y(t)\|_{C^N} = \|(\mathrm{Id} + R_{fx}) \circ \mathcal{F}_x\|_{C^N} \leq M_N (1 + \gamma)(1 + K)^N \leq c_1(K, N)\gamma,$$

where $c_1(K, N) = 2M_N(1 + K)^N$. Now we can estimate the first term in (8.50):

$$\|A_x(y(t))\|_{C^N} \leq M_N \|(\mathcal{P}_x)^{-1}\|_{C^N} \|y(t)\|^N_{C^N} \leq M_N K[1+c_1(K,N)\gamma]^N \leq c_2(K,N)\gamma^N,$$

where $c_2(K, N) = M_N K(1 + c_1(K, N))^N$, and we note that $\mathcal{P}_x^{-1}$ is a polynomial of degree at most $d \leq N$. Combining this with (8.52) and K^{N-1} estimate from (8.45) for the last term in (8.50), we obtain the following estimate for the norm of J_x in (8.49):

$$\|J_x\| \leq c(N)c_2(K,N)\gamma^N \cdot r\,\|t\|^\alpha \|\bar{S}\|_C \cdot K^{N-1} \leq c_3(K,N)\,\gamma^N \cdot r\,\|t\|^\alpha \cdot \|\bar{S}\|_C. \tag{8.54}$$

where $c_3(K, N) = c(N)c_2(K, N)K^{N-1}$ and $c(N)$ is an estimate on the number of terms in J_x.

Now we estimate the main term in (8.49) using (8.38), (8.44), (8.53), and (8.51):

$$\begin{aligned}
&\|D^{(1)}_{t''} (\mathcal{P}_x)^{-1} \circ D^{(N)}_{t'} S_{fx} \circ D^{(1)}_t \mathcal{F}_x\| \leq \\
&\leq \|D^{(1)}_{t''}(\mathcal{P}_x)^{-1}\| \cdot \|D^{(N)} S_{fx}\|_\alpha \|t'\|^\alpha \cdot \|D^{(1)}_t \mathcal{F}_x\|^N \leq \\
&\leq e^{-\chi_1+\varepsilon+\varepsilon'} \cdot \|\bar{S}\|_C \cdot e^{\alpha(\chi_\ell+\varepsilon+\varepsilon')}\|t\|^\alpha \cdot e^{N(\chi_\ell+\varepsilon+\varepsilon')} = e^{-\delta}\|t\|^\alpha \|\bar{S}\|_C,
\end{aligned} \tag{8.55}$$

by the definition of δ (8.42) where $\nu = -(N+\alpha)\chi_\ell + \chi_1$. By (8.46) we have $e^{-\delta} = 1 - 2\theta$. Finally, we estimate (8.49) combining (8.54) and (8.55). For any $\bar{R} \in B_\gamma$,

$$\|D_t^{(N)}([D_{\bar{R}}T]\bar{S})_x\| \le \|t\|^\alpha \cdot \|\bar{S}\|_{\mathcal{C}} \left(1 - 2\theta + c_3(K,N)\,\gamma^N \cdot r\right) \le \|t\|^\alpha \cdot \|\bar{S}\|_{\mathcal{C}}\,(1-\theta),$$

since $r \le \theta/(c_3(K,N)\,\gamma^N)$ by (8.47). Then, for all $\bar{R} \in B_\gamma$ we obtain

$$\begin{aligned}
&\| D_t^{(N)}([D_{\bar{R}}T]\bar{S})_x \| \le \|t\|^\alpha \cdot \|\bar{S}\|_{\mathcal{C}}\,(1-\theta), \quad \text{hence}\\
&\| D^{(N)}([D_{\bar{R}}T]\bar{S})_x \|_\alpha \le (1-\theta)\cdot \|\bar{S}\|_{\mathcal{C}}, \quad \text{and so}\\
&\| [D_{\bar{R}}T]\bar{S} \|_{\mathcal{C}} = \sup_x \, \| D^{(N)}(T(\bar{S}))_x \|_\alpha \le (1-\theta)\cdot \|\bar{S}\|_{\mathcal{C}}.
\end{aligned}$$

Thus $\|D_{\bar{R}}T\| \le 1-\theta$ for all $\bar{R} \in B_\gamma$. Since $\|T(\bar{0})\|_{\mathcal{C}} \le \theta\gamma$ from the definition of γ (8.46), the operator T is a contraction from B_γ to itself. Thus T has a unique fixed point $\bar{H} \in B_\gamma$, which is section of $\mathcal{C}$ bounded in $C^{N,\alpha}$ norm and continuous in C^N norm. The corresponding family of $C^{N,\alpha}$ maps $\mathcal{H}_x = \mathrm{Id} + H_x$ satisfies (8.36), that is, conjugates $\mathcal{P}_x$ and $\mathcal{F}_x$. Then the maps $\mathcal{H}_x$ defined on $B_{x,r}$ can be uniquely extended to $C^{N,\alpha}$ diffeomorphisms on $B_{x,\rho}$, and then on $B_{x,\sigma}$, by the invariance

$$\mathcal{H}_x(t) = (\mathcal{P}_x^k)^{-1} \circ \mathcal{H}_{f^k x} \circ \mathcal{F}_x^k(t)$$

since, for each $t \in B_{x,\sigma}$, we have $\mathcal{F}_x^k(t) \in B_{x,r}$ for some k.

8.5.3 Proof of Part (2): the (Non)uniqueness of $\mathcal{H}$ and $\mathcal{P}$

This essentially follows from the "uniqueness" of the construction. First, starting with $\mathcal{H}_1 = \tilde{\mathcal{H}}$, we inductively construct coordinate changes $\mathcal{H}_k = \{\mathcal{H}_{k,x}\}$ for $k = 1, \ldots, N$ so that the corresponding normal form $\mathcal{P}_{k,x}$ is of sub-resonance type and their Taylor polynomials $\mathcal{H}_{k,x}(t) \sim \sum_{n=1}^N H_{k,x}^{(n)}(t)$ coincide with the Taylor polynomial of $\mathcal{H}$ to order k, that is $H_x^{(n)} = H_{k,x}^{(n)}$ for $n = 1, \ldots, k$. For $\mathcal{H}_1 = \tilde{\mathcal{H}}$ we have $H_x^{(1)} = H_{1,x}^{(1)} = \mathrm{Id}$ and $\mathcal{P}_{1,x}$ is sub-resonance by the assumption.

Suppose $\mathcal{H}_{k-1}$, $k \ge 2$, is constructed and has $H_x^{(n)} = H_{k-1,x}^{(n)}$ for $n = 1, \ldots, k-1$. Then $\mathcal{P}$ and $\mathcal{P}_{k-1}$ have the same terms up to order $k-1$. Hence $H_{k-1,x}^{(k)}$ and $H_x^{(k)}$ satisfy the same equation (8.27) when projected to the factor-bundle $\mathcal{Q}^{(k)}/\mathcal{S}^{(k)}$. Indeed, the Q term defined by (8.26) is composed only of $F^{(i)}$ and terms $H^{(i)}$ and $P^{(i)}$ with $1 < i \le k-1$, which are the same for $\mathcal{H}_{k-1}$ and $\mathcal{H}$. By uniqueness we obtain that

$$H_x^{(k)} = H_{k-1,x}^{(k)} + \Delta_x^{(k)}, \quad \text{where } \Delta_x^{(k)} \in \mathcal{S}_x^{(k)}.$$

Then the coordinate change $\mathcal{H}_{k,x} = (\mathrm{Id} + \Delta_x^{(k)}) \circ \mathcal{H}_{k-1,x}$ has the same Taylor terms as $\mathcal{H}$ up to order k and, since the polynomial $\mathrm{Id} + \Delta_x^{(k)}$ is in $\mathcal{S}_x$, $\mathcal{H}_k$ conjugates $\mathcal{F}$ to a sub-resonance normal form $\mathcal{P}_{k,x} = (\mathrm{Id} + \Delta_{fx}^{(k)}) \circ \mathcal{P}_{k-1,x} \circ (\mathrm{Id} + \Delta_x^{(k)})^{-1}$.

Thus in N steps we obtain the coordinate change

$$\mathcal{H}_{N,x} = G_x \circ \tilde{\mathcal{H}}_x, \quad \text{where} \quad G_x = (\mathrm{Id} + \Delta_x^{(N)}) \circ \cdots \circ (\mathrm{Id} + \Delta_x^{(2)}) \in \mathcal{S}_x.$$

In fact, for $n > d$ we have $\mathcal{S}^{(n)} = 0$ and hence $\Delta^{(n)} = 0$, so that $\mathcal{H}_N = \mathcal{H}_d$. Since $\mathcal{H}_{N,x}$ has the same Taylor terms at 0 as $\mathcal{H}_x$ up to order $N \geq d$, the "moreover" part of (2) yields $\mathcal{H}_N = \mathcal{H}$ and hence $G_x \circ \tilde{\mathcal{H}}_x = \mathcal{H}_{N,x} = \mathcal{H}_x$. The "moreover" part of (2) follows from the uniqueness of the fixed point in the last step of the construction of $\mathcal{H}$.

8.5.4 Proof of Part (1'): Construction of Resonance Normal Form

Now we construct a *polynomial* coordinate change $\mathcal{H}$ that brings the sub-resonance normal form $\mathcal{P}_x = \sum_{n=1}^{d} P_x^{(n)}$ to a resonance normal form $\tilde{\mathcal{P}}_x = \sum_{n=1}^{d} \tilde{P}_x^{(n)}$.

We will inductively construct the terms of polynomial coordinate changes $\mathcal{H}_x = \sum_{n=1}^{d} H_x^{(n)}$. All terms will be in the group $\mathcal{S}$ of sub-resonance polynomials, so the process will stop in d steps and yield sub-resonance of polynomial diffeomorphisms $\mathcal{H}_x$ and resonance normal form $\tilde{\mathcal{P}}_x$. The base case is $n = 1$, where we take $H_x^{(1)} = \mathrm{Id}$, which leaves $\tilde{P}_x^{(1)} = P_x^{(1)} = F_x$ in the resonance form, that is, block triangular.

Now we assume inductively that the terms of degree $k < n$ are constructed so that $H^{(k)}$ is a continuous section of $\mathcal{S}^{(k)}$ and $\tilde{P}_x^{(k)}$ is a resonance polynomial and is continuous in x. As before, we consider the terms of degree n in the conjugacy equation $\mathcal{H}_{fx} \circ \mathcal{P}_x = \tilde{\mathcal{P}}_x \circ \mathcal{H}_x$,

$$P_x^{(n)} + H_{fx}^{(n)} \circ F_x + \sum H_{fx}^{(i)} \circ P_x^{(j)} = F_x \circ H_x^{(n)} + \tilde{P}_x^{(n)} + \sum \tilde{P}_x^{(j)} \circ H_x^{(i)},$$

where the summations are over all i and j such that $ij = n$ and $1 < i,j < n$. Then

$$\tilde{P}_x^{(n)} \circ F_x^{-1} = -F_x \circ H_x^{(n)} \circ F_x^{-1} + H_{fx}^{(n)} + Q_x, \quad \text{where} \tag{8.56}$$

$$Q_x = \left(P_x^{(n)} + \sum_{ij=n,\ 1<i,j<n} H_{fx}^{(i)} \circ P_x^{(j)} - \tilde{P}_x^{(j)} \circ H_x^{(i)} \right) F_x^{-1}.$$

We note that Q_x is composed only of terms $H^{(i)}$ and $\tilde{P}^{(i)}$ with $1 < i < n$, which are already constructed, and terms $P^{(i)}$ with $1 < i \leq n$, which are given. Since composition of sub-resonance polynomials is a sub-resonance polynomial, by the inductive assumption Q_x is a continuous section of $\mathcal{S}^{(n)}$.

We consider splitting $\mathcal{S}_x^{(n)} = \mathcal{R}_x^{(n)} \oplus \mathcal{SS}_x^{(n)}$, where $\mathcal{R}_x^{(n)}$ and $\mathcal{SS}_x^{(n)}$ denote the maps of resonance and strict sub-resonance types respectively. We seek $H_x^{(n)}$ so that the right side of (8.56) is in $\mathcal{R}_x^{(n)}$, and hence so will be $\tilde{P}_x^{(n)}$ when defined by this equation.

Projecting (8.56) to the factor bundle $\mathcal{S}^{(n)}/\mathcal{R}^{(n)}$, we need to solve the equation

$$0 = -F_x \circ \bar{H}_x^{(n)} \circ F_x^{-1} + \bar{H}_{fx}^{(n)} + \bar{Q}_x, \tag{8.57}$$

where $\bar{H}^{(n)}$ and $\bar{Q}$ are the projections of $H^{(n)}$ and Q respectively. We consider the automorphism Φ^{-1} of the bundle $\mathcal{S}^{(n)}$ covering f with fiber maps

$$\Phi_x^{-1} : \mathcal{S}_x^{(n)} \to \mathcal{S}_{fx}^{(n)} \quad \text{where} \quad \Phi_x^{-1}(R) = F_x \circ R \circ F_x^{-1}. \tag{8.58}$$

Since F preserves the splitting $\mathcal{E} = \mathcal{E}^1 \oplus \cdots \oplus \mathcal{E}^\ell$, the resonance and strict sub-resonance types are preserved by Φ^{-1}. We denote by $\bar{\Phi}^{-1}$ the induced automorphism of $\mathcal{S}^{(n)}/\mathcal{R}^{(n)}$ and see that (8.57) becomes

$$\bar{H}_{fx}^{(n)} = \tilde{\Phi}_x^{-1}(\bar{H}_x^{(n)}), \quad \text{where} \quad \tilde{\Phi}_x^{-1}(R) = \bar{\Phi}_x^{-1}(R) - \bar{Q}_x. \tag{8.59}$$

Thus a solution of (8.57) is a $\tilde{\Phi}^{-1}$-invariant section of $\mathcal{S}^{(n)}/\mathcal{R}^{(n)}$.

Lemma 5.5. *The map* $\Phi^{-1} : \mathcal{SS}^{(n)} \to \mathcal{SS}^{(n)}$ *given by* (8.58) *is a contraction over* f*, and hence so is* $\tilde{\Phi}^{-1} : \mathcal{S}^{(n)}/\mathcal{R}^{(n)} \to \mathcal{S}^{(n)}/\mathcal{R}^{(n)}$ *given by* (8.59). *More precisely,* $\|\Phi_x^{-1}(R)\| \le e^{\lambda+(d+2)\varepsilon} \cdot \|R\|$.

Proof. The statement about $\tilde{\Phi}^{-1}$ follows since the linear part $\bar{\Phi}^{-1}$ of $\tilde{\Phi}^{-1}$ is given by Φ^{-1} when $\mathcal{S}^{(n)}/\mathcal{R}^{(n)}$ is naturally identified with $\mathcal{SS}^{(n)}$.

By Lemma 5.1, for polynomials of homogeneous type $s = (s_1, \ldots, s_\ell)$ with $s_1 + \cdots + s_\ell = n$ we have $\|\Phi_x^{-1}(R)\| \le e^{\chi_i - \sum s_j \chi_j + (n+1)\varepsilon}\|R\|$. For all strict sub-resonance homogeneous types we have $\chi_i - \sum_{j=1}^{\ell} s_j \chi_j \le \mu$ by the definition of μ (8.14) and hence for any $R \in \mathcal{SS}_x^{(n)}$ we have $\|\Phi_x(R)\| \le e^{\mu+(n+1)\varepsilon} \cdot \|R\|$. Since $n \le d$ the exponent satisfies $\mu + (n+1)\varepsilon \le \lambda + (d+1)\varepsilon < 0$ since $\varepsilon < \varepsilon_0$ given by (8.15). □

We conclude that $\tilde{\Phi}^{-1}$ is a contraction and hence has a unique continuous invariant section $\bar{H}^{(n)}$. We choose a continuous section $H^{(n)}$ of $\mathcal{S}^{(n)}$ which projects to $\bar{H}^{(n)}$, which is defined uniquely up to a section of $\mathcal{R}^{(n)}$. For example, one can take $H^{(n)}$ in $\mathcal{SS}^{(n)}$. Once $H^{(n)}$ is chosen, we define $\tilde{P}_x^{(n)}$ by Eq. (8.56) and get a continuous section $\tilde{P}^{(n)}$ of $\mathcal{R}^{(n)}$. This completes the inductive step and the construction of $\mathcal{H}$ and $\tilde{\mathcal{P}}$.

8.5.5 Proof of Part (2'): The (Non)uniqueness for Resonance Normal Form

This follows from the "uniqueness" of the construction in the previous section, similarly to the proof of part (2). The process of transition from $\tilde{\mathcal{H}}_x'$ to $\mathcal{H}_x'$ stays in the group of resonance polynomials $\mathcal{R}_x$ and we obtain $\mathcal{H}_x = \mathcal{H}_{d,x} = G_x \circ \tilde{\mathcal{H}}_x'$, where $G_x \in \mathcal{R}_x$.

8.5.6 Proof of Part (3): Centralizer

First we prove that the derivative of $\mathcal{G}$ at zero section, $\Gamma_x = D_0\mathcal{G}_x$, is sub-resonance. Since Γ_x is linear, this is equivalent to the fact that Γ_x preserves the flag of fast sub-bundles associated with the splitting (8.12). Suppose to the contrary that for some $x \in X$ and some $i > j$ we have a unit vector t in $\mathcal{E}_x^j$ such that $t' = \Gamma_x(t)$ has nonzero component $t_i' \neq 0$ in $\mathcal{E}_{gx}^i$. Then we have

$$\|(F_{gx}^n \circ \Gamma_x)(t)\| \geq \|F_{gx}^n(t_i')\| \geq e^{(\chi_i - \varepsilon)n} \, \|t_i'\|.$$

On the other hand, since the extensions and hence their derivatives commute, we have

$$\|(F_{gx}^n \circ \Gamma_x)(t)\| = \|\Gamma_{f^n x}(F_x^n(t))\| \leq \|\Gamma_{f^n x}\| \cdot e^{(\chi_j+\varepsilon)n}\|t\| \leq Ce^{(\chi_j+\varepsilon)n},$$

which is impossible for large n as ε is small enough. Indeed, since $i > j$ we have non sub-resonance relation $-\chi_i + \chi_j < 0$, so by definition (8.13) of $\tilde{\lambda}$ we have $-\chi_i + \chi_j \leq \tilde{\lambda} < 0$, and hence by definition (8.15) of ε_0 we have $\varepsilon_0 \leq -\lambda/(d+2) \leq -\tilde{\lambda}/3 < (\chi_i - \chi_j)/2$. Since $\varepsilon < \varepsilon_0$, this yields $\chi_j + \varepsilon < \chi_i - \varepsilon$.

Similarly, we can further show that Γ_x is resonance, namely preserves the splitting. Using the preceding notations, if $i < j$ we can estimate backward iterates: for $n < 0$ we have

$$e^{(\chi_i+\varepsilon)n} \, \|t_i'\| \leq \|F_{gx}^n(t_i')\| \leq \|(F_{gx}^n \circ \Gamma_x)(t)\| = \|\Gamma_{f^n x}(F_x^n(t))\| \leq \|\Gamma_{f^n x}\| \cdot e^{(\chi_j-\varepsilon)n}\|t\|,$$

which is impossible since $\chi_i + \varepsilon < \chi_j - \varepsilon$. This follows as previously from (8.15) and (8.14) since for a strict sub-resonance $\chi_i < \chi_j$ we have $\chi_i - \chi_j \leq \mu < 0$ and hence $\varepsilon < \varepsilon_0 \leq -\mu/(d+1) \leq -\mu/2 < (-\chi_i + \chi_j)/2$.

Now we consider a new family of coordinate changes

$$\tilde{\mathcal{H}}_x = \Gamma_x^{-1} \circ \mathcal{H}_{gx} \circ \mathcal{G}_x,$$

which also satisfies $\tilde{\mathcal{H}}_x(0) = 0$ and $D_0\tilde{\mathcal{H}}_x = \text{Id}$. A direct calculation shows that

$$\begin{aligned} \tilde{\mathcal{H}}_{fx} \circ \mathcal{F}_x \circ \tilde{\mathcal{H}}_x^{-1} &= \Gamma_{fx}^{-1} \circ \mathcal{H}_{fgx} \circ \mathcal{G}_{fx} \circ \mathcal{F}_x \circ \mathcal{G}_x^{-1} \circ \mathcal{H}_{gx}^{-1} \circ \Gamma_x = \\ &= \Gamma_{fx}^{-1} \circ \mathcal{H}_{fgx} \circ \mathcal{F}_{gx} \circ \mathcal{H}_{gx}^{-1} \circ \Gamma_x = \Gamma_{fx}^{-1} \circ \mathcal{P}_{gx} \circ \Gamma_x = \tilde{\mathcal{P}}_x. \end{aligned}$$

Hence, if $\mathcal{P}_x$ is a sub-resonance polynomial, then so is $\tilde{\mathcal{P}}_x$ as a composition of sub-resonance polynomials. Now part (2) of the theorem gives $\tilde{\mathcal{H}}_x = G_x\mathcal{H}_x$ for some $G_x \in \mathcal{S}_x$ which depends continuously on x. Then the definition of $\tilde{\mathcal{H}}_x$ yields

$$\mathcal{H}_{gx} \circ \mathcal{G}_x = \Gamma_x \circ \tilde{\mathcal{H}}_x = (\Gamma_x G_x) \circ \mathcal{H}_x$$

so that $\mathcal{H}_{gx} \circ \mathcal{G}_x \circ \mathcal{H}_x^{-1} = \Gamma_x G_x$, which is again a sub-resonance polynomial, as claimed.

Similarly, using part (2'), we can obtain that if $\mathcal{P}_x$, and hence $\tilde{\mathcal{P}}_x$, are resonance polynomials, then $\mathcal{H}_{gx} \circ \mathcal{G}_x \circ \mathcal{H}_x^{-1} = \Gamma_x G_x$, where G_x, and hence $\Gamma_x G_x$, are also resonance polynomials for each $x \in X$.

This completes the proof of Theorem 4.3. □

8.5.7 Proof of Corollary 4.4

By part (2) of Theorem 4.3, if we fix a choice of Taylor polynomials of degree d for $\mathcal{H}_x$, then the family $\mathcal{H}_x$ is unique. Then for each $N > d$ we can do the construction in part (1) with this fixed choice of Taylor polynomials and obtain the family of C^N diffeomorphisms $\mathcal{H}_x$. By uniqueness, all these families coincide and hence $\mathcal{H}_x$ are C^∞ diffeomorphisms. For $\tilde{\mathcal{H}}$ in part (1') the smoothness follows since, as we will show, it is a composition of $\mathcal{H}$ with a polynomial diffeomorphism.

8.6 Proof of Theorem 4.6 and Corollary 4.8

The parts (1), (1'), (2), (2'), (3), and (5) of Theorem 4.6 are obtained using Theorem 4.3 as follows. We consider the vector bundle $\mathcal{E} = T\mathcal{W}$ with $\mathcal{E}_x = T_x\mathcal{W}$. To construct extension $\mathcal{F}$ as in Theorem 4.3 we restrict f to the leaves of $\mathcal{W}$ and obtain $\mathcal{F}_x$ by identifying $B(x,\sigma) \subset T_x\mathcal{W}$ with a neighborhood of x in $\mathcal{W}_x$ using an exponential map. It is easy to see that Assumptions 4.1 are satisfied with $N = \lfloor r \rfloor$ and $\alpha = r - \lfloor r \rfloor$.

Hence Theorem 4.3 gives existence of families $\{\mathcal{H}_x\}_{x\in X}$ and $\{\mathcal{H}'_x\}_{x\in X}$ of local normal form coordinates as in (1) and (1') satisfying uniqueness properties (2) and (2'). Then, as in Remark 4.5, they can be extended uniquely to global diffeomorphisms $\mathcal{H}_x \colon \mathcal{W}_x \to \mathcal{E}_x$. We note that the Hölder condition at 0 in Theorem 4.3 implies that $\mathcal{H}_x$ in (1) is globally Hölder along $\mathcal{W}_x$ by part (4). This also implies that $\mathcal{H}'_x$ in (1') is globally Hölder since by (2) it differs from $\mathcal{H}_x$ by a polynomial diffeomorphism.

To prove (3), we similarly restrict g to the leaves of $\mathcal{W}$ and obtain the extension $\mathcal{G}$ commuting with $\mathcal{F}$, so that the result follows from (3) of Theorem 4.3.

The existence of $\{\mathcal{H}_x\}_{x\in X}$ as in part (5) can be obtained by constructing the Taylor terms of $\mathcal{H}_x$ that depend smoothly on x as indicated in Remark 5.3; see [KS16] for more details of this argument.

Part (4) requires a different argument for which we refer to [KS16, KS17]. First one shows inductively that the Taylor polynomial of the transition maps is sub-resonance, and then argues that error term is zero.

The first part of Corollary 4.8 follows directly since the splitting (8.6) is trivial, $d = 1$, and hence there are no nonlinear sub-resonance polynomials. This also means that $\tilde{\lambda}$ and μ are not needed as there are no corresponding relations and we get $\varepsilon_0 = \lambda/3 = -\chi/3$. Hence we need $\varepsilon < \min\{\varepsilon_0/3, \nu/(N+\alpha+1)\}$. The second term is $-\chi(N+\alpha-1)/(N+\alpha+1)$ and smallest for $N = 1$ and gives $-\chi\alpha/(2+\alpha)$, which is less than $\varepsilon_0/3$. This means that we need $\varepsilon < -\chi\alpha/(2+\alpha)$, which yields the interval $(-\chi(1+\frac{\alpha}{2+\alpha}), -\chi(1-\frac{\alpha}{2+\alpha}))$ with endpoint ratio $1+\alpha$.

References

[A] L. Arnold. Random Dynamical Systems. Springer Monographs in Mathematics, 1998.

[AK92] L. Arnold and X. Kedai. Normal forms for random diffeomorphisms. J. Dyn. Differ. Equ. 4, (1992), 445–483.

[BrKo] I. U. Bronstein and A. Ya. Kopanskii. Smooth Invariant Manifolds and Normal Forms. World Scientific, 1994.

[Bu18] C. Butler. Rigidity of equality of Lyapunov exponents for geodesic flows. J. Differ. Geom. 109 (2018), 39–79.

[DWX19] D. Damjanovic, A. Wilkinson, and D. Xu. Pathology and asymmetry: centralizer rigidity for partially hyperbolic diffeomorphisms. arXiv:1902.05201.

[dlLO98] R. de la Llave and R. Obaya. Regularity of the composition operator in spaces of Hölder functions. Discrete and Contin. 5 (1999), 157–184.

[F07] Y. Fang. On the rigidity of quasiconformal Anosov flows. Ergod. Theory Dyn. Syst. 27 (2007), 1773–1802.

[FFH10] Y. Fang, P. Foulon, and B. Hasselblatt. Zygmund strong foliations in higher dimension. J. Mod. Dyn. 4 (2010), 549–569.

[Fe95] R. Feres. The invariant connection of 1/2-pinched Anosov diffeomorphism and rigidity. Pacific J. Math. 171 (1995), 139–155.

[Fe04] R. Feres. A differential-geometric view of normal forms of contractions. In Modern Dynamical Systems and Applications, Cambridge University Press (2004), 103–121.

[FKSp11] D. Fisher, B. Kalinin, and R. Spatzier. Totally non-symplectic Anosov actions on tori and nilmanifolds. Geom. Topol. 15 (2011), 191–216.

[GKS11] A. Gogolev, B. Kalinin, and V. Sadovskaya. Local rigidity for Anosov automorphisms (with Appendix by R. de la Llave). Math. Res. Lett. 18 (2011), 843–858.

[GKS19] A. Gogolev, B. Kalinin, and V. Sadovskaya. Center foliation rigidity for partially hyperbolic toral diffeomorphisms. Math. Ann. 387 (2023), no. 3–4, 1579–1602.

[Gu02] M. Guysinsky. The theory of non-stationary normal forms. Ergod. Theory Dyn. Syst. 22 (2002), 845–862.

[GuKt98] M. Guysinsky and A. Katok. Normal forms and invariant geometric structures for dynamical systems with invariant contracting foliations. Math. Res. Lett. 5 (1998), 149–163.

[KKt01] B. Kalinin and A. Katok. Invariant measures for actions of higher rank abelian groups. Proceedings of Symposia in Pure Mathematics. 69 (2001), 593–637.

[KKt07] B. Kalinin and A. Katok. Measure rigidity beyond uniform hyperbolicity: invariant measures for Cartan actions on tori. J. Mod. Dyn. 1 (2007), 123–146.

[KKtR11] B. Kalinin, A. Katok, and F. Rodriguez-Hertz. Nonuniform measure rigidity. Ann. Math. 174 (2011), 361–400.

[KtL91] A. Katok and J. Lewis. Local rigidity for certain groups of toral automorphisms. Israel J. Math. 75 (1991), 203–241.

[KtR15] A. Katok and F. Rodriguez-Hertz. Arithmeticity and topology of higher rank actions of Abelian groups. J. Mod. Dyn. 10 (2016), 115–152.

[KtSp97] A. Katok and R. Spatzier. Differential rigidity of Anosov actions of higher rank abelian groups and algebraic lattice actions. Tr. Mat. Inst. Steklova 216 (1997), Din. Sist. i Smezhnye Vopr., 292–319; translation in Proc. Steklov Inst. Math. 216 (1997), 287–314.

[KS03] B. Kalinin and V. Sadovskaya. On local and global rigidity of quasiconformal Anosov diffeomorphisms. J. Inst. Math. Jussieu 2 (2003), 567–582.

[KS06] B. Kalinin and V. Sadovskaya. Global rigidity for totally nonsymplectic Anosov $\mathbb{Z}^k$ actions. Geom. Topol. 10 (2006), 929–954.

[KS16] B. Kalinin, V. Sadovskaya. Normal forms on contracting foliations: smoothness and homogeneous structure. Geom. Dedicata, 183 (2016), 181–194.

[KS17] B. Kalinin and V. Sadovskaya. Normal forms for non-uniform contractions. J. Mod. Dyn. 11 (2017), 341–368.

[LL05] W. Li and K. Lu. Sternberg theorems for random dynamical systems. Commun. Pure Appl. Math. 58 (2005), 0941–0988.

[M19] K. Melnick. Nonstationary smooth geometric structures for contracting measurable cocycles. Ergod. Theory Dyn. Syst. 39 (2019), 392–424.

[P] Ya. Pesin. Lectures on Partial Hyperbolicity and Stable Ergodicity. Zurich Lectures in Advanced Mathematics, EMS, 2004.

[S05] V. Sadovskaya. On uniformly quasiconformal Anosov systems. Math. Res. Lett. 12 (2005), 425–441.

[St57] S. Sternberg. Local contractions and a theorem of Poincaré. Am. J. Math. 79 (1957), 809–824.

9

Survey on Entropy-Type Invariants of Subexponential Growth in Dynamical Systems

ADAM KANIGOWSKI*, ANATOLE KATOK†, AND DAREN WEI‡

Measure-theoretic and topological entropy are classical invariants in the theory of dynamical systems. There are several recently developed entropy-type invariants for systems of subexponential growth: sequence entropy, slow entropy, Kakutani invariants, scaled entropy, entropy dimensions and entropy convergence rate. They measure the complexity of zero entropy systems by different approaches. These new invariants and corresponding new theories have many applications and interesting properties. This chapter gives a comprehensive exposition of the slow entropy theory and also discusses some related topics.

9.1 Introduction

9.1.1 Short History of Entropy-Type Invariants

Classification is one of the central problems in ergodic theory and dynamical systems. For a dynamical system, one of its most important and natural characteristics is its complexity. Therefore it is natural to try to classify dynamical systems based on their complexity. The classical and natural way to measure orbit complexity for a dynamical system, continuous or measurable, is to look at the growth of the number of distinguishable orbits segments as a function of time. Distinguishable is the key word here and, according to the setting, it is given a specific meaning. Basically there are three principal settings for the problem corresponding to the following classes of dynamical systems: *topological dynamical systems*, *symbolic dynamical systems* and *measurable dynamical systems*.

Distinguishing the Orbit Segments and Counting. Orbit segments are distinguished with the use of auxiliary tools which can be viewed as a mathematical

* A. K. was partially supported by the NSF grant DMS-1956310,

† A. K. was partially supported by the NSF grant DMS-16-02409.

‡ D. W. was partially supported by the NSF grant DMS-16-02409.

abstraction of the notion of observation. Those tools are accordingly *Bowen metric* in the topological case, *value of the zero coordinate* in the symbolic case, and *entropy* $H(\xi)$[1] *of a partition* in the measurable case. Notice that the Bowen metric at orbit length T is continuous as a function of T, and hence to distinguish orbit segments one needs to put an arbitrary threshold; in the remaining cases the tool is discrete, but in the last one it is rather arbitrary. To refine the device in the topological case, the threshold is set at an arbitrary low fixed level before counting, and in the measurable one, more and more refined partitions are considered, so that the σ-algebra $\mathfrak{B}(\xi)$ becomes more and more representative. In the symbolic case, in principle one may take values of several successive coordinates rather than a single one, but this does not change the outcome for any counting procedure. Similarly, for many "good" continuous and measurable systems, for adequate counting it is sufficient to fix any threshold below a certain fixed number (e.g., the *expansivity constant*) or to consider a single partition (e.g., a *generator*).

For the *counting* we measure the asymptotic growth of orbit segments with the time length. As a natural measure of complexity one uses (n, ϵ)*-spanning set* or (n, ϵ)*-separated set*[2] in the topological setting, *the number of sequences of successive n symbols which appear in elements of a closed invariant set* in the symbolic setting and *entropy* $H(\xi_{-n}^{T})$[3] of the joint partition in the measurable setting.

Topological and Measure-Theoretic Entropy. The exponential growth of orbit complexity is given by the *topological entropy* and *measure-theoretic entropy*. These isomorphism invariants have been extensively studied over the last sixty years to deal with systems possessing exponential orbit growth. The basic connection between the entropies is provided by the Variational Principle for entropy.[4] Based on the Shannon–McMillan–Breiman theorem, one can interpret the variational principle as a precise quantification of the following assertion:

($\mathcal{VP}$) *The global exponential growth rate of orbit complexity can be arbitrarily well approximated by the exponential rates of growth for the sets of statistically typical orbits satisfying various uniform distribution laws.*

Distance between Orbit Segments. The method of distinguishing orbit segments just described assumes uniform accuracy of the measuring device and, in the case of partitions, ignores its discrete nature. If we measure complexity using partitions in the topological setting, measurements for nearby points and close to the boundary of the partition may give different results. These considerations motivate an alternative approach of measuring the distance between orbit segments: using the

[1] See (9.9) for a detailed definition.
[2] See Section 9.2.1 for a detailed discussion.
[3] $\xi_{-n}^{T} = \vee_{i=0}^{n-1} T^{-i}(\xi)$.
[4] See Section 9.2.3.

average in time rather than maximal deviation. As it turns out, the resulting notions do not change in the case of exponential orbit complexity, but make a crucial difference in the subexponential case. The integral or *Hamming metric*[5] between the orbit segments for a map f or a flow Φ in a metric space is defined accordingly as

$$d_n^{H,f}(x,y) = \frac{1}{n}\sum_{i=0}^{n-1} d(f^i(x), f^i(y)) \tag{9.1}$$

and

$$d_t^{H,\Phi}(x,y) = \frac{1}{t}\int_0^t d(\Phi^s(x), \Phi^s(y))\, ds. \tag{9.2}$$

In the case of discrete measurements, which we use for symbolic and measurable systems, one simply assumes that distance between symbols or different elements of the partition ξ is equal to one and uses (9.1). In other words, the Hamming distance between two n-strings of symbols is equal to the proportion of the number of places where the symbols in the strings are different. Let Ω_N^n be the set of n-strings of sequences of symbols for an alphabet with N elements. We use simplified notation for the Hamming metric in this case:

$$d_n^H(\omega, \omega') = \frac{1}{n}\sum_{i=0}^{n-1}(1 - \delta_{\omega_i \omega_i'}). \tag{9.3}$$

Both the original measuring scheme and the average one can be put into an even more general context by considering a "coherent" system of metrics d_n for the sets Ω_N^n of n-strings of sequences of symbols for an alphabet with N elements. Coherence means convexity with respect to concatenation of strings. (The original scheme, of course, corresponds to the trivial choice: $d_n(\omega, \omega') = 1$ for $\omega \neq \omega'$.)

A broad class of metrics can be obtained by fixing a set of elementary operations, and defining the distance as the minimal number of elementary operations needed to transform one string into the other, divided by the length of the strings. For the Hamming metric, elementary operations are just replacements of one symbol by another at the same place. Another important special case is that of *Kakutani metrics* d_n^M or $\bar{f}$*-metric*[6], where an elementary operation consists in crossing out a symbol and inserting another one *at any place*. Basic properties of those metrics are discussed in [38, Section 4]. For the case of Hamming metrics for ergodic measure-preserving transformations this is worked out in detail in [39, Section 1]. The treatment in the Kakutani metric case is quite similar. In both cases the key observation is that one can calculate the size of an ϵ-ball using Stirling's formula and on the exponential in n scale this size goes to zero with ϵ. This also makes the symbolic case straightforward.

[5] See Definition 4.1.2 for general case.

[6] See (9.59) for a continuous version.

Measuring Subexponential Growth. In the case of zero topological entropy, measuring the growth rates for (n, ϵ)-spanning sets or (n, ϵ)-separated sets and the number of sequences of successive n symbols which appear in elements of a closed invariant sets provides a natural approach to the study of orbit complexity. In the symbolic case at least, it is a well-researched subject branching into combinatorics, algorithmic complexity and suchlike; see [17]. We will discuss the topological case later.

However, in the measurable case a difficulty appears. It turns out that, for any aperiodic measure-preserving transformation T, any speed of growth below linear in n appears for the entropy $H(\xi^T_{-n})$ for an appropriately chosen finite partition ξ; see for example [3], although this is not the earliest proof of this fact in the literature. Several solutions have been suggested to overcome this problem.

Before outlining those, let us point out that independently of their success, one cannot hope for a counterpart of variational principle with the standard choice of metric in the topological category, since topological orbit complexity may change for different uniquely ergodic realizations (e.g., irrational rotation and the Denjoy example, [25, Example 2.6.9]). Hence a modification of the approach in the topological setting may also be in order.

Classical entropy theory does not provide any information for systems with slower orbit growth, namely dynamical systems with subexponential orbit growth, for which the invariant is zero. In order to get some non-trivial invariants for such systems, one needs to observe their dynamical properties at different rates.

The earliest successful construction of slow growth invariants for measure-preserving transformations was found by Kushnirenko [48] and he denoted this invariant as the *sequence entropy*. The idea here is to give the system enough time to develop exponential behavior, not against the number of iterates but rather against the number of measurements; namely, instead of taking the standard consecutive sequence of iterates to compute the asymptotic growth of different codes, one calculates the asymptotic growth along a given (convenient) subsequence of consecutive iterates.

Kushnirenko [48] used the sequence entropy to show that the horocycle flow is not isomorphic with its Cartesian square (see also [34]). Moreover, it is shown that this invariant vanishes at all scales if and only if the system has discrete spectrum. Following Kushnirenko's approach, Hulse [29], Newton [53, 54, 55], Krug [47], Dekking [9] and Lemańczyk [49] further developed the measure-theoretic sequence entropy and Goodman [24] modified this construction to the topological category. Sequence entropy is also related to classical ergodic theory properties, such as mixing: Saleski [64], Hulse [29], Huang, Shao and Ye [28]; mild mixing: Zhang [75], Huang, Shao and Ye [28] and weak mixing: Hulse [29], Zhang [75], Huang, Shao and Ye [28]. The crucial idea behind the sequence entropy is that the lack of exponential growth is compensated with the sparseness of the defining subsequence.

On the other hand, in [37, 38] A. Katok suggested to count the number of statistically different orbits (Hamming balls) in a scale which can be adjusted to the system (subexponential, polynomial, logarithmic, etc.). This approach was further developed by Katok and Thouvenot [41], where the classical slow entropy definition was generalized to actions of amenable groups and then was used to give a criterion for smooth realization of $\mathbb{Z}^k$ actions. Then Galatolo [23] generalized this invariant to the topological setting. Later Ferenczi [15] proved that vanishing of the measure-theoretic slow entropy at all scales is equivalent to discrete spectrum in the measurable category; recently, Kanigowski, Vinhage and Wei [34] proved an analogue of this result in the topological category. Hochman [26] studied slow entropy for higher-rank actions and gave a counterexample for smooth realization in the setting of infinite-measure-preserving $\mathbb{Z}^2$-actions. A. Katok, S. Katok and Rodriguez-Hertz [43] studied the connections between the slow entropy and Fried entropy for maximal rank $\mathbb{Z}^k$-actions as well as the lower bounds of these quantities for maximal rank $\mathbb{Z}^k$-actions.

One of the most important differences between slow entropy and sequence entropy is that slow entropy is better adjusted to deal with non-homogeneous systems. For homogeneous systems, computation of both invariants is based on controlled (polynomial) divergence of nearby points (see [34] for more details). Therefore, it can be shown that the metric orbit growth is the same (up to constants) as the topological orbit growth and the second one is easier to compute while dealing with, say, unipotent matrices. However, if the system lacks homogeneity, that is, some rank-one systems, translation flows, smooth flows on surfaces, Liouvillian time changes of linear flows on $\mathbb{T}^2$, time changes of unipotent flows, then the sequence entropy is very difficult to control. Indeed, since one only takes a (sparse) subsequence of iterations, it is almost impossible to control what happens between two consecutive elements of the subsequence. On the other hand, slow entropy tracks entire pieces of orbits and therefore it is easier to control the growth.

The aim of this survey is to give a systematic exposition of slow-entropy-type invariants. Most of the survey is devoted to the slow entropy introduced by Katok and Thouvenot. Besides slow entropy and sequence entropy, we will also describe several other invariants which share some common features with slow entropy, namely Fried entropy ([43]), Kakutani equivalence invariants ([60, 62, 35]), scaled entropy ([76, 77]), entropy dimensions ([8, 18, 1, 10, 11]), entropy convergence rates ([3, 4, 5]) and scaling entropy ([67, 68, 69, 71, 72, 73, 74, 65, 66]). These invariants reflect different features of a given dynamical system from different points of view.

9.1.2 Plan of the Chapter

The structure of this chapter is as follows:

- In §9.2 we review the basic theory of entropy, which includes topological entropy, measure-theoretic entropy and the variational principle.
- In §9.3 we introduce the sequence entropy, describing its properties and connections with the classical entropy. Moreover, we mention spectral theory, the variational principle and provide some examples of the sequence entropy in homogeneous setting.
- In §9.4 we give a systematic introduction of the slow entropy, relations with spectral theory, Kushnirenko's inequality, conditions on vanishing of slow entropy, failure of the variational principle and different upper and lower quantities of slow entropy. Several interesting examples are also provided to illustrate the features of slow entropy; for example, quasi-unipotent flows, rank-one systems, surface flows, AbC constructions and some $\mathbb{Z}^k$-actions.
- In §9.5 we mainly concentrate on several invariants related to slow entropy, namely Fried entropy, Kakutani equivalence invariants, scaled entropy, entropy dimensions and entropy convergence rate. Several examples of these invariants with detailed discussions are also provided in this section.
- In §9.6 we post some open questions related to slow entropy invariants.

Acknowledgments The authors would like to thank Svetlana Katok and Mariusz Lemańczyk for their help, support and careful reading of the first draft of the chapter.

9.2 Classical Entropy Theory

In this section, we will review the classical theory of topological and measure-theoretic entropy, which will help us formulate parallel results for the sequence entropy and the slow entropy. For more details, see, for example, [40].

9.2.1 Topological Entropy

Let (X, d) be a compact metric space and $T\colon (X, d) \to (X, d)$ be a continuous map. For $\epsilon > 0$ let $S_d(\epsilon)$ denote the minimal number of balls of radius ϵ which covers the whole space, $D_d(\epsilon)$ the minimal number of sets with diameter less than or equal to ϵ whose union covers the whole space and $N_d(\epsilon)$ the maximal number of ϵ-separated points in the space. It is clear that these three quantities satisfy the following inequalities:

$$\begin{aligned} D_d(2\epsilon) &\le S_d(\epsilon) \le D_d(\epsilon), \\ N_d(2\epsilon) &\le S_d(\epsilon) \le N_d(\epsilon). \end{aligned} \tag{9.4}$$

In order to measure the complexity of the dynamical system (X, T, d), the Bowen metric is introduced as follows:

$$d_n^T(x,y) := \max\{d(x,y), d(Tx, Ty), \dots, d(T^{n-1}x, T^{n-1}y)\}. \tag{9.5}$$

Combining the quantities defined in the first paragraph and the Bowen metric, the complexity of (X, T, d) with respect to $\epsilon > 0$ is measured by the following quantity:

$$h_d(T, \epsilon) := \limsup_{n\to\infty} \frac{\log S_{d_n^T}(\epsilon)}{n}. \tag{9.6}$$

Notice that $h_d(T, \epsilon)$ is non-decreasing as a function of ϵ and thus we can define the quantity $h_d(T)$ of (X, T, d) as

$$h_d(T) := \lim_{\epsilon\to 0} h_d(T, \epsilon). \tag{9.7}$$

In fact, $h_d(T)$ is independent of the metric d:

Proposition 2.1.1. *If d' is another metric on X which defines the same topology as d, then $h_d(T) = h_{d'}(T)$.*

Thus we have the following:

Definition 2.1.2 (Topological entropy). *The quantity $h_d(T)$ calculated for any metric generating a given topology in X is called the topological entropy of T and is denoted as $h_{top}(T)$.*

Remark 2.1.3. *We point out that the following quantities[7] are equivalent to the definition of topological entropy:*

$$\begin{aligned} h_{top}(T) &= \lim_{\epsilon\to 0} \liminf_{n\to\infty} \frac{\log S_{d_n^T}(\epsilon)}{n} = \lim_{\epsilon\to 0} \lim_{n\to\infty} \frac{\log D_{d_n^T}(\epsilon)}{n} \\ &= \lim_{\epsilon\to 0} \limsup_{n\to\infty} \frac{\log N_{d_n^T}(\epsilon)}{n} = \lim_{\epsilon\to 0} \liminf_{n\to\infty} \frac{\log N_{d_n^T}(\epsilon)}{n}. \end{aligned} \tag{9.8}$$

The following proposition describes some basic features of topological entropy.

Proposition 2.1.4. *We have the following basic properties of topological entropy:*

(1) If the map S is a factor of T, then $h_{top}(S) \le h_{top}(T)$.

(2) If Λ is a closed f-invariant set, then $h_{top}(T\restriction \Lambda) \le h_{top}(T)$.

(3) If $X = \cup_{i=1}^m \Lambda_i$, where Λ_i, $i = 1, 2, \dots, m$ are closed T-invariant sets, then $h_{top}(T) = \max_{1\le i\le m} h_{top}(T_{\restriction\Lambda_i})$.

(4) $h_{top}(T^m) = |m| h_{top}(T)$.

(5) $h_{top}(T \times S) = h_{top}(T) + h_{top}(S)$.

We end this section with the following basic examples.

[7] The reason they are equivalent is the inequalities shown in (9.4).

Example 2.1.5. *1. The topological entropy of any translation T_γ of the torus or any linear flow (T_w^t) on the torus is equal to zero.*

2. The topological entropy of the gradient flow on the round sphere is equal to zero.

3. If $E_m \colon S^1 \to S^1$ is an expanding map of degree m, $E_m(z) = z^m$, then $h_{top}(E_m) = \log|m|$.

4. For a topological Markov chain σ_A, $h_{top}(\sigma_A) = \log|\lambda_A^{\max}|$, where $\lambda_A^{\max}$ is the maximal eigenvalue of A.

9.2.2 Measure-Theoretic Entropy

Instead of considering the complexity in the topological category, similar invariants can be constructed in the measurable category for measure-preserving dynamical systems.

Let $(X, \mathscr{B}, \mu)$ be a probability Borel space and I a finite or countable set of indices. Suppose we have a measurable partition $\xi = \{C_\alpha | \alpha \in I\}$. We define $H(\xi)$ as follows:

$$H(\xi) = H_\mu(\xi) = - \sum_{\alpha \in I, \mu(C_\alpha > 0)} \mu(C_\alpha) \log \mu(C_\alpha). \tag{9.9}$$

The *conditional entropy* of ξ with respect to a partition $\eta = \{D_\alpha | \alpha \in J\}$ is defined as

$$H(\xi|\eta) = -\sum_{\beta \in J} \mu(D_\beta) \sum_{\alpha \in I} \mu(C_\alpha|D_\beta) \log(\mu(C_\alpha|D_\beta)),$$

where $\mu(A|B) = \frac{\mu(A \cap B)}{\mu(B)}$.

Moreover, we introduce the following notions for partitions:

(1) $\xi \le \eta$ if and only if for all $D \in \eta$ there exists a $C \in \xi$ such that $D \subset C$ and we will say η is a *refinement* of ξ and ξ is subordinate to η;
(2) The *joint partition* of ξ and η is defined as $\xi \vee \eta := \{C \cap D | C \in \xi,\ D \in \eta,\ \mu(C \cap D) > 0\}$;
(3) Two partitions ξ and η are *independent* if $\mu(C \cap D) = \mu(C) \cdot \mu(D)$ for all $C \in \xi, D \in \eta$.

We have the following proposition.

Proposition 2.2.1. *Let $\xi = \{C_\alpha | \alpha \in I\}$, $\eta = \{E_\alpha | \alpha \in J\}$, $\zeta = \{E_\alpha | \alpha \in K\}$ be finite or countable measurable partitions of (X, μ). Then:*

(1) $0 < -\log(\sup_{\alpha \in I} \mu(C_\alpha)) \le H(\xi) \le \log \operatorname{card} \xi$; furthermore, if ξ is finite, then $H(\xi) = \log \operatorname{card} \xi$ if and only if all elements of ξ have equal measure.
(2) $0 \le H(\xi|\eta) \le H(\xi)$; $H(\xi|\eta) = H(\xi)$ if and only if ξ and η are independent; $H(\xi|\eta) = 0$ if and only if $\xi \le \eta (mod 0)$. If $\zeta \ge \eta$, then $H(\xi|\zeta) \le H(\xi|\eta)$.

(3) *$H(\xi \vee \eta|\zeta) = H(\xi|\zeta) + H(\eta|\xi \vee \zeta)$. In particular, for $\zeta = \nu$ we obtain*

$$H(\xi \vee \eta) = H(\xi) + H(\eta|\xi). \tag{9.10}$$

(4) *$H(\xi \vee \eta|\zeta) \le H(\xi|\zeta) + H(\eta|\zeta)$; in particular, $H(\xi \vee \eta) \le H(\xi) + H(\eta)$.*

(5) *$H(\xi|\eta) + H(\eta|\zeta) \ge H(\xi|\zeta)$.*

(6) *If λ is another measure on X, then, for every partition ξ measurable for both μ and λ and for any $p \in [0, 1]$,*

$$pH_\mu(\xi) + (1-p)H_\lambda(\xi) \le H_{p\mu+(1-p)\lambda}(\xi).$$

Assume now we have a measure-preserving dynamical system $(X, \mathscr{B}, \mu, T)$. The definition of $H(\xi)$ and Proposition 2.2.1 allow us to measure the complexity of $(X, \mathscr{B}, \mu, T)$ with respect to a partition ξ. If we define the $\xi_{-n}^T = \bigvee_{i=0}^{n-1} T^{-i}(\xi)$, then $H_n(T, \xi)$ as follows:

$$H_n(T, \xi) = -\frac{1}{n} \sum_{C \in \xi_{-n}^T} \mu(C) \log \mu(C). \tag{9.11}$$

By the subadditivity property of $H(\xi)$ (Proposition 2.2.1 (4)), we have

$$(n+m)H_{n+m}(T, \xi) \le nH_n(T, \xi) + mH_m(T, \xi),$$

and thus

$$h(T, \xi) := \lim_{n\to\infty} H_n(T, \xi)$$

exists.

Remark 2.2.2. *$h(T, \xi)$ is also equal to the limit of the conditional entropy:*

$$h(T, \xi) = \lim_{n\to\infty} H(\xi|T^{-1}(\xi_{-n}^T)). \tag{9.12}$$

In order to simplify our notation, we define the *Rokhlin distance* between two partitions ξ and η as

$$\rho(\xi, \eta) = H(\xi|\eta) + H(\eta|\xi). \tag{9.13}$$

Proposition 2.2.3. *The following are some basic properties of $h(T, \xi)$:*

(1) *$0 \le \limsup_{n\to\infty} -\frac{1}{n} \log(\sup_{C\in\xi_{-n}^T} \mu(C)) \le h(T, \xi) \le H(\xi)$.*

(2) *$h(T, \xi \vee \eta) \le h(T, \xi) + h(T, \eta)$.*

(3) *$h(T, \eta) \le h(T, \xi) + H(\eta|\xi)$; in particular, if $\xi \le \eta$, then $h(T, \xi) \le h(T, \eta)$.*

(4) *$|h(T, \xi) - h(T, \eta)| \le \rho(\xi, \eta)$ (the Rokhlin inequality).*

(5) *$h(T, T^{-1}(\xi)) = h(T, \xi)$, and if T is invertible, $h(T, \xi) = h(T, T(\xi))$.*

(6) *$h(T, \xi) = h(T, \vee_{i=0}^{k} T^{-i}(\xi))$ for $k \in \mathbb{N}$, and if T is invertible, $h(T, \xi) = h(T, \vee_{i=-k}^{k} T^{i}(\xi))$ for all $k \in \mathbb{N}$.*

By taking the supremum of $h(T, \xi)$ over all measurable finite entropy partitions, the entropy of $(X, \mathscr{B}, \mu, T)$ is defined as follows:

Definition 2.2.4 (Measure-theoretic entropy). *The measure-theoretic entropy of T with respect to μ is*

$$h_\mu(T) := \sup\{h(T, \xi) \mid \xi \text{ is a measurable partition with } H(\xi) < \infty\}. \tag{9.14}$$

One of the most important features of measure-theoretic entropy is the generator theorem, which is a crucial property for applications.

Definition 2.2.5. *A partition ξ is called a generator for T if $\mathfrak{C} = \{\xi\}$ is a sufficient family. A family $\mathfrak{C}$ of measurable partitions with finite entropy is called sufficient with respect to the measure-preserving transformation T if*

(1) for a noninvertible T, partitions subordinate to partitions of the form $\vee_{i=0}^{k} T^{-i}(\xi)$ ($\xi \in \mathfrak{C}$, $k \in \mathbb{N}$) form a dense subset in the space of all partitions with finite entropy equipped with the Rokhlin metric.

(2) for an invertible T the same holds for partitions subordinate to $\vee_{i=-l}^{l} T^{i}(\xi)$ ($\xi \in \mathfrak{C}$, $l \in \mathbb{N}$).

Theorem 2.2.6. *If ξ is a generator for T, then $h_\mu(T) = h_\mu(T, \xi)$. More generally, $h_\mu(T) = \sup_{\xi \in \mathfrak{C}} h_\mu(T, \xi)$ for any sufficient family $\mathfrak{C}$ of partitions.*

It is worthwhile to notice that the measure-theoretic entropy can be introduced in a similar way as topological entropy. More precisely, the measure-theoretic entropy turns out to be the asymptotic value of the number of balls needed to cover a subset of positive measure instead of the whole space X [39]: let $N(T, n, \epsilon, \delta)$ denote the minimal number of ϵ-balls in the d_n^T-metric that cover a set of measure at least $1 - \delta$.

Theorem 2.2.7 (Katok [39]). *For every $\delta > 0$,*

$$h_\mu(T) = \lim_{\epsilon \to 0} \liminf_{n \to \infty} \frac{\log N(T, n, \epsilon, \delta)}{n} = \lim_{\epsilon \to 0} \limsup_{n \to \infty} \frac{\log N(T, n, \epsilon, \delta)}{n}.$$

We list here some basic properties of measure-theoretic entropy:

Proposition 2.2.8. *Let $T\colon (X, \mu) \to (X, \mu)$ be a measure-preserving transformation of a probability space $(X, \mathscr{B}, \mu)$ and η, ξ be measurable partitions with finite entropy. Then:*

(1) If $S\colon (Y, \nu) \to (Y, \nu)$ is a factor of $T\colon (X, \mu) \to (X, \mu)$ then $h_\mu(S) \leq h_\mu(T)$.
(2) If A is invariant for T and $\mu(A) > 0$, then $h_\mu(T) = \mu(A) h_{\mu_A}(T) + \mu(X \backslash A) h_{\mu_{X\backslash A}}(T)$.

(3) *If* μ, λ *are two invariant probability measures for* T, *then for any* $p \in [0,1]$

$$h_{p\mu+(1-p)\lambda}(T) \geq ph_\mu(T) + (1-p)h_\lambda(T).$$

(4) $h_\mu(T^k) = kh_\mu(T)$ *for any* $k \in \mathbb{N}$. *If* T *is invertible, then* $h_\mu(T^{-1}) = h_\mu(T)$ *and hence* $h_\mu(T^k) = |k|h_\mu(T)$ *for any* $k \in \mathbb{Z}$.

(5) $h_{\mu\times\lambda}(T \times S) = h_\mu(T) + h_\lambda(S)$.

We recall the following simple examples to end this section.

Example 2.2.9. *1. Rotations and linear flows on tori have zero measure-theoretic entropy (with respect to Lebesgue measure* λ*).*[8]

2. For an expanding map E_k *on* S^1*, its measure-theoretic entropy (with respect to Lebesgue measure* λ*) is*

$$h_\lambda(E_k) = \log|k|.$$

3. For the full N-shift σ_N *with Bernoulli measure* $(p_1,\ldots,p_N)$*, its measure-theoretic entropy is*

$$h_\mu(\sigma_N) = -\sum_{i=1}^{N} p_i \log p_i.$$

9.2.3 Variational Principle

Once we have topological entropy and measure-theoretic entropy, it is natural to ask whether there is any relation between these two invariants. Notice first that there are some important differences between these two quantities: the measure-theoretic entropy of the union of two invariants sets is the weighted sums of measure-theoretic entropies, but the topological entropy is the maximum of the two topological entropies. This corresponds to the observation that topological entropy measures the maximal complexity and measure-theoretic entropy measures the statistical complexity (statistics based on a given measure). The classical relation between these two entropies is given by the variational principle:

Theorem 2.3.1 (Variational principle). *If* $f: X \to X$ *is a homeomorphism of a compact metric space* (X,d)*, then*

$$h_{top}(f) = \sup\{h_\mu(f) | \mu \in \mathfrak{M}(f)\}. \tag{9.15}$$

Here $\mathfrak{M}(f)$ *is the set of all* f*-invariant Borel probability measures.*

It is natural to ask whether there always exists a measure that achieves the preceding supremum. In general, this is not true see, e.g., [6], [51] and [63]. In what

[8] The entropy of a flow is defined as the entropy of its time-one map.

follows, we make some comments on the existence and uniqueness of measures of maximal entropy.

Remark 2.3.2. *Expansive maps of compact metric spaces have a measure of maximal entropy.*

Remark 2.3.3. *Measure of maximal entropy might not be unique. The simplest example is to consider a union of several disjoint copies of the same expansive system (which is obviously expansive). In this case, there is more than one measure of maximal entropy.*

Remark 2.3.4. *All transitive topological Markov chains, horseshoes and hyperbolic toral automorphisms have a unique measure of maximal entropy.*

We provide several examples at the end of this section.

Example 2.3.5. *1. By the variational principle, it is easy to see the following formula,*

$$h_{top}(f) = h_{top}(f\restriction_{NW(f)}),$$

where $NW(f)$ is the set of all non-wandering points of f, where we recall that $x \in X$ is a non-wandering point of f if, for any open set $U\ni x$ there exists an $N > 0$ such that $f^N(U) \cap U \neq \emptyset$. The reason we have the preceding equation is that the support of any invariant measure is included in $NW(f)$ and we then use the variational principle.

2. By the previous example it follows that the measure-theoretic entropy of the gradient flow for any function with isolated critical points is zero.

3. By the variational principle, the maximal entropy of a full shift (with N symbols) is $\log N$ and the maximal entropy measure is $\left(\frac{1}{N}, \dots, \frac{1}{N}\right)$.

9.2.4 Kushnirenko's Inequality

In the theory of dynamical systems, one of the most challenging open problems is that of smooth realization of measure-preserving transformations. The only known restriction for the smooth realization problem is the finite entropy condition, which was originally introduced by Kushnirenko for absolutely continuous measures. We will now recall Kushnirenko's inequality. We need to recall several definitions:

Definition 2.4.1. *Let (X, d) be a compact metric space and $b(\epsilon)$ the minimal cardinality of a covering of X by ϵ-balls. Then*

$$D(X) = \limsup_{\epsilon \to 0} \frac{\log b(\epsilon)}{|\log \epsilon|} \in \mathbb{R} \cup \{\infty\}$$

is called the ball dimension of X.

Definition 2.4.2. *Let (X,d) be a compact metric space and $f\colon X \to X$ a Lipschitz map. Then the Lipschitz constant $L(f)$ of f is defined by*

$$L(f) = \sup_{x \neq y} \frac{d(f(x),f(y))}{d(x,y)}.$$

Theorem 2.4.3 (Kushnirenko's inequality). *Let (X,d) be a compact metric space of finite ball dimension $D(X)$ and $f\colon X \to X$ be a Lipschitz map. Then*

$$h_{top}(f) \leq D(X) \max(0, \log L(f)). \tag{9.16}$$

A natural corollary of Kushnirenko's inequality together with the variational principle (see Section 9.2.3) is the following obstruction for smooth realization:

Corollary 2.4.4. *For a C^1 map $f\colon M \to M$ of a compact Riemannian manifold,*

$$h_{top}(f) \leq \max(0, \dim M \log \sup_x \|Df_x\|) < \infty.$$

9.2.5 Mixing, Weak Mixing and Mild Mixing

Sequence entropy and slow entropy are related to some ergodic and spectral properties. We will introduce basic notions in both the measurable and the topological category.

Definition 2.5.1 (Mixing, weak mixing, rigid and mild mixing). *Let $(X, \mathscr{B}, \mu, T)$ be a measure-preserving dynamical system. Then*

(a) Let $U_T\colon L^2(X,\mu) \to L^2(X,\mu)$ denote the Koopman operator *associated with T, namely $U_T(f) = f \circ T$;*
(b) T is mixing *if for all $B, C \in \mathscr{B}$:*

$$\lim_{n\to\infty} \mu(T^{-n}B \cap C) = \mu(B)\mu(C); \tag{9.17}$$

(c) T is weakly mixing *(or has* continuous spectrum*) if, for all $B, C \in \mathscr{B}$,*

$$\lim_{n\to\infty} \frac{1}{n} \sum_{i=0}^{n-1} |\mu(T^{-i}B \cap C) - \mu(B)\mu(C)| = 0; \tag{9.18}$$

(d) T is rigid *if there exists an increasing sequence $\{t_n\}$ such that for any $f \in L^2(X, \mathscr{B}, \mu)$,*

$$T^{t_n} f \to f$$

in L^2;

(e) T *is* mildly mixing *if there are no non-constant rigid functions in* $L^2(X, \mathscr{B}, \mu)$, *where a function* $f \in L^2(X, \mathscr{B}, \mu)$ *is called rigid if there exists an increasing sequence* $\{t_n\}$ *such that* $T^{t_n} f \to f$ *in the* L^2*-topology.*

We have the following inclusions:

$$\text{Mixing systems} \subset \text{Mildly mixing systems} \subset \text{Weakly mixing systems} \subset \text{Ergodic systems}. \tag{9.19}$$

The first inclusion is straightforward from the preceding definitions. The second inclusion is based on the following characterizations of weak mixing and mild mixing:

Theorem 2.5.2 (Furstenberg–Weiss [22]). *T is mildly mixing if, for all ergodic systems $(Y, S, \nu, \mathscr{C})$ (where $\nu(Y)$ might be infinite) not of type I, the product system $(X \times Y, \mathscr{B} \times \mathscr{C}, \mu \times \nu, T \times S)$ is also ergodic.*

We recall that T is weakly mixing if and only if for all **finite** ergodic systems $(Y, S, \nu, \mathscr{C})$, the product system $(X \times Y, T \times S, \mu \times \nu, \mathscr{B} \times \mathscr{C})$ is also ergodic. Moreover, we also recall that all inclusions in (9.19) are strict.

The preceding definitions have their analogs in the topological category:

Definition 2.5.3 (Topological properties: transitivity, mixing, weak mixing and mild mixing). *Let (X, T) be a topological dynamical system:*

(a) (X, T) *is said to be* transitive *if any closed invariant subset is either nowhere dense or the whole space* X*;*
(b) (X, T) *is said to be* topologically weakly mixing *if* $(X \times X, T \times T)$ *is transitive;*
(c) (X, T) *is said to be* topologically mixing *if, for any non-empty open subsets* U *and* V *of* X*, there is* $N \in \mathbb{N}$ *such that* $U \cap T^{-n}V \neq \emptyset$ *for each* $n \geq N$*;*
(d) (X, T) *is said to be* topologically mildly mixing *if, for any transitive system* (Y, S), $(X \times Y, T \times S)$ *is transitive.*

9.3 Sequence Entropy

In this section, we introduce the notion of sequence entropy and describe its basic properties. We also discuss relations between sequence entropy and classical entropy. Several examples are provided to illustrate some interesting properties of sequence entropy.

9.3.1 Measure-Theoretic Sequence Entropy

Notice that a crucial feature of classical entropy is that it is an isomorphism invariant.[9] As a consequence, entropy is a powerful tool in the (isomorphism) classification problem of dynamical systems. For example, by calculating measure-theoretic entropy, we can easily show that Bernoulli shifts $\left(\frac{1}{2}, \frac{1}{2}\right)$ and $\left(\frac{1}{3}, \frac{1}{3}, \frac{1}{3}\right)$ are not isomorphic. However, for systems of zero entropy, classical entropy theory provides no information on classification.[10] In order to distinguish non-isomorphic zero entropy systems, Kushnirenko [48] developed an invariant based on modifying the classical entropy definition by replacing iterates along the sequence of natural numbers by iterates along a fixed subsequence.

Definition 3.1.1 (Sequence entropy, Kushnirenko [48]). *Suppose that $(X, \mathscr{B}, \mu, T)$ is a dynamical system (where T is invertible). For a given integer-valued sequence $A = \{t_1, t_2, \ldots, t_n, \ldots\}$, a measurable partition (finite or countable) ξ satisfying $H(\xi) < \infty$, let*

$$\begin{aligned} h_A(T, \xi) &:= \limsup_{n\to\infty} \frac{1}{n} H\left(\bigvee_{i=1}^{n} T^{-t_i}\xi\right) \\ h_{A,\mu}(T) &:= \sup_{\xi} h_A(T, \xi). \end{aligned} \tag{9.20}$$

We call $h_{A,\mu}(T)$ the (measure-theoretic) sequence entropy of T.

By an analogous argument as for the classical entropy, one shows that $h_{A,\mu}(T)$ is an isomorphism invariant. If we take $t_i = i - 1$, $h_{A,\mu}(T)$ coincides with the classical entropy. However, by adjusting the subsequence to a given dynamical system (e.g., a system of zero entropy), sequence entropy can often provide additional information about the dynamics. We point out an interesting property of sequence entropy. Dekking [9] proved that measure-theoretic sequence entropy does not depend monotonically on A: if A is a subsequence of B, it is possible that $h_{A,\mu}(T) < h_{B,\mu}(T)$.

By adjusting the proofs for classical entropy, one recovers analogous properties of the sequence entropy:

Proposition 3.1.2 (Kushnirenko [48]).

$$|h_{A,\mu}(T, \xi) - h_{A,\mu}(T, \eta)| \leq \rho(\xi, \eta), \tag{9.21}$$

where $\rho(\xi, \eta)$ is defined in (9.13).

We also have a generator-type result:

[9] Topological entropy is an isomorphic invariant due to Proposition 2.1.4 (1), and measure-theoretic entropy is a metric invariant due to Proposition 2.2.8 (1).

[10] For example, the horocycle flow and its Cartesian square.

Proposition 3.1.3 (Kushnirenko [48]). *Let $\xi_1 \leq \xi_2 \leq \ldots \leq \xi_n \leq \ldots$ and $\bigcap_{i=1}^{\infty} \xi_i = \epsilon$ (ϵ is the point partition). Then for any A and T, we have*

$$h_{A,\mu}(T) = \lim_{k\to\infty} h_{A,\mu}(T, \xi_k). \tag{9.22}$$

Once we can control the growth rate of the transformation T, we have:

Proposition 3.1.4 (Kushnirenko [48]). *Suppose that T is a diffeomorphism of a compact m-dimensional Riemannian manifold M^m and $\|T\|$ is the maximum expansion of an $(m-1)$-dimensional non-orientated volume along all $(m-1)$-dimensional tangential directions. Let $A = \{t_1, t_2, \ldots, t_n, \ldots\}$ be a given sequence. If $\|T^{t_n}\| \leq C\lambda^n$ (where $C > 0$ does not depend on n), then $h_{A,\mu}(T) \leq m \log \lambda$.*

However, the sequence entropy behaves differently than the classical entropy when considering Cartesian products:

Proposition 3.1.5 (Kushnirenko [48], Lemańczyk [49], Hulse [30]). *Let $(X, \mathscr{B}, \mu, T)$ and $(X', \mathscr{B}', \mu', T')$ be two dynamical systems. Then*

$$h_{A,\mu\times\mu'}(T \times T') \leq h_{A,\mu}(T) + h_{A,\mu'}(T'). \tag{9.23}$$

Remark 3.1.6. *In [48] it is claimed that one has equality in* (9.23). *Counterexamples to that were given in [49] (see also [30]) where the author showed that the inequality is strict.*

Remark 3.1.7. *As shown by Lemańczyk [49], the formula $h_{A,\mu}(T^k) = kh_{A,\mu}(T)$, $k \geq 2$ is not true in general. This answered a question proposed by Saleski [64].*

Recall that the sequence entropy with the sequence of integers is by definition equal to the classical entropy. It is natural to ask whether there exists a class of more general sequences for which there is a relation between the sequence entropy and the classical entropy. This was shown to be true for large density sequences and systems of positive (classical) entropy. More precisely, E. Krug and D. Newton have the following result to describe the relation between the sequence entropy and the classical entropy:

Definition 3.1.8 (Krug–Newton [47]). *For a given sequence of integers $A = \{t_1, \ldots\}$ and $k, n \in \mathbb{N}$, let $U_A(n,k) = \{t_i + j\colon 1 \leq i \leq n, -k \leq j \leq k\}$. Let $S_A(n,k) = \operatorname{Card} U_A(n,k)$ and define,*[11]

$$K(A) := \lim_{k\to\infty} \overline{\lim_{n\to\infty}} \frac{1}{n} S_A(n,k). \tag{9.24}$$

[11] Notice that $S_A(n,k)$ is an increasing positive function of k; that means $\overline{\lim}_{n\to\infty} \frac{1}{n} S_A(n,k)$ is an increasing nonnegative function of k. By allowing the value ∞, $K(A)$ is well defined.

Theorem 3.1.9 (Krug–Newton [47]). *Let T be any automorphism of (X,μ) and A any sequence of integers. Then $h_{A,\mu}(T) = K(A) \cdot h(T)$, where the right-hand side is to be interpreted as 0 if $K(A) = 0$ and $h(T) = \infty$ and is undefined if $K(A) = \infty$ and $h(T) = 0$.*

Remark 3.1.10. *Notice that the preceding theorem does not provide any information for systems of zero entropy, for which one can hope to take sparser sequences to recover some information. For example, it was shown by Kushnirenko [48] that the time one map u_1 of the horocycle flow satisfies: $1 \leq h_{2^n,\mu}(u_1) \leq 6$ and this result was improved in [34] to $h_{2^n,\mu}(u_1) = 3$. We provide more details in Section 9.3.7.*

9.3.2 Spectral Theory and Measure-Theoretic Sequence Entropy

In this section, we will discuss the relation between the spectrum of T and the measure-theoretic sequence entropy of T.

We start from the following result of Kushnirenko:

Theorem 3.2.1 (Kushnirenko [48]). *An invertible transformation T has discrete spectrum if and only if $h_{A,\mu}(T) = 0$ for any sequence A.*

Remark 3.2.2. *This theorem should be compared with the corresponding result for the slow entropy in Section 9.4.4.*

On the other hand, we have the following result for systems with continuous spectrum:

Theorem 3.2.3 (Pitskel [57]). *Let ξ be a k-element partition of the space (X,μ) and let χ_i be the characteristic function of the i-th element, $i = 1,\dots,k$. Let U_T be the Koopman operator and set $\psi_i := \chi_i - \int \chi_i d\mu$. Let ψ_i, $i = 1,\dots,k$ belong to an invariant subspace $L \subset L^2(X)$ and let the operator U_T have a continuous spectrum on L. Then there exists a sequence A such that*

$$h_{A,\mu}(T,\xi) = H(\xi). \tag{9.25}$$

Recall that for the classical entropy we have the generator theorem. This turns out not to be correct for the sequence entropy. By strengthening Pitskel's result, P. Hulse obtained the following result:

Theorem 3.2.4 (Hulse [29]). *There exists a weakly mixing automorphism T and an increasing sequence of natural numbers A such that $h_{A,\mu}(T) = \infty$ and $h_{A,\mu}(T,\xi) = H(\xi)$ for all ξ with finite entropy. Moreover, for any $\epsilon > 0$, there exists a two-element A-generator ξ for T such that $H(\xi) < \epsilon$.*

9.3.3 Sequence Entropy and Spectral Properties

In this section we will describe further results on connections of the sequence entropy and spectral properties (recall that the first such result was established by Kushnirenko, Theorem 3.2.1). The first two results describe how sequence entropy characterizes weak mixing:

Theorem 3.3.1 (Hulse [29]). *An invertible measure-preserving system* $(X, \mathscr{B}, \mu, T)$ *is weakly mixing if and only if there exists an increasing sequence of natural numbers* A *such that* $h_{A,\mu}(T, \xi) = H(\xi)$ *for all* ξ *with* $H(\xi) < \infty$.

The following result gives yet another characterization of weak mixing:

Theorem 3.3.2 (Zhang [75]). *An invertible measure-preserving system* $(X, \mathscr{B}, \mu, T)$ *is weakly mixing if and only if for any set* $\Gamma \subset \mathbb{N}$ *with positive density, there is a subset* Γ_1 *of* Γ *such that for any finite partition* ξ, $h_{\Gamma_1,\mu}(\xi) = H(\xi)$.

The following results describe the connections between the sequence entropy, mild mixing, mixing and rigidity:

Theorem 3.3.3 (Zhang [75]). *An invertible measure-preserving system* $(X, \mathscr{B}, \mu, T)$ *is rigid if and only if there exists a subset* Γ *of* $\mathbb{N}$ *such that if* $\{F_i\}$ *is any sequence of pairwise disjoint finite subsets of* Γ *and* $s_i = \sum_{a \in F_i} a$, *then* $h_{\{s_i\},\mu}(T) = 0$.

Theorem 3.3.4 (Zhang [75]). *An invertible measure-preserving system* $(X, \mathscr{B}, \mu, T)$ *is mildly mixing if and only if, for any subset* Γ *of* $\mathbb{N}$, *there is a sequence* $\{F_n\}$ *of pairwise disjoint finite subsets of* Γ *such that for any finite partition* ξ *of* X *and* $s_i = \sum_{a \in F_i} a$, $h_{\{s_i\},\mu}(T, \xi) = H(\xi)$.

Theorem 3.3.5 (Saleski [64], Hulse [29]). *An invertible measure-preserving dynamical system* $(X, \mathscr{B}, \mu, T)$ *is mixing if and only if, for any infinite sequence* A, *there exists a subsequence* $B \subset A$ *such that* $h_{B,\mu}(T, \xi) = H(\xi)$ *for all* ξ *with* $H(\xi) < \infty$.

9.3.4 Topological Sequence Entropy

In 1974, seven years after Kushnirenko's definition of measure-theoretic sequence entropy, Goodman [24] introduced topological sequence entropy and investigated its properties. Later, Huang, Shao and Ye [28] studied relations of topological weak mixing and topological sequence entropy, which is a counterpart of the corresponding results in the measurable category, from the previous section. We start this section with the definition of the topological sequence entropy.

Definition 3.4.1 (Topological sequence entropy, Goodman [24]). *Suppose that X is a locally compact metric space and $T\colon X \to X$ is a continuous map. Let $\mathscr{A}(X)$ be the collection of all open covers of X. For a compact subset $K \subset X$ and a given open cover α, let $N(\alpha, K)$ denote the minimal cardinality of a sub-cover of α that covers K. For a given sequence of integers $A = \{t_1, \ldots, t_n, \ldots\}$, we define*

$$\begin{aligned} h_A(T, \alpha, K) &:= \overline{\lim_{n\to\infty}} \frac{1}{n} H\left(\bigvee_{i=1}^{n} T^{-t_i}\alpha\right), \\ h_A(T) &:= \sup_K \sup_{\alpha \in \mathscr{A}(X)} h_A(T, \alpha, K). \end{aligned} \tag{9.26}$$

Remark 3.4.2. *If X is a compact metric space, we will omit the K in the $h_A(T, \alpha, K)$ for simplicity.*

The quantity $h_A(T)$ denotes the topological sequence entropy of (X, T) with respect to the sequence A.

If T is a homeomorphism, then the above definition of topological sequence entropy can be extended to negative integers. Analogously to the measure-theoretic sequence entropy, by taking $t_i = i-1$, $h_A(T)$ coincides with the classical topological entropy.

Recall that the classical topological entropy can also be defined through maximal separated sets (or minimal spanning sets) with respect to the Bowen distance. There is also a parallel version of the topological sequence entropy's definition based on maximal separated sets (or minimal spanning sets).

Definition 3.4.3 (Topological sequence entropy, Goodman [24]). *Suppose that X is a locally compact metric space, $T\colon X \to X$ is a continuous map and $A = \{t_1, \ldots, t_n, \ldots\}$ is a given sequence of integers. For $\alpha \in \mathscr{A}(X)$, we say a set $E \subset X$ is (A, n, α)-separated (with respect to T) if, for any distinct points $x, y \in E$, there is an integer j, $1 \le j \le n$ such that $T^{t_j}x \in U \in \alpha$ and $T^{t_j}y \notin U$. For a compact subset $K \subset X$, let $N(A, n, \alpha, K)$ denote the largest cardinality of a (A, n, α)-separated set in K. Define*

$$\begin{aligned} h_A(T, \alpha, K) &= \overline{\lim_{n\to\infty}} \frac{1}{n} \log N(A, n, \alpha, K), \\ h_A(T) &= \sup_K \sup_{\alpha \in \mathscr{A}(X)} h_A(T, \alpha, K). \end{aligned} \tag{9.27}$$

The quantity $h_A(T)$ denotes the topological sequence entropy of (X, T) with respect to the sequence A.

Remark 3.4.4. *As we have already mentioned, topological sequence entropy can also be defined by minimal spanning sets. We say a set F (A, n, α)-spans another set K if, for each point $x \in K$, there is a point $y \in F$ and there exists $U \in \alpha$ with $T^{t_j}x, T^{t_j}y \in U$ for some*

$j \leq n$. For a compact set $K \subset X$, let $S(A,n,\alpha,K)$ denote the smallest cardinality of a set which (A,n,α)-spans K and denote $S(A,\alpha,K) = \overline{\lim}_{n\to\infty}\frac{1}{n}\log S(A,n,\alpha,K)$. In this situation, define topological sequence entropy as

$$h_A(T,K) := \sup_{\alpha\in\mathscr{A}(X)} S(A,\alpha,K). \tag{9.28}$$

Remark 3.4.5. *In fact, the preceding two definitions are equivalent. Indeed, one can prove that*

$$S(A,n,\alpha,K) \leq N(A,n,\alpha,K) \leq S(A,n,\gamma,K), \tag{9.29}$$

where γ is a refined open cover of α.

Remark 3.4.6. *Similar to the measure-theoretic sequence entropy, Lemańczyk [49] showed that topological sequence entropy also does not depend monotonically on A: it is possible that $h_A(T) < h_B(T)$ if A is a subsequence of B.*

Topological sequence entropy has also the following useful properties:

Proposition 3.4.7 (Generator-type property, Goodman [24]). *Let $\{\alpha_n\}, n = 1,\ldots$ be a sequence of elements of $\mathscr{A}(X)$ for which diam$(\alpha_n) \to 0$. Then $h_A(T,\alpha_n) \to h_A(T)$ as $n \to \infty$.*

Proposition 3.4.8 (Cartesian square, Goodman [24]). *We have*

$$h_A(T\times T) = 2h_A(T). \tag{9.30}$$

Remark 3.4.9. *The preceding equality is not true for general products of two systems, as shown by Lemańczyk [49] (this answered a question of Goodman, [24]). One only has the following inequality:*

$$h_A(T\times S) \leq h_A(T) + h_A(S). \tag{9.31}$$

Remark 3.4.10. *As for iterates of T, Lemańczyk [49] showed that it is not true that $h_A(T^k) = kh_A(T)$, $k \geq 2$ which answered a question by Saleski [64] in the topological category.*

The following two results describe the behavior of the topological sequence entropy under factors or extensions.

Proposition 3.4.11 (Goodman [24]). *Let $\pi : (X,T) \to (Y,S)$ be a homomorphism and suppose each fiber $\pi^{-1}y$, $y \in Y$, has at most n points. Let A be any sequence. Then, $h_A(T) \leq h_A(S) + \log n$.*

Proposition 3.4.12 (Goodman [24]). *For any surjective dynamical system (X,T) and sequence A, we have $h_A(T) = h_A(T^*)$, where (X^*,T^*) is the natural extension of the (X,T).*

9.3.5 Relations of Topological Sequence Entropy and Topological Properties

We first introduce some notation in the topological sequence entropy:

Definition 3.5.1. *Let (X, T) be a topological system.*

(1) (X, T) is called bounded if $h_A(T) < K$ for all sequences A; otherwise, we will say (X, T) is unbounded.

(2) (X, T) is called null if the topological sequence entropy is zero for any sequence.

(3) Let S^1 be the unit circle in the complex plane and $C(X)$ be the collection of all continuous maps from X into K. Given a system (X, T), we define a sequence of groups

$$K = G_1 \subset G_2 \subset \ldots \subset C(X)$$

inductively as follows:

$$G_{n+1} = \{f \in C(X) : f(Tx) = gf(x) \text{ for all } x \in X \text{ and some } g \in G_n\}.$$

Denote by $n(T)$ the least number n for which $G_{n+1} = G_n$.

(4) An admissible cover $\mathcal{U}$ is a finite cover $\mathcal{U} = \{U_1, U_2, \ldots, U_n\}$ such that $\left(\bigcup_{j \neq i} U_j\right)^c$ has non-empty interior for each $i \in \{1, 2, \ldots, n\}$.

At first, a counterexample to the topological version of Kushnirenko's discrete spectrum theorem is described. Recall that by Theorem 3.2.1, vanishing of measure-theoretic sequence entropy for any sequence is equivalent to discrete spectrum. It is natural to conjecture a corresponding topological version, namely

A minimal flow is null if and only if it is equicontinuous.

In fact, this is not the case for topological sequence entropy. In [24], Goodman gave an example, which is a minimal null topological dynamical system that is not distal.[12]

Example 3.5.2 (Goodman [24]). *Let $\alpha \in [0, 1]$ be an irrational number and define $A_0 = \left\{e^{2\pi i\theta} : 0 \leq \theta \leq \frac{1}{2}\right\}$, $A_1 = \left\{e^{2\pi i\theta} : \frac{1}{2} \leq \theta \leq 1\right\}$ and $X = \{x \in \{0, 1\}^{\mathbb{Z}} : \cap_{r=-\infty}^{\infty} e^{2\pi i r\alpha} A_{x_r} \neq \emptyset\}$. We also denote T as the shift given by $(Tx)_n = x_{n+1}$ for $x \in X$. Then (X, T) is a minimal null topological system that is not distal.*

Moreover, the nullness property is quite stable under common operations and also related to unique ergodicity.

[12] Recall that equicontinuity implies distality.

Theorem 3.5.3 (Huang–Li–Shao–Ye [27]). *The property of nullness of a topological dynamical system is stable under factor maps, arbitrary products and inverse limits. Moreover, if a null system is minimal, then it is uniquely ergodic.*

We also have the following characterization of topological weak mixing:

Theorem 3.5.4 (Goodman [24]). *Let (X,T) be an invertible topological dynamical system:*

(1) If (X,T) is non-trivial and topologically weakly mixing, then (X,T) is unbounded;
(2) If (X,T) is minimal, then it is topologically weakly mixing if and only if it has no non-trivial bounded factors;
(3) Assume that (X,T) is a non-trivial bounded system. If (X,T) is minimal, then $n(T) \geq 2$. If (X,T^r) is minimal for all integers $r \neq 0$, then $n(T) \leq 2$.

With the help of admissible covers we can describe topological mixing, topological weak mixing and topological mild mixing as follows:

Theorem 3.5.5 (Huang, Shao and Ye [28]). *Let (X,T) be a topological dynamical system; then we have*

(1) (X,T) is topologically weakly mixing if and only if for any admissible open cover $\mathscr{U}$ of X there is some $A \subset \mathbb{Z}_+$ such that $h_A(T,\mathscr{U}) = \log N(\mathscr{U})$, where $N(\mathscr{U})$ is the cardinality of the $\mathscr{U}$;
(2) (X,T) is topologically mildly mixing if and only if for any admissible open cover $\mathscr{U}$ of X and any IP set F there is some $A \subset F$ such that $h_A(T,\mathscr{U}) = \log N(\mathscr{U})$, where IP set F is defined as

$$F = \left\{ \sum_{i\in\alpha} b_i \colon \alpha \text{ is a finite non-empty subset of } \mathbb{N} \right\},$$

for a sequence of natural numbers b_i;
(3) Assume that (X,T) is topologically mixing. Then, for any admissible open cover $\mathscr{U}$ of X and any infinite $F \subset \mathbb{Z}_+$, there is some infinite $A \subset F$ such that $h_A(T,\mathscr{U}) = \log N(\mathscr{U})$.

9.3.6 Variational Principle for Sequence Entropy

Recall that the connection between classical topological and measure theoretic entropy is given by the variational principle. It is therefore interesting to ask for an analogous property for the sequence entropy. We will discuss this in this section.

Recall that $\mathscr{M}(X,T)$ denotes the collection of all regular probability measures on Borel subsets of X that are invariant under the T.

Definition 3.6.1 (Finite covering dimension, Goodman [24]). *A space X has finite covering dimension at most n if any open cover of X has an open refinement of order at most $n+1$. The order of a collection α of sets is defined to be the maximum number of sets in α with non-empty intersection.*

Theorem 3.6.2 (Goodman [24]). *Let X be a compact Hausdorff space and T be a continuous map from X to X. Suppose X has finite covering dimension. Then, for any sequence A and any measure $\mu \in \mathcal{M}(X, T)$,*

$$h_{A,\mu}(T) \leq h_A(T). \tag{9.32}$$

The opposite inequality also holds under some additional assumptions (see Theorem 3.1.9 by Krug and Newton):

Theorem 3.6.3 (Goodman [24]). *Let X be a compact Hausdorff space and T be a continuous map from X to X. Suppose X has finite covering dimension. Then, for any sequence A, suppose that either $K(A) < \infty$ or $h_A(T) > 0$; then,*

$$h_A(T) = \sup_{\mu \in \mathcal{M}(X,T)} h_{A,\mu}(T). \tag{9.33}$$

Remark 3.6.4. *In [24], a counterexample is provided to show that without the condition in Theorem 3.1.9, we may have*[13]

$$h_A(T) > \sup_{\mu \in \mathcal{M}(X,T)} h_{A,\mu}(T).$$

9.3.7 Sequence Entropy of Quasi-unipotent Systems

In this subsection, one special example related to sequence entropy is introduced, namely the sequence entropy of quasi-unipotent flows. This example is a generalization of the original Kushnirenko's computation for horocycle flows.

9.3.7.1 Introduction to Quasi-unipotent Flows

Let G be a connected Lie group and $\mathfrak{g}$ its Lie algebra. Suppose Γ is a discrete subgroup of G with co-finite volume and μ is the Haar measure on G/Γ, which is induced from Riemannian volume on G. We define a quasi-unipotent flow as follows:

Definition 3.7.1 (Quasi-unipotent flow). *An element $U \in \mathfrak{g}$ is called quasi-unipotent if it can be written as $U = U' + Q$, where $\mathrm{ad}_{U'}^N = 0$ for some N, Q is* ad-compact, *and $[Q, U'] = 0$. A quasi-unipotent element $U \in \mathrm{Lie}(G)$ induces a quasi-unipotent flow on a homogeneous space G/Γ by*

$$\varphi_t(g\Gamma) = \exp(tU)g\Gamma.$$

[13] More precisely, the example is a Chacon system, which has a special spacers sequence.

In order to describe the complexity of quasi-unipotent flows, the following structure is introduced as a basis system for homogeneous dynamics.

Definition 3.7.2. *A* chain *in* $\mathfrak{g}$ *with respect to a quasi-unipotent element U of depth m is a linearly independent set* $\{X_j : 0 \le j \le m\}$ *such that* X_0 *is in the centralizer of U and*

$$\mathrm{ad}_U(X_j) = X_{j-1} \text{ for all } 1 \le j \le m.$$

A double chain *in* $\mathfrak{g}$ *with respect to U of depth m is a linearly independent set* $\{X_{j,i} : 0 \le j \le m, i = 0, 1\}$ *and a number* α *such that* $\mathrm{ad}_Q(X_{j,0}) = -\alpha X_{j,1}$, $\mathrm{ad}_Q(X_{j,1}) = \alpha X_{j,0}$ *for all* $0 \le j \le m$, $X_{0,i}$ *is in the centralizer of* U' *for* $i = 0, 1$ *and*

$$\mathrm{ad}_{U'}(X_{j,i}) = X_{j-1,i} \text{ for all } 1 \le j \le m, i = 0, 1.$$

A chain basis *of* $\mathfrak{g}$ *with respect to U is a basis of chains and double chains. The sequence of depths* $(m_1, \ldots, m_n)$ *of chains and double chains, with each double chain listed twice, is called the* chain structure *of U.*

It follows from the Jordan block decomposition that all quasi-unipotent elements have a chain basis. Thus, for a quasi-unipotent flow $\varphi_t(g\Gamma) = \exp(tU)g\Gamma$ with chain structure $(m_1, m_2, \ldots, m_n)$, we define

$$GR(U) = \sum_{i=1}^{n} \frac{m_i(m_i + 1)}{2}.$$

9.3.7.2 Sequence Entropy of Quasi-unipotent Flows

The sequence entropy of quasi-unipotent flows is described in the following theorem:

Theorem 3.7.3 (Kanigowski–Vinhage–Wei [34]). *For a quasi-unipotent flow* φ_t *on homogeneous space* G/Γ *with chain structure* $(m_1, m_2, \ldots, m_n)$, *Haar measure* μ *and the sequence* $A = \{C\lambda^n\}$ $(C > 0)$, *we have*

$$h_A(\varphi_t) = h_{A,\mu}(\varphi_t) = GR(U) \log \lambda. \tag{9.34}$$

Remark 3.7.4. *This result extends Kushnirenko's original result in [48] by calculating the precise value of the sequence entropy of quasi-unipotent flows. For example, the sequence entropy*[14] *of horocycle flows with respect to the sequence* $\{2^n\}$ *is equal to* 3.

Remark 3.7.5. *It is also worthwhile to point out that the topological sequence entropy equals the measure-theoretic sequence entropy for the Haar measure; that is, the variational principle holds for quasi-unipotent flows; for more details, see [34].*

[14] By taking logarithm function with base 2, which is the same as in Kushnirenko's original paper [48].

9.4 Slow Entropy

In this section, a basic theory of slow entropy is introduced, applications of slow entropy to smooth realization and spectrum are discussed. Moreover, several special features of slow entropy are also discussed such as failure of the variational principle and difference between upper and lower definitions of slow entropy. Many examples of slow entropy are also studied: quasi-unipotent flows, rank-one systems, surface flows and AbC constructions.

9.4.1 Measure-Theoretic Slow Entropy

In this subsection, following [41], the definition of the slow entropy is introduced in the amenable group setting.

Definition 4.1.1. *Suppose that Γ is an amenable discrete group and F is a subset of Γ. Then*

$$\Omega_{N,F} = \{w = (w_r)_{r\in F}; w_r \in \{1,\dots,N\}\}$$

and define a natural projection $\pi_{F,F'} : \Omega_{N,F} \to \Omega_{N,F'}$ for $F' \subset F$.

Suppose $T\colon (X,\mu)\times\Gamma \to (X,\mu)$ is an action of the group Γ by measure-preserving transformations of a Lebesgue space. Let $\xi = \{c_1,\dots,c_N\}$ be a finite measurable partition. We define the coding map as $\phi_{T,\xi} : X \to \Omega_{N,\Gamma}$ if $(\phi_{T,\xi})_r(x) = w_r(x)$ where $T(r)x \in c_{w_r(x)}$. Denote the partial coding $\phi^F_{T,\xi}$ for $F \subset \Gamma$ as $\phi^F_{T,\xi} = \pi_{\Gamma,F}\circ\phi_{T,\xi}$ and call $\phi^F_{T,\xi}$ the *F-name of x with respect to ξ*. The measure induced by a partial coding is $(\phi^F_{T,\xi})_*\mu$. Then, the Hamming metric on $\Omega_{N,F}$ is defined as follows.

Definition 4.1.2 (Hamming metric). *For any finite set $F \subset \Gamma$, the Hamming metric $d^H_F(w,w')$ is defined by*

$$d^H_F(w,w') = \frac{1}{\operatorname{card} F}\sum_{r\in F}(1-\delta_{\omega_r\omega'_r}). \tag{9.35}$$

Now the cover of symbolic space $\Omega_{N,F}$ can be introduced based on the Hamming metric. Denote by $S^H_\xi(T,F,\epsilon,\delta)$ the minimal number of $d^H_F - \epsilon-$balls in $\Omega_{N,F}$, whose union has measure at least $1-\delta$ with respect to $(\phi^F_{T,\xi})_*\mu$. Finally we assume that $\{F_n\}_{n=1}^\infty$ is a Følner sequence of finite subsets of Γ. Then the slow entropy can be defined as follows:

Definition 4.1.3 (Slow entropy, Katok–Thouvenot [41]). *Let $\boldsymbol{a} = \{a_n(t)\}_{n\in\mathbb{N},t>0}$ be a family of positive sequences increasing to infinity and monotone in t. Then define the upper measure-theoretic slow entropy of T with respect to ξ by*

$$\overline{ent}_{a}^{\mu}(T,\xi)=\lim_{\delta\to 0}\lim_{\epsilon\to 0}A(\epsilon,\delta,\xi), \tag{9.36}$$

where $A(\epsilon,\delta,\xi)=\begin{cases}\sup B(\epsilon,\delta,\xi), & \text{if } B(\epsilon,\delta,\xi)\neq\emptyset;\\ 0, & \text{if } B(\epsilon,\delta,\xi)=\emptyset,\end{cases}$ *for*

$$B(\epsilon,\delta,\xi)=\{t>0\colon \limsup_{n\to\infty}\frac{S_{\xi}^{H}(T,F_n,\epsilon,\delta)}{a_n(t)}>0\}.$$

The upper measure-theoretic slow entropy of T is defined by

$$\overline{ent}_{a}^{\mu}(T)=\sup_{\xi}\overline{ent}_{a}^{\mu}(T,\xi). \tag{9.37}$$

Remark 4.1.4. *Similarly, the lower a-entropy can be defined by changing the* lim sup *to* lim inf *in the above and use* $\underline{ent}_{a}^{\mu}(f)$ *to represent the lower a-entropy of T.*

Remark 4.1.5. *By considering the continuous version of the Hamming metric, Definition 4.1.3 can be modified to flows.*

If T is ergodic and the scaling function is exponential: $a_n(t)=\exp(tn)$, then the slow entropy invariant coincides with the classical measure-theoretic entropy. The following theorem explains this phenomenon and also should be understood as a general version of Theorem 2.2.7 in [39] and Proposition 2 in [15].

Theorem 4.1.6 (Katok–Thouvenot [41]). *Let* $\{F_n\}$ *be a Følner sequence in* Γ. *If T is ergodic, then*

$$\lim_{n\to\infty}\lim_{\epsilon\to 0}\lim_{\delta\to 0}\frac{\log S_{\xi}^{H}(T,F_n,\epsilon,\delta)}{|F_n|}=h_{\mu}(T).$$

Recall that in the classical measure-theoretic entropy setting, a very powerful theorem in calculation of entropy is the generator theorem (Theorem 2.2.6). In fact, this is also the case for the slow entropy:

Theorem 4.1.7 (Katok–Thouvenot [41]). *If* ξ *is a generator, then* $\overline{ent}_{a}^{\mu}(T)=\overline{ent}_{a}^{\mu}(T,\xi)$. *More generally, if* ξ_m *is a sufficient family of partitions, then*

$$\overline{ent}_{a}^{\mu}(T)=\sup_{m}\overline{ent}_{a}^{\mu}(T,\xi_m).$$

The upper and the lower slow entropy have different behavior under products:

Proposition 4.1.8 (Ferenczi [15]). *If the scaling function satisfies* $a_n(t+s)=a_n(t)a_n(s)$, *we have:*

$$\begin{aligned}\underline{ent}_{a}^{\mu}(T_1\times T_2)&\geq\underline{ent}_{a}^{\mu}(T_1)+\underline{ent}_{a}^{\mu}(T_2),\\ \overline{ent}_{a}^{\mu}(T_1\times T_2)&\leq\overline{ent}_{a}^{\mu}(T_1)+\overline{ent}_{a}^{\mu}(T_2).\end{aligned} \tag{9.38}$$

At the end of this section, several natural examples are provided to calculate slow entropy:

Example 4.1.9. *Consider the Morse system σ_M generated by the following substitution:*

$$\begin{aligned} a &\to ab, \\ b &\to ba. \end{aligned} \tag{9.39}$$

By calculations in [52], it follows that the upper growth of the system has rate $\frac{10n}{3}$ and lower growth rate is $3n$, which means they are of the same (linear) order.

Thus we have

$$\overline{ent}^{\mu}_{a}(\sigma_M) = \underline{ent}^{\mu}_{a}(\sigma_M) = 1,$$

where $a_n(t) = n^t$.

Example 4.1.10. *Consider the Rudin–Shapiro system σ_{RS} generated by the following substitution:*

$$\begin{aligned} a &\to ab, \\ b &\to ac, \\ c &\to db, \\ d &\to dc. \end{aligned} \tag{9.40}$$

Then this system's upper growth and lower growth coincide and the asymptotic value is $8n$. Thus we have

$$\overline{ent}^{\mu}_{a}(\sigma_{RS}) = \underline{ent}^{\mu}_{a}(\sigma_{RS}) = 1,$$

where $a_n(t) = n^t$.

9.4.2 Topological Slow Entropy and Goodwyn's Theorem

We will follow the ideas from [41] and [23] to define the topological slow entropy. Suppose $K \subset X$ is a compact subset of a locally compact metric space (X, d) and $T\colon (X, d) \times \Gamma \to (X, d)$ is an action of the group Γ by homeomorphisms. Let F be a finite subset of Γ and we define Bowen metric as $d_F^T = \max_{\gamma \in F} d \circ T(\gamma)$. Moreover, we assume that $\{F_n\}_{n=1}^{\infty}$ is a Følner sequence of finite subsets of Γ. We will modify the definition in Section 9.2.1 a little to cover the case where X is non-compact.

Definition 4.2.1 (Minimal separated set and maximal spanning set). *Let $S_{d_F^T,K}(\epsilon)$ be the minimal number of d_F^T-ϵ-balls required to cover K (since K is compact, this is finite); let $N_{d_F^T,K}(\epsilon)$ be the maximal number of d_F^T-ϵ-balls with centers in K that can be placed disjointly in X.*

Then the topological slow entropy can be defined as follows:

Definition 4.2.2 (Topological slow entropy). *Let $\{a_n(t)\}_{n\in\mathbb{N},t>0}$ be a family of positive sequences increasing to infinity and monotone in t. The topological slow entropy of T with respect to $\{a_n(t)\}_{n\in\mathbb{N},t\geq 0}$ is*

$$\overline{ent}_a^{\mathrm{top}}(T) = \sup_K \lim_{\epsilon\to 0} N(\epsilon, K), \tag{9.41}$$

where $N(\epsilon,K) = \begin{cases} \sup N_1(\epsilon,K), & \text{if } N_1(\epsilon,K)\neq\emptyset; \\ 0, & \text{if } N_1(\epsilon,K)=\emptyset, \end{cases}$ *for*

$$N_1(\epsilon,K) = \left\{ t>0\colon \limsup_{n\to\infty} \frac{N_{d^T_{F_n},K}(\epsilon)}{a_n(t)} > 0 \right\}.$$

Equivalently, $\overline{ent}_a^{\mathrm{top}}(T)$ can also be defined as follows:

$$\overline{ent}_a^{\mathrm{top}}(T) = \sup_K \lim_{\epsilon\to 0} S(\epsilon, K), \tag{9.42}$$

where $S(\epsilon,K) = \begin{cases} \sup S_1(\epsilon,K), & \text{if } S_1(\epsilon,K)\neq\emptyset; \\ 0, & \text{if } S_1(\epsilon,K)=\emptyset, \end{cases}$ *for*

$$S_1(\epsilon,K) = \left\{ t>0\colon \limsup_{n\to\infty} \frac{S_{d^T_{F_n},K}(\epsilon)}{a_n(t)} > 0 \right\}.$$

Remark 4.2.3. *If we pick $a_n(t) = e^{nt}$, then topological slow entropy as just defined coincides with the classical topological entropy.*

9.4.2.1 Slow Entropy Goodwyn's Theorem

It is natural to ask for the variational principle for measure-theoretic and topological slow entropy. In fact, the answer is negative and counterexamples are provided in Section 9.4.5. On the other hand, we still have one-sided inequality in the slow entropy setting, namely slow entropy Goodwyn's theorem. Indeed, this is a corollary of Proposition 2 in [41]. We introduce the following definition to formulate the slow entropy Goodwyn's theorem.

Definition 4.2.4. *A metric space X is well-partitionable if for any Borel probability measure μ, compact set K and $\epsilon > 0$, there exists a finite partition $\mathscr{P}$ of K whose atoms have diameter less than ϵ and such that $\mu\left(\bigcup_{\xi\in\mathscr{P}}\partial_\epsilon\xi\right) < \epsilon$, where $\partial_\epsilon\xi = \{y\in X\colon B(y,\epsilon)\cap\xi\neq\emptyset$ but $B(y,\epsilon)\not\subseteq\xi\}$.*

Notice that any smooth manifold is well-partitionable.[15]

[15] In [41] the authors consider a compact space X, but their proof of Proposition 2 generalizes easily to the case where X is well-partitionable.

Theorem 4.2.5 (Slow entropy Goodwyn's Theorem). *Suppose X is well-partitionable and $T\colon (X,d)\times\Gamma\to(X,d)$ is an action of the group Γ by homeomorphisms preserving a non-atomic Borel probability measure μ. Let $\{a_n(t)\}_{n\in\mathbb{N},t>0}$ be a family of sequences increasing to infinity and monotone in t; then*

$$\overline{ent}_a^{\mu}(T)\le\overline{ent}_a^{\mathrm{top}}(T).$$

9.4.3 Generalized Kushnirenko's Inequality

As we have already mentioned at the beginning of Section 9.2.4, slow entropy is a useful tool in the smooth realization problem. In fact, this is also one of the main results of [41]. In [41], the authors not only gave a generalized version of Kushnirenko's inequality in the slow entropy setting, but also constructed an example to distinguish two versions of Kushnirenko's inequality, that is, a $\mathbb{Z}^2$-action that cannot be smoothly realized but satisfies Kushnirenko's inequality (Theorem 2.4.3) and fails the generalized Kushnirenko inequality (Theorem 4.3.2).

The generalized Kushnirenko inequality will be derived from the following proposition.

Proposition 4.3.1 (Katok–Thouvenot [41]). *Assume that the group Γ is generated by a finite set Γ_0 and that the Følner sets F_n consists of elements whose word-length norm with respect to Γ_0 does not exceed a_n. Suppose that the compact metric space X has finite box dimension D with respect to the metric d and Γ acts on X by **bi-Lipschitz homeomorphisms**. Denote by L a common Lipschitz constant for all elements $T(r)$, $r\in\Gamma_0\cup\Gamma_0^{-1}$. Then for any partition ξ with $\mu(\partial\xi)=0$, and some constant $c(\epsilon)$ only depending on ϵ and X, we have*

$$S_{\xi}^{H}(T,F_n,\epsilon,\epsilon)\le c(\epsilon)\exp(Da_n\log L).$$

Combining the preceding proposition with Theorem 4.1.7, the generalized Kushnirenko inequality can be formulated as follows:

Theorem 4.3.2 (Generalized Kushnirenko's inequality, Katok–Thouvenot [41]). *Under the same assumptions as in Proposition 4.3.1, a measure-preserving action of Γ by Lipschitz homeomorphisms in a finite box dimensional space can be smoothly realized if its slow entropy is no more than exponential measured by the diameter of the Følner set in the word-length metric.*

Notice that zero entropy of the action corresponds to the subexponential growth measured by the number of elements in the Følner set. Thus, for any group that is not a finite extension of $\mathbb{Z}$ there is a gap between the classical entropy and slow entropy. In fact, this indicates that zero entropy systems with superexponential growth in terms of diameters of Følner sets can be constructed, and such systems do not have smooth or Lipschitz realization.

9.4.3.1 Failure of Kushnirenko's Inequality

Let $C_n = [0, n-1] \times [0, n-1] \subset \mathbb{Z}^2$ and T_1, T_2 are two commuting measure-preserving transformations generating the $\mathbb{Z}^2$−action T on the space (X, μ). Then by [41], the existence of the following special $\mathbb{Z}^2$-action is guaranteed:

Proposition 4.3.3 (Katok–Thouvenot [41]). *Let ϵ_n, $n = 1, 2, \ldots$ be a sequence of positive numbers decreasing to 0. There exists a measure-preserving ergodic $\mathbb{Z}^2$-action T on the unit interval with Lebesgue measure generated by transformations T_1, T_2 and a generating partition ξ such that*

(1) for some $\epsilon > 0$, $\delta > 0$,

$$\limsup_{n\to\infty} \frac{\log(S_\xi^H(T, C_n, \epsilon, \delta))}{n^2 \epsilon_n} \geq 1;$$

(2) the $\mathbb{Z}^2$-entropy $h_\mu(T) = 0$;

(3) every element of the $\mathbb{R}^2$ suspension action has zero entropy.

As a corollary of the preceding proposition we get the following:

Corollary 4.3.4 (Katok–Thouvenot [41]). *If $\lim_{n\to\infty} n\epsilon_n = \infty$, then the action constructed in Proposition 4.3.3 is not isomorphic to any action by Lipschitz homeomorphisms on a compact metric space of finite box dimension preserving a Borel probability measure. In particular, it is not isomorphic to an action by diffeomorphisms of a compact differentiable manifold.*

The main method used in the construction of the preceding action is the cutting and stacking method by alternating periodic and independent concatenations. Interested readers may go to [41] for details.

Slow entropy can also be used for smooth realization problems in the setting of infinite-measure-preserving systems. By generalizing slow entropy to infinite-measure-preserving systems, M. Hochman [26] produced examples of infinite-measure-preserving $\mathbb{Z}^2$-actions that are not isomorphic to actions by diffeomorphisms on a compact manifold preserving an infinite Borel measure.

9.4.4 Criterion of Total Vanishing of Slow Entropy

Recall that for a dynamical system $(X, \mathscr{B}, \mu, T)$, if its sequence entropy is zero for all subsequences of $\{1, 2, \ldots, \}$, then it has discrete spectrum. As slow entropy is also an invariant that measures the complexity of a given system, it is natural to expect that the total vanishing of slow entropy systems may have some analogous spectral consequences. Indeed, by [15] (measure-theoretic slow entropy) and [34] (topological slow entropy), total vanishing of slow entropy at all scales will restrict the system to being isomorphic to a translation in respectively the measurable and the topological category. This is the content of the two results below:

Theorem 4.4.1 (Ferenczi [15]). *T is measure-theoretically isomorphic to a translation on a compact group if and only if*

$$\overline{ent}_a^{\mu}(T) = 0 \text{ or } \underline{ent}_a^{\mu}(T) = 0 \tag{9.43}$$

with respect to every family of scales $a_n(t)$.

Theorem 4.4.2 (Kanigowski–Vinhage–Wei [34]). *A minimal homeomorphism $T\colon X \to X$ of a compact metric space is topologically conjugated to a translation on a compact abelian group if and only if $\overline{ent}_a^{\mathrm{top}}(f) = 0$ for every family of scales $a_n(t)$.*

9.4.5 Failure of Variational Principle for Slow Entropy

The variational principle is one of the most important results in classical entropy theory. Unfortunately, it does not hold for slow entropy–type invariants. Although we still have Goodwyn's theorem for slow entropy, there are examples for which the topological slow entropy may not be equal to the supremum of measure-theoretic slow entropies, even for uniquely ergodic systems.

A general idea for the construction of counterexamples is based on Theorem 4.4.1 and Theorem 4.4.2: one finds a minimal topological system $f\colon X \to X$, all of whose ergodic measures yield discrete spectrum systems (Kronecker systems) but which is not topologically conjugate to a translation on a compact abelian group. In fact, this is possible by the *approximation-by-conjugation method* first used by Anosov and Katok in [2]. More precisely, we have the following:

Proposition 4.5.1. *Let M be a manifold with a free circle action. Then there exists a uniquely ergodic, volume-preserving, C^∞ diffeomorphism $f\colon M \to M$ which is measurably conjugate to a translation on a torus $\mathbb{T}^d$, $d \geq 1$.*

If $f\colon X \to X$ is a homeomorphism of a compact metric space, let $\mathfrak{M}(f)$ denote the space of invariant measures. We may also use the example due to Furstenberg:

Proposition 4.5.2 (Furstenberg [21]). *There exists a minimal C^∞ diffeomorphism of $\mathbb{T}^2$, with a one-parameter family of measures all of which yield discrete spectrum systems and isomorphic to R_α for some fixed $\alpha \in S^1$.*

As a result we establish the failure of the variational principle for slow entropy:

Corollary 4.5.3 (Kanigowski–Vinhage–Wei [34]). *Let M be a manifold with a free circle action. Then there exists a C^∞ diffeomorphism $f\colon M \to M$ and a family of scales a_χ such that*

$$\sup_{\mu\in\mathfrak{M}(f)} \overline{ent}^{\mu}_{a_\chi}(f) < \overline{ent}^{\mathrm{top}}_{a_\chi}(f).$$

9.4.6 Different Upper and Lower Quantities of Slow Entropy

Notice in the definition of slow entropy we may take lim sup or lim inf in (9.36) to get respectively $\overline{\mathrm{ent}}^{\mu}_{a}$ or $\underline{\mathrm{ent}}^{\mu}_{a}$, respectively. Recall that, for classical entropy, the two preceding quantities coincide. As follows by a result of Cyr and Kra [7], this is not the case for slow entropy:

Theorem 4.6.1 (Cyr–Kra [7]). *Assume $(a_n)_{n\in\mathbb{N}}$ and $(b_n)_{n\in\mathbb{N}}$ are two non-decreasing sequences of positive integers such that $\lim_{n\to\infty} a_n = \infty$, $\lim_{n\to\infty} \frac{1}{n}\log b_n = 0$ and $a_n \le b_n$ for all $n \in \mathbb{N}$. There exists a minimal subshift (X_∞, σ) of topological entropy zero and a σ-invariant, ergodic measure μ supported on X_∞ such that*

$$\begin{aligned} &\sup_{\xi}\lim_{\delta\to 0}\lim_{\epsilon\to 0}\liminf_{n\to\infty}\frac{S^H_\xi(\sigma,n,\epsilon,\delta)}{a_n} \le 1,\\ &\sup_{\xi}\lim_{\delta\to 0}\lim_{\epsilon\to 0}\limsup_{n\to\infty}\frac{S^H_\xi(\sigma,n,\epsilon,\delta)}{b_n} \ge 1. \end{aligned} \tag{9.44}$$

Let

$$s_n(t) := \begin{cases} b_n(2t-1), & 1\le t;\\ a_n(2-2t)+b_n(2t-1), & \frac{1}{2}\le t\le 1;\\ 2ta_n + (1-2t)\frac{a_0}{2}, & 0\le t\le \frac{1}{2}. \end{cases} \tag{9.45}$$

Due to $a_n \le b_n$, $\lim_{n\to\infty} a_n = \infty$ and $\lim_{n\to\infty}\frac{1}{n}\log(b_n) = 0$, it is clear that $\{s_n(t)\}_{n\in\mathbb{N},t>0}$ is a family of positive sequences increasing to infinity for $t>0$, monotone in t for $t\in[0,\infty)$ and $s_n(0)$ is a positive constant sequence. Moreover, $s_n(t)$ satisfies $s_n(\frac{1}{2}) = a_n$ and $s_n(1) = b_n$.

By using the language of slow entropy, Theorem 4.6.1 can be reformulated as follows:

Theorem 4.6.2. *Let $(s_n)_{n\in\mathbb{N}}$ be defined as just shown. There exists a minimal subshift (X_∞,σ) of topological entropy zero and an ergodic measure μ supported on X_∞ such that*

$$\underline{ent}^{\mu}_{s}(\sigma) \le \frac{1}{2} < 1 \le \overline{ent}^{\mu}_{s}(\sigma). \tag{9.46}$$

9.4.7 Slow Entropy of Quasi-unipotent Flows

Recall that quasi-unipotent flows have zero entropy. Therefore, it is a natural idea to use slow entropy to study quasi-unipotent flows. In this section, we mainly state results from [34]. Define G as a connected Lie group and $\mathfrak{g}$ as its Lie algebra. Suppose Γ is a discrete subgroup of G with co-finite volume and μ is the Haar measure on G/Γ that is induced from Riemannian volume on G.

Theorem 4.7.1 (Kanigowski–Vinhage–Wei [34]). *Let φ_t be an ergodic quasi-unipotent flow*[16] *on a homogeneous space G/Γ with chain structure $(m_1,\dots,m_n)$. Then the topological and measure-theoretic slow entropy of φ_t with respect to $a_n(\chi)=n^\chi$ coincide:*

$$\overline{ent}_a^{\mu}(\varphi_t)=\underline{ent}_a^{\mu}(\varphi_t)=\overline{ent}_a^{\text{top}}(\varphi_t)=\underline{ent}_a^{\text{top}}(\varphi_t)=GR(U),$$

where $GR(U)$ is defined as in Section 9.3.7.1.

Remark 4.7.2. *As a corollary to Theorem 4.7.1, one obtains a version of the variational principle for quasi-unipotent flows and slow entropy. Namely, the topological slow entropy of a quasi-unipotent flows is the supremum of the measure-theoretic slow entropies taken over all invariant measures. Recall that by Section 9.4.5 this property is fairly special.*

9.4.7.1 Examples of Quasi-unipotent Flows

We give an example of a quasi-unipotent flow and calculate its slow entropy.

Let $G=\mathrm{SL}(d,\mathbb{R})$, $\Gamma\subset G$ be any lattice, and

$$U=\begin{pmatrix}0&1&&&\\&0&1&&\\&&\ddots&\ddots&\\&&&0&1\\&&&&0\end{pmatrix},\qquad \exp_{alg}(tU)=\begin{pmatrix}1&t&t^2/2&\cdots&t^{d-1}/(d-1)!\\&1&t&\cdots&t^{d-2}/(d-2)!\\&&\ddots&\ddots&\vdots\\&&&1&t\\&&&&1\end{pmatrix}.$$

We call U the *principal* nilpotent element associated to the algebra $\mathfrak{sl}(d,\mathbb{R})$. In the special case of $G=\mathrm{SL}(d,\mathbb{R})$, any nilpotent algebra element is conjugated to a block form element

$$U=\begin{pmatrix}U_1&&&\\&U_2&&\\&&\ddots&\\&&&U_n\end{pmatrix},$$

[16] See Section 9.3.7.1 for quasi-unipotent flow's definition.

where each $U_i \in \mathfrak{sl}(d_i, \mathbb{R})$ is a principal element. Note that this is exactly the Jordan normal form of the matrix U. Call each U_i a *principal block* and the sequence $(\dim(U_1), \ldots, \dim(U_n))$ the *block sequence of* U.

We get the following Corollary of Theorem 4.7.1:

Corollary 4.7.3. *(Kanigowski–Vinhage–Wei [34]) The topological and measure-theoretic slow entropy of a principal unipotent flow equals* $GR(U)$*, where*

$$GR(U) = \sum_{i=1}^{m} \frac{1}{6} k_i(4k_i + 1)(k_i - 1) + \sum_{i=1}^{m-1} \sum_{j=i+1}^{m} \frac{1}{3} k_i \left(k_i^2 + 3k_j^2 - 3k_j - 1\right),$$

where $(k_1, \ldots, k_m)$*,* $k_i \leq k_{i+1}$ *is the block sequence of* U*.*

9.4.8 Slow Entropy of Rank-One Systems

Rank-one systems form a very important class of systems in ergodic theory; they have a quite simple constructive definition and they are a rich source of examples and counterexamples in ergodic theory. We define a rank-one system as follows:

Definition 4.8.1 (β-Rank one). *Let* β *belong to* $(0, 1]$*. A system* (X, T, μ) *is of* β*-rank one if for every partition* α *of* X*, for every positive* ϵ*, there exist two subsets* F *and* A *of* X*, a positive integer* h *and a partition* α' *of* A *such that*

(1) $F, TF, \ldots, T^{h-1}F$ *are disjoint;*
(2) $A = \cup_{j=0}^{h-1} T^j F$*;*
(3) $\mu(A) > \beta$*;*
(4) $|\alpha|_A - \alpha'| < \epsilon$*,*
(5) α' *is refined by the sets* $\{F, TF, \ldots, T^{h-1}F\}$*.*

Moreover, if $\beta = 1$*, we will say* β*-rank-one system is a rank-one system.*

Recall that rank-one systems have zero entropy, thus it is natural to use slow entropy to measure the complexity of rank-one systems. By Ferenczi's results, we have the following general theorem about the upper bound of slow entropy for rank-one systems:

Theorem 4.8.2 (Ferenczi [15]). *For any rank-one system* (X, T, μ) *and* $a_n(t) = n^t$*, we have*

$$\underline{\mathrm{ent}}_a^{\mu}(T) \leq 2.$$

Kanigowski [33] got more precise upper bounds on slow entropy of local rank-one flows:[17]

[17] Slow entropy can be defined analogously for $\mathbb{R}$-actions with the use of a continuous Hamming metric.

Theorem 4.8.3 (Kanigowski [33]). *Let $\beta \in (0,1]$ and $g(n)$ be any sequence of positive numbers such that $\lim_{n\to+\infty} g(n) = +\infty$. Then, for any measure-preserving β-rank-one flow T_t which acts on (X,μ) and $a_n(t) = n(g(n))^t$, we have $\underline{ent}_a^{\mu}(T_t) = 0$.*

It is worthwhile to point out that although the lower *symbolic complexity* can be controlled for a rank-one system with the same bound as earlier, the upper symbolic complexity with respect to the scaling function $a_n(t) = n^t$ can be $+\infty$. For more details, interested readers may go to [14].

For a special rank-one system, namely the Chacon system, the precise value of slow entropy is obtained:

Proposition 4.8.4 (Ferenczi [15]). *For the Chacon system (X,T,μ) and $a_n(t) = n^t$,*

$$\underline{ent}_a^{\mu}(T) = \overline{ent}_a^{\mu}(T) = 1. \tag{9.47}$$

9.4.9 Slow Entropy of Surface Flows

Smooth flows on surfaces play a really important role in dynamical systems. One of the reasons is that dimension 2 is the lowest dimension in which we can have some non-trivial ergodic behavior. We restrict our attention to smooth surface flow with fixed points.[18] Recall that smooth surface flows have zero entropy due to Pesin's formula, thus it is natural to try to compute slow entropy in this class. It turns out that ergodic properties of smooth surface flows are successfully studied via their *special representation*. We will restrict our attention to so-called Arnol'd- and Kochergin-type special flows (defined in what follows) over irrational rotations. Kanigowski [33] computed the precise value of slow entropy of these flows in the scales $a_n(t) = n(\log n)^t$ (for Arnol'd flows) and $a_n(t) = n^t$ (for Kochergin flows). Flows that we consider have the following special representation (with the base automorphism T and the roof function f):

Definition 4.9.1 (Arnol'd flows and Kochergin flows). *Let*

- $T = R_\alpha : \mathbb{T} \to \mathbb{T}$, $R_\alpha x = x + \alpha \mod 1$;
- *f is a $C^2(\mathbb{T}\setminus\{0\})$ function that satisfies:*

$$\lim_{x\to 0^+} \frac{f(x)}{h(x)} = A_1 \text{ and } \lim_{x\to 0^-} \frac{f(x)}{h(1-x)} = B_1, \text{ where } A_1, B_1 > 0; \tag{9.48}$$

$$\lim_{x\to 0^+} \frac{f'(x)}{h(x)} = -A_2 \text{ and } \lim_{x\to 0^-} \frac{f'(x)}{h(1-x)} = B_2, \text{ where } A_2, B_2 > 0; \tag{9.49}$$

[18] If a smooth surface flow has no fixed points, then by the Lefschetz formula, the surface is a two-dimensional torus and the flow is a smooth time-change of a linear flow.

$$\lim_{x\to 0^+} \frac{f''(x)}{h(x)} = A_3 \text{ and } \lim_{x\to 0^-} \frac{f''(x)}{h(1-x)} = B_3, \text{ where } A_3, B_3 > 0. \tag{9.50}$$

- *We call such flows* Kochergin flows *if* $f(x) = x^{-\gamma}$, $0 < \gamma < 1$ *and denote them by* $\mathscr{T}_t^{f,\gamma}$.
- *We call such flows* Arnol'd flow *if* $f(x) = -\log x$ *and we assume additionally that* $A_i + B_i \neq 0, i = 1, 2, 3$ *and denote them as* $\mathscr{T}_t^f$.

The following two full measure sets are needed for the main results:

Definition 4.9.2. *Let* $\{q_n\}$ *be the denominators of* α *and let*

$$\mathscr{D} := \{\alpha \in \mathbb{R}\backslash\mathbb{Q}\colon q_{n+1} \leq C(\alpha)q_n \log q_n (\log n)^2 \text{ for every } n \in \mathbb{N}\},$$

$$K_\alpha := \{n\colon q_{n+1} \leq q_n \log^{\frac{7}{8}} q_n \text{ for every } n \in \mathbb{N}\},$$

and

$$\mathscr{E} := \{\alpha \in \mathbb{R}\backslash\mathbb{Q}\colon \sum_{i\notin K_\alpha} \log^{-\frac{7}{8}} q_i < +\infty\}.$$

Remark 4.9.3. *By Khinchin's theorem [45] we have* $\lambda(\mathscr{D}) = 1$. *By Fayad and Kanigowski [12], we have* $\lambda(\mathscr{E}) = 1$.

Then we have following two theorems:

Theorem 4.9.4 (Kanigowski [33]). *Let* $a_n(t) = n(\log n)^t$. *Then for every* $\alpha \in \mathscr{D} \cap \mathscr{E}$ *and the corresponding Arnol'd flow* $\mathscr{T}_t^f$, *we have* $\underline{\text{ent}}_a^\mu(\mathscr{T}_t^f) = 1$.

Theorem 4.9.5 (Kanigowski [33]). *Let* $a_n(t) = n^t$. *Then for every* $\alpha \in \mathscr{D}$ *and every* $\gamma \in (0,1)$ *for the corresponding Kochergin flow* $\mathscr{T}_t^{f,\gamma}$, *we have* $\underline{\text{ent}}_a^\mu(\mathscr{T}_t^{f,\gamma}) = 1 + |\gamma|$.

9.4.10 Slow Entropy of AbC Constructions

Recall in Section 9.4.5 that the AbC construction helps us to show the failure of the variational principle for slow entropy.

9.4.10.1 Brief Description of the AbC Construction

We give a brief description of the Anosov–Katok construction on a two-dimensional torus $\mathbb{T}^2 = \mathbb{R}^2/\mathbb{Z}^2$. Let $S_t(x,y) = (x+t, y)$.

We use notation

$$\Delta_q = \{(x,y) \in \mathbb{T}^2 : 0 \leq x \leq \frac{1}{q}\}, \quad \eta_q = \{S_{\frac{i}{q}}\Delta_q\}_{i=0}^{q-1}. \tag{9.51}$$

Given any summable sequence of positive real numbers $\{\epsilon_n\}$, the construction proceeds inductively. Assume that we have chosen sequences of integers $\{k_m\}$ and $\{l_m\}$ for $m = 1, 2, 3, \ldots, n-1$, sequences of integers $\{p_m\}, \{q_m\}$, a sequence of rationals $\{\alpha_m\}$, sequences of diffeomorphisms $\{h_m\}, \{H_m\}, \{T_m\}$, sequences of partitions $\{\eta_m\}, \{\xi_{q_m}\}$, and a sequence of sets $\{E_m\}$ for $m = 1, 2, 3, \ldots, n$, such that the following properties are satisfied for $m < n$:

$$
\begin{aligned}
&\alpha_m = \frac{p_m}{q_m}, \quad \alpha_{m+1} = \alpha_m + \beta_m, \quad \beta_m = \frac{1}{k_m l_m q_m^2},\\
&p_{m+1} = k_m l_m p_m + 1, \quad q_{m+1} = k_m l_m q_m^2,\\
&H_{m+1} = h_{m+1} \circ h_m \circ \ldots \circ h_1,\\
&T_{m+1} = H_m^{-1} \circ S_{\alpha_m} \circ H_m,\\
&\xi_{m+1} = H_{m+1}^{-1} \eta_{q_{m+1}},\\
&\mu(E_m) < \epsilon_m,\\
&\|T_{m+1} - T_m\|_{C^m} < \epsilon_m.
\end{aligned}
\tag{9.52}
$$

At the $n+1$-th step we choose k_n, l_n, h_{n+1} and E_{n+1}. Note that this automatically gives us q_{n+1}, p_{n+1} $H_{n+1}, \eta_{q_{n+1}}, \xi_{q_{n+1}}, \beta_{n+1}, \alpha_{n+1}$ and T_{n+1}. We finally define $T := \lim_{n\to\infty} T_n$.

We have the following result regarding the slow entropy of AbC constructions:

Theorem 4.10.1 (Kanigowski–Vinhage–Wei, [36]). *Assume that $\epsilon_{n_0} > \sum_{n=n_0+1}^{\infty} \epsilon_n$ for all n_0 sufficiently large. Then for any scale $a_n(t)$, there exists a diffeomorphism obtained by the AbC construction whose slow entropy is zero.*

Remark 4.10.2. *Here the topological slow entropy cannot vanish at all scales as (non-trivial) AbC constructions are not topologically conjugate to a translation.*

Combining slow entropy Goodwyn's theorem (Theorem 4.2.5) with Theorem 4.10.1, we have the following corollary.

Corollary 4.10.3. *Measure-theoretic slow entropy of an AbC construction can be made zero at arbitrary small scales by adjusting the coefficients of AbC construction.*

9.4.11 Slow Entropy of Higher-Rank Actions

In order to measure the complexity of smooth measure-preserving (higher-rank) actions, we consider the slow entropy invariant. One needs to modify the original slow entropy definition to a specific scale that will work better for higher-rank abelian action:

Definition 4.11.1 (Slow entropy of a $\mathbb{Z}^k$ action [43]). *Let α be an action of $\mathbb{Z}^k$. Given a norm d on $\mathbb{R}^k$, let F_s^d be the intersection of the lattice $\mathbb{Z}^k$ with the ball centered at 0 and radius s. We define the slow entropy of α w.r.t. to the norm d and the partition ξ as*

$$sh(d,\alpha,\xi) = \lim_{\epsilon,\delta\to 0} \limsup_{s\to\infty} \frac{1}{s} \log S_\xi^H(\alpha, F_s^d, \epsilon, \delta).$$

By taking the supremum over all finite measurable partitions, we define

$$sh(d,\alpha) = \sup_\xi sh(d,\alpha,\xi). \tag{9.53}$$

Finally, we define the slow entropy of α as

$$sh(\alpha) = \inf_{d:vol(d)=1} sh(d,\alpha),$$

where $vol(d)$ is the volume of the unit ball in the norm d under the assumption that $\mathbb{Z}^k \subset \mathbb{R}^k$ has co-volume 1.

In order to state results related to slow entropy of Cartan actions, we need to introduce some basic notation related to algebraic number theory. Suppose $A \in \mathrm{SL}(n,\mathbb{Z})$ is a matrix and has an irreducible characteristic polynomial f. Denote the centralizer of A with rational coefficients as $Z_Q(A)$; then $Z_Q(A)$ can be identified with the field $K = \mathbb{Q}(\lambda)$, where λ is an eigenvalue of A. Denote this identification by $\gamma: p(A) \to p(\lambda)$ with $p \in \mathbb{Q}(x)$. Then define O_K as the ring of integers in K and U_K as the group of units in O_K. We state the result on slow entropy of Cartan actions:

Theorem 4.11.2 (A. Katok–S. Katok–Rodriguez Hertz [43]). *Let α be a Cartan action on the torus $\mathbb{T}^n$. Then*

$$sh(\alpha) = C(n)R^{\frac{1}{n-1}}, \tag{9.54}$$

with $R = kR_K$, where R_K is the regulator, $k = [U_K : \gamma(\alpha)]$, and

$$\frac{n-1}{2} \le C(n) \le n-1. \tag{9.55}$$

9.5 Other Entropy-Type Invariants

In this section, we will introduce other invariants related to slow entropy: Fried entropy, entropy convergence rate, Kakutani invariant and entropy dimension.

9.5.1 Fried Average Entropy

In this section we will define Fried average entropy for higher-rank actions and then introduce some relations between slow entropy and Fried average entropy. We will follow [20] and [43] in this section. We define the entropy function of a one-parameter subgroup of $\mathbb{R}^k$ action which preserves a probability ergodic measure μ first:

Definition 5.1.1. *The entropy function of an $\mathbb{R}^k$ action α, denoted by h_μ^α, associates to $t^v = (t_1, \dots, t_k) \in \mathbb{R}^k$ the value of the measure-theoretic entropy of $\alpha(t^v)$: $h_\mu^\alpha(t^v) = h_\mu(\alpha(t^v))$.*

We now define the Fried average entropy of an $\mathbb{R}^k$ action:

Definition 5.1.2 (The Fried average entropy, Fried [20]). *Let α be an $\mathbb{R}^k$ action with a fixed volume element. The* Fried average entropy *of α, denoted as $h^F(\alpha)$, is defined as*

$$h^F(\alpha) = \frac{2^k}{k!vol(B(h_\mu^\alpha))},$$

where $B(h_\mu^\alpha)$ is the unit ball with respect to the entropy function.

If α is a $\mathbb{Z}^k$ action, then the Fried average entropy of α is defined as the Fried average entropy of its suspension.

Here are some basic properties of Fried average entropy:

Proposition 5.1.3 ([20],[43]). *(a) Let α_1 be an action of $\mathbb{R}^{k_1}$ on M_1, α_2 be an action of $\mathbb{R}^{k_2}$ on M_2, preserving the measures μ_1 and μ_2 respectively, and $\alpha = \alpha_1 \times \alpha_2$ be the product action of $\mathbb{R}^{k_1+k_2}$ with the measure $\mu = \mu_1 \times \mu_2$. Then*

$$h^F(\alpha) = h^F(\alpha_1) \cdot h^F(\alpha_2).$$

(b) For $k = 1$, $h^F(\alpha)$ is equal to the usual entropy of the flow, namely the entropy of its time-one map.

(c) The Fried average entropy is equal to zero if the entropy function is a semi-norm.

(d) The Fried average entropy of a $\mathbb{Z}^k$ action is independent of the choice of generators.

(e) The Fried average entropy of the restriction of a $\mathbb{Z}^k$ action α to a finite index subgroup $A \subset \mathbb{Z}^k$ is equal to h_F^α multiplied by the index of the subgroup A.

Recall that α is a maximal rank action, if it is an action of $\mathbb{Z}^a \times \mathbb{R}^b$ with $a + b = n - 1$ on a $n+b$-dimensional manifold. We have the following dichotomy for $k \geq 2$:

Theorem 5.1.4 (A. Katok–S. Katok–Rodriguez Hertz [43]). *Suppose α is a maximal-rank $\mathbb{Z}^k$ action preserving μ. Then either μ is absolutely continuous and the Fried average entropy is positive or the Fried average entropy of the action is equal to zero.*

For Cartan actions, we have the following quantitative description of Fried average entropy. We define Cartan action first.

Definition 5.1.5 (Cartan action, Hurder [31]). *Let $\mathscr{A}$ be a free abelian group with a given set of generators $\Delta = \{\gamma_1, \ldots, \gamma_n\}$. Then (ϕ, Δ) is a Cartan C^r-action on the n-manifold X if*

(a) $\phi\colon \mathscr{A} \times X \to X$ is a C^r-action on X;
(b) each $\gamma_i \in \Delta$ is ϕ-hyperbolic, namely $\phi(\gamma_i)$ is an Anosov diffeomorphism of X, and $\phi(\gamma_i)$ has one-dimensional, strong stable foliation $\mathscr{F}_i^{ss}$;
(c) the tangential distributions $E_i^{ss} = T\mathscr{F}_i^{ss}$ are pairwise transverse with their internal direct sum $E_1^{ss} \oplus \ldots \oplus E_N^{ss} \cong TX$.

Moreover, if $\phi\colon \Gamma \times X \to X$ is an Anosov C^r-action on a manifold X, we say that ϕ is a Cartan lattice action if there is a subset of commuting hyperbolic elements $\Delta = \{\gamma_1, \ldots, \gamma_n\} \subset \Gamma$, which generate an abelian subgroup $\mathscr{A}$, such that the restriction of $\phi|\mathscr{A}$ is an abelian Cartan C^r-action on X.

Then we have

Theorem 5.1.6 (Lower bound of Fried average entropy, A. Katok–S. Katok–Rodriguez Hertz [43]). *The Fried average entropy $h^F(\alpha)$ of a Cartan action α of a given rank $n-1 \geq 2$ is bounded away from zero by a positive function that grows exponentially with n,*

$$h^F(\alpha) > 0.000752 \exp(0.244n). \tag{9.56}$$

Moreover,

$$h^F(\alpha) \geq 0.089.$$

Combining the main result in [44] with the preceding theorem, we have the following characterization of weakly mixing maximal rank $\mathbb{Z}^k$ actions:

Corollary 5.1.7 (A. Katok–S. Katok–Rodriguez Hertz [43]). *The Fried average entropy of a weakly mixing maximal rank action of $\mathbb{Z}^k$, $k \geq 2$, is either equal to zero or is greater than* 0.089; *furthermore the lower bound grows exponentially with k as* (9.56).

At the end of this section, we will establish the connection between Fried average entropy and slow entropy of higher rank actions. The first relation is for affine actions on the torus:

Theorem 5.1.8 (A. Katok–S. Katok–Rodriguez Hertz [43]). *For an affine action α on a torus, $sh(\alpha) = 0$ if and only if $h^F(\alpha) = 0$.*

If we assume α is a weakly mixing maximal rank action, we get the following formulas explicitly establishing the relation between slow entropy and Fried average entropy:

Theorem 5.1.9 (A. Katok–S. Katok–Rodriguez Hertz [43]). *Suppose α is a weakly mixing maximal rank $\mathbb{Z}^k$ action for $k \geq 2$. Then we have*

$$sh(\alpha) = c(k)\binom{2k}{k}^{\frac{1}{k}}(h^F(\alpha))^{\frac{1}{k}}, \tag{9.57}$$

where $c(k) \in [\frac{k}{4}, \frac{k}{2}]$.

Remark 5.1.10. *Theorem 5.1.9 implies that the slow entropy is either equal to zero or uniformly bounded away from zero. Moreover, the lower bound grows linearly with respect to the rank.*

9.5.2 Kakutani Equivalence Invariant

Kakutani equivalence relation was first introduced by Kakutani [32] in 1943. Later this equivalence relation was developed by Katok [37], [38] and Feldman [13] by introducing a property named standardness (zero entropy loosely Bernoulli or loosely Kronecker), which is invariant under Kakutani equivalence.

Later, Ratner introduced a Kakutani invariant in [60] and [62] to study Kakutani equivalence quantitatively. In this section we will concentrate on this invariant and related results. The general idea of this invariant is to use the $\bar{f}$-metric instead of the Hamming metric to measure the complexity of a given system.

9.5.2.1 Definition and Basic Theorems

Let T and S be two ergodic transformations on $(X, \mathscr{B}, \mu)$ and $(Y, \mathscr{C}, \nu)$, respectively. We call T and S *Kakutani equivalent*, denoted as $T \overset{K}{\sim} S$, if there exist $A \subset X$ and $B \subset$ with $\mu(A)\nu(B) > 0$ such that $(T|_A, A, \mu_A)$ and $(S|_B, B, \nu_B)$ are measurably isomorphic.

Kakutani equivalence can also be defined through *time changes*. For a flow T_t acting on $(X, \mathscr{B}, \mu)$ and a function $\alpha \in L^1_+(X, \mathscr{B}, \mu)$, the flow T^α_t is called a *time change* of T_t if $T^\alpha_t(x) = T_{u(t,x)}(x)$, where $u(t,x)$ satisfies $\int_0^{u(t,x)} \alpha(T_s x)ds = t$. Then two ergodic measure-preserving flows T_t, S_t acting on $(X, \mathscr{B}, \mu)$ and $(Y, \mathscr{C}, \nu)$ are *Kakutani equivalent* if S_t is isomorphic to T^α_t for some $\alpha \in L^1_+(X, \mathscr{B}, \mu)$.

In 1977, Katok introduced a hierarchy structure among Kakutani equivalence classes by a binary relation called *majorization*. An ergodic flow S_t is majorized by an ergodic flow T_t, which is denoted as $S_t \overset{M}{\prec} T_t$, if there exists an ergodic flow $\tilde{T}_t$ which is Kakutani equivalent to T_t such that S_t is a factor of $\tilde{T}_t$. It is clear that if $T_t \overset{K}{\sim} S_t$, then we have $S_t \overset{M}{\prec} T_t$ and $T_t \overset{M}{\prec} S_t$, which guarantees that majorization defines a quasi-partial order among all Kakutani equivalence classes.[19]

[19] We do not know whether $S_t \overset{M}{\prec} T_t$ and $T_t \overset{M}{\prec} S_t$ imply $S_t \overset{K}{\sim} T_t$.

Among all Kakutani equivalence classes, there exists a special equivalence class majorized by all other classes which is called the *standard class* and can be characterized geometrically as follows: a flow T_t is *standard* or *loosely Bernoulli of zero entropy* if it is Kakutani equivalent to an irrational linear flow[20] on $\mathbb{T}^2$.

From now on, we will assume that T_t is an ergodic flow acting on a Lebesgue space $(X, \mathscr{B}, \mu)$, $\mathcal{P}$ is a finite measurable partition of X and $I_T(x) = \{T_s x : s \in [0, T]\}$. Suppose that $x \in X$ and we denote $\mathcal{P}(x)$ as the atom of $\mathcal{P}$ containing x.

Definition 5.2.1 (($\epsilon, \mathcal{P}$)-matchable, Ratner [62]). *For $x, y \in X$, $\epsilon > 0$ and $T > 1$, $I_T(x)$ and $I_T(y)$ are called $(\epsilon, \mathcal{P})$-matchable if there exist subsets $A = A(x,y), A' = A'(x,y) \subset [0,T]$ with $l(A) > (1-\epsilon)T$, $l(A') > (1-\epsilon)T$ and an increasing absolutely continuous map $h = h(x,y)$ from A onto A' such that $\mathcal{P}(T_t x) = \mathcal{P}(T_{h(t)} y)$, for all $t \in A$ and the derivative h', satisfies*

$$|h'(t) - 1| < \epsilon, \quad \forall t \in A. \tag{9.58}$$

We call h an $(\epsilon, \mathcal{P})$-matching from $I_T(x)$ onto $I_T(y)$.

With the preceding definition, Kakutani invariants can be defined in the following process (following Ratner):

Definition 5.2.2 (Kakutani invariant, Ratner [62]). *We have the following:*

(1) Define

$$\bar{f}_T(x, y, \mathcal{P}) = \inf\{\epsilon : I_T(x) \text{ and } I_T(y) \text{ are } (\epsilon, \mathcal{P})\text{-matchable}\}. \tag{9.59}$$

(2) Denote $B_T(x, \epsilon, \mathcal{P}) = \{y \in X : \bar{f}_T(x, y, \mathcal{P}) < \epsilon\}$ the $(T, \mathcal{P})$-ball of radius $\epsilon > 0$ centered at $x \in X$, $T > 1$.

(3) A family $\alpha_T(\epsilon, \mathcal{P})$ of $(T, \mathcal{P})$-balls of radius $\epsilon > 0$ is called $(\epsilon, T, \mathcal{P})$-cover of X if we have $\mu(\cup \alpha_T(\epsilon, \mathcal{P})) > 1 - \epsilon$.

(4) Denote $K_T(\epsilon, \mathcal{P}) = \inf |\alpha_T(\epsilon, \mathcal{P})|$ where $|A|$ denote the number of elements in A and inf is taken over all $(\epsilon, T, \mathcal{P})$-covers of X.

(5) Let $\mathcal{F}$ denote the family of all non-decreasing functions from $\mathbb{R}^+$ onto itself, converging to $+\infty$.

(6) For $u \in \mathcal{F}$, we denote

$$\begin{aligned} \beta(u, \epsilon, \mathcal{P}) &= \liminf_{T \to \infty} \frac{\log K_T(\epsilon, \mathcal{P})}{u(T)}; \\ e(u, \mathcal{P}) &= \limsup_{\epsilon \to 0} \beta(u, \epsilon, \mathcal{P}); \\ e(T_t, u) &= \sup_{\mathcal{P}} e(u, \mathcal{P}). \end{aligned} \tag{9.60}$$

We shortly document some basic properties of $e(T_t, u)$ in the following:

[20] Notice that standardness implies ergodicity. By Katok's results in [38], all irrational linear flows on $\mathbb{T}^2$ are Kakutani equivalent.

Theorem 5.2.3 (Generator theorem, Ratner [60], [62]). *Let T_t be an ergodic measure-preserving flow on $(X, \mathscr{B}, \mu)$ and let $\mathcal{P}_1 \leq \mathcal{P}_2 \leq \ldots$ be an increasing sequence of finite measurable partitions of X such that $\vee_{n=1}^{\infty} \mathcal{P}_n$ generates the σ-algebra $\mathscr{B}$. Then $e(T_t, u) = \sup_m e(u, \mathcal{P}_m)$ for all $u \in \mathcal{F}$.*

Theorem 5.2.4 (Ratner [60], [62]). *Let T_t and $\tilde{T}_t$ be two ergodic Kakutani equivalent measure-preserving flows on $(X, \mathscr{B}, \mu)$ and $(\tilde{X}, \tilde{\mathscr{B}}, \tilde{\mu})$ respectively. Then*

$$e(T_t, u) = e(\tilde{T}_t, u)$$

for all $u \in U$ with

$$\lim_{t\to\infty} \frac{u(at)}{u(t)} = 1 \text{ for all } a > 0.$$

Remark 5.2.5. *Theorem 5.2.4 guarantees $e(T_t, u)$ is a Kakutani invariant for some certain functions u. A simple example which satisfies the assumptions of Theorem 5.2.4 is $u(t) = \log t$.*

9.5.2.2 Application of Kakutani Invariants

With the help of Kakutani invariants, we can give another characterization of standard flows, which should be understood as a quantitative version of majorization characterization:

Theorem 5.2.6 (Ratner [60], [62]). *A zero-entropy ergodic measure-preserving flow T_t is standard if and only if $e(T_t, u) = 0$ for all $u \in \mathcal{F}$.*

Recall that Ratner proved that the horocycle flow is standard and its Cartesian product is not standard in [58] and [59]. Moreover, Ratner gave the following estimates of the Kakutani invariant of the n-fold Cartesian product of the horocycle flow.

Theorem 5.2.7 (Ratner [60], [62]). *Let $h_t^{(n)} = h_t \times \cdots \times h_t$ $n = 1, 2, \ldots$, be the n-fold Cartesian product of the horocycle flow on $(\mathrm{SL}(2, \mathbb{R})/\Gamma)^n$ with cocompact lattice Γ and $u(t) = \log t$, $t > 1$. Then*

$$3n - 3 \leq e(h_t^{(n)}, u) \leq 3n - 2,$$

for all $n = 1, 2, \ldots$.

Combining Theorem 5.2.6 and Theorem 5.2.7, it is clear that among the n-fold Cartesian product of the horocycle flow, only the horocycle flow itself is standard.

Inspired by M. Ratner's results on the horocycle flow, it is natural to consider the Kakutani equivalence for general unipotent flows, this has even been considered by M. Ratner in [61]. Based on multiscale analysis and structure of semisimple

Lie groups, Kanigowski, Vinhage and Wei obtained the following result about the Kakutani invariant for unipotent flows. Recall that $GR(U)$ is defined in Theorem 4.7.1.

Theorem 5.2.8 (Kanigowski–Vinhage–Wei [35]). *Let G be a semisimple linear Lie group and $\phi_t = L_{\exp(tU)}$ be a unipotent flow on G/Γ. If Γ is cocompact, we have*

$$e(\phi_t, \log) = GR(U) - 3.$$

For finite volume Γ, we have

$$GR(U) - 4 \leq e(\phi_t, \log) \leq GR(U) - 3.$$

Moreover, if $GR(U) = 3$, then (ϕ_t) is standard.

By a direct computation, we get $3n - 4 \leq e(h_t^{(n)}, \log) \leq 3n - 3$. This generalizes M. Ratner's results [60] and [62] to any lattice in $(\mathrm{SL}(2,\mathbb{R}))^k$. If we assume additionally that the lattice is cocompact, then $e(h_t^{(n)}, \log) = 3n - 3$. Moreover, combining Theorem 5.2.8 with semisimple Lie group structure, we can use the following corollaries to provide a solution to M. Ratner's ICM Problem 1 in [61] (see also [46]):

Corollary 5.2.9 (Kanigowski–Vinhage–Wei [35]). *The only ergodic unipotent flows on compact quotients of linear semisimple Lie groups which are standard are of the form* $\phi_t = \begin{pmatrix} 1 & t \\ 0 & 1 \end{pmatrix} \times \mathrm{id}$ *acting on* $(\mathrm{SL}(2,\mathbb{R}) \times G')/\Gamma$, *where Γ is irreducible.*

Corollary 5.2.10 (Kanigowski–Vinhage–Wei [35]). *Let G be a linear semisimple Lie group with* $\dim G > 3$, *and G/Γ be a compact homogeneous space of G.*

(a) There are ergodic unipotent flows on G/Γ which are not standard.
(b) If G is simple, no unipotent flow on G/Γ is standard.
(c) If G has real rank at least two, there are two unipotent flows on G/Γ (which are not identity, but not necessarily ergodic) which are not Kakutani equivalent.
(d) If $G \cong \mathrm{SL}(d,\mathbb{R})$, then there are at least $d-1$ flows on G/Γ which are pairwise non-Kakutani equivalent.

9.5.3 Scaled Entropy

Scaled entropy, which was introduced by Pesin and Zhao ([76], [77]), is an invariant defined through the Carathéodory approach. This invariant also measures the complexity of a given system. Although scaled entropy may be quite different from slow entropy, there exist some relations which can be used to give an intuitive understanding of scaled entropy based on slow entropy. In the remaining part of this section, we give a short introduction of scaled entropy based on [76], [77].

9.5.3.1 Scaled Topological Entropy

Let (X, d) be a compact metric space and $T\colon X \to X$ be a continuous transformation.

(i) Let $\mathbf{a} = \{a_n\}_{n\geq 1}$ be a sequence of positive numbers which we call a *scaled sequence* if it is monotonically increasing to infinity.

(ii) Given $\epsilon > 0$, a scaled sequence $\mathbf{a} = \{a_n\}_{n\geq 1}$, a subset $Z \subset X$ and $\alpha, N > 0$, we define

$$M(Z,\alpha,N,\epsilon,\mathbf{a}) = \inf\Big\{\sum_i \exp(-\alpha a_{n_i})\colon \cup_i B_{n_i}(x_i,\epsilon) \supset Z,\ \ x_i \in X \text{ and } n_i \geq N \ \ \forall i\Big\}, \tag{9.61}$$

where $B_n(x_i, \epsilon) = \{y \in X\colon d_n^T(x, y) < \epsilon\}$ and d_n^T is the Bowen metric defined in (9.5).

(iii) As $M(Z, \alpha, N, \epsilon, \mathbf{a})$ is monotonically increasing with respect to N, its limit exists as $N \to \infty$ and we define $m(Z, \alpha, N, \epsilon, \mathbf{a}) = \lim_{N\to\infty} M(Z, \alpha, N, \epsilon, \mathbf{a})$.

(iv) Then define $E_Z(T, \mathbf{a}, \epsilon)$ and $E_Z(T, \mathbf{a})$ as

$$\begin{aligned} E_Z(T,\mathbf{a},\epsilon) &= \inf\{\alpha\colon m(Z,\alpha,\epsilon,\mathbf{a}) = 0\} = \sup\{\alpha : m(Z,\alpha,\epsilon,\mathbf{a}) = +\infty\}; \\ E_Z(T,\mathbf{a}) &= \lim_{\epsilon\to 0} E_Z(T,\mathbf{a},\epsilon). \end{aligned} \tag{9.62}$$

We call $E_Z(T, \mathbf{a})$ the *scaled topological entropy* of T on the set Z.

(v) For $Z \subset X$, let $N(Z, n, \epsilon)$ be the smallest number of Bowen's ball $B_n(x, \epsilon)$ that are needed to cover the set Z. Then for any scaled sequence $\mathbf{a} = \{a_n\}_{n\geq 1}$, we define the following two quantities:

$$\begin{aligned} \underline{E}_Z(T,\mathbf{a},\epsilon) &= \liminf_{n\to\infty} \frac{1}{a_n} \log N(Z,n,\epsilon), \\ \overline{E}_Z(T,\mathbf{a},\epsilon) &= \limsup_{n\to\infty} \frac{1}{a_n} \log N(Z,n,\epsilon). \end{aligned} \tag{9.63}$$

By taking $\epsilon \to 0$, we define *lower scaled topological entropy* $\underline{E}_Z(T, \mathbf{a})$ and *upper scaled topological entropy* $\overline{E}_Z(T, \mathbf{a})$ as follows:

$$\begin{aligned} \underline{E}_Z(T,\mathbf{a}) &= \lim_{\epsilon\to 0} \underline{E}_Z(T,\mathbf{a},\epsilon), \\ \overline{E}_Z(T,\mathbf{a}) &= \lim_{\epsilon\to 0} \overline{E}_Z(T,\mathbf{a},\epsilon). \end{aligned} \tag{9.64}$$

Remark 5.3.1. *Scaled entropy $E_Z(T,\mathbf{a})$, lower scaled topological entropy $\underline{E}_Z(T,\mathbf{a})$ and upper scaled topological entropy $\overline{E}_Z(T,\mathbf{a})$ are invariant under a topological conjugacy. Moreover, we have the following inequalities that describe the relation of these three invariants:*

$$E_Z(T,\mathbf{a}) \le \underline{E}_Z(T,\mathbf{a}) \le \overline{E}_Z(T,\mathbf{a}). \tag{9.65}$$

9.5.3.2 Scaled Metric Entropy

If μ is a T-invariant measure, we can define *scaled metric entropy* for a given scaled sequence $\mathbf{a}$ as:

$$E_\mu(T,\mathbf{a}) = \lim_{\epsilon\to 0}\lim_{\delta\to 0}\inf\{E_Z(T,\mathbf{a},\epsilon) : \mu(Z) > 1-\delta\}, \tag{9.66}$$

where $E_Z(T,\mathbf{a},\epsilon)$ is defined in (9.62).

Moreover, define *lower scaled metric entropy* $\underline{E}_\mu(T,\mathbf{a})$ and *upper scaled metric entropy* $\overline{E}_\mu(T,\mathbf{a})$ as

$$\begin{aligned}\underline{E}_\mu(T,\mathbf{a}) &= \lim_{\epsilon\to 0}\lim_{\delta\to 0}\inf\{\underline{E}_Z(T,\mathbf{a},\epsilon)\colon \mu(Z) > 1-\delta\},\\ \overline{E}_\mu(T,\mathbf{a}) &= \lim_{\epsilon\to 0}\lim_{\delta\to 0}\inf\{\overline{E}_Z(T,\mathbf{a},\epsilon)\colon \mu(Z) > 1-\delta\},\end{aligned} \tag{9.67}$$

where $\underline{E}_Z(T,\mathbf{a},\epsilon)$ and $\overline{E}_Z(T,\mathbf{a},\epsilon)$ are defined in (9.63).

Remark 5.3.2. *Same as in the topological case, scaled metric entropy, lower scaled metric entropy and upper scaled metric entropy are invariant under a topological conjugacy. Moreover, we have*

$$E_\mu(T,\mathbf{a}) \le \underline{E}_\mu(T,\mathbf{a}) \le \overline{E}_\mu(T,\mathbf{a}). \tag{9.68}$$

9.5.3.3 Scaled Entropy's Properties

By [77], we can describe the scaled metric entropy based on slow entropy language. Let $\mathbf{a} = \{a_n\}_{n\ge 1}$ be a scaled sequence and μ is a T-invariant ergodic measure; we have

$$\begin{aligned}\overline{S}(T,\mu) &\le \overline{E}_\mu(T,\mathbf{a}),\\ \underline{S}(T,\mu) &\le \underline{E}_\mu(T,\mathbf{a}),\end{aligned} \tag{9.69}$$

where $\overline{S}(T,\mu)$ and $\underline{S}(T,\mu)$ are defined as

$$\begin{aligned}\overline{S}(T,\mu) &= \sup_\xi \lim_{\epsilon\to 0}\lim_{\delta\to 0}\limsup_{n\to\infty}\frac{\log S_\xi^H(T,n,\epsilon,\delta)}{a_n},\\ \underline{S}(T,\mu) &= \sup_\xi \lim_{\epsilon\to 0}\lim_{\delta\to 0}\liminf_{n\to\infty}\frac{\log S_\xi^H(T,n,\epsilon,\delta)}{a_n},\end{aligned} \tag{9.70}$$

for ξ from the collection of all finite measurable partitions and $S_\xi^H(T,n,\epsilon,\delta)$ from Section 9.4.1.

9.5.4 Entropy Dimensions

Similar to slow entropy, entropy dimension is also an invariant which is finer than the classical entropy. We give a basic introduction here, and all material in this section is contained in [8], [18], [1], [10], and [11].

Definition 5.4.1 (Topological entropy dimension). *Let X be a compact metric space and T be a continuous endormorphism of X. For an open cover $\mathcal{U}$ denote by $N(\mathcal{U})$ the minimal number of elements among all subcovers of $\mathcal{U}$ and $\mathcal{U}_0^{n-1} = \mathcal{U} \vee T^{-1}\mathcal{U} \ldots \vee T^{-n+1}\mathcal{U}$. Then the upper entropy dimension of (X, T) w.r.t. an open cover is*

$$\begin{aligned} \overline{D}(\mathcal{U}) &= \inf\{s \geq 0\colon \limsup_{n\to\infty} \frac{1}{n^s} \log(N(\mathcal{U}_0^{n-1})) = 0\} \\ &= \sup\{s \geq 0\colon \limsup_{n\to\infty} \frac{1}{n^s} \log N(\mathcal{U}_0^{n-1}) = \infty\}. \end{aligned} \tag{9.71}$$

The upper entropy dimension of (X, T) is defined as

$$\overline{D}(X, T) = \sup_{\mathcal{U}} \overline{D}(\mathcal{U}).$$

The lower entropy dimension of (X, T) w.r.t. an open cover $\mathcal{U}$ is defined as

$$\begin{aligned} \underline{D}(\mathcal{U}) &= \inf\{s \geq 0\colon \liminf_{n\to\infty} \frac{1}{n^s} \log(N(\mathcal{U}_0^{n-1})) = 0\} \\ &= \sup\{s \geq 0\colon \liminf_{n\to\infty} \frac{1}{n^s} \log N(\mathcal{U}_0^{n-1}) = \infty\}. \end{aligned} \tag{9.72}$$

The lower entropy dimension of (X, T) is defined to be

$$\underline{D}(X, T) = \sup_{\mathcal{U}} \underline{D}(\mathcal{U}).$$

If $\overline{D}(X, T) = \underline{D}(X, T)$, we denote it as $D(X, T)$ and call it the entropy dimension of (X, T).

We also have measure-theoretic entropy dimension:

Definition 5.4.2 (Measure-theoretic entropy dimension). *Let $(X, \mathcal{B}, \mu, T)$ be a measure-preserving system, α a finite measurable partition of X and $\epsilon > 0$. Denote $K(n, \epsilon)$ as the minimal cardinality of a set of Hamming balls*[21] *with radius ϵ in d_n^H distance that covers a set of measure at least $1 - \epsilon$. Then define*

$$\overline{D}(\alpha, \epsilon) = \sup\{s \in [0, 1]\colon \limsup_{n\to\infty} \frac{\log K(n, \epsilon)}{n^s} > 0\},$$

$$\overline{D}(\alpha) = \lim_{\epsilon\to 0} \overline{D}(\alpha, \epsilon),$$

$$\overline{D}_\mu(X, T) = \sup_{\alpha} \overline{D}(\alpha).$$

21 See 4.1.2 for more details.

and

$$\underline{D}(\alpha,\epsilon) = \sup\{s \in [0,1]\colon \liminf_{n\to\infty} \frac{\log K(n,\epsilon)}{n^s} > 0\},$$

$$\underline{D}(\alpha) = \lim_{\epsilon\to 0} \underline{D}(\alpha,\epsilon),$$

$$\underline{D}_\mu(X,T) = \sup_\alpha \underline{D}(\alpha).$$

The dimensions $\overline{D}_\mu(X,T)$ and $\underline{D}_\mu(X,T)$ are called upper and lower entropy dimensions of $(X,\mathscr{B},\mu,T)$. We denote them as $D_\mu(X,T)$ if they are equal and called entropy dimension.

Remark 5.4.3. *In [8], de Carvalho showed that if the topological entropy is finite, then both the upper and lower entropy dimensions are less than or equal to 1 and if the topological entropy is finite and positive, then both the upper and lower entropy dimensions are equal to 1. From this result, we know that entropy dimension should be mainly used for zero entropy systems.*

Remark 5.4.4. *In [10] and [11], Dou, Huang and Park defined topological and measure-theoretic entropy dimension via the dimension of entropy-generating sequences. Moreover, they also established Goodwyn's inequality and many interesting properties of entropy dimension.*

Remark 5.4.5. *Ahn, Dou and Park [1] proved that there exists a system such that topological entropy dimension of a topological dynamical system (X,T) is strictly larger than the supremum of its measure-theoretic entropy dimensions, which demonstrates the failure of the variational principle for entropy dimension.*

As with slow entropy, entropy dimension also has the product property in the non-classical version:

Proposition 5.4.6 (Ferenczi–Park [18]).

$$\overline{D}_\mu(X\times Y, T_1\times T_2) = \max\{\overline{D}_\mu(X,T_1), \overline{D}_\mu(Y,T_2)\}. \tag{9.73}$$

Remark 5.4.7. *It is also worthwhile to notice that measure-theoretic entropy dimension has a flexibility phenomenon. Namely for any $\alpha \in (0,1)$, there exists system $(X,T,\mathscr{B},\mu)$ such that $D_\mu(X,T)=\alpha$. For more details, we refer to [18] for reference.*

Remark 5.4.8. *It is also shown by Ferenczi and Park [18] that the upper and lower measure-theoretic entropy dimensions may be different, namely there exists a system $(X,T,\mathscr{B},\mu)$ such that $\overline{D}_\mu(X,T)=1$ and $\underline{D}_\mu(X,T)=0$.*

9.5.5 Entropy Convergence Rate

Instead of considering the Hamming or Bowen metric, Blume ([3], [4], [5]) created a new isomorphism invariant, called *entropy convergence rate*, for measure-preserving systems by using a direct approach to study zero entropy systems, namely only changing the scaling function of the original entropy function. The idea here is straightforward: if there is no effective upper sublinear bound among all partitions, one looks at the lower bound instead. Unlike Kushnirenko's sequence entropy, where vanishing of all invariants characterizes systems with discrete spectrum, Blume shows that there is a universal lower bound, somewhat slower than logarithmic for all aperiodic transformations; for rank-one systems it is at most logarithmic and for mixing rank-one systems the logarithmic lower bound is sharp. In order to avoid an obvious problem of missing more complicated parts of the orbit structure, this approach needs to be modified; for example, one may consider the slowest growth rate of joint entropy only among generating partitions.

Let $(X, T, \mathscr{B}, \mu)$ be a probability space with a measure-preserving endormorphism T and α be a finite partition of X; then we define $\alpha^{n-1} = \vee_{i=1}^{n} T^{-i}\alpha$. By defining $f(x) = -x\log_2 x$ for $x \in [0, 1]$, the entropy of a partition α is defined as $H(\alpha) = \sum_{A\in\alpha} f(\mu(A))$.

Definition 5.5.1 (Entropy convergence rate, Blume [3]). *Let $\{a_n\}_{n\geq 1}$ be a monotonically increasing sequence with $\lim_{n\to\infty} a_n = \infty$ and $c \in (0, \infty)$. If $\mathcal{P}$ is a class of partitions of X, then we say that (X, T) is of type $(LS \geq c)$ for $(\{a_n\}_{n\geq 1}, \mathcal{P})$ if*

$$\limsup_{n\to\infty} \frac{H(\alpha^{n-1})}{a_n} \geq c, \text{ for all } \alpha \in \mathcal{P};$$

and (X, T) is of type $(LI \geq c)$ for $(\{a_n\}_{n\geq 1}, \mathcal{P})$ if

$$\liminf_{n\to\infty} \frac{H(\alpha^{n-1})}{a_n} \geq c, \text{ for all } \alpha \in \mathcal{P}.$$

Remark 5.5.2. *The types $(LS < \infty)$, $(LS = \infty)$, $(LI < \infty)$, $(LI = \infty)$, $(LI > 0)$, $(LS \leq c)$, $(LI \leq c)$ are defined analogously.*

Remark 5.5.3. *By definition, we have following relations:*

$$\begin{aligned} &(LI \geq c) \Rightarrow (LS \geq c), \quad (LS \leq c) \Rightarrow (LI \leq c), \\ &(LI > 0) \Rightarrow (LS > 0), \quad (LS < \infty) \Rightarrow (LI < \infty). \end{aligned} \tag{9.74}$$

In order to state results about the entropy convergence type of a measure-preserving system, we need to specify the class of partitions under consideration. In the following text, we will mainly consider these two cases:

(1) $P(X) = \{\alpha(E) \mid \alpha(E) = \{E, X \setminus E\}$ for some $E \in \mathscr{B}$ with $0 < \mu(E) < 1\}$;

(2) $R(X) = \{\alpha(E) \mid E \in \mathscr{B}$ and $\lim_{n\to\infty} \max\{\mu(A) | A \in \alpha(E)_0^{n-1}\} = 0\}$.

Now we have enough preparation to introduce related results:

Theorem 5.5.4 (Blume [3]). *If (X, T) is an aperiodic measure-preserving system and $\{a_n\}_{n=1}^{\infty}$ is a positive monotone increasing sequence with $\lim_{n\to\infty} \frac{a_n}{n} = 0$, then there exists an $E \subset \mathscr{B}$ such that*

$$\lim_{n\to\infty} \frac{H(\alpha(E)^{n-1})}{a_n} = \infty.$$

Remark 5.5.5. *By definition, it is easy to see that this implies nonexistence of aperiodic systems of types $(LS < \infty)$, $(LS \leq c)$ or $(LI \leq c)$.*

However, if we assume additionally that T is completely ergodic, then we can describe the entropy convergence rate for general monotonically increasing sequence a_n:

Theorem 5.5.6 (Blume [4]). *If T is a completely ergodic transformation on* $([0, 1], \mathscr{B}, \mu)$, *then there exists a positive monotonically increasing sequence $\{a_k\}_{k\geq 1}$ with $\lim_{k\to\infty} a_k = \infty$ such that $\forall E \in \mathscr{B}$ with $\mu(E) \in (0, 1)$, we obtain*

$$\liminf_{k\to\infty} \frac{H(\alpha(E)^{k-1})}{a_k} \geq 1.$$

On the other hand, if we keep the aperiodicity condition but add some conditions on the sequence a_n and the partition $\alpha(E)$, we will obtain almost the same result:

Theorem 5.5.7 (Blume [3]). *Let (X, T) be an aperiodic measure-preserving system and let $g\colon [0, \infty) \to \mathbb{R}$ be a positive monotonely increasing function with $\int_1^{\infty} \frac{g(x)}{x^2} dx < \infty$. If $E \in \mathscr{B}$ is such that*

$$\lim_{n\to\infty} \max\{\mu(A) | A \in \alpha(E)^{n-1}\} = 0,$$

then

$$\limsup_{n\to\infty} \frac{H(\alpha(E)^{n-1})}{g(\log_2 n)} = \infty.$$

Moreover, if we replace aperiodicity by complete ergodicity, we will have the following characterization of completely ergodic systems based on the function g:

Corollary 5.5.8 (Blume [3]). *Let (X, T) be completely ergodic and let g be given as in Theorem 5.5.7. If $E \subset \mathscr{B}$ is such that $0 < \mu(E) < 1$, then*

$$\limsup_{n\to\infty} \frac{H(\alpha(E)^{n-1})}{g(\log_2 n)} = \infty.$$

Recall that so far we have always considered g satisfying $\int_1^\infty \frac{g(x)}{x^2}dx < \infty$. If we consider some g with some fast growth, namely $\int_1^\infty \frac{g(x)}{x^2}dx = \infty$, then we obtain:

Theorem 5.5.9 (Blume [4]). *If T is a completely ergodic transformation on $([0,1], \mathscr{B}, \mu)$, then there exists a positive concave function g defined on $[1,\infty)$ with $\int_1^\infty \frac{g(x)}{x^2}dx = \infty$, such that for all $E \in \mathscr{B}$ with $\mu(E) \in (0,1)$, we obtain*

$$\limsup_{k\to\infty} \frac{H(\alpha(E)^{k-1})}{g(\log_2 k)} \geq 1.$$

Blume also used entropy convergence rate to characterize rank-one systems:

Theorem 5.5.10 (Blume [3]). *If (X,T) is a rank-one system, then there is an $\alpha \in P(X)$ such that*

$$\liminf_{n\to\infty} \frac{H(\alpha^{n-1})}{\log_2 n} \leq 2.$$

If we assume additionally that (X,T) is mixing and $E \in \mathscr{B}$ with $0 < \mu(E) < 1$, then

$$\limsup_{n\to\infty} \frac{H(\alpha(E)^{n-1})}{\log_2 n} > 0.$$

Remark 5.5.11. *By Theorem 5.5.10 it follows that no rank-one system can have a convergence rate of type $(LI \geq c)$ for any sequence a_n that grows faster than $\log_2 n$. Recall that Ferenczi (Theorem 4.8.2) proved that the rank-one system has growth rate at most n^2. It is worthwhile to compare Theorem 5.5.10 with Theorem 4.8.2, as they show that the complexity of the rank-one system has the same bounds.*

Remark 5.5.12. *By considering information function $I_\alpha(x) = -\log_2 \mu(\alpha(x))$, we can formalize pointwise version results parallel to Theorem 5.5.7 and Corollary 5.5.8.*

9.5.6 Scaling Entropy

By using dynamics of functions of several variables, A. Vershik [67, 68, 69] introduced another measure-theoretic invariant called scaling entropy. Many applications [70, 71, 72, 73, 74, 65, 66] have been found since it was introduced. We document some basic definitions and related properties here.

Let ρ be a semimetric on a standard probability space (X,μ); then

- ρ is *measurable* if it is measurable on (X^2, μ^2) as a function of two variables. In this case, we say the triple (X, μ, ρ) is a *semimetric triple*;
- ρ is *admissible* if it is separable on some subset $X_1 \subset X$ of full measure. In this case, we say the triple $(X,' mu, \rho)$ is an *admissible semimetric triple*.

For every $\epsilon > 0$, let $k \in \mathbb{N}$ be the least natural number for which the space X can be partitioned intro disjoint measurable subsets $X_0, \ldots, X_k$ such that $\mu(X_0) < \epsilon$ and $\mathrm{diam}_\rho(X_j) < \epsilon$ for $j = 0, \ldots, k$. Then we define the *ϵ-entropy* of a semimetric triple (X, μ, ρ) as

$$\mathbb{H}_\epsilon(X, \mu, \rho) = \log_2 k.$$

Let $T : (X, \mu) \circlearrowleft$ be an ergodic automorphism. We define the averages of ρ along T as

$$T^n_{av}\rho = \frac{1}{n}\sum_{k=0}^{n-1} T^k \rho.$$

Then we can define the scaling sequence for ρ as follows:

Definition 5.6.1. *Let (X, μ, T) be an ergodic system and ρ be an admissible semimetric on (X, μ). A sequence $\{h_n\}$ of positive numbers is called a* scaling sequence for ρ *if*

$$\limsup_n \frac{\mathbb{H}_\epsilon(X, \mu, T^n_{av}\rho)}{h_n} < +\infty, \text{ for all } \epsilon > 0,$$
$$\liminf_n \frac{\mathbb{H}_\epsilon(X, \mu, T^n_{av}\rho)}{h_n} > 0, \text{ for sufficiently small } \epsilon > 0.$$

We denote the class of all scaling sequences for ρ by $\mathcal{H}(X, \mu, T, \rho)$.

By [70], $\mathcal{H}(X, \mu, T, \rho)$ is independent of the semimetric:

Theorem 5.6.2. *Let ρ and $\tilde{\rho}$ be two summable admissible generating semimetrics on (X, μ). Then $\mathcal{H}(X, \mu, T, \rho) = \mathcal{H}(X, \mu, T, \tilde{\rho})$.*

As a result, the scaling entropy sequence can be defined as follows:

Definition 5.6.3. *A sequence $\{h_n\}$ of positive numbers is called a scaling entropy sequence of an ergodic dynamical system (X, μ, T) if $\{h_n\} \in \mathcal{H}(X, \mu, T, \rho)$ for some summable admissible generating*[22] *semimetric ρ.*

9.6 Open Questions

In this section we propose several open questions related to slow entropy–type invariants.

[22] We say a measurable semimetric ρ on (X, μ) is a generating semimetric for (X, mu, T) if there exists a full measure subset $X' \subset X$ such that for every $x, y \in X'$, there exists $k \in \mathbb{Z}$ such that $T^k\rho(x,y) > 0$.

9.6.1 Slow Entropy and Spectral Properties

Recall that a measure-preserving transformation T on $(X, \mathscr{B}, \mu)$ is said to have *countable Lebesgue spectrum* if there exists an infinite orthonormal set $\{1, \phi_0, \dots\} \subset L^2(X)$ such that $\phi_{ik} = U_T^k(\phi_i) \in 1^\perp$ are all pairwise distinct and orthogonal, where U_T is the Koopman operator induced from T. It will be interesting to see if there is any relation between slow entropy and Lebesgue spectrum:

Question 6.1.1. *Does countable Lebesgue spectrum imply any below restrictions on the slow entropy?*

Moreover, one can ask the following question for rigid systems (recalling that rigidity implies singular spectrum):

Question 6.1.2. *Is it possible to have the upper slow entropy for a rigid transformation positive with respect to $a_n(t) = n^t$?*

9.6.2 Slow Entropy and Finite Rank Systems

Recall that a system is a finite rank system if, instead of approximating it with one sequence of towers (which ϵ-refines every partition), one is allowed finitely many such towers.[23] We have the following question:

Question 6.2.1. *What about the finite rank systems' slow entropy?*

It is worth pointing out that in a general rank-one situation, Ferenczi [15] and Kanigowski [33] obtain estimates of upper bounds of lower slow entropy; see Theorem 4.8.2 and Theorem 4.8.3 for more details.

Recall that a very natural class of finite rank systems is given by interval exchange transformations (IETs). More precisely, suppose $n > 0$, π is a permutation defined on $\{1, \dots, n\}$ and $\lambda = (\lambda_1, \dots, \lambda_n)$, where $\sum_{i=1}^n \lambda_i = 1$ and $\lambda_i > 0$ for $1 \le i \le n$. Then a map $T_{\pi,\lambda}\colon [0,1] \to [0,1]$ is an (π, λ)-interval exchange transformation if, for $x \in [a_i, a_i + \lambda_i)$, we have

$$T_{\pi,\lambda}(x) = x - a_i + a_i',$$

where $a_i = \sum_{1\le j\le i} \lambda_j$, $a_i' = \sum_{1\le j\le \pi(i)} \lambda_{\pi^{-1}(j)}$ for $1 \le i \le n$.

It is interesting to study growth rate for IETs:

Question 6.2.2. *For an interval exchange transformation, is it possible that its upper (lower) slow entropy is 1 in the scale $a_n(t) = n^t$?*

[23] For a more precise definition, see [16].

However, if we restrict ourselves in the rank-one situation by adding some additional ergodic conditions, we may ask the following simple version of Question 6.2.1[24]:

Question 6.2.3. *What about the slow entropy of Ornstein's [56] mixing rank-one system? While it must satisfy Theorem 4.8.3, does it have an infinite upper slow entropy when* $a_n(t) = n^{kt}$ *for any k?*

9.6.3 Realization of Slow Entropy

One of the important combinatorial tools in ergodic theory is fast approximation. (See [42] for more details.) Thus it is reasonable to consider the realization of ergodic systems with certain given slow entropy by fast approximation:

Question 6.3.1. *Is it possible to have a system constructed by fast periodic approximation with any subexponential slow entropy (upper or lower)?*

9.6.4 Other Related Questions

Recall that slow entropy is designed to measure the complexity of zero entropy ergodic systems. Thus the estimates of the slow entropy of certain representative systems would be quite interesting and meaningful:

Question 6.4.1. *Can we get estimates of the slow entropy for billiards*[25] *in polygons and polyhedra?*

Question 6.4.2. *Can we get estimates of the slow entropy of Gaussian*[26] *systems?*

Finally, we recall the following (well-known) question on a relation between classical entropy and sequence entropy:

Question 6.4.3. *Does there exist a zero-entropy system whose sequence entropy is positive with respect to the sequence* $\{n^2\}$? *One can ask the same question for the sequence* $\{p(n)\}$ *where p is a polynomial.*

References

[1] Y. Ahn, D. Dou, K. K. Park, *Entropy dimension and variational principle* (English summary), Studia Math. 199 (2010), 295–309.

24 See conjecture in [15] for more explanations.
25 For detailed discussion about billiards in polygons, we refer to [19].
26 For detailed discussion of Gaussian systems, we refer to [50].

[2] D. V. Anosov, A. B. Katok, *New examples in smooth ergodic theory: ergodic diffeomorphisms*, (Russian), Trudy Moskov. Mat. Obšč. 23 (1970), 3–36.
[3] F. Blume, *Possible rates of entropy convergence*, Ergod. Theory Dyn. Syst. 17 (1997), 45–70.
[4] F. Blume, *Minimal rates of entropy convergence for completely ergodic systems*, Israel J. Math. 108 (1998), 1–12.
[5] F. Blume, *The rate of entropy convergence*, doctoral dissertation, University of North Carolina at Chapel Hill (1995).
[6] J. Buzzi, *Intrinsic ergodicity of smooth interval maps*, Israel J. Math. 100 (1997), 125–161.
[7] V. Cyr, B. Kra, *Realizing ergodic properties in zero entropy subshifts*, arXiv:1902.08645.
[8] M. de Carvalho, *Entropy dimension of dynamical systems*, Portugal. Math. 54 (1997), 19–40.
[9] F. M. Dekking, *Some examples of sequence entropy as an isomorphism invariant*, Trans. Amer. Math. Soc. 259 (1980), 167–183.
[10] D. Dou, W. Huang, K. K. Park, *Entropy dimension of topological dynamical systems*, Trans. Amer. Math. Soc. 363 (2011), 659–680.
[11] D. Dou, W. Huang, K. K. Park, *Entropy dimension of measure preserving systems*, Trans. Amer. Math. Soc. 371 (2019), 7029–7065.
[12] B. Fayad, A. Kanigowski, *Multiple mixing for a class of conservative surface flows*, Invent. Math. 203.2 (2016), 555–614.
[13] J. Feldman, *New K-automorphisms and a problem of Kakutani*, Israel J. Math. 24 (1976), 16–38.
[14] S. Ferenczi, *Rank and symbolic complexity*, Ergod. Theory Dyn. Syst. 16 (1996), 663–682.
[15] S. Ferenczi, *Measure-theoretic complexity of ergodic systems*, Israel J. Math. 100 (1997), 189–207.
[16] S. Ferenczi, *Systems of finite rank*, Colloq. Math. 73 (1997), 35–65.
[17] S. Ferenczi, *Complexity of sequences and dynamical systems*, Combinatorics and number theory (Tiruchirapalli, 1996). Discrete Math. 206 (1999), 145–154.
[18] S. Ferenczi, K. K. Park, *Entropy dimensions and a class of constructive examples*, Discrete Contin. Dyn. Syst. 17 (2007), 133–141.
[19] G. Forni, C. Matheus, *Introduction to Teichmüller theory and its applications to dynamics of interval exchange transformations, flows on surfaces and billiards*, J. Mod. Dyn. 8 (2014), 271–436.
[20] D. Fried, *Entropy for smooth abelian actions*, Proc. Amer. Math. Soc. 87 (1983), 111–116.
[21] H. Furstenberg, *Strict ergodicity and transformation of the torus*, Amer. J. Math. 83 (1961), 573–601.
[22] H. Furstenberg, B. Weiss, *The finite multipliers of infinite ergodic transformations*, in N. G. Markley, J. C. Martin, and W. Perrizo (eds.), The Structure of Attractors in Dynamical Systems. Lecture Notes in Mathematics, 668, Springer, Berlin, 1978, pp. 127–132.
[23] S. Galatolo, *Global and local complexity in weakly chaotic dynamical systems*, Discrete Contin. Dyn. Syst. 9 (2003), 1607–1624.
[24] T. N. T. Goodman, *Topological sequence entropy*, Proc. London Math. Soc. (3) 29 (1974), 331–350.
[25] B. Hasselblatt, A. Katok, *Principal structures*, in Handbook of Dynamical Systems, Vol. 1, Elsevier, 2002, pp. 1–203.

[26] M. Hochman, *Slow entropy and differentiable models for infinite-measure preserving $\mathbb{Z}^k$ actions*, Ergod. Theory Dyn. Syst. 32 (2012), 653–674.
[27] W. Huang, S. M. Li, S. Shao, X. Ye, *Null systems and sequence entropy pairs*, Ergod. Theory Dyn. Syst. 23 (2003), 1505–1523.
[28] W. Huang, S. Shao, X. Ye, *Mixing via sequence entropy*, Algebraic and topological dynamics, 101–122, Contemp. Math., 385, American Mathematical Society, Providence, RI, 2005.
[29] P. Hulse, *Sequence entropy and subsequence generators*, J. London Math. Soc. (2) 26 (1982), 441–450.
[30] P. Hulse, *Counterexamples to the product rule for entropy*, Dyn. Syst. 24 (2009), 81–95.
[31] S. Hurder, *Rigidity for Anosov actions of higher rank lattices*, Ann. of Math. (2) 135 (1992), 361–410.
[32] S. Kakutani, *Induced measure preserving transformations*, Proc. Imp. Acad. Tokyo 19, (1943), 635–641.
[33] A. Kanigowski, *Slow entropy for some smooth flows on surfaces*, Israel J. Math. 226 (2018), 535–577.
[34] A. Kanigowski, K. Vinhage, D. Wei, *Slow entropy of parabolic flows*, Comm. Math. Phys. 370 (2019), 449–474.
[35] A. Kanigowski, K. Vinhage, D. Wei, *Kakutani equivalence of unipotent flows*, preprint, Duke Math. J. 170 (2021), no. 7, 1517–1583.
[36] A. Kanigowski, K. Vinhage, D. Wei, unpublished notes.
[37] A. B. Katok, *Time change, monotone equivalence, and standard dynamical systems*, Dokl. Akad. Nauk SSSR 223 (1975), 789–792.
[38] A. B. Katok, *Monotone equivalence in ergodic theory*, Izv. Akad. Nauk SSSR Ser. Mat. 41 (1977), 104–157, 231.
[39] A. Katok, *Lyapunov exponents, entropy and periodic orbits for diffeomorphisms*, Inst. Hautes Études Sci. Publ. Math. (1980), 137–173.
[40] A. Katok, B. Hasselblatt, *Introduction to the modern theory of dynamical systems*, Encyclopedia of Mathematics and Its Applications, 54. Cambridge University Press, Cambridge, 1995.
[41] A. Katok, J.-P. Thouvenot, *Slow entropy type invariants and smooth realization of commuting measure-preserving transformations*, Ann. Inst. H. Poincaré Probab. Statist. 33 (1997), 323–338.
[42] A. Katok, *Combinatorial constructions in ergodic theory and dynamics*, University Lecture Series, 30. American Mathematical Society, Providence, RI, 2003.
[43] A. Katok, S. Katok, F. Rodriguez Hertz, *The Fried average entropy and slow entropy for actions of higher rank abelian groups*, Geom. Funct. Anal. 24 (2014), 1204–1228.
[44] A. Katok, F. Rodriguez Hertz, *Arithmeticity and topology of smooth actions of higher rank abelian groups*, J. Mod. Dyn. 10 (2016), 135–172.
[45] A. Ya. Khinchin, *Continued fractions*, University of Chicago Press, 1964.
[46] D. Y. Kleinbock, N. Shah, A. Starkov, *Dynamics of subgroup actions on homogeneous spaces of Lie groups and applications to number theory: handbook of dynamical systems*, Vol. 1A, 813–930, North-Holland, Amsterdam, 2002.
[47] E. Krug, and D. Newton, *On sequence entropy of automorphisms of a Lebesgue space*, Z. fur Angew. Math. Phys. 24 (1972), 211–214.
[48] A. G. Kushnirenko, *Metric invariants of entropy type*, Uspehi Mat. Nauk 22 1967 no. 5 (137), 57–65.
[49] M. Lemańczyk, *The sequence entropy for Morse shifts and some counterexamples*, Studia Math. 82 (1985), 221–241.

[50] M. Lemańczyk, F. Parreau, J.-P. Thouvenot, *Gaussian automorphisms whose ergodic self-joinings are Gaussian*, Fund. Math. 164 (2000), 253–293.
[51] M. Misiurewicz, *Diffeomorphism without any measure with maximal entropy*, Bull. Acad. Polon. Sci. Sér. Sci. Math. Astronom. Phys. 21 (1973), 903–910.
[52] B. Mossé, *Reconnaissabilité des substitutions et complexité des suites automatiques*, Bull. Soc. Math. France 124 (1996), 329–346.
[53] D. Newton, *On sequence entropy. I*, Math. Syst. Theory 4 (1970), 119–125.
[54] D. Newton, *On sequence entropy. II*, Math. Syst. Theory 4 (1970), 126–128.
[55] D. Newton, *A note on sequence entropy*, Bull. London Math. Soc. 18 (1986), 24–28.
[56] D. S. Ornstein, *On the root problem in ergodic theory*, Proceedings of the Sixth Berkeley Symposium on Mathematical Statistics and Probability, Vol. II: Probability theory, pp. 347–356. University of California Press, Berkeley, CA, 1972.
[57] B. S. Pitskel, *Certain properties of A-entropy*, Mat. Zametki 5 (1969), 327–334.
[58] M. Ratner, *Horocycle flows are loosely Bernoulli*, Israel J. Math. 31 (1978), 122–132.
[59] M. Ratner, *The Cartesian square of the horocycle flow is not loosely Bernoulli*, Israel J. Math. 34 (1979), 72–96.
[60] M. Ratner, *Some invariants of Kakutani equivalence*, Israel J. Math. 38 (1981), 231–240.
[61] M. Ratner, *Interactions between ergodic theory, Lie groups, and number theory*, Proceedings of the International Congress of Mathematicians, Vol. 1, 2 (Zürich, 1994), 157–182, Birkhäuser, Basel, 1995.
[62] M. Ratner, *Time change invariants for measure preserving flows*, Modern theory of dynamical systems, 263–273, Contemp. Math., 692, American Mathematics Society, Providence, RI, 2017.
[63] S. Ruette, *Mixing C^r maps of the interval without maximal measure*, Israel J. Math. 127 (2002), 253–277.
[64] A. Saleski, *Sequence entropy and mixing*, J. Math. Anal. Appl. 60 (1977), 58–66.
[65] G. A. Veprev, *Scaling entropy of unstable systems*, Zap. Nauchn. Sem. S.–Peterburg. Otdel. Mat. Inst. Steklov. (POMI) 498 (2020), Teoriya Predstavleniĭ, Dinamicheski Sistemy, Kombinatornye Metody. XXXI, 5–17.
[66] G. A. Veprev, *The scaling entropy of a generic action*, Zap. Nauchn. Sem. S.–Peterburg. Otdel. Mat. Inst. Steklov. (POMI) 507 (2021), Teoriya Predstavleniĭ, Dinamicheski Sistemy, Kombinatornye Methody. XXXIII, 5–14; translation in J. Math. Sci. (NY) 261 (2022), 595–600.
[67] A. M. Vershik, A. D. Gorbul'skiǐ, *Scaled entropy of filtrations of σ-algebras*, Teor. Veroyatn. Primen. 52 (2007), 446–467; translation in Theory Probab. Appl. 52 (2008), 493–508.
[68] A. M. Vershik, *Dynamics of metrics in measure spaces and their asymptotic invariants*. Markov Process. Relat. Fields 16 (2010), 169–184.
[69] A. M. Vershik, *Scaled entropy and automorphisms with a pure point spectrum*, Algebra i Analiz 23 (2011), 111–135; translation in St. Petersburg Math. J. 23 (2012), 75–91.
[70] A. M. Vershik, P. B. Zatitskiy, F. V. Petrov, *Geometry and dynamics of admissible metrics in measure spaces*, Cent. Eur. J. Math. 11 (2013), 379–400.
[71] P. B. Zatitskiy, *On a scaling entropy sequence of a dynamical system*, Translation of Funktsional. Anal. i. Prilozhen. 48 (2014), 70–74. Funct. Anal. Appl. 48 (2014), 291–294.
[72] P. B. Zatitskiy, *Scaling entropy sequence: invariance and examples*, J. Math. Sci. (NY) 209 (2015), 890–909.

[73] P. B. Zatitskiĭ, F. V. Petrov, *On the subadditivity of a scaling entropy sequence*, Zap. Nauchn. Sem. S.-Peterburg. Otdel. Mat. Inst. Steklov. (POMI) 436 (2015), Teoriya Predstavleniĭ, Dinamicheskie Sistemy, Kombinatornye Metody. XXV, 167–173; translation in J. Math. Sci. (NY) 215 (2016), 734–737.

[74] P. B. Zatitskiĭ, *On the possible growth rate of a scaling entropy sequence*, Zap. Nauchn. Sem. S.-Peterburg. Otdel. Mat. Inst. Steklov. (POMI) 436 (2015), Teoriya Predstavleniĭ, Dinamicheskie Sistemy, Kombinatornye Metody. XXV, 136–166; translation in J. Math. Sci. (NY) 215 (2016), 715–733.

[75] Q. Zhang, *Sequence entropy and mild mixing*, Canad. J. Math. 44 (1992), 215–224.

[76] Y. Zhao, Y. Pesin, *Scaled entropy for dynamical systems*, J. Stat. Phys. 158 (2015), 447–475.

[77] Y. Zhao, Y. Pesin, *Erratum to: Scaled entropy for dynamical systems*, J. Stat. Phys. 162 (2016), 1654–1660.

10

On the Smooth Realization Problem and the AbC Method

PHILIPP KUNDE*

We give an overview of the Approximation by Conjugation method introduced by Dmitri Anosov and Anatole Katok and its contributions to the smooth realization problem. In particular, we present various versions of the method, recent results, and related open problems.

"This topic is very close to my heart."
– Anatole Katok

Introduction

An important question in Ergodic Theory dating back to the foundational paper [Ne32] of von Neumann is the so-called smooth realization problem.

Smooth realization problem. *Are there smooth versions to the objects and concepts of abstract ergodic theory? Does every ergodic measure-preserving transformation have a smooth model?*

By a smooth model one means a smooth diffeomorphism of a compact manifold preserving a measure equivalent to the volume element which is isomorphic to the measure-preserving transformation. (If one waives the condition that the diffeomorphism should preserve a smooth measure, then Lind and Thouvenot showed that every finite-entropy measure-preserving transformation can be realized as a linear automorphism of the 2-torus with respect to a suitable invariant Borel probability measure [LT78]. There it is also shown that every finite entropy ergodic transformation is isomorphic to a Lebesgue measure-preserving homeomorphism of $\mathbb{T}^2$.) More specifically, one can ask

* P. K. acknowledges financial support from a DFG Forschungsstipendium under Grant No. 405305501.

- about the existence of volume-preserving diffeomorphisms with specific ergodic properties, and
- what ergodic properties, if any, are imposed upon a dynamical system by the fact that it should be smooth.

The only known general restriction is due to A. G. Kushnirenko, who proved that such a diffeomorphism must have finite entropy [Ku65]. In other words, no measure-preserving transformation with finite entropy is known not to be realizable on any manifold. There are restrictions in low dimension: any circle diffeomorphism with invariant smooth measure is conjugated to a rotation and any weakly mixing surface diffeomorphism of positive metric entropy is Bernoulli by Pesin theory [Pe77]. Besides that, no restrictions are known. On the other hand, there is a lack of general results on the smooth realization problem. According to A. Katok, it is one of the five most resistant problems in dynamics [Ka04].

One of the most powerful tools of constructing smooth volume-preserving diffeomorphisms of entropy zero with prescribed ergodic or topological properties is the so-called *approximation by conjugation* method (also known as the *AbC* method or Anosov–Katok method) developed by D. Anosov and A. Katok in [AK70]. In particular, it provided the first example of an ergodic diffeomorphism on the disc $\mathbb{D}^2$. In fact, on every smooth compact connected manifold M of dimension $d \geq 2$ admitting a non-trivial circle action $\mathcal{R} = \{R_t\}_{t\in\mathbb{S}^1}$ this method enables the construction of smooth diffeomorphisms with specific ergodic properties (e.g., weakly mixing ones in [AK70, Section 5] and uniquely ergodic ones in [FH77]) or non-standard smooth realizations of measure-preserving systems (e.g., [AK70, Section 6], [FSW07] and [Be13]). These diffeomorphisms are constructed as limits of conjugates $T_n = H_n \circ R_{\alpha_{n+1}} \circ H_n^{-1}$ with $\alpha_{n+1} = \frac{p_{n+1}}{q_{n+1}} = \alpha_n + \frac{1}{k_n \cdot l_n \cdot q_n^2} \in \mathbb{Q}$ and $H_n = H_{n-1} \circ h_n$, where the h_n's are measure-preserving diffeomorphisms satisfying $R_{\frac{1}{q_n}} \circ h_n = h_n \circ R_{\frac{1}{q_n}}$. In each step the conjugation map h_n and the parameter k_n are chosen such that the diffeomorphism T_n imitates the desired property with a certain precision. In a final step of the construction, the parameter l_n is chosen large enough to guarantee closeness of T_n to T_{n-1} in the C^∞-topology, and so the convergence of the sequence $(T_n)_{n\in\mathbb{N}}$ to a limit diffeomorphism is provided. The power and flexibility of the method comes from the fact that this convergence is not connected with any convergence of the conjugating diffeomorphisms H_n.

This short outline of the AbC construction already indicates that it is rather technically involved. We give a more detailed description of the general scheme of the method in the next section. It is a striking anecdote that, although containing several elaborate constructions, the draft of the seminal paper [AK70] was written at one single weekend (see [Ka17] which also contains further historic background information).

In many ways, the AbC method can be seen as a smooth version of the concept of *periodic approximation in ergodic theory* [KS67] which was developed by A. Katok and A. Stepin in their dissertations awarded with the prestigious prize of the Moscow Mathematical Society for young mathematicians. The concept has numerous applications in measurable dynamics and helped to solve several long-standing problems dating back to von Neumann and Kolmogorov. A detailed exposition can be found in [Ka03].

The flexibility of the AbC method also allows a multitude of interesting applications. Currently, the AMS publication database lists about a hundred citations. So, it is an impossible task to review all of them and to draw attention to them as they would deserve. Instead of that we focus on some recent results and emphasize related open problems. The reader should also consult the nice survey article [FK04] describing several other applications as well as problems which are still open apart from Problem 7.11 therein (see Section 10.4). Moreover, we refer to the very recent survey article [FK18] on quasi-periodic dynamics describing several results and questions on pseudo-rotations of the disc[1] and instability of elliptic equilibria in smooth Hamiltonian systems, where the AbC method has been applied fruitfully (see, e.g., [FS17]).

10.1 The Scheme of the AbC Method

Unless specified explicitly, in the following M always denotes a smooth compact connected manifold of dimension $d \geq 2$ admitting a smooth non-trivial circle action $\mathcal{R} = \{R_t\}_{t \in \mathbb{S}^1}$ preserving a smooth volume ν. In this connection, we note that for any given smooth volume form $\hat{\nu}$ on M we can obtain an $\mathcal{R}$-invariant volume form $\nu = \int_0^1 (R_t)_* \hat{\nu}$ by averaging $\hat{\nu}$ over the action.

Typically, constructions on $\mathbb{S}^1 \times [0, 1]^{d-1}$ with Lebesgue measure and standard actions

$$R_t(\theta, r_1, \ldots, r_{d-1}) = (\theta + t, r_1, \ldots, r_{d-1})$$

can be transferred to a general compact connected smooth manifold of dimension $d \geq 2$ with a non-trivial circle action (see, e.g., [FSW07, Proposition 1] or [GK20, Section 3.1] for a detailed exposition). So, it usually suffices to do constructions on $\mathbb{S}^1 \times [0, 1]^{d-1}$. To alleviate notation we present the scheme of the AbC method for $d = 2$.

[1] For $k \in \mathbb{N} \cup \{\infty\} \cup \{\omega\}$ a C^k pseudo-rotation of the disc $\mathbb{D}^2$ is an orientation- and area-preserving C^k diffeomorphism of $\mathbb{D}^2$ that preserves the origin, leaves invariant the boundary $\partial\mathbb{D}^2$ of the disk and with no other periodic point than the origin. The ergodic AbC diffeomorphisms constructed in [AK70] are C^∞ pseudo-rotations and provided the first examples of pseudo-rotations which are not topologically conjugated to rigid rotations.

For this purpose, we introduce partitions $\xi_{kq,s} = \left\{ \Delta^{i,j}_{kq,s} \,\middle|\, 0 \le i < kq,\ 0 \le j < s \right\}$, where

$$\Delta^{i,j}_{kq,s} = \left[\frac{i}{kq}, \frac{i+1}{kq}\right) \times \left[\frac{j}{s}, \frac{j+1}{s}\right).$$

An AbC diffeomorphism is constructed inductively as the limit of a sequence $T_n = H_n \circ R_{\alpha_{n+1}} \circ H_n^{-1}$ with $\alpha_{n+1} \in \mathbb{Q}$ and conjugation maps $H_n = H_{n-1} \circ h_n$. In each step one builds a conjugation map h_n and a new rational number $\alpha_{n+1} = \frac{p_{n+1}}{q_{n+1}}$ with p_{n+1}, q_{n+1} relatively prime using sequences of positive integers $\{k_n\}_{n\in\mathbb{N}}$, $\{l_n\}_{n\in\mathbb{N}}$, and $\{s_n\}_{n\in\mathbb{N}}$ that are defined over the course of the construction. Moreover, we fix a decreasing sequence $\{\varepsilon_n\}_{n\in\mathbb{N}}$ of positive numbers satisfying $\sum_{n\in\mathbb{N}} \varepsilon_n < \infty$.

In the first step, one can pick an arbitrary $\alpha_1 = \frac{p_1}{q_1} \in \mathbb{Q}$ with p_1, q_1 relatively prime and let $H_0 = \mathrm{id}$.

At step $n+1$ of the induction process we assume that we have already defined H_{n-1}, $\{\alpha_m\}_{m=1}^{n}$, $\{k_m\}_{m=1}^{n-1}$, $\{l_m\}_{m=1}^{n-1}$, and $\{s_m\}_{m=1}^{n-1}$. In particular, we know $T_{n-1} = H_{n-1} \circ R_{\alpha_n} \circ H_{n-1}^{-1}$. Then we proceed in the inductive steps in the following order.

1. We choose positive integers k_n and s_n as well as an R_{1/q_n}-equivariant permutation $\mathfrak{h}_n$ of the partition $\xi_{k_n q_n, s_n}$. The choice of the parameters and the combinatorics of $\mathfrak{h}_n$ depend on the specific properties (e.g., weak mixing) that we are targeting to prove for the limit transformation. In particular, we require that k_n and s_n are sufficiently large such that

$$\operatorname{diam}\left(H_{n-1}\left(\Delta^{i,j}_{k_n q_n, s_n}\right)\right) < \frac{1}{2^n}$$

   for all $0 \le i < k_n q_n$, $0 \le j < s_n$. Hereby, we guarantee that $\{H_{n-1}\left(\xi_{k_n q_n, s_n}\right)\}_{n\in\mathbb{N}}$ yields a generating sequence of partitions. Notice also that the R_{1/q_n}-equivariance of our permutation $\mathfrak{h}_n$ is needed so that we can satisfy the upcoming condition (10.1).
2. We construct a volume-preserving C^∞ diffeomorphism h_n satisfying

$$R_{\frac{1}{q_n}} \circ h_n = h_n \circ R_{\frac{1}{q_n}} \tag{10.1}$$

   such that h_n coincides with the identity in a neighborhood of the boundary of $M = \mathbb{S}^1 \times [0,1]$ and with the action of $\mathfrak{h}_n$ on the "inner kernels"

$$\tilde{\Delta}^{i,j}_{k_n q_n, s_n} = \left[\frac{i+\varepsilon_n}{k_n q_n}, \frac{i+1-\varepsilon_n}{k_n q_n}\right) \times \left[\frac{j+\varepsilon_n}{s_n}, \frac{j+1-\varepsilon_n}{s_n}\right)$$

   of all the partition elements $\Delta^{i,j}_{k_n q_n, s_n} \in \xi_{k_n q_n, s_n}$. Such a diffeomorphism can be obtained by "Moser's trick" (see, e.g., [AK70, Lemma 1.1] or [FW19a, Section 6.2]).

3. We set
$$\alpha_{n+1} = \alpha_n + \frac{1}{k_n l_n q_n^2}, \tag{10.2}$$
where we choose the parameter $l_n \in \mathbb{N}$ sufficiently large such that
$$\begin{aligned} d_{C^n}(T_n, T_{n-1}) &= d_{C^n}\left(H_n \circ R_{\alpha_{n+1}} \circ H_n^{-1}, H_{n-1} \circ R_{\alpha_n} \circ H_{n-1}^{-1}\right) \\ &= d_{C^n}\left(H_n \circ R_{\alpha_{n+1}} \circ H_n^{-1}, H_{n-1} \circ R_{\alpha_n} \circ h_n \circ h_n^{-1} \circ H_{n-1}^{-1}\right) \\ &\overset{(*)}{=} d_{C^n}\left(H_n \circ R_{\alpha_{n+1}} \circ H_n^{-1}, H_{n-1} \circ h_n \circ R_{\alpha_n} \circ h_n^{-1} \circ H_{n-1}^{-1}\right) \\ &\overset{(**)}{\leq} C(n) \cdot \left(|||H_n|||_{n+1}\right)^{n+1} \cdot |\alpha_{n+1} - \alpha_n| \\ &= C(n) \cdot \left(|||H_n|||_{n+1}\right)^{n+1} \cdot \frac{1}{k_n l_n q_n^2} \\ &< \varepsilon_n. \end{aligned}$$
Here, we exploited the commutation relation (10.1) in $(*)$ and a generalized mean value theorem in $(**)$, where $|||H|||_n = \max\{\|H\|_{C^n}, \|H^{-1}\|_{C^n}\}$ with C^n norm the C^n norm $\|H\|_{C^n}$ given by
$$\max\left\{\max_{z\in M} |D_{\vec{a}} H_i(z)| \;\middle|\; i = 1, \ldots, d,\ \vec{a} \in \mathbb{N}_0^d \text{ multi-index with } 0 \leq \sum_{j=1}^{d} a_j \leq n\right\}$$
and the constant $C(n)$ depends only on the order of derivatives and the geometry of the manifold. Moreover, we observe that by our definition in Eq. (10.2) we have
$$p_{n+1} = p_n k_n l_n q_n + 1 \text{ and } q_{n+1} = k_n l_n q_n^2,$$
i.e., p_{n+1} and q_{n+1} are relatively prime.

This completes the construction at stage $n+1$. Notice that in this way, we can guarantee that $\{T_n\}_{n\in\mathbb{N}}$ is a Cauchy sequence and converges to a limit diffeomorphism in $\text{Diff}^{\infty}(M, \nu)$.

It is even possible to keep this limit diffeomorphism within any given C^∞-neighborhood of the initial element R_{α_1} or, by applying a fixed diffeomorphism g as H_0 first, of $g \circ R_{\alpha_1} \circ g^{-1}$. So the construction can be carried out in a neighborhood of any diffeomorphism conjugate to an element of the action. Thus, $\mathcal{A}(M) = \overline{\left\{h \circ R_t \circ h^{-1} : t \in \mathbb{S}^1, h \in \text{Diff}^\infty(M, \nu)\right\}}^{C^\infty}$ is a natural space for the produced diffeomorphisms. The method often even allows one to prove genericity theorems in $\mathcal{A}(M)$. Moreover, we will consider the restricted space $\mathcal{A}_\alpha(M) = \overline{\left\{h \circ R_\alpha \circ h^{-1} : h \in \text{Diff}^\infty(M, \nu)\right\}}^{C^\infty}$ for a prescribed $\alpha \in \mathbb{S}^1$.

In their pioneering paper [AK70] Anosov and Katok constructed diffeomorphisms with specific ergodic properties as well as non-standard smooth realizations.

Theorem 1.1 ([AK70]). *(a) The set of weakly mixing diffeomorphisms is residual (i.e., it contains a dense G_δ-set) in $\mathcal{A}(M)$ in the $\mathrm{Diff}^\infty(M)$-topology.*
(b) There is an ergodic diffeomorphism $T \in \mathrm{Diff}^\infty(M,\nu)$ metrically isomorphic to the circle rotation by $\alpha = \lim_{n\to\infty} \alpha_n$.
(c) For any positive integer h there is an ergodic diffeomorphism $T \in \mathrm{Diff}^\infty(M,\nu)$ metrically isomorphic to some translation on $\mathbb{T}^h$.
(d) There is an ergodic diffeomorphism $T \in \mathrm{Diff}^\infty(M,\nu)$ metrically isomorphic to some translation on $\mathbb{T}^\infty$.

We stress that until that point not even the existence of an ergodic area-preserving diffeomorphism of the disc was known. Moreover, part (d) provides the earliest example of a smooth realization of a transformation whose natural domain is not a finite-dimensional manifold.

10.2 Quantitative Version of the AbC Method

In their foundational paper [AK70] Anosov and Katok proved that the set of weakly mixing diffeomorphisms is residual in $\mathcal{A}(M)$ in the $\mathrm{Diff}^\infty(M)$-topology. Actually, it follows from the proof that the result holds true in $\mathcal{A}_\alpha(M)$ for a G_δ-set of $\alpha \in \mathbb{S}^1$. However, the proof does not give a full description of the set of $\alpha \in \mathbb{S}^1$. Such an investigation is started in a landmark paper by B. Fayad and M. Saprykina.

Theorem 2.1 ([FS05]). *If $\alpha \in \mathbb{S}^1$ is Liouville, then the set of weakly mixing diffeomorphisms is residual in the $C^\infty(M)$-topology in $\mathcal{A}_\alpha(M)$. In case of the disc $\mathbb{D}^2$ or the annulus $\mathbb{A} = \mathbb{S}^1 \times [0,1]$, the constructed diffeomorphisms will have the rotation R_α as their restriction to the boundary.*

Here, we recall that an irrational number α is called *Liouville* if, for every $C > 0$ and every positive integer n, there are infinitely many pairs of integers (p,q) with $q > 1$ such that

$$0 < \left|\alpha - \frac{p}{q}\right| < \frac{C}{q^n}.$$

So, a Liouville number can be approximated very well by a sequence of rational numbers. If an irrational number is not Liouville, it is called *Diophantine* (i.e., it satisfies that there are constants $C > 0$ and $\tau > 0$ such that for any pair (p, q) of integers we have $\left|\alpha - \frac{p}{q}\right| > \frac{C}{q^{1+\tau}}$).

In connection with this, "Herman's last geometric theorem" says that for a C^∞-diffeomorphism T of $\mathbb{D}^2$ or $\mathbb{A}$, whose restriction to the boundary has Diophantine rotation number, the boundary is accumulated by a set of positive measure of T-invariant curves [FK09]. In particular, T cannot be ergodic. Combining this with Theorem 2.1, one obtains the following dichotomy.

Corollary 2.2. *Let M be $\mathbb{D}^2$ or $\mathbb{A}$. Then $\alpha \in \mathbb{R} \setminus \mathbb{Q}$ is Diophantine if and only if there is no ergodic C^∞-diffeomorphism T of M whose restriction to the boundary has rotation number α.*

The proof of Theorem 2.1 relies on a more explicit construction of the conjugation maps compared to the original Anosov–Katok method such that one can deduce norm estimates of the form

$$|||H_n|||_k \leq q_n^{2k}.$$

Exploiting the commutation relation $h_n \circ R_{\alpha_n} = R_{\alpha_n} \circ h_n$ and a generalized mean value theorem thus allows one to estimate the C^n distance between T_n and T_{n-1} as

$$\begin{aligned}
d_{C^n}(T_n, T_{n-1}) &= d_{C^n}\left(H_n \circ R_{\alpha_{n+1}} \circ H_n^{-1}, H_{n-1} \circ R_{\alpha_n} \circ H_{n-1}^{-1}\right) \\
&= d_{C^n}\left(H_n \circ R_{\alpha_{n+1}} \circ H_n^{-1}, H_n \circ R_{\alpha_n} \circ H_n^{-1}\right) \\
&\leq C(n) \cdot |||H_n|||_{n+1}^{n+1} \cdot |\alpha_{n+1} - \alpha_n| \\
&\leq C(n) \cdot q_n^{2\cdot(n+1)^2} \cdot 2 \cdot |\alpha - \alpha_n|,
\end{aligned}$$

where the constant $C(n)$ only depends on n and the geometry of the manifold. Since α is assumed to be Liouville, one can make this distance small enough to guarantee that $(T_n)_{n\in\mathbb{N}}$ is a Cauchy sequence and converges in the C^∞-topology to a limit diffeomorphism $T \in \mathcal{A}_\alpha(M)$.

This variant of the AbC method is called the *quantitative version*. It also allows one to carry out other Anosov–Katok constructions in the restricted space $\mathcal{A}_\alpha(M)$ for a prescribed Liouville number α; see for example [GK18]. The quantitative version of the AbC method is also used to find non-standard smooth realizations of Liouville rotations.

Theorem 2.3 ([FSW07]). *For every Liouville number α there exists an ergodic $T \in \mathrm{Diff}^\infty(M, \mu)$ measure-theoretically isomorphic to the circle rotation R_α.*

In the case of the torus $\mathbb{T}^d$, $d \geq 2$, the result can be strengthened to obtain a uniquely ergodic C^∞-diffeomorphism T [FSW07, Theorem 2].

It is not known if any rotation by a Diophantine rotation number allows a non-standard smooth realization (see also [FK04, Problem 7.9]).

Problem 2.4. *Given a Diophantine circle rotation, is there an ergodic volume-preserving C^∞-diffeomorphism on a manifold of dimension greater than 1 that is measure-theoretically isomorphic to the rotation?*

Due to "Herman's last geometric theorem," this seems to be very hard to achieve in case of the disc. In fact, an open question by Herman asks for Diophantine α if every element of $\mathcal{A}_\alpha(\mathbb{D}^2)$ is smoothly conjugate to the rigid rotation $S_\alpha(\theta, r) = (\theta + 2\pi\alpha, r)$. In [FK09] this question is answered affirmatively for diffeomorphisms that have rotation number α on the boundary and are sufficiently close, in a differentiable sense, to the rigid rotation S_α.

More generally, one can ask if there are any maps with interesting behavior in $\mathcal{A}_\alpha(M)$ for α Diophantine. Even the following starting point seems to require new ideas beyond the AbC method.

Problem 2.5. *Let α be Diophantine. Are there topologically transitive (or even ergodic area-preserving) C^∞-diffeomorphisms in $\mathcal{A}_\alpha(\mathbb{T}^2)$?*

We refer to Section 10.8 for further comments on pseudo-rotations of the torus.

10.3 Topological Version of the AbC Method

In another version (called *topological version* in [FK04]) of the AbC method one tries to exercise control over every orbit of the initial $\mathbb{S}^1$-action, while the original construction by Anosov and Katok was only able to exercise control over almost every orbit of the $\mathbb{S}^1$-action $\mathcal{R} = \{R_t\}_{t\in\mathbb{S}^1}$. Thus, in the topological versions one needs that every orbit of $\mathcal{R}$ spend most of its time inside zones on which one has clear control over the mapping behavior of the conjugation map (i.e., apart from so-called "turbulence zones" resulting from smoothly realizing the combinatorial description of the conjugation map). This requires a careful design and study of the geometry of such zones. Such topological constructions deal with minimality and the number of ergodic invariant measures (e.g., unique ergodicity), for instance. The topological version was announced in [Ka72] and independently developed in [FH77]. Therein A. Fathi and M. Herman apply the Baire category theorem on the space $\mathcal{A}(M)$ to prove the existence of minimal and uniquely ergodic diffeomorphisms.

Theorem 3.1 ([FH77]). *On any compact and connected smooth d-dimensional manifold M without boundary ($d \geq 2$) admitting a locally free*[2] *circle action the set of minimal and uniquely ergodic C^{∞}-diffeomorphisms is a dense G_{δ}-set in $\mathcal{A}(M)$ in the $C^{\infty}(M)$-topology.*

Related to the previous section, one can pose the following problem.

Problem 3.2. *Find such a result in the restricted spaces $\mathcal{A}_{\alpha}(M)$ for any Liouville number $\alpha \in \mathbb{S}^1$.*

In [FSW07], for every Liouville number α this restricted space $\mathcal{A}_{\alpha}(\mathbb{T}^d)$ on the torus $\mathbb{T}^d$ is considered: Fayad, Saprykina and Windsor construct a smooth diffeomorphism that is uniquely ergodic and measure-theoretically isomorphic to the rigid rotation R_{α} on $\mathbb{S}^1$. Concerning this number of ergodic invariant measures, for any $r \in \mathbb{Z}^+$ Windsor [Win01] constructed minimal diffeomorphisms with exactly r ergodic invariant measures in $\mathcal{A}(M)$ on any compact and connected smooth manifold M without boundary, where M is of dimension at least 2 and admits a free C^{∞}-action. While conversely a uniquely ergodic transformation on a compact metric space preserving a Borel measure is minimal on the support of the measure (e.g., [HK95, Proposition 4.1.18]), the first example that minimality does not imply unique ergodicity is due to Markov (see [NS60, Section 9.35]). In the analytic category Furstenberg constructed skew-products admitting uncountably many ergodic measures [Fu61]. In fact, these counterexamples carry great meaning in the history of Ergodic Theory: they showed that the so-called *quasi-ergodic hypothesis* (i.e., each orbit is dense in each surface of constant energy) does not imply the equality of space means and time means and so helped to find the right notion of ergodicity.

Special cases of manifolds with boundary are treated as well. By a fixed-point theorem of Bourgain, any area- and orientation-preserving diffeomorphism of the disc $\mathbb{D}^2$ and the annulus $\mathbb{A}$, respectively, has at least three ergodic invariant measures, namely the Lebesgue measure μ on the manifold, the Dirac-δ-measures at the fixed points of the rotations and the Lebesgue measures on the boundary components. These measures are called the *natural measures*. In [FK04], Fayad and Katok constructed diffeomorphisms with this minimal number of ergodic invariant measures. In fact, they proved that the set of such diffeomorphisms is a residual subset in the closure $\mathcal{A}'(\mathbb{D}^2)$ in the C^{∞}-topology of the conjugates of rotations with conjugacies fixing every point of the boundary and the fixed points of the action by rotations.

A stronger property than unique ergodicity is distributional unique ergodicity (sometimes also called strong unique ergodicity): a C^{∞}-diffeomorphism T with

[2] A group action is said to be locally free if the stabilizer at every point is discrete.

only one (up to multiplication by a constant) T-invariant distribution is *distributionally uniquely ergodic*, DUE for short. Ergodic translations on tori are the archetypal examples of DUE diffeomorphisms. In [AFK15] Avila, Fayad, and Kocsard use the AbC method to construct some new DUE systems.

Theorem 3.3 ([AFK15]). *Let P be either a compact nilmanifold (i.e., $P = N/\Gamma$ with N a nilpotent connected and simply connected Lie group and $\Gamma < N$ a uniform lattice) or a homogeneous space of compact type (i.e., $P = G/H$, where G is a compact Lie group and $H < G$ a closed subgroup). Then there exist DUE diffeomorphisms on $M := \mathbb{S}^1 \times P$.*

Hereby, they provided counterexamples to a conjecture by G. Forni that tori are the only closed manifolds supporting DUE diffeomorphisms. In fact, it is even possible to prove the existence of real-analytic DUE diffeomorphisms on $M = \mathbb{S}^1 \times P$ [AFK15, Section 3.2].

10.4 Realization of Spectral Properties

To a measure-preserving transformation $T\colon (X, \mu) \to (X, \mu)$ one can associate the *Koopman operator* $U_T\colon L^2(X, \mu) \to L^2(X, \mu)$ defined by $U_T f = f \circ T$ and study its properties. One of the main problems in the spectral theory of dynamical systems at the interface of unitary operator theory and ergodic theory is the following question:

Question 4.1. *What are possible spectral properties for a Koopman operator associated with a measure-preserving transformation?*

More specifically, one can search for transformations possessing specific essential values $\mathcal{M}_{U_T}$ of the spectral multiplicities:

Question 4.2. *Given a subset $E \subset \mathbb{N} \cup \{\infty\}$, is there an ergodic transformation T such that $\mathcal{M}_{U_T} = E$?*

These two problems are open and no restrictions (except for the obvious ones) are known. However, there is impressive progress concerning Question 4.2 (see [Da13] for a survey on spectral multiplicities of ergodic actions) and there exist two standard points of view: to consider the spectrum of T (and in particular $\mathcal{M}_{U_T}$) either on $L^2\,(X, \mu)$ or on the orthogonal complement $L^2_0\,(X, \mu)$ of the constant functions. In [KL95] it was proved that all possible subsets of $\mathbb{N} \cup \{\infty\}$ can be realized as $\mathcal{M}_{U_T}$ for some ergodic transformation T in the first case (since 1 is always an eigenvalue because of the constant functions, "possible" means any subset of $\mathbb{N} \cup \{\infty\}$ with 1 as an element). In the second case, the Cartesian powers of a transformation provide a good opportunity for the construction of examples with the infimum of essential spectral multiplicities larger than 1. Although it seemed very unlikely that these Cartesian powers have finite maximal spectral multiplicity, this is the generic case: independently, Ageev and Ryzhikov proved the celebrated result that for a generic automorphism T the Cartesian square $T \times T$ has homogeneous spectrum of multiplicity 2 (see [Ag99] and [Ry99], respectively).

In the smooth category there are only a few results in this direction. Explicitly, Danilenko asks which subsets $E \neq \{1\}$ admit a smooth ergodic transformation T with $\mathcal{M}_{U_T} = E$ [Da13, Section 10]. Blanchard and Lemanczyk showed that every set E containing 1 as well as $\operatorname{lcm}(e_1, e_2)$ for $e_1, e_2 \in E$ is realizable as the set of essential spectral multiplicities for a Lebesgue measure-preserving analytic diffeomorphism of a finite-dimensional torus [BL93]. Answering Problem 7.11 in [FK04] affirmatively, the author proved the subsequent result using the quantitative version of the AbC method:

Theorem 4.3 ([Ku16]). *Let α be a Liouville number. Then the set of smooth weakly mixing diffeomorphisms, that have a homogeneous spectrum of multiplicity 2 for $T \times T$ and a maximal spectral type disjoint with its convolutions, is residual in $\mathcal{A}_\alpha(M)$ in the Diff$^\infty(M)$-topology.*

The second part of this theorem is linked to a conjecture of Kolmogorov, later reiterated by Rokhlin and Fomin (after verifying that the property holds for all dynamical systems known at that time, especially large classes of systems of probabilistic origin like Gaussian ones), namely that every ergodic transformation possesses the so-called *group property*, namely the maximal spectral type σ is symmetric and dominates its square $\sigma * \sigma$. This conjecture is an analogue of the well-known group property of the set of eigenvalues of an ergodic automorphism and was proven to be false. Indeed, in [St66] A. M. Stepin gave the first example of a dynamical system without the group property. Oseledets constructed an analogous weakly mixing example [Os69]. Later, Stepin showed that for a generic transformation all convolutions σ_0^k, $k \in \mathbb{N}$, of the maximal spectral type σ_0 on $L_0^2(X, \mu)$ are mutually singular (see [St87]).

Ageev was even able to show that for the n-th power $T^n = T \times \cdots \times T$ of a generic measure-preserving transformation T it holds $M(T^n) = \{n, n \cdot (n-1), \ldots, n!\}$ [Ag99, Theorem 2]. He also proved for every $n \in \mathbb{N}$ the existence of an ergodic transformation with homogeneous spectrum of multiplicity n in the orthogonal complement of the constant functions [Ag05, Theorem 1], solving Rokhlin's problem on homogeneous spectrum in ergodic theory.

Problem 4.4. *Is it possible to find a similar result in the smooth category? Which subsets E admit a smooth ergodic transformation T with $\mathcal{M}_{U_T} = E$?*

10.5 The AbC Method in Symplectic Dynamics

We start by recalling that a *symplectic manifold* is a smooth manifold M equipped with a closed nondegenerate differential 2-form ω, called the *symplectic form*. Since ω is nondegenerate, every differentiable function $H\colon M \to \mathbb{R}$ determines a unique vector field X_H by requiring $dH(Y) = \omega(X_H, Y)$ for every vector field Y on M. Then X_H is called the *Hamiltonian vector field* with *Hamiltonian* H.

In a much cited paper [Ka73], Katok uses a variant of the AbC method to construct smooth Hamiltonian systems that are arbitrarily close in a differentiable sense to a given integrable Hamiltonian system satisfying some degeneracy conditions and that are ergodic on each level $H^{-1}(c)$ of constant energy with respect to the invariant measure induced by the symplectic form.

Problem 5.1. *Can one obtain further ergodic properties like weak mixing?*

The difficulty is to make the construction on various surfaces of constant energy coherent. In [Ka73] Katok also applies the construction to obtain examples of nondegenerate Finsler metrics[3] on spheres and projective spaces which are arbitrarily close to the standard Riemannian metric such that the corresponding geodesic flows have only two ergodic components which cover the space up to a set of arbitrarily small measure.

To state the next results, we recall that a diffeomorphism $T\colon (M,\omega) \to (M',\omega')$ between two symplectic manifolds is called a *symplectomorphism* if $T^*\omega' = \omega$. We denote the group of symplectomorphisms of a symplectic manifold (M,ω) by $\mathrm{Symp}(M,\omega)$, and $\mathrm{Symp}_0(M,\omega)$ is the connected component of the identity map. Inspired by the Baire Category approach to the AbC method from [FH77], L. Hernández-Corbato and F. Presas prove the existence of minimal symplectomorphisms.

Theorem 5.2 ([HP19]). *If a symplectic manifold (M,ω) admits a locally free $\mathbb{S}^1$-action by symplectomorphisms, then there exists $\psi \in \mathrm{Symp}_0(M,\omega)$ such that ψ is minimal.*

As mentioned in [HP19], it would be more ambitious to obtain the result in the Hamiltonian symplectomorphism group. Since a Hamiltonian $\mathbb{S}^1$-action is never locally free, this seems to require a more elaborate construction in the spirit of the original Anosov–Katok constructions in [AK70] instead of the Baire Category approach.

Problem 5.3. *Can one obtain the statement in the Hamiltonian symplectomorphism group?*

[3] Generalizing the notion of a Riemannian metric, a *Finsler metric* is a continuous nonnegative function $F\colon TM \to [0,\infty)$ such that for each point $x \in M$ we have $F(v+w) \leq F(v) + F(w)$ for all $v, w \in T_xM$, $F(\lambda v) = \lambda F(v)$ for all $\lambda \geq 0$, and $F(v) > 0$ unless $v = 0$.

While every symplectic manifold must be of even dimension, contact geometry is in many ways an odd-dimensional counterpart. A *contact manifold* (M, ξ) is a smooth manifold M of odd dimension $2n+1$ which is equipped with a differential 1-form A satisfying $\xi = \ker A$ that is non-degenerate in the sense that the wedge product $A \wedge (dA)^{\wedge^n}$ does not vanish. Then a diffeomorphism $T\colon M \to M$ of a contact manifold (M, A) is called a *contactomorphism* if $T_*\xi = \xi$ (or, equivalently, if $T^*A = \varphi \cdot A$ with a nonzero C^∞-function $\varphi\colon M \to \mathbb{R}$). As earlier, we denote the group of contactomorphisms of a contact manifold (M, ξ) by $\mathrm{Cont}(M, \xi)$ and the connected component of the identity by $\mathrm{Cont}_0(M, \xi)$, which is known to correspond to the Hamiltonian contactomorphism group (i.e., any contact vector field is Hamiltonian). In this setting, Hernández-Corbato and Presas can obtain ψ also to be uniquely ergodic.

Theorem 5.4 ([HP19]). *If a contact manifold (M, ξ) admits a locally free $\mathbb{S}^1$-action by contactomorphisms, then there exists $\psi \in \mathrm{Cont}_0(M, \xi)$ such that ψ is minimal and uniquely ergodic.*

Problem 5.5. *In all these settings, one can ask about more subtle ergodic properties.*

This is related to the question in [GG18] if projective spaces $\mathbb{CP}^n$, $n \geq 2$, admit dynamically interesting Hamiltonian pseudo-rotations. Here, a Hamiltonian diffeomorphism on $\mathbb{CP}^n$ is called a pseudo-rotation if it has exactly $n+1$ periodic points which is the minimal possible number of fixed points. In particular, one can ask about ergodicity or, more generally, a finite number of ergodic measures. In [LRS22], F. Le Roux and S. Seyfaddini use the AbC method to prove that toric[4] symplectic manifolds admit Hamiltonian pseudo-rotations with a finite and minimal set of ergodic measures which consists of the measure induced by the symplectic volume form and the Dirac measures supported at the fixed points of the torus action. In some sense, this is a symplectic analogue to the construction in [FK04, Section 3] of area-preserving C^∞ diffeomorphisms of the disk with a minimal number of ergodic invariant measures (see Section 10.3).

In the converse direction, symplectic methods prove useful in analyzing pseudo-rotations on $\mathbb{D}^2$. In [Ba15] B. Bramham applied pseudoholomorphic curves techniques to show that for a dense subset of Liouville rotation numbers a pseudo-rotation on $\mathbb{D}^2$ is C^0-rigid (hence, it cannot be mixing). Recently, Albers, Geiges, and Zehmisch have shown that a large class of AbC diffeomorphisms on $\mathbb{D}^2$, which equal a rigid rotation near the boundary, embed into a closed 3-manifold as

[4] A $2n$-dimensional closed and connected symplectic manifold is called *toric* if it admits an effective Hamiltonian action of $\mathbb{T}^n$.

Poincaré return maps on a global surface of section[5] for a Reeb flow[6] [AGZ22]. As mentioned by the authors, this might allow one to use global symplectic methods for studying these pseudo-rotations.

10.6 Ergodic Properties of the Derivative Extension

As mentioned earlier, Anosov and Katok proved that the set of weakly mixing diffeomorphisms is generic in $\mathcal{A}(M)$ in the $C^\infty(M)$-topology. In extension of it, R. Gunesch and A. Katok constructed weakly mixing diffeomorphisms preserving a measurable Riemannian metric in [GK00]. Using the quantitative version of the AbC method as described in Section 10.2 one can prove the denseness of volume-preserving diffeomorphisms, that are weakly mixing and preserve a measurable Riemannian metric, in $\mathcal{A}_\alpha(M)$ for every prescribed Liouville number $\alpha \in \mathbb{S}^1$ [GK18]. The existence of a measurable invariant metric for a diffeomorphism is equivalent to the existence of an invariant measure for the projectivized derivative extension which is absolutely continuous in the fibers. Hence, it is a natural question to study the ergodic properties of the projectivized derivative extension with respect to such a measure. Actually, the constructions in [GK00] as well as [GK18] are as non-ergodic as possible: their projectivized derivative extensions are isomorphic to the direct product of the diffeomorphism in the base with the trivial action in the fibers so that each ergodic component intersects almost every fiber in a single point. The author was also able to realize the other extreme possibility by constructing diffeomorphisms whose differential is ergodic with respect to such a smooth measure in the projectivization of the tangent bundle.

Theorem 6.1 ([Ku20]). *There exists a volume-preserving diffeomorphism in $\mathcal{A}(M)$, whose projectivized derivative extension is ergodic with respect to a measure in the projectivization of the tangent bundle which is absolutely continuous in the fibers.*

This construction provides the only known examples of measure-preserving diffeomorphisms whose differential is ergodic with respect to a smooth measure in the projectivization of the tangent bundle. In fact, it is possible to extend the result in such a way that the diffeomorphism is weakly mixing and lies in the restricted space $\mathcal{A}_\alpha(M)$ for any prescribed Liouville number α. It is an interesting challenge to study further ergodic properties of the projectivized derivative extension with respect to such a measure.

5 A *global surface of section* for the flow of a smooth non-singular vector field X on a closed 3-dimensional manifold M is an embedded compact surface $\Sigma \subset M$ such that each component of the boundary $\partial\Sigma$ is a periodic orbit of X and such that the interior $\mathrm{Int}(\Sigma)$ is transverse to X and the orbit of X through any point in $M \setminus \partial\Sigma$ intersects $\mathrm{Int}(\Sigma)$ in forward and backward time.

6 Let $(M, \xi = \ker A)$ be a contact manifold. The *Reeb vector field* R of A is defined uniquely by the conditions $dA(R, \cdot) = 0$ and $A(R) = 1$.

Problem 6.2. *Is it possible to do the constructions in such a way that the projectivized derivative extension is weakly mixing with respect to such a measure?*

10.7 Diffeomorphisms of the Circle

Let F be the group of orientation-preserving C^∞-diffeomorphisms of the circle. Furthermore, for $\alpha \in \mathbb{S}^1$ we consider the subspace F_α of F consisting of all the C^∞-diffeomorphisms of the circle with rotation number α. If α is irrational, for any $f \in F_\alpha$ there is a unique orientation-preserving homeomorphism H_f of the circle such that $f = H_f \circ R_\alpha \circ H_f^{-1}$ and $H_f(0) = 0$ by the celebrated Denjoy's theorem. Yoccoz proved that the subspace O_α of F_α of all the diffeomorphisms, for which H_f are C^∞-diffeomorphisms, is C^∞-dense in F_α ([Yo95]). If α is Diophantine, then we even have $F_\alpha = O_\alpha$.

One can use the AbC method to show that, for Liouvillean α, several subspaces of F_α which are defined by intermediate regularities of H_f are also C^∞-dense in F_α. To obtain interesting results one has to drop the requirement in the AbC construction that the conjugation maps h_n are volume-preserving. Instead of this, one chooses them to distort lengths in a controllable way. Prototype results can be found in [HK95, Section 12.6] and [Ma12] where it is shown that the following spaces are C^∞-dense in F_α for Liouvillean rotation number α:

- $G_{0,\text{sing}}$: H_f is singular and is not d-Hölder for any $d \in (0,1)$.
- $G_{1,\text{ac}}$: H_f is absolutely continuous and is d-Hölder for any $d \in (0,1)$, but is not bi-Lipschitz.
- For $k \in \mathbb{N}$ G_k: H_f is a C^k-diffeomorphism, but is not a C^{k+1}-diffeomorphism.

Further dense properties in the space of circle diffeomorphisms are studied in [Ma12] and [Ku18].

The AbC method is also used to provide examples that the assumption of hyperbolicity[7] of the measure is crucial in the Eckmann–Ruelle conjecture. This conjecture states that for a $C^{1+\alpha}$ diffeomorphism, a hyperbolic ergodic measure μ and μ-a.e. point, the pointwise dimension exists. Here, we recall that the upper and lower pointwise dimensions at a point x of a Borel measure μ on a metric space X are defined as

$$\overline{d}_\mu(x) = \limsup_{r\to 0} \frac{\log \mu(B(x,r))}{\log r} \quad \text{and} \quad \underline{d}_\mu(x) = \liminf_{r\to 0} \frac{\log \mu(B(x,r))}{\log r}.$$

If $\overline{d}_\mu(x) = \underline{d}_\mu(x)$, then one defines this value as the *pointwise dimension* $d_\mu(x)$. The conjecture was answered affirmatively in [BPS99]. Extending earlier joint

[7] A measure is called *hyperbolic* if all its Lyapunov exponents are different from zero.

work with B. Kalinin [KS02], V. Sadovskaya obtained the following result via the quantitative version of the AbC method.

Theorem 7.1 ([Sa08]). *Let α be a Liouville number and let $0 \leq s \leq 1$. There exists a C^∞ circle diffeomorphism f with rotation number α such that for its unique invariant measure μ:*

1. $\underline{d}_\mu(x) = s$ *and* $\overline{d}_\mu(x) = 1$ *for μ-a.e.* $x \in \mathbb{S}^1$.
2. $\dim_H(\mu) = \underline{\dim}_B(\mu) = s$ *and* $\overline{\dim}_B(\mu) = 1$.

For the sake of completeness, we recall the following definitions of Hausdorff, upper and lower box dimensions of a Borel probability measure μ:

$$\begin{aligned}
\dim_H(\mu) &= \inf\{\dim_H X \mid \mu(X) = 1\},\\
\underline{\dim}_B(\mu) &= \lim_{\varepsilon \to 0} \inf\left\{\underline{\dim}_B(X) \mid \mu(X) > 1 - \varepsilon\right\},\\
\overline{\dim}_B(\mu) &= \lim_{\varepsilon \to 0} \inf\left\{\overline{\dim}_B(X) \mid \mu(X) > 1 - \varepsilon\right\}.
\end{aligned}$$

The next problem is stated in [Sa08].

Problem 7.2. *Does there exist a circle diffeomorphism with irrational rotation number whose invariant measure has upper pointwise dimension less than one on a set of positive measure?*

10.8 Pseudo-Rotations of the Torus

A homeomorphism f on $\mathbb{T}^d$ homotopic to the identity is called a *pseudo-rotation*, if there exists a vector $\omega \in \mathbb{R}^d$ and a lift $F: \mathbb{R}^d \to \mathbb{R}^d$ of f, such that for all $z \in \mathbb{R}^d$

$$\lim_{n \to \infty} \frac{F^n(z) - z}{n} = \omega.$$

The pseudo-rotation is called *totally irrational* if $\omega = (\omega_1, \ldots, \omega_d) \in \mathbb{R}^d$ is totally irrational, namely if $\omega_1, \ldots, \omega_d, 1$ are rationally independent. We say a pseudo-rotation f has *bounded mean motion*, with constant $c \geq 0$, if $\|F^n(z) - z - n\omega\| \leq c$ for all $n \in \mathbb{Z}$ and $z \in \mathbb{R}^d$. In [J09] T. Jäger obtained a classification for pseudo-rotations of $\mathbb{T}^2$ with bounded mean motion analogous to Poincaré's celebrated classification of circle homeomorphisms.

Norton and Sullivan asked the following question: if f is a diffeomorphism of $\mathbb{T}^2$, $h: \mathbb{T}^2 \to \mathbb{T}^2$ is a continuous map homotopic to the identity, and $hf = S_\omega h$, where $S_\omega: \mathbb{T}^2 \to \mathbb{T}^2$ is a translation by a totally irrational vector $\omega \in \mathbb{R}^2$, are there natural geometric conditions (e.g., smoothness) on f that force h to be a homeomorphism? In [WZ18] J. Wang and Z. Zhang give a negative answer to this question in the C^∞

category. On the one hand, they apply the AbC method to construct a C^∞ area-preserving totally irrational pseudo-rotation of $\mathbb{T}^d$ with bounded mean motion that is not topologically conjugate to a translation. On the other hand, they use a theorem by Jäger that if f is a minimal totally irrational pseudo-rotation on $\mathbb{T}^d$ homotopic to the identity with bounded mean motion, then f is semi-conjugate to a totally irrational translation on $\mathbb{T}^d$ and the semi-conjugacy is homotopic to the identity [J09]. Combining these two steps, they obtain the following theorem.

Theorem 8.1 ([WZ18]). *For any $d \geq 2$, there exists a C^∞ volume-preserving and minimal map $f\colon \mathbb{T}^d \to \mathbb{T}^d$ which is semi-conjugate to a translation by a map homotopic to the identity, but is not topologically conjugate to a translation.*

Recently, in [WY21] Wang and Yang extended the result to the real-analytic category using the AbC method and the concept of the block-slide type of maps (see also Section 10.10.1).

We also mention that working on $\mathbb{T}^2$ or, more generally, on any manifold admitting a $\mathbb{T}^2$ action $\{R_{\alpha,\beta}\}_{(\alpha,\beta)\in\mathbb{T}^2}$ allows to use a *two-dimensional version* of the AbC method. Here, one constructs a diffeomorphism T as a limit of diffeomorphisms $T_n = H_n \circ R_{\alpha_{n+1},\alpha'_{n+1}} \circ H_n^{-1}$ with $\alpha_{n+1}, \alpha'_{n+1} \in \mathbb{Q}$ and conjugation maps $H_n = H_{n-1} \circ h_n$ satisfying $h_n \circ R_{\alpha_n,\alpha'_n} = R_{\alpha_n,\alpha'_n} \circ h_n$. This two-dimensional version of the AbC method gives more flexibility in the construction, especially for approximating periodic processes with towers of unequal height. It was used in [GK20] to construct the first example of a smooth diffeomorphism T of zero-entropy whose Cartesian square $T \times T$ is loosely Bernoulli.

Theorem 8.2 ([GK20]). *Let $d \geq 2$ and let M be a d-dimensional smooth, compact and connected manifold, possibly with boundary, admitting a smooth effective torus action $\left\{R_{\alpha,\beta}\right\}_{(\alpha,\beta)\in\mathbb{T}^2}$ preserving a smooth volume μ. Then there is a weakly mixing C^∞-diffeomorphism $T \in \mathcal{B} := \overline{\left\{h \circ R_{\alpha,\beta} \circ h^{-1} \,:\, h \in \mathrm{Diff}^\infty(M,\mu), (\alpha,\beta) \in \mathbb{T}^2\right\}}^{C^\infty}$ of topological entropy 0 such that $T \times T$ is loosely Bernoulli. Furthermore, the set of such diffeomorphisms is residual in $\mathcal{B}$ in the $C^\infty(M)$-topology.*

Here, we recall that an ergodic automorphism T of entropy zero is called *loosely Bernoulli* (sometimes also *standard* or *loosely Kronecker*) if it is Kakutani equivalent[8] to an irrational circle rotation. If T is an ergodic automorphism such that $T \times T$ is loosely Bernoulli, then T must itself be loosely Bernoulli. However, the converse is not true. Most notably, M. Ratner showed that for any (non-identity) transformation T in the horocycle flow, T is loosely Bernoulli, but $T \times T$ is not loosely Bernoulli.

8 Two ergodic measure-preserving automorphisms T and S are said to be *Kakutani equivalent* if they are isomorphic to cross-sections of the same ergodic measure-preserving flow.

Problem 8.3. *Can one construct diffeomorphisms with loosely Bernoulli Cartesian square on manifolds admitting a $\mathbb{S}^1$-action?*

10.9 Coding AbC Maps and Anti-classification Results

Another major question in Ergodic Theory was asked in von Neumann's pioneering paper [Ne32]: the so-called *isomorphism problem* asks to classify the ergodic measure preserving transformations (MPT's) up to measure-theoretic isomorphisms. It has been a guiding light for directions of research within ergodic theory and has been solved only for some special classes of transformations (e.g., transformations with pure point spectrum). In particular, Kolmogorov entropy classifies a large class of highly random systems like Bernoulli shifts. This result is an example of a classification result. Informally, a *classification* is a method of determining isomorphism between transformations by computing other invariants for which equivalence is easy to determine. Many properties of transformations like mixing of various types or finite rank have been characterized and studied in connection with this problem, but the general problem remained intractable.

Starting in the late 1990s, so-called *anti-classification* results have been established and have demonstrated in a rigorous way that classification is not possible. In their landmark paper [FRW11] M. Foreman, D. Rudolph and B. Weiss showed that the measure-isomorphism relation for ergodic MPT's is a complete analytic set.[9] In particular, it is not a Borel set. Informally speaking, this result says that determining isomorphism between ergodic transformations is inaccessible to countable methods that use countable amounts of information.

One way to show that not all finite ergodic measure preserving transformations have a smooth model would be to show that their classification is easier than the general classification problem. But recently Foreman and Weiss were able to show that the variety of ergodic transformations that have smooth models is rich enough that one can upgrade their aforementioned anti-classification result to the C^∞ category ([FW19b], [FW22]).

Theorem 9.1 ([FW22]). *Let M be either $\mathbb{T}^2$, $\mathbb{D}^2$, or $\mathbb{A}$ and μ be the Lebesgue measure. Then the measure-isomorphism relation among pairs $(S, T) \in \mathit{Diff}^\infty(M, \mu) \times \mathit{Diff}^\infty(M, \mu)$ is a complete analytic set and hence not a Borel set with respect to the C^∞-topology.*

Thus, von Neumann's classification problem is impossible even for C^∞-diffeomorphisms of compact surfaces.

[9] A subset B of a Polish space is *analytic* if and only if it is the continuous image of a Borel subset of a Polish space. Moreover, if $\mathcal{S}$ is a collection of sets and $B \in \mathcal{S}$, then B is called *complete* for Borel reductions if and only if every $A \in \mathcal{S}$ is Borel reducible to B. Here, for Polish spaces X and Y a Borel function $f : X \to Y$ is called a *Borel reduction* from $A \subseteq X$ to $B \subseteq Y$ if and only if for all $x \in X$: $x \in A$ if and only if $f(x) \in B$.

The AbC method is a key ingredient in transferring the anti-classification result from [FRW11] to the smooth category: Foreman and Weiss found a way to code smooth diffeomorphisms on $M = \mathbb{T}^2, \mathbb{A}, \mathbb{D}^2$ constructed by an untwisted version of the AbC method (i.e., the conjugation map h_n maps its fundamental domain $\left[0, \frac{1}{q_n}\right] \times [0, 1]$ into itself) into some symbolic systems (the so-called *uniform circular systems*) and showed that all these circular systems can be realized smoothly [FW19a].

In joint work with S. Banerjee we show that these circular systems can even be realized as real-analytic diffeomorphism on $\mathbb{T}^2$. Hereby, we are also able to obtain the analogous anti-classification result for real-analytic diffeomorphisms of the 2-torus [BKpp]. Hence, von Neumann's classification problem is impossible even when restricting to real-analytic diffeomorphisms of $\mathbb{T}^2$. It is an interesting problem to extend the result to other manifolds (see also the next section for a brief outline of challenges in the real-analytic category).

In [FW19a, Section 7.7.] Foreman and Weiss express the following problem.

Problem 9.2. *Find symbolic representations for other versions of the AbC method.*

This seems to be particularly interesting for twisted AbC constructions of weakly mixing diffeomorphisms. Given such symbolic representations of the AbC constructions of weakly mixing diffeomorphisms, one might be able to deduce an anti-classification result for weakly mixing diffeomorphisms.

We also mention that the equivalence relation of isomorphism on ergodic MPT's is known to be turbulent [FW04]. Hence, no generic class can have a complete set of algebraic invariants.

Problem 9.3. *Is the equivalence relation of isomorphism on ergodic diffeomorphisms of compact manifolds turbulent?*

10.10 Real-Analytic AbC Constructions

While the AbC method is one of the most powerful tools of constructing C^∞ diffeomorphisms with prescribed ergodic or topological properties, there are great challenging differences in the real-analytic case as discussed in [FK04, Section 7.1] as well as [BK19, Section 6.3]: since maps with very large derivatives in the real domain or its inverses are expected to have singularities in a small complex neighborhood, for a real analytic family S_t, $0 \leq t \leq t_0$, $S_0 = \mathrm{id}$, the family $h \circ S_t \circ h^{-1}$ is expected to have singularities very close to the real domain for any $t > 0$. So, the domain of analyticity for maps of our form $T_n = H_n \circ R_{\alpha_{n+1}} \circ H_n^{-1}$ will shrink at any step of the construction and the limit diffeomorphism will not be real-analytic. Thus, it is necessary to find conjugation maps of a special form which may be inverted more or less explicitly in such a way that one can guarantee analyticity of the map and its inverse in a large complex domain. In fact, in all to date real-analytic AbC-constructions one needs the diffeomorphisms T_n to be analytic on such a fixed domain rather than h_n. This is a stronger assumption and we do not know how to do real-analytic constructions without such an assumption so far.

10.10.1 Real-Analytic AbC Constructions on the Torus

First of all, we will give a brief description of real-analytic diffeomorphisms on the torus as well as the concept of *block-slide type of map* introduced in [Ba17]. Afterwards, we will state some recent results obtained in this setting. Additionally, we will also mention some open problems.

10.10.1.1 The Topology of Real-Analytic Diffeomorphisms on the Torus

Any real-analytic diffeomorphism on $\mathbb{T}^d$ homotopic to the identity admits a lift to a map from $\mathbb{R}^d$ to $\mathbb{R}^d$ and has the form $F(x_1,\dots,x_d) = (x_1 + f_1(x_1,\dots,x_d),\dots,x_d + f_d(x_1,\dots,x_d))$, where $f_i\colon \mathbb{R}^d \to \mathbb{R}$ are $\mathbb{Z}^d$-periodic real-analytic functions. Any real-analytic $\mathbb{Z}^d$-periodic function defined on $\mathbb{R}^d$ can be extended to some complex neighborhood[10] of $\mathbb{R}^d$ as a holomorphic function. For a fixed $\rho > 0$, let $\Omega_\rho := \{(z_1,\dots,z_d) \in \mathbb{C}^d\colon |\mathrm{Im}(z_1)| < \rho,\dots,|\mathrm{Im}(z_d)| < \rho\}$ and for a function f defined on this set, put $\|f\|_\rho := \sup_{(z_1,\dots,z_d)\in\Omega_\rho} |f(z_1,\dots,z_d)|$. We define $C^\omega_\rho(\mathbb{T}^d)$ to be the space of all $\mathbb{Z}^d$-periodic real-analytic functions on $\mathbb{R}^d$ that extends to a holomorphic function on Ω_ρ and $\|f\|_\rho < \infty$. Then we define $\mathrm{Diff}^{\,\omega}_\rho(\mathbb{T}^d,\mu)$ to be the set of all measure-preserving real-analytic diffeomorphisms of $\mathbb{T}^d$ homotopic to the identity, whose lift $F(x) = (x_1 + f_1(x),\dots,x_d + f_d(x))$ to $\mathbb{R}^d$ satisfies $f_i \in C^\omega_\rho(\mathbb{T}^d)$ and we also require the lift $\tilde{F}(x) = (x_1 + \tilde{f}_1(x),\dots,x_d + \tilde{f}_d(x))$ of its inverse to $\mathbb{R}^d$ to satisfy $\tilde{f}_i \in C^\omega_\rho(\mathbb{T}^d)$. The metric d_ρ in $\mathrm{Diff}^{\,\omega}_\rho(\mathbb{T}^d,\mu)$ is defined by

$$d_\rho(f,g) = \max\{\tilde{d}_\rho(f,g), \tilde{d}_\rho(f^{-1},g^{-1})\}, \quad \text{where}$$
$$\tilde{d}_\rho(f,g) = \max_{i=1,\dots,d}\{\inf_{n\in\mathbb{Z}} \|f_i - g_i + n\|_\rho\}.$$

10.10.1.2 The Concept of Block-Slide Type Maps and Their Analytic Approximations

As mentioned earlier, finding real-analytic conjugation maps whose complexification is entire is central to current versions of the real-analytic AbC method. In this section we present a way to get those in the case of the torus $\mathbb{T}^d$, $d \geq 2$.

We recall that a *step function* on the unit interval is a finite linear combination of indicator functions on intervals. We define, for $1 \leq i,j \leq d$, and $i \neq j$, the following piecewise continuous map $\mathfrak{h}\colon \mathbb{T}^d \to \mathbb{T}^d$ defined by

[10] We identify $\mathbb{R}^d$ inside $\mathbb{C}^d$ via the natural inclusion $(x_1,\dots,x_d) \mapsto (x_1 + i0,\dots,x_d + i0)$.

$$\mathfrak{h}(x_1,\ldots,x_d) := (x_1,\ldots,x_{i-1},x_i + s(x_j) \mod 1, x_{i+1},\ldots,x_d),$$

where s is a step function on the unit interval. We refer to any finite composition of maps of this kind as a *block-slide type of map* on the torus. The nomenclature is motivated from the fact that a finite composition of maps defined above has the effect of sliding solid blocks on the torus. These maps can be approximated extremely well by measure-preserving real analytic diffeomorphisms outside a set of arbitrarily small measure. This can be achieved because step function can be approximated well by real-analytic functions whose complexification is entire:

Lemma 10.1 ([Ba17], Lemma 4.7, [Ku17], Lemma 3.6). *Let k and N be two positive integers and $\beta = (\beta_0,\ldots,\beta_{k-1}) \in [0,1)^k$. Assume k is even. Consider a step function of the form*

$$\tilde{s}_{\beta,N} \colon [0,1) \to \mathbb{R} \quad \textit{defined by} \quad \tilde{s}_{\beta,N}(x) = \sum_{i=0}^{kN-1} \tilde{\beta}_i \chi_{[\frac{i}{kN},\frac{i+1}{kN})}(x).$$

Here $\tilde{\beta}_i := \beta_j$, where $j := i \mod k$. Then, given any $\varepsilon > 0$ and $\delta > 0$, there exists a periodic real-analytic function $s_{\beta,N} \colon \mathbb{R} \to \mathbb{R}$ satisfying the following properties:

1. *Entirety: The complexification of $s_{\beta,N}$ extends holomorphically to $\mathbb{C}$.*
2. *Proximity criterion: $s_{\beta,N}$ is L^1 close to $\tilde{s}_{\beta,N}$. We can say more:*

$$\sup_{x\in[0,1)\setminus F} |s_{\beta,N}(x) - \tilde{s}_{\beta,N}(x)| < \varepsilon.$$

3. *Periodicity: $s_{\beta,N}$ is $1/N$ periodic. More precisely, the complexification will satisfy*

$$s_{\beta,N}(z + n/N) = s_{\beta,N}(z) \qquad \textit{for all } z \in \mathbb{C} \textit{ and } n \in \mathbb{Z}.$$

4. *Bounded derivative: The derivative is small outside a set of small measure,*

$$\sup_{x\in[0,1)\setminus F} |s'_{\beta,N}(x)| < \varepsilon.$$

Here, $F = \cup_{i=0}^{kN-1} I_i \subset [0,1)$ is a union of intervals centered around $\frac{i}{kN}$, $i = 1,\ldots,kN-1$ and $I_0 = [0,\frac{\delta}{2kN}] \cup [1-\frac{\delta}{2kN},1)$ and $l(I_i) = \frac{\delta}{kN}$ for all i.

This lemma enables us to approximate a block-slide type of map on the torus by a measure-preserving real-analytic diffeomorphism:

Proposition 10.2. *Let $\mathfrak{h} \colon \mathbb{T}^d \to \mathbb{T}^d$ be a block-slide type of map which commutes with $R_{1/q}$ for some natural number q. Then for any $\varepsilon > 0$ and $\delta > 0$, there exists a real-analytic diffeomorphism $h \in \mathrm{Diff}^{\omega}_{\infty}(\mathbb{T}^d,\mu)$ satisfying the following conditions:*

1. *Proximity property: There exists a set $E \subset \mathbb{T}^d$ such that $\mu(E) < \delta$ and*

$$\sup_{x \in \mathbb{T}^d \setminus E} \|h(x) - \mathfrak{h}(x)\| < \varepsilon.$$

2. *Commutativity property: $h \circ R_{1/q} = R_{1/q} \circ h$.*

10.10.1.3 Recent Results and Open Problems

The first successful application of an AbC method in the real-analytic category was exhibited in [Sa03], where M. Saprykina constructed uniquely ergodic real-analytic diffeomorphisms on $\mathbb{T}^2$ using conjugation maps of the explicit form

$$h_n(x, y) = (x, y + c_n \cdot \sin(2\pi q_n x))$$

with a sufficiently fast increasing sequence $(c_n)_{n \in \mathbb{N}}$.

Recently, in joint work with S. Banerjee we are able to find real-analytic counterparts of several AbC constructions using the real-analytic approximation of block-slide type of maps. In particular, we prove the following non-standard realization result:

Theorem 10.3 ([BK19], Theorem B). *For any $\rho > 0$ and any two integers $h \geq 1$ and $d \geq 2$ there exists an ergodic real-analytic diffeomorphism $T \in \mathrm{Diff}^{\omega}_{\rho}(\mathbb{T}^d, \mu)$ which is metrically isomorphic to an ergodic translation of $\mathbb{T}^h$.*

As a corollary we get

Corollary 10.4. *For any $\rho > 0$ and any two integers $h \geq 1$ and $d \geq 2$ there exists an ergodic real-analytic diffeomorphism $T \in \mathrm{Diff}^{\omega}_{\rho}(\mathbb{T}^d, \mu)$ such that T has a discrete spectrum generated (over $\mathbb{Z}$) by h linearly independent eigenvalues.*

This disproves a long-standing conjecture of Kolmogorov in [Ko57] stating that on a d-dimensional real-analytic manifold an ergodic real-analytic diffeomorphism preserving an analytic measure may have a discrete spectrum with only d distinct eigenvalues.

We can also produce more exotic examples by showing that there are minimal but not uniquely ergodic measure-preserving real-analytic diffeomorphisms.

Theorem 10.5. *For any $\rho > 0$ and any natural number $r \geq 1$, there exists a real-analytic diffeomorphism $T \in \mathrm{Diff}^{\omega}_{\rho}(\mathbb{T}^2, \mu)$ which is minimal and has exactly r ergodic invariant measures each of which is absolutely continuous with respect to the Lebesgue measure.*

This theorem parallels the previously mentioned result of Windsor in the smooth category ([Win01]).

We have also been able to find some real-analytic counterparts for the results presented in Section 10.6.

Theorem 10.6 ([Ku17]). *For any $\rho > 0$ and any integer $d \geq 2$, there are weakly mixing real-analytic diffeomorphisms $T \in \mathrm{Diff}^{\omega}_{\rho}(\mathbb{T}^d, \mu)$ preserving a measurable Riemannian metric.*

Hereby, we solved [GK00, Problem 3.9] in the affirmative. As discussed in Section 10.6, it is an interesting challenge to study ergodic properties of the projectivized derivative extension with respect to a smooth measure in the projectivization of the tangent bundle.

Problem 10.7. *Is it possible to do the real-analytic constructions in such a way that the projectivized derivative extension is ergodic with respect to such a measure?*

In another recent joint work with S. Banerjee we get a real-analytic counterpart of the smooth realization of uniform circular systems in [FW19a]. As already indicated in Section 10.9, this allows us to transfer the anti-classification result of [FRW11] to the real-analytic category and to conclude that the measure-isomorphism relation among pairs of ergodic real-analytic diffeomorphisms is not a Borel set [BKpp, Theorem B]. More precisely, we show the following anti-classification result.

Theorem 10.8 ([BKpp]). *For every $\rho > 0$ the measure-isomorphism relation among pairs (S, T) of ergodic $\mathrm{Diff}^{\omega}_{\rho}(\mathbb{T}^2, \mu)$-diffeomorphisms is a complete analytic set and hence not a Borel set.*

The real-analytic realization of circular systems also allows one to construct an uncountable family of ergodic diffeomorphisms in $\mathrm{Diff}^{\omega}_{\rho}(\mathbb{T}^2, \mu)$ which are pairwise not Kakutani-equivalent [BKpp, Theorem D]. Hereby, one can produce real-analytic non-Bernoulli diffeomorphisms with property K on manifolds of dimension greater than 4 and show the existence of uncountably many real-analytic diffeomorphisms measure-theoretically K, with the same entropy, and pairwise not isomorphic (not even Kakutani equivalent).

In connection with Kakutani equivalence one can ask if there is a real-analytic analogue of Theorem 8.2.

Problem 10.9. *Is there a real-analytic diffeomorphism T of zero entropy whose Cartesian square $T \times T$ is loosely Bernoulli?*

10.10.2 Real-Analytic AbC Constructions on Odd-Dimensional Spheres

Up to now, tori and odd-dimensional spheres are the only manifolds where real-analytic AbC constructions have been carried out successfully. An odd-dimensional sphere $\mathbb{S}^{2n-1}$ sits naturally inside $\mathbb{C}^n$. Hence, we can describe

$$\mathbb{S}^{2n-1} := \left\{(z_1,\dots,z_n) \in \mathbb{C}^n \mid |z_1|^2 + \cdots + |z_n|^2 = 1\right\}$$

with effective circle action given by the Hopf fibration

$$\phi^t\colon \mathbb{S}^{2n-1} \to \mathbb{S}^{2n-1},\ \phi^t(z_1,\dots,z_n) = \mathrm{e}^{2\pi \mathrm{i} t} \cdot (z_1,\dots,z_n).$$

Moreover, we have a large collection of well-behaved analytic flows:

$$\begin{aligned}
\xi_j^t(z_1,\dots,z_n) &= \left(z_1,\dots,z_{j-1},\mathrm{e}^{2\pi \mathrm{i} t} z_j, z_{j+1},\dots,z_n\right),\\
\tau_j^t(z_1,\dots,z_n) &= \left(z_{1,t}, z_2,\dots,z_{j-1}, z_{j,t}, z_{j+1},\dots,z_n\right),\\
\text{where } z_{1,t} &= \frac{1}{2}\left(\left(\mathrm{e}^{2\pi \mathrm{i} t}+1\right) z_1 + \left(\mathrm{e}^{2\pi \mathrm{i} t}-1\right) z_j\right),\\
z_{j,t} &= \frac{1}{2}\left(\left(\mathrm{e}^{2\pi \mathrm{i} t}-1\right) z_1 + \left(\mathrm{e}^{2\pi \mathrm{i} t}+1\right) z_j\right).
\end{aligned}$$

In particular, this collection of flows has the transitivity property (i.e., any point on the sphere can be moved to any other point by riding orbits of these flows for finite time). Constructing the conjugating diffeomorphisms for the AbC method using a finite composition of diffeomorphisms of the type

$$g_{1,i,j,q}(z_1,\dots,z_n) = \xi_i^{A\psi_{j,q}(z_1,\dots,z_n)},\quad g_{2,i,j,q}(z_1,\dots,z_n) = \xi_i^{A\chi_{j,q}(z_1,\dots,z_n)},$$

where $A > 0$, $\psi_{j,q}(z_1,\dots,z_n) = \mathrm{Re}\left(z_j^q\right)$, $\chi_{j,q}(z_1,\dots,z_n) = \mathrm{Re}\left((z_1 - z_j)^q\right)$, Fayad and Katok were able to construct uniquely ergodic real-analytic diffeomorphisms on $\mathbb{S}^{2n-1}$, $n \geq 2$ ([FK14]). Even minimal real-analytic diffeomorphisms on $\mathbb{S}^3$ have not been known before.

Remark. As explained in [FK14, Remark 1] the line of argument can be extended to compact Lie groups and some of their homogeneous spaces with an action of $\mathbb{S}^1$ by left translation.

One can aim at even more subtle ergodic properties. In [Fapp], G. Farré combines the aforementioned constructions from [FK14] and the criterion for weak mixing from [FS05] to build real-analytic diffeomorphisms of odd-dimensional spheres that are weakly mixing with respect to the volume. Additionally, we can ask about a non-standard realization result (cf. [FK04, Problem 7.4]):

Problem 10.10. *Does there exist a real-analytic volume-preserving diffeomorphism of $\mathbb{S}^{2n-1}$, $n \geq 2$, which is measurably isomorphic to an irrational rotation of the circle?*

10.10.3 Real-Analytic AbC Method on Other Manifolds

Constructing real-analytic dynamical systems on arbitrary real-analytic manifolds using the AbC method still continues to remain an intractable problem (see also [BK19, Section 6.3]). Motivated from the successful approach on odd-dimensional spheres, the key to such constructions might be finite collections of flows on the manifold that have the "transitivity" property. Then the conjugation maps h_n are constructed by taking composition of time one maps of time changes of these flows. If these flows had enough analyticity, then the diffeomorphisms constructed out of them will have enough analyticity for the AbC construction. So, it seems natural to understand what potentially could constitute a class of such "nice" manifolds supporting existence of such flows.

On manifolds where the $\mathbb{S}^1$-action is not free or if it has a boundary, for example in the case of a disk, one can consider the flow of the Hamiltonian $H(x,y) = xy(1 - x^2 - y^2)$ and observe that it has the so-called transitivity property but not enough analyticity. In fact, in this case one can check that the conjugation maps h_n constructed in this manner do not even belong to any fixed complex domain containing the manifold. As a result we do not know if there exists a zero-entropy real-analytic ergodic diffeomorphism on the disc:

Problem 10.11. *Does there exist a zero-entropy ergodic area-preserving real-analytic diffeomorphism on the disc $\mathbb{D}^2$?*

Not even a transitive example is known so far. Notice that a positive answer would negate Birkhoff's rigidity question of whether a real-analytic pseudo-rotation of angle α is analytically conjugated to the rotation S_α on the disc.

On general real-analytic principal $\mathbb{S}^1$-bundles one can construct flows and diffeomorphisms h_n together with their inverses belonging to some fixed complex domain containing the manifold. But then again we cannot guarantee that the composition T_n and its inverse can be extended to some complex domain independent of n. In this case and also the previous one, one can construct vector fields with enough analyticity (in fact, polynomial for the disk), but when one integrates such a vector field, one can guarantee that the resulting flows are real-analytic but they are not analytic enough in the sense that they do not extend to the whole of $\mathbb{C}^n$. The problem here is created by the fact that we are integrating this vector field over a non-compact domain and Picard–Lindelof in general does not guarantee that the flow exists for a fixed time interval over the whole domain. This situation does not arise when the flows or their vector fields are in a certain sense "affine" as in the case of torus (translation) and the odd-dimensional spheres (linear). So one needs some new ideas when working with manifolds where the flows are nonlinear.

Acknowledgments The author would like to take the opportunity to express his deepest gratitude to Anatole Katok for inspiring discussions and warm encouragement. He also thanks Shilpak Banerjee and an anonymous referee for helpful comments on this survey article.

References

[Ag99] O. N. Ageev: *On ergodic transformations with homogeneous spectrum*. J. Dyn. Control Syst. 5 (1999), 149–152.

[Ag05] O. N. Ageev: *The homogeneous spectrum problem in ergodic theory*. Invent. Math. 160 (2005), 417–446.

[AGZ22] P. Albers, H. Geiges, and K. Zehmisch: *Pseudorotations on the 2-disc and Reeb flows on the 3-sphere*. Ergod. Theory Dyn. Syst. 42 (2022), 402–436.

[AK70] D. V. Anosov and A. Katok: *New examples in smooth ergodic theory. Ergodic diffeomorphisms*. Trudy Moskov. Mat. Obsc. 23 (1970), 3–36.

[AFK15] A. Avila, B. Fayad, and A. Kocsard: *On manifolds supporting distributionally uniquely ergodic diffeomorphisms*. J. Differ. Geom. 99 (2015), 191–213.

[Ba17] S. Banerjee: *Non-standard real-analytic realization of some rotations of the circle*. Ergod. Theory Dyn. Syst. 37 (2017), 1369–1386.

[BK19] S. Banerjee and P. Kunde: *Real-analytic AbC constructions on the torus*. Ergod. Theory Dyn. Syst. 39 (2019), 2643–2688.

[BKpp] S. Banerjee and P. Kunde: *Real-analytic realization of Uniform Circular Systems and some applications*. Journal d'Analyse Math., 148 (2022), 399–455.

[BPS99] L. Barreira, Ya. Pesin, and J. Schmeling: *Dimension and product structure of hyperbolic measures*. Ann. Math. 149 (1999), 755–783.

[Be13] M. Benhenda: *Non-standard smooth realization of translations on the torus*. J. Mod. Dyn. 7 (2013), 329–367.

[BL93] F. Blanchard and M. Lemanczyk: *Measure-preserving diffeomorphisms with an arbitrary spectral multiplicity*. Topol. Methods Nonlinear Anal. 1 (1993): 275–294.

[Ba15] B. Bramham: *Pseudo-rotations with sufficiently Liouvillean rotation number are C^0-rigid*. Invent. Math. 199 (2015), 561–580.

[Da13] A. Danilenko: *A survey on spectral multiplicities of ergodic actions*. Ergod. Theory Dyn. Syst. 33 (2013), 81–117.

[Fapp] G. Farré: *Analytic weakly mixing diffeomorphisms on odd dimensional spheres*. Discrete and Continuous Dynamical Systems, 43, 1, (2023), 461–481.

[FH77] A. Fathi and M. R. Herman: *Existence de diffemorphismes minimaux*. Dynamical systems, Vol. I Warsaw. Astérisque 49 (1977), 37–59.

[FK04] B. Fayad and A. Katok: *Constructions in elliptic dynamics*. Ergod. Theory Dyn. Syst. 24 (2004), 1477–1520.

[FK14] B. Fayad and A. Katok: *Analytic uniquely ergodic volume preserving maps on odd spheres*. Comment. Math. Helvetici 89 (2014), 963–977.

[FK09] B. Fayad and R. Krikorian: *Herman's last geometric theorem*. Ann. Scient. Ecole. Norm. Sup. 42 (2009), 193–219.

[FK18] B. Fayad and R. Krikorian: *Some questions around quasi-periodic dynamics*. International Congress of Mathematicians 2018, Rio de Janeiro, Brazil, (2019) 1909–1932.

[FS05] B. Fayad and M. Saprykina: *Weak mixing disc and annulus diffeomorphisms with arbitrary Liouville rotation number on the boundary*. Ann. Scient. Ecole. Norm. Sup. 38 (2005), 339–364.

[FS17] B. Fayad and M. Saprykina: *KAM-instability of elliptic fixed points in arbitrary frequency*. AMS Contemporary Mathematics 692, Modern Theory of Dynamical Systems, 2017.

[FSW07] B. Fayad, M. Saprykina, and A. Windsor: *Nonstandard smooth realizations of Liouville rotations*. Ergod. Theory Dyn. Syst. 27 (2007), 1803–1818.

[FRW11] M. Foreman, D. J. Rudolph, and B. Weiss: *The conjugacy problem in ergodic theory*. Ann. Math. 173 (2011), 1529–1586.

[FW04] M. Foreman and B. Weiss: *An anti-classification theorem for ergodic measure preserving transformations*. J. Eur. Math. Soc. 6 (2004), 277–292.

[FW19a] M. Foreman and B. Weiss: *A symbolic representation of Anosov–Katok systems*. J. Anal. Math. 137 (2019), 603–661.

[FW19b] M. Foreman and B. Weiss: *From odometers to circular systems*. J. Mod. Dyn. 15 (2019), 345–423.

[FW22] M. Foreman and B. Weiss: *Measure preserving diffeomorphisms of the torus are unclassifiable*. J. Eur. Math. Soc. 24 (2022), 2605–2690.

[Fu61] H. Furstenberg: *Strict ergodicity and transformation of the torus*. Amer. J. Math. 83 (1961), 573–601.

[GK20] M. Gerber and P. Kunde: *A smooth zero-entropy diffeomorphism whose product with itself is loosely Bernoulli*. J. Anal. Math. 141 (2020), 521–583.

[GG18] V. Ginzburg and B. Gürel: *Hamiltonian pseudo-rotations of projective spaces*. Invent. Math. 214 (2018), 1081–1130.

[GK00] R. Gunesch and A. Katok: *Construction of weakly mixing diffeomorphisms preserving measurable Riemannian metric and smooth measure*. Discrete Contin. Dynam. Systems 6 (2000), 61–88.

[GK18] R. Gunesch and P. Kunde: *Weakly mixing diffeomorphisms preserving a measurable Riemannian metric with prescribed Liouvillean rotation behavior*. Discrete Contin. Dynam. Systems 38 (2018), 1615–1655.

[HK95] B. Hasselblatt and A. Katok: *Introduction to the Modern Theory of Dynamical Systems*. Cambridge University Press, Cambridge, 1995.

[HP19] L. Hernández-Corbato and F. Presas: *The conjugation method in symplectic dynamics*. Rev. Mat. Iberoam. 35 (2019), 1–22.

[J09] T. Jäger: *Linearization of conservative toral homeomorphisms*. Invent. Math. 176 (2009), 601–616.

[KS02] B. Kalinin and V. Sadovskaya: *On pointwise dimension of nonhyperbolic measures*. Ergod. Theory Dyn. Syst. 22 (2002), 1783–1801.

[Ka72] A. Katok: *Minimal diffeomorphisms on principal $\mathbb{S}^1$-bundles* (Russian). Abstracts of sixth All-Union Topological Conference, Metsniereba Tbilisi, page 63, 1972.

[Ka73] A. Katok: *Ergodic perturbations of degenerate integrable Hamiltonian systems* (Russian). Izv. Akad. Nauk SSSR Ser. Mat. 37 (1973), 539–576.

[Ka03] A. Katok: *Combinatorical Constructions in Ergodic Theory and Dynamics*. American Mathematical Society, Providence, 2003.

[Ka04] A. Katok: *Five most resistant problems in dynamics*. MSRI-Evans Lecture, Berkeley, September 2004.

[Ka17] A. Katok: *Dmitry Viktorovich Anosov: His life and mathematics*. Modern Theory of Dynamical Systems: A Tribute to Dmitry Victorovich Anosov, Contemporary Mathematics, AMS, 692, 2017, 1–21.

[KS67] A. Katok and A. Stepin: *Approximations in ergodic theory*. Uspehi Mat. Nauk 22 (1967), 81–106.

[Ko57] A. N. Kolmogorov: *Theorie generale des systemes dynamiques et mecanique classique. (French)* 1957 Proceedings of the International Congress of Mathematicians, Amsterdam, Vol. 1: pp. 315–333, Erven P. Noordhoff N.V., Groningen; North-Holland Publishing Co., Amsterdam, 1957.

[Ku16] P. Kunde: *Smooth diffeomorphisms with homogeneous spectrum and disjointness of convolutions*. J. Mod. Dyn. 10 (2016), 439–481.

[Ku17] P. Kunde: *Real-analytic weak mixing diffeomorphism preserving a measurable Riemannian metric*. Ergod. Theory Dyn. Syst. 37 (2017), 1547–1569.

[Ku18] P. Kunde: *Further dense properties of the space of circle diffeomorphisms with a Liouville rotation number*. J. Dyn. Differ. Equ. 30 (2018), 1145–1160.

[Ku20] P. Kunde: *Weakly mixing diffeomorphism with ergodic derivative extension*. Comment. Math. Helv. 95 (2020), 1–25.

[Ku65] A. G. Kushnirenko: *An upper bound of the entropy of classical dynamical systems*. Sov. Math. Dokl. 6 (1965), 360–362.

[KL95] J. Kwiatkowski and M. Lemanczyk: *On the multiplicity function of ergodic group extensions II*. Studia Math. 116 (1995), 207–214.

[LRS22] F. Le Roux and S. Seyfaddini: *The Anosov–Katok method and pseudorotations in symplectic dynamics*. J. Fixed Point Theory Appl. 24 (2022), Article 36, 1–39.

[LT78] D. A. Lind and J. P. Thouvenot: *Measure-preserving homeomorphisms of the torus represent all finite entropy ergodic transformations*. Math. Systems Theory 11 (1978), 275–282.

[Ma12] S. Matsumoto: *Dense properties of the space of the circle diffeomorphisms with a Liouville number*. Nonlinearity 25 (2012), 1495–1511.

[NS60] V. V. Nemytskii and V. V. Stepanov: *Qualitative Theory of Differential Equations*. Princeton University Press, Princeton, 1960.

[Ne32] J. von Neumann: *Zur Operatorenmethode in der klassischen Mechanik*. Ann. Math. 33 (1932), 587–642.

[Os69] V. I. Oseledets: *An automorphism with simple continuous spectrum not having the group property*. Mat. Zametki 5 (1969), 323–326.

[Pe77] Ya. Pesin: *Characteristic Lyapunov exponents and smooth ergodic theory*. Uspehi Mat. Nauk. 32 (1977), 55–112.

[Ry99] V. V. Ryzhikov: *Transformations having homogeneous spectra*. J. Dyn. Control Syst. 5 (1999), 145–148.

[Sa08] V. Sadovskaya: *Dimensional characteristics of invariant measures for circle diffeomorphisms*. Ergod. Theory Dyn. Syst. 29 (2009), 1979–1992.

[Sa03] M. Saprykina: *Analytic non-linearizable uniquely ergodic diffeomorphisms on the two-torus*. Ergod. Theory Dyn. Syst. 23 (2003), 935–955.

[St66] A. M. Stepin: *Spectral properties of ergodic dynamical systems with locally compact time*. Dokl. Akad. Nauk SSSR 169 (1966), 773–776.

[St87] A. M. Stepin: *Spectral properties of generic dynamical systems*. Math. USSR-Izv. 29 (1987), 159–192.

[WY21] J. Wang and H. Yang: *A Question of Norton–Sullivan in the Analytic Case*. International Mathematics Research Notices (2021), 17201–17219.

[WZ18] J. Wang and Z. Zhang: *The rigidity of pseudo-rotations on the two-torus and a question of Norton-Sullivan*. Geom. Funct. Anal. 28 (2018), 1487–1516.

[Win01] A. Windsor: *Minimal but not uniquely ergodic diffeomorphisms*. In: *Smooth ergodic theory and its applications*. American Mathematical Society, Providence, RI, 2001.

[Yo95] J.-C. Yoccoz: *Centralisateurs et conjugasion différentiable des difféomorphismes du cercle*. Astérisque 231 (1995), 89–242.

11

On Some Spectral Problems in Ergodic Theory

MARIUSZ LEMAŃCZYK

We survey selected advances in the spectral theory of dynamical systems, present perspectives and discuss some open problems.

11.1 Introduction

Historically, spectral invariants were the first which rigorously distinguished between intuitively different measure-theoretic dynamical systems. Over decades a vision of spectral classification of dynamical systems considerably evolved: from a belief in serious restrictions imposed by dynamics to a present belief that it is rather common to have little restrictions (compared to general and well-known classification of unitary operators on separable Hilbert spaces). This, in practice, is observed in the prevalence of constructions and, using Anatole Katok's expression (shared by the author of the article): "a weak hope for theorems." Spectral theory not only provides invariants for dynamics, but also serves as an "engine" for complicated measure-theoretic constructions of dynamical systems with interesting (sometimes "exotic") properties; perhaps [27] is the most illustrative article in this direction. Moreover, by regarding measures on product spaces as relevant Markov operators of L^2-spaces, spectral theory is closely related to the joining theory (in particular, Furstenberg's disjointness theory) of dynamical systems – the most modern tool in contemporary (abstract) ergodic theory.

Many interesting spectral questions turned out to be difficult and remain unsolved – our current understanding of the spectral classification of dynamical systems is definitely unsatisfactory. In this article we discuss classical invariants of spectral theory like the maximal spectral type and the multiplicity function, but also concentrate on selected latest achievements and discuss perspectives and open problems concerning spectral properties of flows, especially in the smooth context. The present material is based on the article [35] written jointly with Anatole Katok.

It also uses the survey articles by A.I. Danilenko [12] and myself [46]. For a nice introduction to the spectral theory of dynamical system, see the monographs [8], [50] and [56].

11.2 Koopman Operator and Spectral Theory

Assume that $(X, \mathcal{B}, \mu)$ is a probability, standard Borel space (non-atomic!), so up to a Borel isomorphism, we may think of $X = [0, 1]$ considered with the σ-algebra of Borel sets and Lebesgue measure. In particular, $L^2(X, \mathcal{B}, \mu)$ is separable. Let $T\colon (X, \mathcal{B}, \mu) \to (X, \mathcal{B}, \mu)$ be an invertible, measure-preserving transformation; we will write $T \in \mathrm{Aut}(X, \mathcal{B}, \mu)$. We can now define the associated *Koopman operator* $U_T\colon L^2(X, \mathcal{B}, \mu) \to L^2(X, \mathcal{B}, \mu)$ by the following formula:

$$U_T f := f \circ T \text{ for } f \in L^2(X, \mathcal{B}, \mu). \tag{1}$$

In the *spectral theory of dynamical systems* we study properties of U_T, that is, properties of T that are stable under **spectral isomorphism** in $\mathrm{Aut}(X, \mathcal{B}, \mu)$.[1] Note that ergodicity, weak mixing and mixing **are** spectral properties while entropy is not (see Example 4.1) and whether mixing of all orders is so, remains an open problem. Note also that $\mathrm{Aut}(X, \mathcal{B}, \mu)$ is a Polish group, when we consider the strong operator topology on it. Indeed, the group $\mathrm{Aut}(X, \mathcal{B}, \mu)$ is closed under pointwise convergence: if $U_{T_n} f \to Vf$ for each $f \in L^2(X, \mathcal{B}, \mu)$ (and V is unitary), then V is also Koopman.

Before we enter into the subject, let us briefly discuss the "Koopman" operators on L^p-spaces for $p \neq 2$, which are defined by the same formula as (1) for $f \in L^p(X, \mathcal{B}, \mu)$. On the one hand, based on the classical fact that there are not too many isometries of $L^p([0, 1])$ for $p \neq 2$, the following result shows that this problem is too strong and in fact, L^p-spectral classification of measure-preserving automorphisms is the same as metric classification:

Proposition 2.1 ([28], [46]). *Assume that* $1 \leq p < +\infty$, $p \neq 2$. *Let* $T \in \mathrm{Aut}(X, \mathcal{B}, \mu)$ *and* $S \in \mathrm{Aut}(Y, \mathcal{C}, \nu)$ *be ergodic. Assume that* $V\colon L^p(X, \mathcal{B}, \mu) \to L^p(Y, \mathcal{C}, \nu)$ *is an isometry such that* $V \circ U_T = U_S \circ V$. *Then* T *and* S *are measure-theoretically isomorphic.*[2]

On the other hand, there are interesting open problems when $p \neq 2$, from which perhaps the following is the best known:

[1] If $T_i \in \mathrm{Aut}(X_i, \mathcal{B}_i, \mu_i)$, $i = 1, 2$, then a unitary operator $V\colon L^2(X_1, \mathcal{B}_1, \mu_1) \to L^2(X_2, \mathcal{B}_2, \mu_2)$ establishes a spectral isomorphism of T_1 and T_2 if $V \circ U_{T_1} = U_{T_2} \circ V$.

[2] The proof of the preceding result uses Lamperti's theorem to obtain that $(Vf)(y) = j(y) \cdot f(Jy)$, where $J\colon Y \to X$ is non-singular. Then the equivariance yields $j(y)f(TJy) = j(Sy)f(JSy)$; take $f = 1$, to obtain that $j = const$ by the ergodicity of S. Finally, we note that the image $J_*(\nu)$ is a T-invariant, ergodic measure satisfying $J_*(\nu) \ll \mu$, and we use the ergodicity of T to conclude.

Problem 1 (Thouvenot's problem, 1986). *Is it true that for each ergodic $T \in \mathrm{Aut}(X, \mathcal{B}, \mu)$ there exists $f \in L^1(X, \mathcal{B}, \mu)$ such that $L^1(X, \mathcal{B}, \mu) = \overline{\mathrm{span}}\{f \circ T^k : k \in \mathbb{Z}\}$ for some $f \in L^1(X, \mathcal{B}, \mu)$?*[3]

Iwanik [26], in 1991, proved that for each Bernoulli automorphism T and $1 < p < +\infty$, each $n \geq 1$ and all $f_1, \dots, f_n \in L^p(X, \mathcal{B}, \mu)$, we have

$$\overline{\mathrm{span}}\{f_j \circ T^k : k \in \mathbb{Z}, j = 1, \dots, n\} \neq L^p(X, \mathcal{B}, \mu)$$

which we can interpret as "Bernoulli automorphisms have infinite multiplicity" in all L^p spaces whenever $1 < p < +\infty$.

11.3 Classification of Unitary Operators on Separable Hilbert Spaces

The spectral classification of unitary operators on separable Hilbert spaces is based on the following:

Theorem 3.1 (Herglotz's theorem). *If $U : H \to H$ is a unitary operator on a Hilbert space H and $f \in H$, then the sequence $\mathbb{Z} \ni k \mapsto \langle U^k f, f \rangle$ is positive definite and therefore there exists a unique (positive, finite, Borel) measure σ_f on $\mathbb{S}^1$ such that*

$$\widehat{\sigma}_f(k) := \int_{\mathbb{S}^1} z^k \, d\sigma_f(z) = \langle U^k f, f \rangle \text{ for each } k \in \mathbb{Z}.$$

σ_f is called the *spectral measure* of f. Let us consider the following example: $H_\sigma := L^2(\mathbb{S}^1, \sigma)$, where σ is a (positive, Borel) finite measure on the circle, and $V_\sigma : H_\sigma \to H_\sigma$, $(V_\sigma f)(z) = z f(z)$ which is clearly unitary. We then have $H_\sigma = \mathbb{Z}(\mathbf{1}) := \overline{\mathrm{span}}\{V_\sigma^k \mathbf{1} : k \in \mathbb{Z}\}$ and one says that H_σ is equal to the **cyclic** space $\mathbb{Z}(\mathbf{1})$, where $\mathbf{1}(z) = 1$. V_σ is an example of a unitary operator (defined on a separable Hilbert space) with *simple spectrum*. Returning to the abstract setting, if H is a (separable) Hilbert space and U is a unitary operator on it such that $H = \mathbb{Z}(f)$ for some $f \in H$, then the map

$$U^k f \mapsto V_{\sigma_f}^k(\mathbf{1}) = z^k, \ k \in \mathbb{Z},$$

extends to a (linear) isometry intertwining U and V_{σ_f}. If H is cyclic, we say that U has *simple spectrum*.

Theorem 3.2 (Spectral theorem). *Assume that $U : H \to H$ is unitary on a separable Hilbert space H. There exists a decomposition, called a* spectral decomposition,[4] $H = \bigoplus_{n \geq 1} \mathbb{Z}(f_n)$ *such that*

$$\sigma_{f_1} \gg \sigma_{f_2} \gg \dots \tag{2}$$

[3] The problem is also open in the category of Bernoulli automorphisms.

[4] Note that, if a Hilbert space admits such a decomposition, then it **must** be separable.

For any other spectral decomposition $H = \bigoplus_{n\geq 1} \mathbb{Z}(f_n')$, *we have*

$$\sigma_{f_n} \equiv \sigma_{f_n'} \text{ for each } n \geq 1. \tag{3}$$

The sequence (2) is called a *spectral sequence*. The type[5] of σ_{f_1} is called the *maximal spectral type* of U and is denoted by σ_U. To obtain a spectral decomposition we choose any cyclic space $\mathbb{Z}(f)$ and extend it to a maximal cyclic space, say $\mathbb{Z}(f_1)$. Such an extension exists by separability, and it has the property that its spectral measure σ_{f_1} dominates all other spectral measures. Then we pass to $\mathbb{Z}(f_1)^{\perp}$ and repeat the previous argument. Again, it is separability which yields the existence of a spectral decomposition.

Then, we define $M_U(z) := \sum_{n\geq 1} \mathbf{1}_{\text{supp}\frac{d\sigma_{f_n}}{d\sigma_{f_1}}}(z)$, which is called the *spectral multiplicity* function of U; it is defined σ_U-a.e. and takes values in $\mathbb{N} \cup \{+\infty\}$, where $\mathbb{N} = \{1, 2, \ldots\}$. Note that, when we know σ_U and M_U, then we can reconstruct a spectral sequence (2) (up to equivalence of measures). It follows that:

- Any sequence $\sigma_1 \gg \sigma_2 \ldots$ can be realized as a spectral sequence of some U (indeed, take $U = V_{\sigma_1} \oplus V_{\sigma_2} \oplus \ldots$).
- Equivalently, given σ and $M\colon \mathbb{S}^1 \to \mathbb{N}\cup\{+\infty\}$ which is measurable and defined σ-a.e., we can find U so that $\sigma = \sigma_U$ and $M = M_U$.
- Two unitary operators are (spectrally) isomorphic if and only if they have the same spectral sequence (2) (up to equivalence of spectral measures); equivalently, if and only if they have the same maximal spectral types and the same multiplicity functions.

Remark 3.3. For Koopman operators, we will use the notation σ_T, M_T instead of σ_{U_T}, M_{U_T}, respectively.

11.4 Basic Questions of Spectral Theory of Koopman Operators

One of the classical unsolved problems of spectral theory is the following:

Problem 2. *Which sequences* $\sigma_1 \gg \sigma_2 \gg \ldots$ *appear as spectral sequences of Koopman operators* $U_T|_{L^2_0(X,\mathcal{B},\mu)}$ *for an ergodic* $T \in \text{Aut}(X, \mathcal{B}, \mu)$*?*[6]

Instead of considering Problem 2, whose solution is hard to see in a reasonable perspective, we can extract the following two natural questions:

[5] By the *type* of σ we mean the class of measures equivalent (i.e., having the same null sets) to σ. In fact, unless confusion may arise, in what follows we will not distinguish between a spectral measure and its type. Note, for example, that the topological support of a measure σ is the same for all measures equivalent to σ.

[6] By an obvious reason, we consider only the U_T-invariant subspace $L^2_0(X, \mathcal{B}, \mu)$ which is the orthocomplement of constant functions. Ergodicity is assumed to avoid "cheap" tricks with repetitions of the same space, say $L^2(X, \mathcal{B}, \mu) \oplus L^2(X, \mathcal{B}, \mu)$ when, for example, we consider the spectral multiplicity problem.

Problem 3. *Which measures appear as the maximal spectral type of an ergodic automorphism?*

Problem 4. *Which subsets of* $\mathbb{N}$ *(see Remark 4.1) appear as the set* $\mathrm{essval}(M_T)$[7] *of essential values of an ergodic automorphism T? (Such sets are called* Koopman realizable.*)*

Example 4.1. (i) Assume that T is Bernoulli. Denote by $\lambda = \lambda_{\mathbb{S}^1}$ the Lebesgue measure of $\mathbb{S}^1$. Then $\lambda \equiv \lambda \equiv \ldots$ is a spectral sequence of U_T, so $\sigma_{U_T} = \sigma_T = [\lambda]$ and $\mathrm{essval}(M_T) = \{+\infty\}$. It follows that all Bernoulli automorphisms are spectrally isomorphic.
(ii) If T is an irrational rotation, then its spectral sequence is given by $\sigma \gg 0 \equiv 0 \equiv \ldots$, where $\sigma_T = \sigma = \sum_{\ell \neq 0}^{\infty} \frac{1}{2^{|\ell|}} \delta_{e^{2\pi i \ell \alpha}}$ and $\mathrm{essval}(M_T) = \{1\}$. More generally, ergodic automorphisms with discrete spectrum have simple spectrum and purely atomic maximal spectral type.

Remark 4.1. Of course, in Problem 4, we should consider subsets of $\mathbb{N} \cup \{+\infty\}$ as possible sets of essential values of the multiplicity function. However, as all constructions which follow will have singular maximal spectral types, by considering their product with a Bernoulli shift, we only add $+\infty$ as the new essential value (corresponding to Lebesgue measure); see [12].

We would like to emphasize that spectral problems are essentially zero-entropy problems because of the classical Rokhlin–Sinai theorem: *If* $T \in \mathrm{Aut}(X, \mathcal{B}, \mu)$*, then in the orthocomplement of* $L^2(\pi(T))$*, where* $\pi(T) \subset \mathcal{B}$ *stands for the Pinsker* σ*-algebra,*[8] *the spectrum is countable Lebesgue.*[9] For a generalization of this result to actions of other groups, see [17] (the result was also obtained independently by Thouvenot (unpublished)).

11.5 Maximal Spectral Type

11.5.1 Restrictions

The following result describes restrictions on the maximal spectral type of an ergodic automorphism T:

Proposition 5.1. *(i) The topological support* $\mathrm{supp}(\sigma_T)$ *of* σ_T *is* $\mathbb{S}^1$.
(ii) The measure $\widetilde{\sigma_T}(A) := \sigma_T(\overline{A})$ *is equivalent to* σ_T.
(iii) If $e^{2\pi i\alpha}$ *is an eigenvalue of* U_T*, then the measure* $\sigma_{T,\alpha}(A) := \sigma_T(e^{2\pi i\alpha} \cdot A)$ *is equivalent to* σ_T.

7 Formally, $n \in \mathrm{essval}(M_T)$ if $\sigma_T(M_T^{-1}(\{n\})) > 0$.
8 The Pinsker σ-algebra is the largest σ-algebra such that the corresponding factor automorphism has zero entropy.
9 There are zero entropy automorphism with countable Lebesgue spectrum: time-one map of horocycle flows (see Section 11.8.1), or the even factor of a Gaussian automorphism given by $\sigma \perp \lambda$ with $\sigma * \sigma \equiv \lambda$ (see, e.g., [47]).

To show the symmetry of the maximal spectral type in (ii), note that each Koopman operator preserves the space of real functions. To obtain (iii) note that, if $f \in L_0^2(X, \mathcal{B}, \mu)$ and g is an eigenfunction corresponding to an eigenvalue $c \in \mathbb{S}^1$, then $|g| = 1$, so $fg \in L^2(X, \mathcal{B}, \mu)$ and, for each $n \in \mathbb{Z}$, we have

$$\int (fg) \circ T^n \cdot \overline{(fg)}\, d\mu = c^n \int f \circ T^n \cdot \bar{f}\, d\mu,$$

so $\sigma_{fg} = \delta_c * \sigma_f$. Finally, (i) requires the following observations: $\operatorname{supp}(\sigma_T) = \{z \in \mathbb{C} \colon z \cdot Id - U_T \text{ is not a bijection}\}$ (i.e., the topological support of σ_T is equal to the Gelfand spectrum of U_T), then (because of the normality of U_T) $\{z \in \mathbb{C} \colon z \cdot Id - U_T \text{ is not a bijection}\} = \{z \in \mathbb{C} \colon z \text{ is an approximative eigenvalue}\}$, so the problem is reduced to solving "inequalities" $\|U_T(f) - zf\|_{L^2} < \epsilon$ (more precisely: for each $z \in \mathbb{S}^1$ and $\epsilon > 0$ we seek $f \in L^2(X, \mathcal{B}, \mu)$ of norm 1), which can be done using Rokhlin's lemma and considering $\sum_{i=0}^{h-1} z^i \mathbf{1}_{T^i F}$ for a Rokhlin tower $F, TF, \ldots, T^{h-1}F$ which almost fills up the whole space.

11.5.2 Exponentials of Measures as Maximal Spectral Types

If σ is a continuous symmetric measure on the circle and $U = \exp(V_\sigma) := \bigoplus_{\ell \geq 0} V_\sigma^{\odot \ell}$, then U is a Koopman operator via the classical Gaussian construction. Indeed, take a centered, real stationary Gaussian process $(X_n)_{n \in \mathbb{Z}}$ with spectral measure σ, namely $\mathbb{E}(X_n \cdot X_0) = \widehat{\sigma}(n)$ for each $n \in \mathbb{Z}$, with joint distribution μ_σ on $\mathbb{R}^{\mathbb{Z}}$ and consider the shift T on $(\mathbb{R}^{\mathbb{Z}}, \mu_\sigma)$. The following is classical:

Theorem 5.2. *We have* $\sigma_{U_{T_\sigma}} = \sum_{n \geq 1} \frac{1}{2^n} \sigma^{*n}$.[10] *In particular,* $\sum_{n \geq 1} \frac{1}{2^n} \sigma^{*n}$ *is a measure of maximal spectral type for an ergodic (in fact, weakly mixing) automorphism.*

We also recall the following (for a new proof, see [42]):

Theorem 5.3 (Girsanov's theorem, 1950s). *Either* $\operatorname{essval}(M_{U_{T_\sigma}}) = \{1\}$ *or* $M_{U_{T_\sigma}}$ *has to be unbounded.*

We can have $\operatorname{essval}(M_{U_{T_\sigma}}) = \{1, +\infty\}$ (take $\sigma \perp \lambda$ with $\sigma * \sigma \equiv \lambda$). In 2010, Danilenko and Ryzhikov [15] proved that **every** multiplicative subsemigroup of $\mathbb{N}$ has a Gaussian "realization." However, they also proved that, in general, $\operatorname{essval}(M_{T_\sigma})$ need not have such a multiplicative structure.[11] More than that, Ryzhikov in [63] showed that for an **arbitrary** $E \subset \mathbb{N}$, the set $E \cup \{+\infty\}$ is the set of essential values for the multiplicity function of a (mixing) Gaussian

[10] σ^{*n} means the convolution $\sigma * \ldots * \sigma$ (n times).

[11] For example, the double-factorials set $\{(2n-1)!! \colon n \in \mathbb{N}\}$ has a Gaussian "realization." This shows that Proposition 6.4.4, claiming the multiplicativity of multiplicity function in the Gaussian case, given in the book by Katok-Thouvenot [37] (also in [58]), is false.

automorphism. (It is unknown whether this can be done without ∞ as an essential value.)

Remark 5.4. Gaussian systems are also useful to show that conditions (i)–(iii) for a measure σ are not sufficient to realize σ as the maximal spectral type of a Koopman operator. Indeed, we can find a (continuous, with full topological support) so-called Kronecker measure σ, namely, a measure satisfying: For each $f \in L^2(\mathbb{S}^1, \sigma)$, $|f| = 1$, and $\epsilon > 0$ there exists $k \in \mathbb{Z}$ such that $\|f - z^k\|_{L^2(\sigma)} < \epsilon$. Then the famous Foiaş–Stratila theorem from 1967 [19] tells us that whenever $T \in \mathrm{Aut}(X, \mathcal{B}, \mu)$ is ergodic and a real-valued $f \in L^2(X, \mathcal{B}, \mu)$ satisfies $\sigma_f \equiv \sigma$, then the process $(f \circ T^n)_{n \in \mathbb{Z}}$ has to be Gaussian; then f will be in the first chaos of the corresponding L^2-space, each σ^{*j} is also a spectral measure and also $\sigma \perp \sum_{j \geq 2} \frac{1}{2^j} \sigma^{*j}$, so σ cannot be a measure of maximal spectral type.

Note that, essentially, **our knowledge about the maximal spectral types of Koopman operators has not changed since the 1960s**!

11.6 Spectral Multiplicity

11.6.1 Maximal Spectral Multiplicity

Because of the limited number of examples, for quite a long time there had been a belief that the maximal spectral multiplicity (that is, the essential supremum of the multiplicity function M_T) is either 1 or $+\infty$. However, in 1966 Oseledets [52] proved that: *There exists an ergodic* $T \in \mathrm{Aut}(X, \mathcal{B}, \mu)$ *such that* $1 < \mathrm{esssup}(M_T) < +\infty$. In fact, in [52] he introduced the concept of interval exchange transformations (IETs), showed that IETs of d intervals have maximal spectral multiplicity $\leq d - 1$ and used the idea of double group extension of such to create a relevant example. Refining Oseledets' construction, in 1983, Robinson (a student of Katok) proved the following:

Theorem 6.1 ([58]). *For each* $n \geq 1$, *there exists an ergodic* T *such that* $\mathrm{esssup}(M_T) = n$.[12]

Recall that for $T \in \mathrm{Aut}(X, \mathcal{B}, \mu)$, $\mathrm{essval}(M_T)$ stands for the set of essential values of the multiplicity function M_T.

General multiplicity conjecture: Each subset $E \subset \mathbb{N}$ is realizable as $\mathrm{essval}(M_T)$. (4)

Recall also that, generally, we are interested in answering the question: which sequences $\sigma_1 \gg \sigma_2 \gg \ldots$ are realizable as spectral sequences of Koopman

[12] Robinson also used double group extension constructions; moreover, Katok–Stepin's theory of periodic approximation to apply a generic type arguments has been used.

operators; cf. Problem 2. In such a sequence we have either $\equiv$ or $\gg$ (without equivalence). It is not hard to see that $n \in \mathrm{essval}(M_U)$ if and only if $\sigma_{f_n} \gg \sigma_{f_{n+1}}$ (which, at this moment, means that that we have absolute continuity without equivalence). We can hence reformulate (4) as follows:

Problem 5. *Which sequences* $(s_n)_{n\geq 1} \in \{\equiv, \gg\}^J$ *are Koopman realizable* ($J \in \mathbb{N}$ *or* $J = \mathbb{N}$)*?*

11.6.2 Essential Values of the Multiplicity Function $1 \in E$

The problem of Koopman realization of a subset E containing 1 has been solved in a series of papers by Robinson (1986), Goodson–Kwiatkowski–Lemańczyk–Liardet (1992) and Kwiatkowski (Jr.)–Lemańczyk (1995), the latter with the final result:

Theorem 6.2 ([59],[24],[45]). *Each subset* $1 \in E \subset \mathbb{N}$ *is Koopman realizable, namely there exists an ergodic* T *such that* $\mathrm{essval}(M_T) = E$.

While [59] still used the idea of double group extension, in [24], there is only one group extension step to create some symmetries. We now provide more detail. Let $T \in \mathrm{Aut}(X, \mathcal{B}, \mu)$ and G be a compact Abelian group. Assume that $\phi\colon X \to G$ a cocycle (a measurable function). We define the corresponding group extension:

$$T_\phi\colon X \times G \to X \times G,\ T_\phi(x, g) = (Tx, \phi(x) + g)$$

(considered with product measure $\mu \otimes m_G$, where m_G stands for the Haar measure of G). We have

$$L^2(X \times G, \mu \otimes m_G) = \sum_{\chi \in \widehat{G}} L^2(X, \mu) \otimes \chi, \tag{5}$$

where $\widehat{G}$ denotes the group of characters of G. Note that the (closed) subspaces in the Fourier decomposition (5) are U_{T_ϕ}-invariant: $U_{T_\phi}(L^2(X, \mu) \otimes \chi) = L^2(X, \mu) \otimes \chi$. Assume that $S \in C(T)$, that is, $S \in \mathrm{Aut}(X, \mathcal{B}, \mu)$ with $ST = TS$, and for some continuous $v \in \mathrm{Aut}(G)$ we can solve the equation

$$\phi(Sx) - v(\phi(x)) = \xi(Tx) - \xi(x)$$

for a measurable $\xi : X \to G$. It is not hard to see that the preceding equation is equivalent to saying that the (measure-preserving) $S_{\xi,v}(x, g) := (Sx, \xi(x) + v(g))$ is an element of the centralizer $C(T_\phi)$ of T_ϕ. However, the T_ϕ-invariant subspaces in (5) are (in general) no longer $S_{\xi,v}$-invariant as

$$U_{S_{\xi,v}}(L^2(X, \mu) \otimes \chi) = L^2(X, \mu) \otimes (\chi \circ v).$$

It easily follows that the lengths of the orbits of the dual automorphism $\widehat{v}$ on $\widehat{G}$ yield a lower bound on the multiplicity. Finally, note that instead of T_ϕ, we can consider its so-called *natural factor* $T_{\phi+H}$ acting on $X \times G/H$ by the formula $(x, g+H) \mapsto (Tx, \phi(x)+g+H)$ for a closed subgroup $H \subset G$. Passing to $L^2(X \times G/H)$, its spectral decomposition (5) yields *fewer* subspaces of the form $L^2(X,\mu) \otimes \chi$ and the key observation in [45] was the following algebraic fact:

Lemma 6.3 ([45]). *For each set $E \subset \mathbb{N}$ containing 1 there exists a countable (discrete) group $\mathscr{G}$, an algebraic automorphism V of $\mathscr{G}$ and a subgroup $\mathscr{H}$ such that*

$$E = L(\mathscr{G}, V, \mathscr{H}) := \{|\{V^j\chi : j \in \mathbb{Z}\} \cap \mathscr{H}| : \chi \in \mathscr{H}\}.$$

The data we need for a relevant construction are obtained from the lemma by duality (in particular, $\mathcal{H}$ is the annihilator of H in $\mathcal{G} = \widehat{G}$).

11.6.3 Essential Values of the Multiplicity Function and Rokhlin's Problem

Let us recall first the classical

Rokhlin's homogeneous spectrum problem: For each $n \geq 2$, is there an ergodic automorphism T such that $\mathrm{essval}(M_T) = \{n\}$? (6)

Essential ideas toward a solution of this problem are due to Katok who already in the mid 1980s proved that for a generic[13] automorphism T, $\mathrm{essval}(M_{T\times T}) \subset \{2,4\}$ (this was obtained via Katok's linked approximation theory). He also formulated the following:

Katok's conjecture: Generically, we have $\mathrm{essval}(M_{T^{\times n}}) = \{n, n(n-1), \ldots, n!\}$ (7)

(note that for $n = 2$ this yields a positive answer to Rokhlin's question). Katok's conjecture was proved in 1999 by Ageev [2], [3] in the general case and Ryzhikov [60] for $n = 2$ (see also Anosov's presentation in [5]). In 2005, Ageev fully answered Rokhlin's question (6):

Theorem 6.4 ([4]). *For each $n \geq 2$ there is an ergodic $T \in \mathrm{Aut}(X, \mathcal{B}, \mu)$ with homogenous spectrum of multiplicity n.*

To obtain the preceding result, Ageev exploited the new idea of using actions of non-Abelian groups and showing that a fixed "direction" automorphism fulfills our spectral requirement. The original proof of Ageev was simplified in [61] and [9]. In fact, Danilenko in 2006, combining methods leading to Theorems 6.2 and 6.4, obtained the following:

[13] Generic is meant topologically: a dense G_δ set in the Polish group $\mathrm{Aut}(X, \mathcal{B}, \mu)$.

Theorem 6.5 ([9]). *For each $n \geq 2$, and $1 \in E \subset \mathbb{N}$, there is an ergodic $T \in \mathrm{Aut}(X, \mathcal{B}, \mu)$ such that* $\mathrm{essval}(M_T) = n \cdot E$.

Remark 6.6. An interesting problem is to find Koopman realizations of the constructions from Theorem 6.5 and others in the smooth category – see Question 5 in Section 10 [12]. I was informed by A. Danilenko that the question on the smooth realization of the homogenous problem for $n = 2$ was originally asked by A. Katok. See a solution of this problem in the analytic category by Kunde [43] and Banerjee–Kunde [7].

11.6.4 Essential Values of the Multiplicity Function $2 \in E$

The realization of subsets containing 2 has been accomplished in three papers, one by Katok-Lemańczyk in 2009 and two by Danilenko in 2010 and 2012, the latter two with the final result:

Theorem 6.7 ([35], [10], [11]). *Each set $2 \in E \subset \mathbb{N}$ is Koopman realizable.*[14]

The main idea behind the proof of the preceding result is to combine techniques giving a solution of Rokhlin's problem (6) for $n = 2$ and returning to the Oseledets double group extension idea. Moreover, the technique of weak limits[15] is exploited to obtain simplicity of the spectrum of tensor product operators of the form $e^V \otimes W$ with simultaneous control of the homogenous multiplicity for the $W \otimes W$. It is also important that Danilenko [10] was able to prove Lemma 6.3 without the awkward assumption $1 \in E$.

Remark 6.8. For other sets which are Koopman realizable: $\{k, \ell, k\ell\}$, $\{k, \ell, m, k\ell, km, \ell m, k\ell m\}$, and so on: see [61], [64].

11.7 Related Results

11.7.1 Infinite-Measure-Preserving Automorphisms

In this context we study the L^2-space of a standard Borel space $(X, \mathcal{B}, \mu)$ with μ infinite (and σ-finite). We consider here Koopman operators on the **whole** $L^2(X, \mathcal{B}, \mu)$, as the constant functions are no longer integrable. Somewhat surprisingly, Danilenko and Ryzhikov proved the following result:

Theorem 7.1 ([14],[15]). *For each subset $E \subset \mathbb{N} \cup \{+\infty\}$ there exists an ergodic infinite-measure-preserving automorphism T such that* $\mathrm{essval}(M_T) = E$.[16]

[14] The papers [10] and [11] are about weakly mixing and mixing realizations, respectively.
[15] A wide use of this technique has been originated by the Moscow school.
[16] The papers [14] and [15] are about weakly mixing and mixing realizations, respectively.

11.7.2 Flows

The spectral theory for flows is similar to that for automorphisms: we assume that a flow $\mathcal{T} = (T_t)$ acting on a standard Borel probability space $(X, \mathcal{B}, \mu)$ is *measurable*, so the associated Koopman one-parameter group

$$\mathbb{R} \ni t \mapsto U_{T_t}(f) = f \circ T_t$$

is continuous for each $f \in L^2(X, \mathcal{B}, \mu)$. Bochner's theorem tells us that the function $t \mapsto \int f \circ T_t \cdot \bar{f}\, d\mu$ is positive definite, so there exists a unique (positive, finite, Borel) measure σ_f on $\widehat{\mathbb{R}} = \mathbb{R}$[17] such that

$$\widehat{\sigma_f}(t) := \int_{\mathbb{R}} e^{2\pi i t s}\, d\sigma_f(s) = \int_X f \circ T_t \cdot \bar{f}\, d\mu, \text{ for each } t \in \mathbb{R}.$$

The cyclic space $\mathbb{R}(f)$ is defined as the smallest, closed (T_t)-invariant subspace containing f, that is, $\mathbb{R}(f) = \overline{\text{span}}\{f \circ T_t \colon t \in \mathbb{R}\}$. As before, we obtain that on the unitary level all is determined by the maximal spectral type $\sigma_{\mathcal{T}}$ (a measure on $\mathbb{R}$) and the multiplicity function $M_{\mathcal{T}} \colon \mathbb{R} \to \mathbb{N} \cup \{+\infty\}$ defined $\sigma_{\mathcal{T}}$-a.e.

Interesting questions arise when we study relations between spectral properties of the flow and the Koopman operators corresponding to nonzero time automorphisms T_t. Some answers belong to folklore. As a sample, consider the multiplicity problem of time-automorphisms, in which the crucial object turns out to be the (Borel) group

$$H(\sigma_{\mathcal{T}}) = \{t \in \mathbb{R} : \delta_t * \sigma_{\mathcal{T}} \equiv \sigma_{\mathcal{T}}\}.$$

Now, if $t \in H(\sigma_{\mathcal{T}})$, then T_t has necessarily uniform countable multiplicity (and its maximal spectral type is given by the image $(e^{2\pi i t \cdot})_*(\sigma_{\mathcal{T}})$); see for example [48]. In particular, for flows having Lebesgue measure as the maximal spectral type, we obtain that each nonzero time automorphism has countable Lebesgue spectrum. One more sample of such a result is the observation from [23] that, if a flow $\mathcal{T}$ has simple singular spectrum, then a typical time automorphism has also simple spectrum.

Another result of the folklore kind is to consider induced (unitary) representations, namely when we pass from unitary operators ($\mathbb{Z}$-actions) to one-parameter groups of unitary operators ($\mathbb{R}$-actions). In fact, if $T \in \text{Aut}(X, \mathcal{B}, \mu)$ then the induced (from U_T) representation is also Koopman and given by the classical construction of the suspension flow (special flow under the constant function 1). However, inducing introduces always $1 \in \mathbb{R}$ as an eigenvalue (for the one-parameter group). We hence obtain the following:

Corollary 7.2 (e.g., [13]). *For each $E \subset \mathbb{N}$ which is Koopman realizable (i.e., $E =$* essval(T)*) there is an ergodic flow $\mathcal{T}$ such that* essval$(\mathcal{T}) = E \cup \{1\}$.

[17] The natural isomorphism between $\widehat{\mathbb{R}}$ and $\mathbb{R}$ is given by $t \mapsto e^{2\pi i t \cdot}$, for $t \in \mathbb{R}$.

However, the ergodic flow in this corollary cannot be replaced by a weakly mixing flow, and to obtain stronger (and more natural) results, we must leave the case of folklore results and rather try to adapt proofs. In [13], some methods from the $\mathbb{Z}$-action case have been adapted to $\mathbb{R}$-actions and they lead to the extensions of Theorems 6.2 and 6.7 to flows. Moreover, Danilenko and Solomko [16] answered Rokhlin's question (6) positively in the class of $\mathbb{R}$-actions.

It would be interesting to understand some continuity in possible variations of spectral properties of Koopman operators corresponding to nonzero time automorphisms. For example, in [48] it has been proved that the function

$$t \mapsto \operatorname{esssup}(M_{T_t})$$

is of second Baire class, answering a question of Thouvenot from the 1990s.

Remark 7.3. See also the article [62] by Ryzhikov about "unusual" behavior of the maximal spectral multiplicity for powers of a weakly mixing automorphism.

11.8 Other Spectral Problems Raised by Anatole Katok: Time Changes of Flows

11.8.1 Time Changes of Horocycle Flows

Let Γ be a lattice in $SL(2,\mathbb{R})$. We consider $X = SL(2,\mathbb{R})/\Gamma$ with the image μ (which is finite) of a Haar measure of $SL(2,\mathbb{R})$. The action

$$h_t(x\Gamma) := \begin{bmatrix} 1 & t \\ 0 & 1 \end{bmatrix} x\Gamma \text{ for } x \in SL(2,\mathbb{R})$$

is the corresponding *horocycle flow*. Horocycle flows have Lebesgue spectrum of infinite multiplicity (Parasyuk [54]). Given a function $v\colon X \to \mathbb{R}^+$, we can consider the corresponding time-change flow $(h_t^v)_{t\in\mathbb{R}}$.[18] Smooth time changes of horocycle flows are mixing (Kushnirenko [44], Marcus [49]).

In 2006, Katok and Thouvenot [37] formulated the following:

Katok–Thouvenot's Conjecture: All flows obtained by a sufficiently smooth time change of horocycle flows have countable Lebesgue spectrum. (8)

The maximal spectral-type Lebesgue part was proved by Forni and Ulcigrai in 2012 [20], and (the absolute continuity of the maximal spectral type) independently

[18] If (R_t) is a (measurable) flow on $(Z,\mathcal{D},\kappa)$ and $v\colon Z \to \mathbb{R}^+$, $v \geq \varepsilon_0 > 0$, is integrable, then, for κ-a.e. $z \in Z$ and all $t \in \mathbb{R}$, there is a unique solution $u = u(t,z)$ of $\int_0^u v(R_s z)\,ds = t$. Then we set $R_t^v(z) = R_{u(t,z)}(z)$ and we obtain a new flow (R_t^v) which preserves the measure $\left(\frac{v}{\int v\,d\kappa}\right)d\kappa$. Note that special flows over an automorphism T are time changes of the suspension of T.

by Tiedra de Aldecoa [66]. However, the problem of multiplicity turned out to require a new approach as the standard representation theory of $SL(2, \mathbb{R})$ (which is used to prove infinite multiplicity for horocycle flows themselves) cannot be used after a time change. Finally, countable Lebesgue spectrum for smooth time changes has been proved by Fayad, Forni and Kanigowski (in 2019) [18].

11.8.2 Time Changes of Linear Flow on the Torus

We now consider special flows over irrational rotations, so let $Tx = x + \alpha$ for $x \in \mathbb{T}$. Let $f : \mathbb{T} \to \mathbb{R}^+$ be piecewise smooth, with the sum of jumps different from zero. The weak-mixing property of the special flow T^f was already proved by von Neumann in 1932(!) [51]. In what follows, we refer to these flows as von Neumann flows. Recently, they became of particular interest after a discovery in [22] that, if α has bounded partial quotients, then they enjoy a (variation of) celebrated Ratner's property. This latter property[19] was discovered by Ratner in the 1980s in [57] for horocycle flows. In 2004, A. Katok formulated the following conjecture:

Katok's Conjecture: Von Neumann's special flows have finite multiplicity. (9)

This conjecture is still open.[20] However, Kanigowski and Solomko in 2016 [31] proved that these flows have no finite rank, so there is no "easy" way to prove possible finite maximal spectral multiplicity by seeing a kind of measurable approximation by finitely many towers.

11.9 Summarizing Comments and Questions

This section is in large part (except for the last paragraphs) taken, sometimes *in extenso*, from [35]. Returning to the multiplicity problem, we notice that in all known constructions, the appearance of nonsimple finite multiplicity spectrum is due to some symmetries:

- symmetry of double skew products with a group structure in the second extension, first noticed by Oseledets [52], originally systematically explored by Robinson and further developed by Goodson–Kwiatkowski–Lemańczyk–Liardet [24],
- the obvious symmetry of the Cartesian powers, first used in the unpublished version of Katok's notes (cf. [34]) which has circulated since mid-eighties, and brought to the final form by Ageev [3] and Ryzhikov [60] and

[19] This property roughly describes a polynomial way of divergence of nearby points which are not in the same orbit.

[20] When the roof function f is smooth, and α admits a fast approximation by rationals, then T^f has simple spectrum – this follows from Katok–Stepin in 1967 [36], as such flows enjoy fast periodic approximation.

- symmetry involving a certain non-Abelian finite extension of a cyclic group discovered by Ageev [4].

These symmetries, especially when they meet together in constructions, lead to considerable progress in our understanding of the multiplicity function of Koopman operators, but the following cases certainly need new ideas:

Problem 6. *Are the simplest unsolved cases* $\{3,4\}$, $\{3,5\}$, $\{3,7\}$ *realizable?*

Let us now come back to classical questions of spectral theory:

Problem 7. *Can the maximal spectral type be absolutely continuous but not Lebesgue?*

Problem 8. *Can the maximal spectral type* $\sigma = \sigma_T$ *for* U_T *be absolutely continuous with respect to its convolution* $\sigma * \sigma$ *but not equivalent to it?*

Recall that Kolmogorov's group property conjecture of the spectrum[21] was disproved in Katok and Stepin's article [36] in 1967, but see also Oseledets [53] and Stepin [65]. Notice that for $\sigma = \sigma_T$ there are three known possibilities:

- σ is equivalent to $\sigma * \sigma$, as for Lebesgue spectrum or for Gaussian systems;
- σ and $\sigma * \sigma$ are mutually singular, as for a generic measure preserving transformation T;
- σ and $\sigma * \sigma$ have a common part, but neither is absolutely continuous with respect to the other, as for $T \times T$ for a generic T (as $\sigma_{T\times T} = \sigma_T + \sigma_T^{*2}$, $\sigma_{T\times T} * \sigma_{T\times T} = \sigma_T^{*2} + \sigma_T^{*3} + \sigma_T^{*4}$, and typically all convolutions of the maximal spectral type are mutually singular).

Note that Problem 8 might be related to Problem 7 as for each absolutely continuous measure σ, σ^{*n} is equivalent to Lebesgue measure for n large enough.

Returning to Proposition 5.1 (i), it is natural to ask the following:

Problem 9. *Is it true that all spectral types in a spectral sequence of a measure preserving transformation with continuous spectrum have full topological support?*

Frączek (unpublished) answered this question positively for some group extensions of rotations.

The following problem is closely related to the famous Banach problem on the existence of an automorphism with simple Lebesgue spectrum.

Problem 10. *Does there exist an ergodic measure-preserving transformation whose maximal spectral type is absolutely continuous but whose spectrum is not Lebesgue with countable multiplicity?*

21 Kolmogorov's group property for $T \in \mathrm{Aut}(X, \mathcal{B}, \mu)$ states that $\sigma_T \ll \sigma_T * \sigma_T$.

The difference from Problem 7 is that it is conceivable that the maximal spectral type is Lebesgue while not all others are. More specifically, one can ask whether Lebesgue maximal spectral type implies homogenous spectrum.

An account of research around the Banach problem was published in 2008 [46]: the main achievement consisted in exhibiting constructions with all possible **even** multiplicities of Lebesgue component in the spectrum. Not much has changed since then. Recall also that Guenais [25] proved that there is a generalized Morse sequence for which the associated dynamical system (subshift) has a Lebesgue component[22] if and only if there exists a sequence of L^1-ultraflat[23] trigonometric polynomials $P_{n_k}(z) = \sum_{j=-n_k}^{n_k} a_j^{(k)} z^j$ ($|z| = 1$) with $a_j^{(n)} = \pm 1$. Recently, on arXiv there appeared the paper [6] in which the authors prove the Littlewood conjecture on the existence of flat (in the uniform sense) trigonometric polynomials: there are $c, C > 0$ such that for each $n \geq 1$ we can find a polynomial $Q_n(z) = \sum_{j=-n}^{n} b_j^{(n)} z^j$ with $b_j^{(n)} = \pm 1$, such that $c\sqrt{n} \leq |Q_n(z)| \leq C\sqrt{n}$ for each $|z| = 1$. This remarkable progress does not seem, however, to imply the existence of a sequence of L^1-ultraflat trigonometric polynomials with coefficients ± 1.

While there was no substantial progress on the Banach problem in the framework of Koopman operators, there were advances on a similar problem for Koopman flows. Prikhodko in [55] published a construction of a rank-one flow having Lebesgue spectrum. As rank one implies simplicity of the spectrum, the result yields a solution of the Banach problem for Koopman flows. To do this, he proved the following version of Littlewood conjecture: *For all* $[a, b] \subset \mathbb{R}^+$ *and* $n \geq 1$, *one can find polynomials* $P_n(t) = \sum_{j=0}^{n-1} e^{2\pi i w_j^{(n)} t}$ *for some real* $w_j^{(n)}$, *such that* $\|P_n\|_{L^1([a,b])}/\sqrt{n} \to 1$. However, although the article was published in 2013, some details were rather cryptic and no further developments of methods/results/ideas from [55] have appeared so far.[24] Of course, as we have noticed in Section 11.7.2, a solution of the Banach problem for flows does not imply it for automorphisms (as all nonzero time automorphisms have countable Lebesgue spectrum). **It seems to be an important task for the ergodic community to convincingly explain the status of the Banach problem.**[25]

22 All subshifts given by generalized Morse sequences have simple spectrum (and they have a discrete component in the spectrum). Therefore, if ultraflat polynomials with coefficients ± 1 do exist, then there are also dynamical systems with Lebesgue component of multiplicity 1.

23 That is, $\|P_{n_k}\|_{L^1}/\sqrt{n_k} \to 1$ when $k \to \infty$.

24 This is to be compared with a solution of Rokhlin's problem for $n = 2$, after which Anosov's book [5] appeared and explained all details.

25 It would also be highly desirable to explain the status of [1] claiming solution of the original Banach problem, namely of the existence of an ergodic (conservative) **infinite**-measure-preserving automorphism whose Koopman operator has simple Lebesgue spectrum, posted firstly on arXiv in 2015.

Lebesgue spectrum (plus a discrete component) is characteristic for algebraic systems (affine transformations, nil-systems, horocycle flows, or more generally, unipotent actions), we have no continuous singular part in the spectrum. Whether one can expect some restrictions on spectral types by looking at smooth systems in dimension 2, say, flows on surfaces or billiards, is definitely less clear. Typically, these dynamical systems have special representations as special flows over irrational rotations or (more generally) over interval exchange transformations. Their roof functions have singularities which correspond to singularities of the original flows on surfaces. For the simplest case, when the roof function is of bounded variation, mixing is excluded, as shown by Kochergin [39] and Katok [33]. In fact, when the base is an irrational rotation, the flows are **spectrally disjoint** with all mixing flows as shown by Frączek–Lemańczyk [21]. The absence of mixing persists for flows with symmetric logarithmic singularities over a typical IET as proved by Ulcigrai [67].[26] On the other hand, mixing is "typical" in the asymmetric case, see for example Kochergin [41] and Khanin-Sinai [38]. Also, mixing appears in case of so-called power singularities [40]. While it was rather expected (including by the author of this article) that the spectrum in dimension 2 is singular, Fayad–Forni–Kanigowski [18] showed that for some class of special flows with power singularities (and over irrational rotations) the spectrum is countable Lebesgue. One can of course speculate that flows with such singularities are simply either horocycle flows or their smooth time changes in "different coordinates,"[27] but it follows from [29] and [32] that special flows with power singularities are not isomorphic to horocycle flows.[28] Moreover, in the recent paper of Kanigowski–Lemanczyk–Ulcigrai [30] it was shown that (some class of) flows under a function with logarithmic singularities are disjoint with horocycle flows. **A full understanding of the spectral theory of smooth flows on surfaces seems to be one of the main challenges in ergodic theory in forthcoming years.**

11.10 Acknowledgments

Research of the author was supported by a Narodowe Centrum Nauki grant UMO-2019/33/B/ST1/00364.

I would like to thank A. Danilenko, K. Frączek, A. Kanigowski, J. Kułaga-Przymus, V. V. Ryzhikov and Y. Tomilov for useful remarks and comments on the first version of the manuscript.

26 Such flows come from some multi-Hamiltonian flows on surfaces of genus ≥ 2.

27 The general theory of loosely Bernoulli yields that any horocycle flow has a special representation over an irrational rotation under a continuous function.

28 It is the so-called slow entropy that distinguishes them.

References

[1] H. El Abdalaoui, *Ergodic Banach problem on simple Lebesgue spectrum and flat polynomials*, arXiv:1508.06439v4.

[2] O. N. Ageev, *On ergodic transformations with homogenous spectrum*, J. Dyn. Control Syst. **5** (1999), 149–152, 619–622; translation in Math. Notes **65** (1999), 516–519.

[3] O. N. Ageev, *On the spectrum of Cartesian powers of classical automorphisms*, (Russian) Mat. Zametki **68** (2000), 643–647; English translation in Math. Notes **68** (2000), no. 5–6, 547–551.

[4] O. N. Ageev, *The homogeneous spectrum problem in ergodic theory*, Invent. Math. **160** (2005), 417–446.

[5] D. W. Anosov, *On Spectral Multiplicities in Ergodic Theory*, in Current Problems in Mathematics 3, Rossiĭskaya Akademiya Nauk, MatematicheskiĭInstitut im. V. A. Steklova, Moscow, 2003. (Available electronically at www.mi.ras.ru/spm/pdf/003.pdf.)

[6] P. Balister, B. Bollobas, R. Morris, J. Suhasrabudhe, M. Tiba, *Flat Littlewood polynomials exist*, Ann. of Math. (2) 192 (2020), no. 3, 977–1004.

[7] S. Banerjee, P. Kunde, *Real-analytic diffeomorphisms with homogeneous spectrum and disjointness of convolutions*, arXiv:1711.02537.

[8] I. P. Cornfeld, S. V. Fomin, Y. G. Sinai, *Ergodic Theory*, Springer, 1982.

[9] A. I. Danilenko, *Explicit solution of Rokhlin's problem on homogenous spectrum and applications*, Ergod. Theory Dyn. Syst., **26** (2006), 1467–1490.

[10] A. I. Danilenko, *On new spectral multiplicities for ergodic maps*, Studia Math. **197** (2010), 57–68.

[11] A. I. Danilenko, *New spectral multiplicities for mixing transformations*, Ergod. Theory Dyn. Syst. **32** (2012), 517–534.

[12] A. I. Danilenko, *A survey on spectral multiplicities of ergodic actions*, Ergod. Theory Dyn. Syst. **33** (2013), 81–117.

[13] A. I. Danilenko, M. Lemańczyk, *Spectral multiplicities for ergodic flows*, Discrete Contin. Dyn. Syst. **33** (2013), 4271–4289.

[14] A. I. Danilenko, V. V. Ryzhikov, *Spectral multiplicities for infinite measure preserving transformations*, Funct. Anal. Appl. **44** (2010), 161–170.

[15] A. I. Danilenko, V. V. Ryzhikov, *Mixing constructions with infinite invariant measure and spectral multiplicities*, Ergod. Theory Dyn. Syst. **31** (2011), 853–873.

[16] A. I. Danilenko, A. V. Solomko, *Ergodic Abelian Actions with Homogenous Spectrum*, (Contemporary Math. 532), American Mathematical Society, Providence, RI, 2010, pp. 137–148.

[17] A. Dooley, V.Y. Golodets, *The spectrum of completely positive entropy actions of countable amenable groups*, J. Funct. Anal. **196** (2002), 1–18.

[18] B. Fayad, G. Forni, A. Kanigowski, *Lebesgue spectrum of countable multiplicity for conservative flows on the torus*, J. Amer. Math. Soc. 34 (2021), no. 3, 747–813.

[19] C. Foiaş, S. Stratila, *Ensembles de Kronecker dans la théorie ergodique*, Comptes rendus de l'Académie des Sciences, sr. A–B **267** (1968), A166–A168.

[20] G. Forni, C. Ulcigrai, *Time-changes of horocycle flows*, J. Mod. Dyn. **6** (2012), 251–273.

[21] K. Frączek, M. Lemańczyk, *A class of special flows over irrational rotations which is disjoint from mixing flows*, Ergod. Theory Dyn. Syst. **24** (2004), 1083–1095.

[22] K. Frączek, M. Lemańczyk, *On mild mixing of special flows over irrational rotations under piecewise smooth maps*, Ergod. Theory Dyn. Syst. **26** (2006), 719–738.

[23] K. Frączek, M. Lemańczyk, *On the self-similarity problem for flows*, Proc. London Math. Soc. **99** (2009), 658–696.

[24] G. R. Goodson, J. Kwiatkowski, M. Lemańczyk, P. Liardet, *On the multiplicity function of ergodic group extensions of rotations*, Studia Math. **102** (1992), 157–174.

[25] M. Guenais, *Morse cocycles and simple Lebesgue spectrum*, Ergod. Theory Dyn. Syst. **19** (1999), 437–446.

[26] A. Iwanik, *The problem of L^p-simple spectrum for ergodic group automorphisms*, Bull. Soc. Math. France **119** (1991), 91–96.

[27] A. del Junco, M. Lemańczyk, *Generic spectral properties of measure-preserving maps and applications*, Proc. Amer. Math. Soc. **115** (1992), 725–736.

[28] A. G. Kachurovskii, *A property of operators generated by ergodic automorphisms*, (Russian) Optimizatsiya **47** (64) (1990), 122–125.

[29] A. Kanigowski, *Slow entropy for smooth flows on surfaces*, Israel J. Math. **226** (2018), 535–577.

[30] A. Kanigowski, M. Lemańczyk, C. Ulcigrai, *On disjointness of some parabolic flows*, Invent. Math. 221 (2020), no. 1, 1–111.

[31] A. Kanigowski, A. V. Solomko, *On the rank of von Neumann special flows*, Ergod. Theory Dyn. Syst. **38** (2018), 2245–2256.

[32] A. Kanigowski, K. Vinhage, D. Wei, *Slow entropy for some parabolic flows*, Comm. Math. Physics. **370** (2019), 449–474.

[33] A. Katok, *Interval exchange transformations and some special flows are not mixing*, Israel J. Math. **35** (1980), 301–310.

[34] A. Katok, *Combinatorial constructions in ergodic theory and dynamics*, AMS, University Lecture Series, **30**, 2003, iv+121pp.

[35] A. Katok, M. Lemańczyk, *Some new cases of realization of spectral multiplicity function for ergodic transformations*, Fundam. Math. **206** (2009), 185–215 (special volume dedicated to M. Misiurewicz).

[36] A. B. Katok, A. M. Stepin, *Approximations in ergodic theory*, (Russian) Uspehi Mat. Nauk **22** (1967) (137), 81–106.

[37] A. Katok, J.-P. Thouvenot, *Spectral properties and combinatorial constructions in ergodic theory*, Handbook of Dynamical Systems, Vol. 1B, 649–743, Elsevier, 2006.

[38] K. M. Khanin, Ya. G. Sinai, *Mixing of some classes of special flows over rotations of the circle*, Funct. Anal. Appl. **26** (1992), 155–169.

[39] A. V. Kočergin, *On the absence of mixing in special flows over the rotation of a circle and in flows on a two-dimensional torus*, (Russian) Dokl. Akad. Nauk SSSR **205** (1972), 949–952.

[40] A. V. Kochergin, *Mixing in special flows over rearrangment of segments and in smooth flows on surfaces*, Math-USSR Acad. Sci. **25** (1975), 441–469.

[41] A. V. Kochergin, *A mixing special flow over a rotation of the circle with an almost Lipschitz function*, Sb. Math. **193** (2002), 359–385.

[42] J. Kułaga, F. Parreau, *Disjointness properties for Cartesian products of weakly mixing systems*, Colloq. Math. **128** (2012), 153–177.

[43] P. Kunde, *Smooth diffeomorphisms with homogeneous spectrum and disjointness of convolutions*, J. Mod. Dyn. **10** (2016), 439–481.

[44] A. G. Kushnirenko, *Spectral properties of some dynamical systems with polynomial divergence of orbits*, Vestnik Moskovskogo Univ. **1–3** (1974), 101–108.

[45] J. Kwiatkowski (Jr.), M. Lemańczyk, *On the multiplicity function of ergodic group extensions. II*, Studia Math. **116** (1995), 207–215.

[46] M. Lemańczyk, *Spectral Theory of Dynamical Systems*, Encyclopedia of Complexity and System Science, Springer (2009), 8554–8575.

[47] M. Lemańczyk, F. Parreau, J.-P. Thouvenot, *Gaussian automorphisms whose ergodic self-joinings are Gaussian*, Fundam. Math. **164** (2000), 253–293.
[48] M. Lemańczyk and F. Parreau, *Special flows over irrational rotation with simple convolution property*, unpublished preprint from 2008, www-users.mat.umk.pl/mlem /publications.php.
[49] B. Markus, *Ergodic properties of horocycle flows for surfaces of negative curvature*, Ann. Math. (2), **105** (1977), 81–105.
[50] M. G. Nadkarni. *Spectral theory of dynamical systems*, Berlin, 1998.
[51] J. von Neumann, *Zur Operatorenmethode in der Klassichen Mechanik*, Ann. Math. **33** (1932), 587–642.
[52] V. I. Oseledets, *On the spectrum of ergodic automorphisms*, Sov. Math. Dokl. **7**, (1966), 776–779.
[53] V. I. Oseledets, *Automorphism with simple and continuous spectrum without the group property* (Russian) Mat. Zametky **5** (1969), 323–326.
[54] O. S. Parasyuk, *Flows of horocycles on surfaces of constant negative curvature*, Uspekhi Mat. Nauk (N.S.) **8** (1953), 125–126 (In Russian).
[55] A. A. Prikhodko, *Littlewood polynomials and their applications to the spectral theory of dynamical systems*, Mat. Sb. **204** (2013), 135–160; English translation in Sb. Math. **204** (2013).
[56] W. Parry, *Topics in Ergodic Theory*, Cambridge Tracts in Mathematics, 75, Cambridge University Press, 1981.
[57] M. Ratner, *Horocycle flows, joinings and rigidity of products*, Ann. Math. **118** (1983), 277–313.
[58] E. A. Robinson, *Ergodic measure-preserving transformations with arbitrary finite multiplicities*, Invent. Math. **72** (1983), 299–314.
[59] E. A. Robinson, *Transformations with highly nonhomogenous spectrum of finite multiplicity*, Israel J. Math. **56** (1986), 75–88.
[60] V. V. Ryzhikov, *Transformations having homogeneous spectra*, J. Dyn. Control Syst. **5** (1999), 145–148.
[61] V. V. Ryzhikov, *Spectral multiplicities and asymptotic operator properties of actions with invariant measure*, Sb. Math. **200** (2009), 1833–1845.
[62] V. V. Ryzhikov, *Spectral multiplicities for powers of weakly mixing transformations*, Mat. Sb. **203** (2012), 149–160.
[63] V. V. Ryzhikov, *On spectral multiplicities of Gaussian actions*, arXiv:1406.3321.
[64] A. V. Solomko, *New spectral multiplicities for ergodici actions*, Studia Math. **208** (2012), 229–247.
[65] A. M. Stepin, *Spectral properties of generic dynamical systems*, (Russian) Izv. Akad. Nauk SSSR Ser. Mat. **50** (1986), 801–834.
[66] R. Tiedra de Aldecoa, *Spectral analysis of time changes for horocycle flows*, J. Mod. Dyn. **6** (2012), 275–285.
[67] C. Ulcigrai, *Absence of mixing in area-preserving flows on surfaces*, Ann. Math. (2) **173** (2011), 1743–1778.

12

Area-Preserving Surface Diffeomorphisms with Polynomial Decay of Correlations Are Ubiquitous

YAKOV PESIN, SAMUEL SENTI, AND FARRUH SHAHIDI*

Dedicated to the memory of Anatole Katok, our friend and mentor

We show that any smooth compact connected and oriented surface admits an area-preserving $C^{1+\beta}$ diffeomorphism with nonzero Lyapunov exponents which is Bernoulli and has polynomial decay of correlations. We establish both upper and lower polynomial bounds on correlations. In addition, we show that this diffeomorphism satisfies the Central Limit Theorem and has the Large Deviation Property. Finally, we show that the diffeomorphism we constructed possesses a unique hyperbolic Bernoulli measure of maximal entropy with respect to which it has exponential decay of correlations.

12.1 Introduction

A classical problem in smooth dynamics known as the *smooth realization problem* asks whether there is a diffeomorphism f of a compact smooth manifold M which has a prescribed collection of ergodic properties with respect to a *natural* invariant measure μ such as the Riemannian volume (or a more general smooth measure, i.e., a measure that is equivalent to volume). Other interesting measures to consider include the measure of maximal entropy. A yet more interesting but substantially more difficult version of the smooth realization problem is to construct a volume-preserving diffeomorphism f with prescribed ergodic properties on any given smooth manifold M. Starting with the basic ergodic property – ergodicity – Anosov and Katok [1] constructed an example of a volume-preserving *ergodic* C^∞ map with some additional metric properties. Katok [12] gave an example of area-preserving C^∞ diffeomorphism with nonzero Lyapunov exponents on any surface which is *Bernoulli* (see the definitions in the next section). Later Brin, Feldman, and Katok [4] and then Brin [3] extended this result by constructing a volume-preserving C^∞ diffeomorphism, which is Bernoulli, on any Riemannian manifold

* Y. P. is supported by the NSF grant DMS-2153053.

of dimension $\geq$ 5. In this example the map has all but one nonzero Lyapunov exponents. Finally, Dolgopyat and Pesin [6] constructed a volume-preserving C^∞ Bernoulli diffeomorphism with *nonzero Lyapunov exponents* on any Riemannian manifold of dimension ≥ 2.

It is natural to ask if a compact smooth manifold admits a volume-preserving Bernoulli diffeomorphism with nonzero Lyapunov exponents that enjoys other important statistical properties such as exponential or polynomial decay of correlations (that is, rate of mixing), the Central Limit Theorem, and the Large Deviations property (all three with respect to a *natural* class of observables, e.g., functions which are Hölder continuous).

In one-dimensional dynamics the famous Mauneville–Pomeau map [19] (with some modifications) provide some examples of a map with an indifferent fixed point preserving a measure which is absolutely continuous with respect to the (one-dimensional) Lebesgue measure. With respect to this measure the decay of correlations is polynomial, the Central Limit Theorem is satisfied, and the map has Large Deviation (with respect to the class of Hölder continuous observables; see [8, 10, 15]).

In the two-dimensional case several examples have been constructed of area-preserving maps on the 2-torus with polynomial decay of correlations (and in some cases with sharp polynomial lower and upper bounds); see [5, 7, 16, 15]. Also, starting from the work of Young, [24, 25], some general techniques for obtaining polynomial decay of correlations for maps admitting the Young tower have been developed by Gouëzel [8] and Sarig [22]; see also [18, 23]. Also in [11] polynomial lower and upper bounds were obtained for almost Anosov diffeomorphisms. This class of diffeomorphisms was introduced by Hu in [10] and they preserve the Sinai–Ruelle–Bowen measure (which may not be volume) and the local dynamics of these maps is quite different than that of the Katok map.

In the present chapter we show that any surface admits an area-preserving $C^{1+\beta}$ diffeomorphism with nonzero Lyapunov exponents which is Bernoulli and has polynomial decay of correlations – more precisely, it allows polynomial lower and upper bounds. It also satisfies the Central Limit Theorem and has Large Deviation.

Interestingly enough, the map we construct also has the unique measure of maximal entropy with exponential decay of correlations. Thus we show that any surface allows a $C^{1+\beta}$ diffeomorphism with the unique measure of maximal entropy with respect to which it has nonzero Lyapunov exponents, is Bernoulli, has exponential decay of correlations, and satisfies the Central Limit Theorem.

While our proof follows the basic scheme of Katok construction in [12], we have to make many substantial changes, which we outline here. Starting with a linear automorphism of the 2-torus T^2 which has four fixed points, we slow down trajectories in sufficiently small neighborhoods of these points and then *correct* the resulting map to obtain an area-preserving diffeomorphism f_{T^2}. We stress that while in the Katok construction the slowdown function can be chosen to be infinitely and arbitrarily flat at 0 (guaranteeing that f_{T^2} is C^∞), to ensure that f_{T^2} has polynomial decay of correlations, we have to choose the slowdown function to be polynomial at 0.[1] This results in f_{T^2} to be of class $C^{2+2\kappa}$. It also has nonzero Lyapunov exponents and is Bernoulli.

To show that f_{T^2} has polynomial decay of correlations we represent this map as a Young diffeomorphism and study the symbolic map on the corresponding Young tower (see Section 12.5). This map preserves the measure that is the lift of the area (see, [23, 24]). The decay of correlations of this symbolic map has been studied extensively (see for, example, [8, 18, 22, 23, 24, 25]) and is tied to the decay of the tail of the return time (see Proposition 10.2). Thus, to establish polynomial upper and lower bounds on the decay of correlations we need to obtain both upper and lower bounds on decay of the tail. This requires a deep understanding of the behavior of trajectories in the slowdown domain and is done in Sections 12.6–12.8 which constitute the technically most difficult part of the work.

Our next step is to carry over the map of the torus to a map on a given surface. To achieve this we follow the approach in [12] and obtain a $C^{2+2\kappa}$ area-preserving Bernoulli diffeomorphism f_{D^2} of the two-dimensional disk with nonzero Lyapunov exponents which is identity on the boundary of the disk. In the original Katok's construction the map f_{D^2} is C^∞ and is infinitely and arbitrarily flat at the boundary of the disk,[2] which allows one to apply some standard results to carry over f_{D^2} to a diffeomorphism f_M of a surface M. In our case f_{D^2} is only finitely flat at the boundary of the disk and we develop a specific construction of a diffeomorphism from the interior of the disk onto an open simply connected and dense subset of M, which extends to a homeomorphism from the closed disk onto M and is area preserving. This diffeomorphism moves f_{D^2} to an area-preserving Bernoulli diffeomorphism f_M of the surface with nonzero Lyapunov exponents; see Section 12.9.

Our next step is to use the conjugacy map and a representation of f_{T^2} as a Young diffeomorphism to obtain a similar representation for f_M; see Section 12.10. Now, to obtain an upper polynomial bound for decay of correlation we use the results in [25] and [18] to choose an appropriate class of observables, which includes all Hölder continuous functions on the surface. To obtain a lower bound we use the result in [23] (which is a hyperbolic version of the result in [8]) and the class of Hölder continuous observables on the surface which vanish inside small neighborhoods of the fixed points; see Section 12.10.

We stress that the exponents in our polynomial lower and upper bounds are different. This is a result of our estimates on the behavior of trajectories of the Katok map in the slowdown domain (see Section 12.6) and may be an artifact of

[1] We believe that the rate of decay of correlations of f_{T^2} is determined by the rate of decay of the slowdown function at 0.

[2] This is due to the fact that the slowdown function can be chosen arbitrarily flat at 0.

our techniques. However, there is a particular class of observables for which our method gives the same exponent in the polynomial lower and upper bounds; see Statement 3(b)(ii) of the Main Theorem 3.1.

Representing the map f_M as a Young tower also allows us to establish the Central Limit Theorem using results in [8, 14] as well as Polynomial Large Deviation using results in [17].

The chapter is organized as follows. After we provide some definitions in the next section we state our Main Theorem in Section 12.3. In Section 12.4 we construct the map f_{T^2} on the 2-torus and state some of its properties, including its class of smoothness. In Section 12.5 we recall the definition of Young diffeomorphisms and describe a representation of f_{T^2} as a Young diffeomorphism. The proof of the main result, Theorem 3.1, occupies Sections 12.6 through 12.10. In Section 12.6, we prove some technical results that establish new crucial properties of the slow-down map. In Sections 12.7 and 12.8 we obtain polynomial respectively lower and upper bounds on the tail of the return time for the Katok map f_{T^2}. In Section 12.9 we show how to carry over the Katok map of the torus to a diffeomorphism of a given surface. Finally, in Section 12.10 we complete the proof of the main result.

Acknowledgments. The first and the second authors would like to thank Bernoulli Center (CIB, Lausanne, Switzerland) where part of the work was done for their hospitality. S. S. was supported by the CNPQ grant. The authors also would like to thank the referee for valuable comments and suggestions.

12.2 Definitions and Notations

Let X be a measurable space and $T: X \to X$ a measurable invertible transformation preserving a measure μ. For the reader's convenience we recall the definitions of some properties of the map which are of interest to us in the chapter.

12.2.1 The Bernoulli Property

We say that (T, μ) has the *Bernoulli property* if it is metrically isomorphic to the Bernoulli shift (σ, κ) associated to some Lebesgue space (Y, ν), so that ν is metrically isomorphic to the Lebesgue measure on an interval together with at most countably many atoms and κ is given as the direct product of $\mathbb{Z}$ copies of ν on $Y^{\mathbb{Z}}$.

12.2.2 Decay of Correlations

Let $\mathcal{H}_1$ and $\mathcal{H}_2$ be two classes of real-valued functions on X called *observables*. For $h_1 \in \mathcal{H}_1$ and $h_2 \in \mathcal{H}_2$, define the correlation function

$$\mathrm{Cor}_n(h_1, h_2) := \int h_1(T^n(x))h_2(x)\, d\mu - \int h_1(x)\, d\mu \int h_2(x)\, d\mu.$$

We say that T has *polynomial decay of correlations* (more precisely, *polynomial upper bound on correlations*) with respect to classes $\mathcal{H}_1$ and $\mathcal{H}_2$ if there exists $\gamma_1 > 0$ such that, for any $h_1 \in \mathcal{H}_1$, $h_2 \in \mathcal{H}_2$, and any $n > 0$,

$$|\mathrm{Cor}_n(h_1, h_2)| \leq Cn^{-\gamma_1},$$

where $C = C(h_1, h_2) > 0$ is a constant.

We say that T admits a *polynomial lower bound on correlations* with respect to classes $\mathcal{H}_1$ and $\mathcal{H}_2$ of observables if there exists $\gamma_2 > 0$ such that, for any $h_1 \in \mathcal{H}_1$, $h_2 \in \mathcal{H}_2$, and any $n > 0$,

$$|\mathrm{Cor}_n(h_1, h_2)| \geq C'n^{-\gamma_2},$$

where $C' = C'(h_1, h_2) > 0$ is a constant.

We say that T has *exponential decay of correlations* with respect to classes $\mathcal{H}_1$ and $\mathcal{H}_2$ if there exists $\gamma_3 > 0$ such that, for any $h_1 \in \mathcal{H}_1$, $h_2 \in \mathcal{H}_2$, and any $n > 0$,

$$|\mathrm{Cor}_n(h_1, h_2)| \leq C''e^{-\gamma_3 n},$$

where $C'' = C''(h_1, h_2) > 0$ is a constant.

12.2.3 The Central Limit Theorem

We say that T satisfies the *Central Limit Theorem (CLT)* with respect to a class $\mathcal{H}$ of observables on X if there exists $\sigma > 0$ such that for any $h \in \mathcal{H}$ with $\int h\, d\mu = 0$ the sum

$$\frac{1}{\sqrt{n}} \sum_{i=0}^{n-1} h(f^i(x))$$

converges in law to a normal distribution $N(0, \sigma)$.

12.2.4 Large Deviation

We say that T has *Polynomial Large Deviation* with respect to a class $\mathcal{H}$ of observables on X if there is $\beta > 0$ such that for any $h \in \mathcal{H}$, any $\varepsilon > 0$, and any sufficiently large $n > 0$,

$$\mu\Big(\Big|\frac{1}{n}\sum_{i=0}^{n-1} h(T^i(x)) - \int h\Big| > \varepsilon\Big) < Kn^{-\beta},$$

where $K = K(\varepsilon, \beta, h) > 1$ is a constant.

12.2.5 Lyapunov Exponents

Let $f\colon M \to M$ be a diffeomorphism of a compact smooth Riemannian manifold M. Given a point $x \in M$ and a vector $v \in T_xM$, the number

$$\chi(x,v) := \limsup_{n\to\infty} \frac{1}{n} \log \|df_x^n v\|$$

is called the *Lyapunov exponents* of v at x. One can show that for every $x \in M$ the function $\chi(x,\cdot)$ takes on finitely many values which we denote by $\chi_1(x) \le \cdots \le \chi_p(x)$, where $p = \dim M$. The functions $\chi_i(x)$, $i = 1,\dots,p$ are Borel measurable and f-invariant.

If μ is an f-invariant measure, then for μ-almost every $x \in M$ and any $v \in T_xM$,

$$\chi(x,v) = \lim_{n\to\infty} \frac{1}{n} \log \|df_x^n v\|.$$

We say that f has *nonzero Lyapunov exponents* with respect to μ or that μ is *hyperbolic* if for μ-almost every x we have that $\chi_i(x) \neq 0$, $i = 1,\dots,p$ and that $\chi_1(x) < 0$ while $\chi_p(x) > 0$.

Note that if μ is ergodic, then $\chi_i(x)$ is a constant for all $i = 1,\dots,p$ and μ-almost every $x \in M$, which we denote by $\chi_i(\mu)$.

12.3 Main Results

Let M be a smooth compact connected oriented surface with area m. Without loss of generality we assume that $m(M) = 1$. Given $\rho > 0$, let $C^\rho := C^\rho(M)$ be the class of all Hölder continuous functions on M with exponent ρ.

Consider a nested sequence of subsets $\{M_j\}$ that exhausts M that is $M_1 \subset M_2 \subset \cdots \subset M$ and $\bigcup_{j\ge1} M_j = M$. Given such a sequence, let $\mathcal{G} = \mathcal{G}(\{M_j\})$ be the class of observables $h \subset C^\rho$ for which there is $k = k(h)$ such that $\mathrm{supp}(h) \subset M_k$.

Theorem 3.1. *Let M be a compact smooth connected and oriented surface. There are numbers $\beta > 0$, $\rho > 0$, $\gamma_2 > \gamma_1 > 0$, and a $C^{1+\beta}$ diffeomorphism f of M preserving area m and satisfying the following conditions:*

(1) *f has the Bernoulli property with respect to m;*
(2) *f has nonzero Lyapunov exponents almost everywhere with respect to m;*
(3) *f admits polynomial upper and lower bounds on correlations with respect to m; more precisely:*
   (a) *for any $h_i \in C^\rho$, $i = 1,2$*

$$|Cor_n(h_1,h_2)| \le C_1 n^{-\gamma_1},$$

where $C_1 = C_1(\|h_1\|_{C^\rho}, \|h_2\|_{C^\rho}) > 0$;

(b) *if* $\int h_1 dm \int h_2 dm > 0$, *then there is a nested sequence of subsets* $\{M_j\}$ *which exhausts* M *such that for any* $h_i \in \mathcal{G}(\{M_j\})$, $i = 1, 2$,

$$|Cor_n(h_1, h_2)| \geq C_2 n^{-\gamma_2},$$

where $C_2 = C_2(\|h_1\|_{C^\rho}, \|h_2\|_{C^\rho}) > 0$;

(4) *the map* f *satisfies the CLT for the class of observables* $h \in C^\rho$, $\int h\, dm = 0$ *with* $\sigma = \sigma(h)$ *given by*

$$\sigma^2 = -\int h^2 dm + 2\sum_{n=0}^{\infty}\int h \cdot h \circ f^n dm,$$

where $\sigma > 0$ *if and only if* h *is not cohomologous to zero, namely,* $h \circ f \neq g \circ f - g$ *for any measurable function* g;

(5) *the map* f *has Polynomial Large Deviation with respect to the class* C^ρ *of observables with the constant* K *of the form* $K = K(\|h\|_{C^\rho})\varepsilon^{-2\beta}$. *In addition, for an open and dense subset of observables in* C^ρ *and sufficiently small* $\varepsilon > 0$,

$$n^{-\beta} < m\left(\left|\frac{1}{n}\sum_{i=0}^{n-1} h(f^i(x)) - \int h\right| > \varepsilon\right)$$

for infinitely many n;

(6) f *has a unique measure of maximal entropy (MME) with respect to which it has the Bernoulli property, nonzero Lyapunov exponents almost everywhere, exponential decay of correlations, and satisfies the CLT with respect to the class* C^ρ *of observables.*

12.4 A Slowdown Map of the 2-Torus

12.4.1 The Definition of a Slowdown Map

Consider the automorphism of the two-dimensional torus $\mathbb{T}^2 = \mathbb{R}^2/\mathbb{Z}^2$ given by the matrix $A := \begin{pmatrix} 5 & 8 \\ 8 & 13 \end{pmatrix}$. It has four fixed points $x_1 = (0, 0)$, $x_2 = (\frac{1}{2}, 0)$, $x_3 = (0, \frac{1}{2})$, and $x_4 = (\frac{1}{2}, \frac{1}{2})$. For $i = 1, 2, 3, 4$ consider the disk $D_r^i = \{(s_1, s_2) : {s_1}^2 + {s_2}^2 \leq r^2\}$ of radius r centered at x_i and set $D_r = \bigcup_{i=1}^4 D_r^i$. Here (s_1, s_2) is the coordinate system obtained from the eigendirections of A and originated at x_i. Let $\lambda > 1$ be the largest eigenvalue of A. There are $r_2 > r_1 > r_0$ such that

$$D_{r_0}^i \subset A(D_{r_1}^i), \quad A(D_{r_1}^i) \cup A^{-1}(D_{r_1}^i) \subset D_{r_2}^i \tag{12.1}$$

and the disks $D_{r_2}^i$ are pairwise disjoint. Fix i and consider the system of differential equations in $D_{r_1}^i$

$$\frac{ds_1}{dt} = s_1 \log \lambda, \quad \frac{ds_2}{dt} = -s_2 \log \lambda. \tag{12.2}$$

Observe that $A|_{D^i_{r_1}}$ is the time-one map of the local flow generated by this system.

We choose a number $0 < \alpha < 1$, $0 < r_0 < 1$, and a function $\psi : [0,1] \to [0,1]$ satisfying:

(K1) ψ is of class C^∞ everywhere on $(0,1]$ but at the origin;
(K2) $\psi(u) = 1$ for $r_0 \le u \le 1$;
(K3) $\psi'(u) > 0$ for $0 < u < r_0$;
(K4) $\psi(u) = (u/r_0)^\alpha$ for $0 \le u \le \frac{r_0}{2}$.

Using the function ψ, we slow down trajectories of the flow by perturbing the system (12.2) in $D^i_{r_1}$ as follows

$$\begin{aligned} \frac{ds_1}{dt} &= s_1\psi({s_1}^2+{s_2}^2)\log\lambda \\ \frac{ds_2}{dt} &= -s_2\psi({s_1}^2+{s_2}^2)\log\lambda. \end{aligned} \tag{12.3}$$

This system of differential equations generates a local flow g^i_t and we denote by g^i the time-one map of this flow. The choices of ψ, r_0, and r_1 (see (12.1)) guarantee that the domain of g^i contains $D^i_{r_1}$. Furthermore, g^i is of class C^∞ in $D^i_{r_1} \setminus \{x_i\}$ and it coincides with A in some neighborhood of the boundary $\partial D^i_{r_1}$. Therefore, the map

$$G(x) = \begin{cases} A(x) & \text{if } x \in \mathbb{T}^2 \setminus D_{r_1}, \\ g^i(x) & \text{if } x \in D^i_{r_1} \end{cases} \tag{12.4}$$

defines a homeomorphism of the torus $\mathbb{T}^2$, which is a C^∞ diffeomorphism everywhere except at the fixed points x_i. Since $0 < \alpha < 1$, we have that

$$\int_0^1 \frac{du}{\psi(u)} < \infty.$$

This implies that the map G preserves the probability measure

$$d\nu = q_0^{-1} q \, dm, \tag{12.5}$$

where m is the area and the density q is a positive C^∞ function that is infinite at x_i and is defined by

$$q(s_1, s_2) := \begin{cases} (\psi({s_1}^2+{s_2}^2))^{-1} & \text{if } (s_1,s_2) \in D^i_{r_1}, \\ 1 & \text{in } \mathbb{T}^2 \setminus D_{r_1} \end{cases}$$

and

$$q_0 := \int_{\mathbb{T}^2} q \, dm.$$

We further perturb the map G by a coordinate change ϕ in $\mathbb{T}^2$ to obtain an area-preserving map. To achieve this, define a map ϕ in $D^i_{r_1}$ by the formula

$$\phi(s_1, s_2) := \frac{1}{\sqrt{q_0(s_1{}^2 + s_2{}^2)}} \left(\int_0^{s_1{}^2+s_2{}^2} \frac{du}{\psi(u)} \right)^{1/2} (s_1, s_2) \tag{12.6}$$

and set $\phi = \mathrm{Id}$ in $\mathbb{T}^2 \setminus D_{r_1}$. Clearly, ϕ is a homeomorphism and is a C^∞ diffeomorphism outside the points x_1, x_2, x_3, x_4. One can show that ϕ transfers the measure ν into the area and that the map $f_{\mathbb{T}^2} = \phi \circ G \circ \phi^{-1}$ is a homeomorphism and is a C^∞ diffeomorphism outside the points x_1, x_2, x_3, x_4. It is called a *slowdown map* (see [12] and also [2]). The following proposition describes some basic properties of this map.

Proposition 4.1 ([12],[2]). *The map $f_{\mathbb{T}^2}$ has the following properties:*

(1) *It is topologically conjugated to A via a homeomorphism H.*
(2) *It admits two transverse invariant continuous stable and unstable distributions $E^s(x)$ and $E^u(x)$ and for almost every point x with respect to area m it has two nonzero Lyapunov exponents, positive in the direction of $E^u(x)$ and negative in the direction of $E^s(x)$. Moreover, the only invariant measure with zero Lyapunov exponents is the atomic measure supported on the fixed points x_i.*
(3) *It admits two continuous, uniformly transverse, invariant foliations with smooth leaves which are the images under the conjugacy map of the stable and unstable foliations for A respectively.*
(4) *For every $\varepsilon > 0$ one can choose $r_0 > 0$ such that*

$$|\int_{D^i_{r_0}} \log |Df_{\mathbb{T}^2}|E^u|\, dm - \log \lambda| < \varepsilon.$$

(5) *It is ergodic with respect to the area m.*

The following proposition establishes regularity of the map $f_{\mathbb{T}^2}$.

Proposition 4.2. *The map $f_{\mathbb{T}^2}$ is of class of smoothness $C^{2+2\kappa}$, where $\kappa = \frac{\alpha}{1-\alpha}$.*

Proof. Fix $i \in \{1, 2, 3, 4\}$ and consider the vector field in $D^i_{r_1}$ given by the right-hand side of (12.2). It is Hamiltonian with respect to the area and the Hamiltonian function $H_1(s_1, s_2) = s_1 s_2 \log \lambda$. The vector field given by (12.3) is obtained from (12.2) by a time change and hence is also Hamiltonian with respect to the measure ν (see (12.5)) and the same Hamiltonian function. The map $f_{\mathbb{T}^2}$ is conjugate via φ (see (12.6)) to the time-one map of the flow generated by (12.3). Since $\phi_* \nu = m$, $f_{\mathbb{T}^2}$ is the time-one map of the flow which is Hamiltonian with respect to the area and the Hamiltonian function $H_2 = H_1 \circ \phi^{-1}$. Using (12.6), we find that H_2 can be given in $D^i_{r_1}$ as follows (see [12]):

$$H_2(s_1, s_2) = \frac{s_1 s_2 h(\sqrt{s_1^2 + s_2^2})}{s_1^2 + s_2^2} \log \lambda,$$

where by (K4), $h(u) = u^{\frac{2}{1-\alpha}}$ and $u = s_1^2 + s_2^2$. To prove that $f_{\mathbb{T}^2}$ is of the desired class of smoothness we will show that the Hamiltonian H_2 has Hölder continuous partial derivatives of second order with Hölder exponent 2κ. To this end we consider the function $g(x,y) = xy(x^2+y^2)^\kappa$ with $\kappa = \frac{\alpha}{1-\alpha}$ and show that g has Hölder continuous partial derivatives of second order with Hölder exponent 2κ. Note that g is of class C^∞ except for $(x,y) = (0,0)$, so we only need to show Hölder continuity of partial derivatives at the origin. Note also that the function g is symmetric, so we only show that $\frac{\partial^2 g}{\partial x^2}$ and $\frac{\partial^2 g}{\partial x \partial y}$ are Hölder continuous. Since $\frac{\partial g}{\partial x}(0,0) = \lim_{\Delta x \to 0} \frac{g(\Delta x,0) - g(0,0)}{\Delta x} = 0$, we have that

$$\frac{\partial g}{\partial x} = \begin{cases} y(x^2+y^2)^\kappa + 2\kappa x^2 y(x^2+y^2)^{\kappa-1}, & (x,y) \neq (0,0); \\ 0 & (x,y) = (0,0). \end{cases}$$

Note that

$$\frac{\partial^2 g}{\partial x \partial y}(0,0) = \lim_{\Delta y \to 0} \frac{\frac{\partial g}{\partial x}(0,\Delta y) - \frac{\partial g}{\partial x}(0,0)}{\Delta y} = \lim_{\Delta y \to 0} \frac{\Delta y (\Delta y)^{2\kappa} - 0}{\Delta y} = 0$$

and hence,

$$\frac{\partial^2 g}{\partial x \partial y} = \begin{cases} (1+2\kappa)(x^2+y^2)^\kappa + \\ \quad +4(\kappa-1)x^2y^2(x^2+y^2)^{\kappa-2} & (x,y) \neq (0,0); \\ 0 & (x,y) = (0,0). \end{cases}$$

Since the function $\frac{\partial^2 g}{\partial x \partial y}$ is differentiable for all $(x,y) \neq (0,0)$, it is Hölder continuous for all pairs of nonzero points (x,y). It remains to show Hölder continuity for pairs of points one of which is zero. It is easy to see that

$$\left| \frac{\partial^2 g}{\partial x \partial y}(x,y) - \frac{\partial^2 g}{\partial x \partial y}(0,0) \right| \leq K(x^2+y^2)^\kappa = Kd((x,y),(0,0))^{2\kappa},$$

where $K > 0$ and d denotes the usual distance. Thus $\frac{\partial^2 g}{\partial x \partial y}$ is Hölder continuous with Hölder exponent 2κ.

Now we consider $\frac{\partial^2 g}{\partial x^2}$. Observe that $\frac{\partial^2 g}{\partial x^2}(0,0) = 0$ and

$$\frac{\partial^2 g}{\partial x^2} = \begin{cases} 6\kappa xy(x^2+y^2)^{\kappa-1} + 4(\kappa-1)x^3y(x^2+y^2)^{\kappa-2} & (x,y) \neq (0,0); \\ 0, & (x,y) = (0,0). \end{cases}$$

It is easy to see that

$$|\frac{\partial^2 g}{\partial x^2}(x,y) - \frac{\partial^2 g}{\partial x^2}(0,0)| \leq K(x^2+y^2)^{\kappa} = Kd((x,y),(0,0))^{2\kappa}.$$

Hence, $\frac{\partial^2 g}{\partial x^2}$ is Hölder continuous with Hölder exponent 2κ. □

12.5 Proof of Theorem 3.1: Representing $f_{\mathbb{T}^2}$ as a Young Diffeomorphism

12.5.1 Young Diffeomorphisms

Let $f\colon M \to M$ be a $C^{1+\epsilon}$ diffeomorphism of a compact smooth Riemannian manifold M. Following [24] we describe a collection of conditions on the map f.[3]

An embedded C^1-disk $\gamma \subset M$ is called an *unstable disk* (respectively, a *stable disk*) if for all $x,y \in \gamma$ we have that $d(f^{-n}(x),f^{-n}(y)) \to 0$ (respectively, $d(f^n(x),f^n(y)) \to 0$) as $n \to +\infty$. A collection of embedded C^1 disks $\Gamma^u = \{\gamma^u\}$ is called a *continuous family of unstable disks* if there exists a homeomorphism $\Phi\colon K^s \times D^u \to \cup\gamma^u$ satisfying:

- $K^s \subset M$ is a Borel subset and $D^u \subset \mathbb{R}^d$ is the closed unit disk for some $d < \dim M$;
- $x \to \Phi|\{x\} \times D^u$ is a continuous map from K^s to the space of C^1 embeddings of D^u into M which can be extended to a continuous map of the closure $\overline{K^s}$;
- $\gamma^u = \Phi(\{x\} \times D^u)$ is an unstable disk.

A *continuous family of stable disks* is defined similarly.

We allow the sets K^s to be non-compact in order to deal with overlaps which appear in most known examples including the Katok map.

A set $\Lambda \subset M$ has *hyperbolic product structure* if there exists a continuous family $\Gamma^u = \{\gamma^u\}$ of unstable disks γ^u and a continuous family $\Gamma^s = \{\gamma^s\}$ of stable disks γ^s such that

- $\dim \gamma^s + \dim \gamma^u = \dim M$;
- the γ^u-disks are transversal to γ^s-disks with an angle uniformly bounded away from 0;
- each γ^u-disk intersects each γ^s-disk at exactly one point;
- $\Lambda = (\cup\gamma^u) \cap (\cup\gamma^s)$.

A subset $\Lambda_0 \subset \Lambda$ is called an *s-subset* if it has hyperbolic product structure and is defined by the same family Γ^u of unstable disks as Λ and a continuous subfamily $\Gamma_0^s \subset \Gamma^s$ of stable disks. A *u-subset* is defined analogously.

[3] Our requirements are slightly different than those in [24] since we do not assume the inducing domain Λ to be compact.

We define the *s-closure* $scl(\Lambda_0)$ of an *s-subset* $\Lambda_0 \subset \Lambda$ by

$$scl(\Lambda_0) := \bigcup_{x \in \overline{\Lambda_0 \cap \gamma^u}} \gamma^s(x) \cap \Lambda$$

and the *u-closure* $ucl(\Lambda_1)$ of a given *u-subset* $\Lambda_1 \subset \Lambda$ similarly:

$$ucl(\Lambda_1) := \bigcup_{x \in \overline{\Lambda_1 \cap \gamma^s}} \gamma^u(x) \cap \Lambda.$$

Assume the map f satisfies the following conditions:

(Y1) There exists $\Lambda \subset M$ with hyperbolic product structure, a countable collection of continuous subfamilies $\Gamma_i^s \subset \Gamma^s$ of stable disks and positive integers τ_i, $i \in \mathbb{N}$ such that the s-subsets

$$\Lambda_i^s := \bigcup_{\gamma \in \Gamma_i^s} (\gamma \cap \Lambda) \subset \Lambda \tag{12.7}$$

are pairwise disjoint and satisfy:
(a) *invariance*: for every $x \in \Lambda_i^s$

$$f^{\tau_i}(\gamma^s(x)) \subset \gamma^s(f^{\tau_i}(x)),\ f^{\tau_i}(\gamma^u(x)) \supset \gamma^u(f^{\tau_i}(x)),$$

where $\gamma^{u,s}(x)$ denotes the (un)stable disk containing x;
(b) *Markov property*: $\Lambda_i^u := f^{\tau_i}(\Lambda_i^s)$ is a u-subset of Λ such that for all $x \in \Lambda_i^s$

$$f^{-\tau_i}(\gamma^s(f^{\tau_i}(x)) \cap \Lambda_i^u) = \gamma^s(x) \cap \Lambda,$$
$$f^{\tau_i}(\gamma^u(x) \cap \Lambda_i^s) = \gamma^u(f^{\tau_i}(x)) \cap \Lambda.$$

(Y2) The sets Λ_i^u are pairwise disjoint.

For any $x \in \Lambda_i^s$ define the *inducing time* by $\tau(x) := \tau_i$ and the *induced map* $\tilde{f}\colon \bigcup_{i \in \mathbb{N}} \Lambda_i^s \to \Lambda$ by

$$\tilde{f}|_{\Lambda_i^s} := f^{\tau_i}|_{\Lambda_i^s}.$$

(Y3) There exists $0 < a < 1$ such that for any $i \in \mathbb{N}$ we have:
(a) For $x \in \Lambda_i^s$ and $y \in \gamma^s(x)$,

$$d(\tilde{f}(x), \tilde{f}(y)) \le a\, d(x,y);$$

(b) For $x \in \Lambda_i^s$ and $y \in \gamma^u(x) \cap \Lambda_i^s$,

$$d(x,y) \le a\, d(\tilde{f}(x), \tilde{f}(y)).$$

For $x \in \Lambda$ let $\mathrm{Jac}f(x) = \det |Df|_{E^u(x)}|$ and $\mathrm{Jac}\tilde{f}(x) = \det |D\tilde{f}|_{E^u(x)}|$ denote the Jacobian of $Df|_{E^u(x)}$ and $D\tilde{f}|_{E^u(x)}$ respectively.

(Y4) There exist $c > 0$ and $0 < b < 1$ such that:
(a) For all $n \geq 0$, $x \in \tilde{f}^{-n}(\bigcup_{i \in \mathbb{N}} \Lambda_i^s)$ and $y \in \gamma^s(x)$ we have

$$\left| \log \frac{\mathrm{Jac}\tilde{f}(\tilde{f}^n(x))}{\mathrm{Jac}\tilde{f}(\tilde{f}^n(y))} \right| \leq cb^n;$$

(b) For any $i_0, \ldots, i_n \in \mathbb{N}$, $\tilde{f}^k(x), \tilde{f}^k(y) \in \Lambda_{i_k}^s$ for $0 \leq k \leq n$ and $y \in \gamma^u(x)$ we have

$$\left| \log \frac{\mathrm{Jac}\tilde{f}(\tilde{f}^{n-k}(x))}{\mathrm{Jac}\tilde{f}(\tilde{f}^{n-k}(y))} \right| \leq cb^k.$$

(Y5) For every $\gamma^u \in \Gamma^u$ one has

$$\mu_{\gamma^u}(\gamma^u \cap \Lambda) > 0, \quad \mu_{\gamma^u}\left(\overline{(\Lambda \setminus \cup \Lambda_i^s) \cap \gamma^u}\right) = 0,$$

where μ_{γ^u} is the leaf volume on γ^u.

(Y6) There exists $\gamma^u \in \Gamma^u$ such that

$$\sum_{i=1}^{\infty} \tau_i \mu_{\gamma^u}(\Lambda_i^s \cap \gamma^u) < \infty.$$

It is shown in [21] that the Katok map is a Young diffeomorphism. We will briefly outline the argument.

12.5.2 A Tower Representation of the Automorphism A

Consider a finite Markov partition $\tilde{\mathcal{P}}$ for the automorphism A. Recall that by definition of Markov partitions, $\tilde{P} = \mathrm{int}\,\tilde{P}$ for any $\tilde{P} \in \tilde{\mathcal{P}}$. Let $\tilde{P} \in \tilde{\mathcal{P}}$ be a partition element which does not intersect any of the disks $D_{r_0}^i$, $i = 1, 2, 3, 4$. Given $\delta > 0$, we can always choose the Markov partition $\tilde{\mathcal{P}}$ in such a way that diam $(\tilde{P}) < \delta$. For a point $x \in \tilde{P}$ denote by $\tilde{\gamma}^s(x)$ (respectively, $\tilde{\gamma}^u(x)$) the connected component of the intersection of $\tilde{P}$ with the stable (respectively, unstable) leaf of x, which contains x. We say that $\tilde{\gamma}^s(x)$ and $\tilde{\gamma}^u(x)$ are *full-length* stable and unstable curves through x.

Given $x \in \tilde{P}$, let $\tilde{\tau}(x)$ be the first return time of x to int$\tilde{P}$. For all x with $\tilde{\tau}(x) < \infty$ denote by

$$\tilde{\Lambda}^s(x) = \bigcup_{y \in \tilde{U}^u(x) \setminus \tilde{A}^u(x)} \tilde{\gamma}^s(y),$$

where $\tilde{U}^u(x) \subseteq \tilde{\gamma}^u(x)$ is an interval containing x and open in the induced topology of $\tilde{\gamma}^u(x)$, and $\tilde{A}^u(x) \subset \tilde{U}^u(x)$ is the set of points which either lie on the boundary of the Markov partition or never return to the set $\tilde{P}$. Note that $\tilde{A}^u(x)$ has zero one-dimensional Lebesgue measure in $\tilde{\gamma}^u(x)$. One can choose $\tilde{U}^u(x)$ such that

(1) for any $y \in \tilde{\Lambda}^s(x)$ we have $\tilde{\tau}(y) = \tilde{\tau}(x)$;
(2) for any $y \in \tilde{P}$ such that $\tilde{\tau}(y) = \tilde{\tau}(x)$ we have $y \in \tilde{\Lambda}^s(x)$.

Moreover, the image under $A^{\tilde{\tau}(x)}$ of $\tilde{\Lambda}^s(x)$ is a u-subset containing $A^{\tilde{\tau}(x)}(x)$. It is easy to see that for any $x, y \in \tilde{P}$ with finite first return time the sets $\tilde{\Lambda}^s(x)$ and $\tilde{\Lambda}^s(y)$ either coincide or are disjoint. Thus we have a countable collection of disjoint sets $\tilde{\Lambda}_i^s$ and numbers $\tilde{\tau}_i$ which give a representation of the automorphism A as a Young diffeomorphism for which the set

$$\tilde{\Lambda} = \bigcup_{i \geq 1} \tilde{\Lambda}_i^s$$

is the base of the tower, the sets $\tilde{\Lambda}_i^s$ are the s-sets and the numbers $\tilde{\tau}_i$ are the inducing times; see [21] for details.

12.5.3 A Tower Representation for the Slowdown Map $f_{\mathbb{T}^2}$

Applying the conjugacy map H, one obtains the element $P = H(\tilde{P})$ of the Markov partition $\mathcal{P} = H(\tilde{\mathcal{P}})$. Since the map H is continuous, given ε, there is $\delta > 0$ such that diam $(P) < \varepsilon$ for any $P \in \mathcal{P}$ provided diam $(\tilde{P}) < \delta$. Further, we obtain the set $\Lambda = H(\tilde{\Lambda})$, which has direct product structure given by the full-length stable $\gamma^s(x) = H(\tilde{\gamma}^s(x))$ and unstable $\gamma^u(x) = H(\tilde{\gamma}^u(x))$ curves. We thus obtain a representation of the slowdown map as a Young diffeomorphism for which

(1) $\Lambda_i^s = H(\tilde{\Lambda}_i^s)$ are s-sets;
(2) $\Lambda_i^u = H(\tilde{\Lambda}_i^u) = f_{\mathbb{T}^2}^{\tau_i}(\Lambda_i^s)$ are u-sets;
(3) $\tau_i = \tilde{\tau}_i$ – the inducing times – are the first return times of points in Λ_i^s to Λ;
(4) $\tilde{f}(x) = f_{\mathbb{T}^2}^{\tau(x)}(x)$ is the induced map.

Note that for all x with $\tau(x) < \infty$

$$\Lambda^s(x) = \bigcup_{y \in U^u(x) \setminus A^u(x)} \gamma^s(y),$$

where $U^u(x) = H(\tilde{U}^u(x)) \subseteq \gamma^u(x)$ is an interval containing x and open in the induced topology of $\gamma^u(x)$, and $A^u(x) = H(\tilde{A}^u(x)) \subset U^u(x)$ is the set of points which either lie on the boundary of the Markov partition or never return to the set P. Note that $A^u(x)$ has zero one-dimensional Lebesgue measure in $\gamma^u(x)$.

In what follows we will always assume that a Markov partition and the slowdown domain are chosen such that the following statement holds.

Proposition 5.1. *Given $Q > 0$, one can choose a Markov partition $\mathcal{P}$ and the number r_0 in the construction of the map $f_{\mathbb{T}^2}$ such that*

(1) *there is a partition element P for which $f^j_{\mathbb{T}^2}(x) \notin D_{r_0}$ for any $0 \le j \le Q$ and for any point x for which either $x \in \Lambda$ or $x \notin f_{\mathbb{T}^2}(D_{r_0})$ while $f^{-1}_{\mathbb{T}^2}(x) \in D^i_{r_0}$ for some $i = 1, 2, 3, 4$.*
(2) *if P_i is the element of the Markov partition containing x_i, $i = 1, 2, 3, 4$, then $x_i \in D^i_{r_0} \subset \mathrm{Int}\, P_i$.*

To prove this proposition, observe that it holds for the automorphism A and hence, it remains to apply the conjugacy homeomorphism H.

Proposition 5.2 ([21], Proposition 6.2). *There exists $Q > 0$ such that the collection of s-subsets Λ^s_i satisfies Conditions (Y1)–(Y6).*

12.5.4 Lifting the Slowdown Map to the Tower

We define *Young tower* with the base Λ by setting

$$\hat{Y} = \{(x, k) \in \Lambda \times \mathbb{N} \colon 0 \le k < \tau(x)\}$$

and the *tower map* $\hat{f}_{\mathbb{T}^2} \colon \hat{Y} \to \hat{Y}$ by $\hat{f}_{\mathbb{T}^2}(x, k) = (x, k + 1)$ if $k < \tau(x) - 1$ and $\hat{f}_{\mathbb{T}^2}(x, k) = (\tilde{f}(x), 0)$ if $k = \tau(x) - 1$ where $\tilde{f} \colon \Lambda \to \Lambda$ is the induced map. The map $\hat{f}_{\mathbb{T}^2}$ is the lift of the slowdown map to the tower and it preserves the *lift measure* $\hat{m} = m \times \text{counting}/(\int_\Lambda \tau)$. We have that $\hat{f}_{\mathbb{T}^2}$ is a measurable bijection from $\Lambda^s_i \times \{k - 1\}$ to $\Lambda^s_i \times \{k\}$ for all $1 \le k < \tau_i - 1$ and from $\Lambda^s_i \times \{\tau_i - 1\}$ to $\Lambda \times \{0\}$.

12.6 Proof of Theorem 3.1: Technical Lemmas

We establish here several technical results on the solutions of the nonlinear systems of differential equations (12.3). Throughout this section we fix a number $0 < \alpha < 1$ and $i \in \{1, 2, 3, 4\}$ and for simplicity we drop the index i in the notation of the disk D^i_r. We also set $f := f_{\mathbb{T}^2}$.

Lemma 6.1 ([21], Lemma 5.1). *For $s = (s_1, s_2) \in D_{\frac{r_0}{2}}$ let*

$$d_{i,j} = d_{i,j}(s_1, s_2) := \frac{\partial^2}{\partial s_i \partial s_j} s_2 \psi(s_1^2 + s_2^2).$$

Then

$$\max_{i,j=1,2} |d_{i,j}| \le \frac{6\alpha}{r_0^\alpha}(s_1^2+s_2^2)^{\alpha-\frac{1}{2}}.$$

Consider a solution $s(t) = (s_1(t), s_2(t))$ of Equation (12.3) with an initial condition $s(0) = (s_1(0), s_2(0))$. Assume it is defined on the *maximal* time interval $[0, T]$ for which $f^{-1}(s(0)) \notin D_{\frac{r_0}{2}}$ and $f(s(T)) \notin D_{\frac{r_0}{2}}$ but $s(t) \in D_{\frac{r_0}{2}}$ for all $0 \le t \le T$. In particular, $s_1(t) \neq 0$ and $s_2(t) \neq 0$. Setting $T_1 = \frac{T}{2}$ we have that $s_1(t) \le s_2(t)$ for all $0 \le t \le T_1$ and $s_1(t) \ge s_2(t)$ for all $T_1 \le t \le T$. The following statement provides effective lower and upper bounds on the functions $s_1(t)$ and $s_2(t)$. For the proof see Lemma 5.2 in the erratum to the paper [21].

Lemma 6.2. *The following statements hold:*

$$|s_2(t)| \ge |s_2(a)|\left(1+2^\alpha C_1 s_2^{2\alpha}(a)(t-a)\right)^{-\frac{1}{2\alpha}}, \quad 0 \le a \le t \le T_1;$$

$$|s_2(t)| \le |s_2(a)|\left(1+C_1 s_2^{2\alpha}(a)(t-a)\right)^{-\frac{1}{2\alpha}}, \quad 0 \le a \le t \le T;$$

$$|s_1(t)| \ge |s_1(b)|\left(1+2^\alpha C_1 s_1^{2\alpha}(b)(b-t)\right)^{-\frac{1}{2\alpha}}, \quad T_1 \le t \le b \le T;$$

$$|s_1(t)| \le |s_1(b)|\left(1+C_1 s_1^{2\alpha}(b)(b-t)\right)^{-\frac{1}{2\alpha}}, \quad 0 \le t \le b \le T,$$

where $C_1 = \frac{2\alpha \log\lambda}{r_0^\alpha}$ *is a constant.*

Consider another solution $\tilde{s}(t) = (\tilde{s}_1(t), \tilde{s}_2(t))$ of Equation (12.3) satisfying an initial condition $\tilde{s}(0) = (\tilde{s}_1(0), \tilde{s}_2(0))$. For $i = 1, 2$, we set

$$\Delta s_i(t) = \tilde{s}_i(t) - s_i(t).$$

For the proof of the next result see Lemma 5.3 in the erratum to the paper [21].

Lemma 6.3. *Fix* $0 < \mu < 1$ *and assume that* $s_1(t) \neq 0$, $s_2(t) \neq 0$ *for all* $0 \le t \le T$, *and that*

(1) $\Delta s_2(t) > 0$ *and* $|\Delta s_1(t)| \le \mu \Delta s_2(t)$ *for* $t \in [0, T]$;
(2) $\left|\frac{\Delta s_2}{s_2}(0)\right| < \frac{1-\mu}{72}$.

Then

$$\Delta s_2(t) \le \frac{\Delta s_2(0)}{s_2(0)} s_2(t)\left(1+2^\alpha C_1 s_2^{2\alpha}(0)\, t\right)^{-\beta'}, \quad 0 \le t \le T_1;$$

$$\Delta s_2(t) \le \frac{\Delta s_2(T_1)}{s_1(T_1)} s_1(t)\left(\frac{1+2^\alpha C_1 s_1^{2\alpha}(b)(b-t)}{1+2^\alpha C_1 s_1^{2\alpha}(b)(b-T_1)}\right)^{\beta'}, \quad T_1 \le t \le b \le T,$$

where $\beta' = \frac{1-\mu}{2^{\alpha+2}}$ *and* C_1 *is the constant in Lemma 6.2. In addition,*

$$\|\Delta s(T)\| \leq \sqrt{1+\mu^2}\,\frac{s_1(T)}{s_2(0)}\,\|\Delta s(0)\|.$$

Given $0 < \alpha < 1$ and $0 < \mu < 1$, denote by

$$\gamma = \frac{1}{2\alpha} + 2^{\alpha-1}(1+\mu) + \frac{1-\mu}{6}, \quad \gamma' = \frac{1}{2\alpha} + \frac{1-\mu}{2^{\alpha+2}}. \tag{12.8}$$

It is easy to see that $\gamma > \gamma' > 2$ for all $0 < \alpha < \frac{1}{4}$ and $0 < \mu < \frac{1}{2}$.

In what follows till the end of the next section we denote by C_2 through C_{10} positive constants that are independent of the time t and the choice of the solution $(s_1(t), s_2(t))$.

Lemma 6.4. *Under the assumptions of Lemma 6.3 for any* $0 \leq t \leq T_1$ *we have*

$$\Delta s_2(t) \leq C_2 \Delta s_2(0) t^{-\gamma'}.$$

Proof. By Lemma 6.2 (the second estimate), one has

$$s_2(t) \leq s_2(0)(1 + C_1 s_2^{2\alpha}(0) t)^{-\frac{1}{2\alpha}}.$$

Therefore, Lemma 6.3 implies that

$$\begin{aligned}\Delta s_2(t) &\leq \frac{\Delta s_2(0)}{s_2(0)} s_2(t)\left(1 + 2^\alpha\, C_1\, s_2^{2\alpha}(0)\, t\right)^{-\beta'} \\ &\leq \Delta s_2(0)\left(1 + C_1\, s_2^{2\alpha}(0)\, t\right)^{-\gamma'}.\end{aligned}$$

Since $(1 + C_1 s_2^{2\alpha}(0))t > C_1 s_2^{2\alpha}(0)t$, we have

$$\Delta s_2(t) \leq \Delta s_2(0)(C_1 s_2^{2\alpha}(0))^{-\gamma'} t^{-\gamma'}$$

and the desired estimate follows, since $s_2(0)$ is of order r_0. □

Lemma 6.5. *Under the assumptions of Lemma 6.3 we have*

$$\begin{aligned}\Delta s_2(t) &\geq \frac{\Delta s_2(0)}{s_2(0)} s_2(t)\left(1 + C_1\, s_2^{2\alpha}(0)\, t\right)^{-\beta}, && 0 \leq t \leq T_1; \\ \Delta s_2(t) &\geq \frac{\Delta s_2(T_1)}{s_1(T_1)} s_1(t)\left(1 + C_1\, s_1^{2\alpha}(T_1)(t - T_1)\right)^{-\beta_1}, && T_1 \leq t \leq T,\end{aligned}$$

where $\beta = (1+\mu)2^{\alpha-1} + \frac{1-\mu}{6}$ *and* $\beta_1 = \beta + \frac{2^\alpha}{\alpha}$.

Proof. Let $s_1 = s_1(t)$, $s_2 = s_2(t)$, $u := s_1^2 + s_2^2$, and $\tilde{u} = \tilde{s}_1^2 + \tilde{s}_2^2$. Assume $s_1(t)$ and $s_2(t)$ are strictly positive (the proof in the case when $s_1(t)$ and $s_2(t)$ are strictly negative follows by symmetry). By Equation (12.3), we have

$$\begin{aligned}\frac{d}{dt}\Delta s_2(t) &= \frac{d}{dt}\tilde{s}_2(t) - \frac{d}{dt}s_2(t) = -(\log\lambda)\,(\tilde{s}_2\psi(\tilde{u}) - s_2\psi(u))\\ &= -\log\lambda\left(\frac{\partial}{\partial s_1}\Big(s_2\psi(u)\Big)\Delta s_1 + \frac{\partial}{\partial s_2}\Big(s_2\psi(u)\Big)\Delta s_2\right)\\ &\quad - \frac{\log\lambda}{2}\sum_{i,j=1,2} d_{i,j}(\xi_1,\xi_2)(\Delta s_i)(\Delta s_j)\end{aligned} \tag{12.9}$$

for some $\xi = (\xi_1, \xi_2)$ for which ξ_i lies between $s_i(t)$ and $\tilde{s}_i(t)$ for $i = 1, 2$ (see Lemma 6.1 for the definition of $d_{i,j}(\xi_1, \xi_2)$). Note that

$$\frac{\partial}{\partial s_1}\Big(s_2\psi(u)\Big) = 2s_1s_2\psi', \quad \frac{\partial}{\partial s_2}\Big(s_2\psi(u)\Big) = 2s_2^2\psi' + \psi$$

and hence,

$$\begin{aligned}\frac{d}{dt}\left(\frac{\Delta s_2}{s_2}\right) &= \frac{1}{s_2}\left(\frac{d}{dt}\Delta s_2\right) - \frac{\Delta s_2}{s_2^2}\left(\frac{ds_2}{dt}\right)\\ &= -\log\lambda\left(2\psi'(s_1\Delta s_1 + s_2\Delta s_2) + \frac{\Delta s_2}{s_2}\psi\right)\\ &\quad + (\log\lambda)\frac{\Delta s_2}{s_2}\psi - \frac{\log\lambda}{2}\sum_{i,j=1,2} d_{i,j}(\xi_1,\xi_2)\frac{\Delta s_i\Delta s_j}{s_2}\\ &= -\frac{2\alpha\log\lambda}{r_0^\alpha}(s_1^2 + s_2^2)^{\alpha-1}(s_1\Delta s_1 + s_2\Delta s_2)\\ &\quad - \frac{\log\lambda}{2}\sum_{i,j=1,2} d_{i,j}(\xi_1,\xi_2)\frac{\Delta s_i\Delta s_j}{s_2}.\end{aligned}$$

Note that for $0 \le t \le T_1$ we have $0 < s_1(t) \le s_2(t)$. Since $|\Delta s_1| \le \mu\Delta s_2 < \Delta s_2$, we have

$$s_1\Delta s_1 + s_2\Delta s_2 \le (s_1\mu + s_2)\Delta s_2 \le (1+\mu)s_2\Delta s_2$$

and Lemma 6.1 yields

$$\sum_{i,j=1,2} d_{i,j}(\xi_1,\xi_2)\Delta s_i\Delta s_j \le \frac{24\,\alpha}{r_0^\alpha}(\xi_1^2 + \xi_2^2)^{\alpha-\frac{1}{2}}(\Delta s_2)^2. \tag{12.10}$$

Therefore, we have

$$\begin{aligned}\frac{d}{dt}\left(\frac{\Delta s_2}{s_2}\right) &\ge -(1+\mu)\frac{2\alpha\log\lambda}{r_0^\alpha}(s_1^2 + s_2^2)^{\alpha-1}s_2^2\frac{\Delta s_2}{s_2}\\ &\quad - \frac{12\alpha\log\lambda}{r_0^\alpha}s_2^{2\alpha}\left(\frac{\xi_1^2+\xi_2^2}{s_2^2}\right)^{\alpha-\frac{1}{2}}\left(\frac{\Delta s_2}{s_2}\right)^2.\end{aligned}$$

Using again the fact that $0 < s_1(t) \le s_2(t),\ 0 < t < T_1$, we obtain

$$\frac{d}{dt}\left(\frac{\Delta s_2}{s_2}\right) \ge -(1+\mu)\frac{\alpha \log \lambda}{r_0^\alpha} 2^\alpha (s_2)^{2\alpha} \frac{\Delta s_2}{s_2} - \frac{12\alpha \log \lambda}{r_0^\alpha} s_2^{2\alpha} \left(\frac{\xi_1^2 + \xi_2^2}{s_2^2}\right)^{\alpha - \frac{1}{2}} \left(\frac{\Delta s_2}{s_2}\right)^2.$$

Let $\chi = \chi(t) = \frac{\Delta s_2}{s_2}(t)$. Then the preceding inequality can be written as

$$\frac{d\chi}{dt} \ge -\frac{\alpha \log \lambda}{r_0^\alpha} s_2^{2\alpha} \chi \left((1+\mu)2^\alpha + 12 \left(\frac{\xi_1^2 + \xi_2^2}{s_2^2}\right)^{\alpha - \frac{1}{2}} \chi\right).$$

Following arguments in [21] (see page 352) one can derive from here that $\left(\frac{\xi_1^2+\xi_2^2}{s_2^2}\right)^{\alpha-\frac{1}{2}} \le 2$ and that

$$\frac{d\chi}{dt} \ge -A \frac{\alpha \log \lambda}{r_0^\alpha} s_2^{2\alpha}(t) \chi(t),$$

where $A = (1+\mu)2^\alpha + \frac{1-\mu}{3}$. By Gronwall's inequality (applied to $-\chi(t)$) and the second inequality in Lemma 6.2, we obtain that

$$\begin{aligned}
\chi(t) &\ge \chi(0) \exp\left(-A \frac{\alpha \log \lambda}{r_0^\alpha} \int_0^t s_2^{2\alpha}(\tau)\, d\tau\right) \\
&\ge \chi(0) \exp\left(-A \frac{\alpha \log \lambda}{r_0^\alpha} \int_0^t s_2^{2\alpha}(0)(1 + C_1 s_2^{2\alpha}(0)\,\tau)^{-1}\, d\tau\right) \\
&= \chi(0) \exp\left(-A \frac{\alpha \log \lambda}{r_0^\alpha} \frac{1}{C_1} \log(1 + C_1 s_2^{2\alpha}(0)\, t\,)\right) \\
&= \chi(0) \exp\left(-\frac{A}{2} \log(1 + C_1 s_2^{2\alpha}(0)\, t\,)\right) \\
&= \chi(0) \left(1 + C_1 s_2^{2\alpha}(0)\, t\,\right)^{-\beta},
\end{aligned}$$

where $\beta = \frac{A}{2} = (1+\mu)2^{\alpha-1} + \frac{1-\mu}{6}$.

In the case $s_2 < 0$ one can show using an argument similar to the preceding one that the same estimate for $\chi(t)$ holds but with exponent $(1+\mu)2^{\alpha-1} - \frac{1-\mu}{6} < \beta$. This completes the proof of the first estimate.

To prove the second estimate, using (12.9), we obtain that

$$\frac{d}{dt}\left(\frac{\Delta s_2}{s_1}\right) = -\log \lambda \left(2 s_2 \psi' \Delta s_1 + (2 s_2^2 \psi' + \psi) \frac{\Delta s_2}{s_1}\right) - \log \lambda \psi \frac{\Delta s_2}{s_1} - \frac{\log \lambda}{2} \sum_{i,j=1,2} d_{i,j}(\xi_1, \xi_2) \frac{\Delta s_i \Delta s_j}{s_1}.$$

By the assumption $|\Delta s_1| \le \mu \Delta s_2$ and positivity of s_1, s_2, ψ', and Δs_2, we obtain

$$\frac{d}{dt}\left(\frac{\Delta s_2}{s_1}\right) \ge -\log\lambda(2\mu s_1 s_2 \psi' + 2s_2^2\psi' + 2\psi)\frac{\Delta s_2}{s_1} - \frac{\log\lambda}{2}\sum_{i,j=1,2} d_{i,j}(\xi_1,\xi_2)\frac{\Delta s_i \Delta s_j}{s_1}.$$

Since $s_2(t) \le s_1(t)$ on $[T_1, T]$, we have

$$2\mu s_1 s_2\psi' + 2s_2^2\psi' + 2\psi = 2\psi'(\mu s_1^2 + s_1^2 + \frac{1}{\alpha}2s_1^2) \le 2^\alpha \frac{\alpha\mu + \alpha + 2}{r_0^\alpha} s_1^{2\alpha}.$$

Using this fact along with the estimate (12.10), we find that

$$\frac{d}{dt}\left(\frac{\Delta s_2}{s_1}\right) \ge -\log\lambda 2^\alpha \frac{\alpha\mu + \alpha + 2}{r_0^\alpha} s_1^{2\alpha}\frac{\Delta s_2}{s_1} - \frac{12\alpha\log\lambda}{r_0^\alpha} s_1^{2\alpha}\left(\frac{\xi_1^2 + \xi_2^2}{s_1^2}\right)^{\alpha - \frac{1}{2}}\left(\frac{\Delta s_2}{s_1}\right)^2.$$

Let $\tilde{\chi} = \tilde{\chi}(t) = \frac{\Delta s_2}{s_1}(t)$. Then the above inequality can be written as

$$\frac{d\tilde{\chi}}{dt} \ge -\frac{\log\lambda}{r_0^\alpha} s_1^{2\alpha}\tilde{\chi}\left(2^\alpha(\alpha\mu + \alpha + 2) + 12\alpha\left(\frac{\xi_1^2 + \xi_2^2}{s_1^2}\right)^{\alpha-\frac{1}{2}}\tilde{\chi}\right). \tag{12.11}$$

It is shown in [21] (see page 354) that

$$\left(\frac{\xi_1^2 + \xi_2^2}{s_1^2}\right)^{\alpha - \frac{1}{2}} \le \begin{cases} (1 - \tilde{\chi})^{2\alpha - 1} & 0 < \alpha \le \frac{1}{2}, \\ 2^{\alpha - \frac{1}{2}}(1 + \tilde{\chi})^{2\alpha - 1} & \frac{1}{2} \le \alpha < 1 \end{cases} \tag{12.12}$$

and that $\tilde{\chi} = \frac{\Delta s_2}{s_2}$ is positive and decreasing (or negative and increasing). Observing that $s_1(T_1) = s_2(T_1)$ and using Assumption (2) of the lemma, we obtain

$$0 \le \tilde{\chi}(T_1) = \frac{\Delta s_2(T_1)}{s_1(T_1)} = \frac{\Delta s_2(T_1)}{s_2(T_1)} \le \frac{\Delta s_2(0)}{s_2(0)} < \frac{1 - \mu}{72}.$$

$$\tilde{\chi}(t) = \frac{\Delta s_2(t)}{s_1(t)} \le \frac{\Delta s_2(T_1)}{s_1(T_1)}(1 + 2^\alpha C_1 s_1^{2\alpha}(t)(t - T_1))^{-\beta} \le \frac{\Delta s_2(T_1)}{s_1(T_1)}. \tag{12.13}$$

Setting $B = 2^\alpha(\alpha\mu + \alpha + 2) + \frac{1-\mu}{3}\alpha$ and combining (12.11), (12.12), and (12.13), we obtain that

$$\left.\frac{d\tilde{\chi}}{dt}\right|_{t=T_1} \ge -\frac{B\log\lambda}{r_0^\alpha} s_1^{2\alpha}(T_1)\tilde{\chi}(T_1). \tag{12.14}$$

Therefore, Gronwall's inequality and the fourth inequality in Lemma 6.2 now yield

$$\begin{aligned}
\tilde{\chi}(t) &\geq \tilde{\chi}(T_1)\exp\left(-B\frac{\log\lambda}{r_0^\alpha}\int_{T_1}^t s_1^{2\alpha}(\tau)\,d\tau\right)\\
&\geq \tilde{\chi}(T_1)\exp\left(-B\frac{\log\lambda}{r_0^\alpha}\int_{T_1}^t s_1^{2\alpha}(t)(1+C_1s_1^{2\alpha}(0)\,(t-\tau))^{-1}\,d\tau\right)\\
&= \tilde{\chi}(T_1)\exp\left(-B\frac{\log\lambda}{r_0^\alpha}\frac{1}{C_1}\log(1+C_1s_1^{2\alpha}(t)\,(t-T_1)\,)\right)\\
&= \tilde{\chi}(T_1)\exp\left(-B\log(1+C_1s_1^{2\alpha}(t)\,(t-T_1)\,)\right)\\
&= \tilde{\chi}(T_1)\left(1+C_1s_1^{2\alpha}(t)\,(t-T_1)\,\right)^{-\beta_1},
\end{aligned}$$

where $\beta_1 = \frac{B}{2\alpha}$. It follows that

$$\Delta s_2(t) \geq \frac{\Delta s_2(T_1)}{s_1(T_1)}s_1(t)\left(1+C_1\,s_1^{2\alpha}(T_1)\,(t-T_1)\right)^{-\beta_1},\ T_1\leq t\leq T.$$

In the case $s_1 < 0$ one can show using an argument similar to the preceding one that the same estimate for $\tilde{\chi}(t)$ holds but with exponent $2^{\alpha-1}(1+\mu)+\frac{2^\alpha}{\alpha}-\frac{1-\mu}{6} < \beta_1$. This completes the proof of the second estimate. □

Lemma 6.6. *Under the assumptions of Lemma 6.3 for all* $0 \leq t \leq T_1$ *we have*

$$\Delta s_2(t) \geq C_3\Delta s_2(0)t^{-\gamma}.$$

Proof. By assumption, Lemma 6.5 holds and the first estimate together with the first inequality in Lemma 6.2 yield

$$\begin{aligned}
\Delta s_2(t) &\geq \frac{\Delta s_2(0)}{s_2(0)}s_2(t)(1+C_1s_2^{2\alpha}(0)t)^{-\beta}\\
&\geq \Delta s_2(0)(1+C_1s_2^{2\alpha}(0)t)^{-(\frac{1}{2\alpha}+\beta)}.
\end{aligned}$$

Since $1+C_1s_2^{2\alpha}(0)t \leq (1+C_1s_2^{2\alpha}(0))t$ for $t \geq 1$, the preceding implies

$$\Delta s_2(t) > \Delta s_2(0)(1+C_1s_2^{2\alpha}(0))^{-(\frac{1}{2\alpha}+\beta)}t^{-(\frac{1}{2\alpha}+\beta)}.$$

It remains to observe that $s_2(0)$ is of order r_0 and so $(1+C_1s_2^{2\alpha}(0))^{-(\frac{1}{2\alpha}+\beta)}$ is a constant and $\gamma = \frac{1}{2\alpha}+\beta$.

For $0 < t \leq 1$ the orbit stays bounded away from the region of perturbation and so the inequality holds for some constant. □

Lemma 6.7. *Under the assumptions of Lemma 6.3:*

$$C_4\Delta s_2(T_1) \geq \Delta s_2(T) \geq C_5\Delta s_2(T_1).$$

Proof. First, we prove the lower bound. The fourth inequality in Lemma 6.2 with $t = T_1$ and $b = T$ yields

$$s_1(T) \geq s_1(T_1)(1 + C_1 s_1^{2\alpha}(T)(T - T_1))^{\frac{1}{2\alpha}}.$$

This and the second estimate in Lemma 6.5 imply that

$$\begin{aligned}\Delta s_2(T) &\geq \frac{\Delta s_2(T_1)}{s_1(T_1)} s_1(T_1)\left(1 + C_1 s_1^{2\alpha}(T)(T - T_1)\right)^{\frac{1}{2\alpha}}\\ &\quad\times \left(1 + C_1 s_1^{2\alpha}(T_1)(T - T_1)\right)^{-\beta_1}\\ &\geq \Delta s_2(T_1)\left(1 + C_1 s_1^{2\alpha}(T_1)(T - T_1)\right)^{\frac{1}{2\alpha} - \beta_1},\end{aligned}$$

where we recall that $s_1(T) \geq s_1(T_1)$ and $\beta_1 = 2^{\alpha-1}(1+\mu) + \frac{1-\mu}{6} + \frac{2^\alpha}{\alpha}$, $0 < \alpha < \frac{1}{4}$, and $0 < \mu < \frac{1}{2}$. It is easy to see that $\beta_1 > \frac{1}{2\alpha}$ and hence, to complete the proof of the lower bound it suffices to show that $s_1^{2\alpha}(T_1)(T - T_1)$ is bounded above by a constant that is independent of T_1, T, and the choice of the solution $(s_1(t), s_2(t))$.

Since $s_1(T_1) = s_2(T_1)$, applying the second inequality in Lemma 6.2 with $a = 0$ and $t = T_1$, we obtain that

$$s_1^{2\alpha}(T_1) \leq s_2^{2\alpha}(0)\left(1 + C_1 s_2^{2\alpha}(0) T_1\right)^{-1}.$$

This implies that

$$\begin{aligned}s_1^{2\alpha}(T_1)(T - T_1) &\leq \frac{s_2^{2\alpha}(0)(T - T_1)}{1 + C_1 s_2^{2\alpha}(0) T_1}\\ &\leq \frac{s_2^{2\alpha}(0)(T - T_1)}{C_1 s_2^{2\alpha}(0) T_1} = \frac{T - T_1}{C_1 T_1}.\end{aligned}$$

Since $\frac{T-T_1}{T_1}$ is of order 1 regardless of the choice of the solution $(s_1(t), s_2(t))$, this completes the proof of the lower bound.

To prove the upper bound we use the existence of invariant stable and unstable cones at every point x in the disk D_r^i, $i = 1, 2, 3, 4$. More precisely, let $K^-(x)$ (respectively, $K^+(x)$) be the cone at x around vertical (respectively, horizontal) line of angle $\frac{\pi}{4}$. One can show (see [12], Proposition 4.1) that these families of cones are invariant under the flow g_t^i (given by Equation (12.3)), that is,

$$dg_\tau^i(K^+(x)) \subset K^+(g_\tau^i(x)), \quad d(g_\tau^i)^{-1}(K^-(x)) \subset K^-((g_\tau^i)^{-1}(x)).$$

Moreover, let us fix positive numbers a and b and consider the region

$$\{(s_1, s_2)\colon |s_1 s_2| \leq a,\ |s_1| \leq b,\ |s_2| \leq b\} \subset D_r^i$$

and a piece of hyperbola

$$\{s_1 s_2 = \varepsilon,\ 0 \leq s_1 \leq b,\ 0 \leq s_2 \leq b\} \text{ for } 0 < \varepsilon \leq a.$$

One can show (see [12], Proposition 4.1) that for every (s_1, s_2) on this hyperbola every angle inside the cone $K^-(x)$ is contracted under $(dg_t^i)^{-1}$ by the factor $d = C\frac{a^2}{b^4}$, where $C > 0$ is a constant independent of x, a, and b (note that given $b > 0$, one can choose a so small as to ensure that $d < 1$).

Let $s(t) = (s_1(t), s_2(t))$ and $\tilde{s}(t) = (\tilde{s}_1(t), \tilde{s}_2(t))$ be two solutions of Equation (12.3) satisfying the assumptions of Lemma 6.3. It follows from what was said earlier that for any $T_1 < t < T$ and any $\tau \geq 0$ such that $T_1 \leq t - \tau$ we have that $\Delta s(t) \subset K^-(s(t))$ and $\|\Delta s(t-\tau)\| \leq \|\Delta s(t)\|$. Since $\Delta s(t) = \Delta s_1(t) + \Delta s_2(t)$ and, by assumption, $|\Delta s_1(t)| \leq \mu \Delta s_2(t)$, choosing $t = T$ and $\tau = T - T_1$, we obtain the desired upper bound. This completes the proof of the lemma. □

Consider a set Λ_i^s and note that it consists of full-length s-curves. Let us fix one of these curves, say σ.

Lemma 6.8. *Assume that σ enters the slowdown disk $D_{\frac{r_0}{2}}$ at time n, so that the intersection $f^n(\sigma) \cap D_{\frac{r_0}{2}}$ is not empty. Assume that σ then exits $D_{\frac{r_0}{2}}$ at time m, $m > n > 1$. Then*

$$C_6(m-n)^{-\gamma} \leq \frac{L(f^m(\sigma))}{L(f^n(\sigma))} \leq C_7(m-n)^{-\gamma'},$$

where γ, γ' are as in (12.8), and L denotes the length of the curve.

Proof. Let x and y be the endpoints of the curve σ. For $k \geq 0$ set $x_k = f^k(x)$ and $y_k = f^k(y)$. It is easy to see that there is $K_0 > 0$ such that for all $k \geq 1$,

$$K_0^{-1} d(x_k, y_k) \leq L(f^k(\sigma)) \leq K_0 d(x_k, y_k), \tag{12.15}$$

where d denotes the usual distance.

Let $s, \tilde{s}\colon [0, N] \to R^2$ be the solutions of Equation (12.2) with initial conditions $s(0) = x_n$ and $\tilde{s}(0) = y_n$ respectively. Also, define $\Delta s_i(t) = \tilde{s}_i(t) - s_i(t)$, $i = 1, 2$ and $\Delta s = (\Delta s_1, \Delta s_2)$. Note that there is $K_1 > 0$ such that, for all $n, m > 0$ and $n \leq j \leq m$,

$$K_1^{-1} ||\Delta s(j)|| \leq d(x_j, y_j) \leq K_1 ||\Delta s(j)||. \tag{12.16}$$

In what follows we will use Lemma 6.3 and we need to check the assumptions of this Lemma. Assumption (1) is satisfied, since y is contained in the stable cone at x. Assumption (2) requires $d(x_i, y_i)$, $i = n_0, m_0$ to be sufficiently small. In view of Proposition 6.1 and (12.16), this can be ensured, if we choose the number r_0 in the construction of the slowdown map sufficiently small to guarantee that Q is sufficiently large.

We have that $f^j(\sigma) \subset D_{\frac{r_0}{2}} \cap \{(s_1, s_2)\colon s_2 \geq s_1\}$ for $n \leq j \leq \frac{n+m}{2}$ and $f^j(\sigma) \subset D_{\frac{r_0}{2}} \cap \{(s_1, s_2)\colon s_2 < s_1\}$ for $\frac{n+m}{2} < j < m$. Applying Lemmas 6.3, 6.6 with $0 < t \leq \frac{n+m}{2}$

and $\frac{n+m}{2} < t \leq m$ and using the lower bound in Lemma 6.7 as well as (12.15) and (12.16), we obtain that

$$\begin{aligned} L(f^m(\sigma)) &\geq K_0^{-1} d(x_m, y_m) \geq K_0^{-1} K_1^{-1} ||\Delta s(m-n)|| \\ &\geq K_0^{-1} K_1^{-1} \Delta s_2(m-n) > K_0^{-1} K_1^{-1} C_5 \Delta s_2(\frac{m-n}{2}) \\ &\geq K_0^{-1} K_1^{-1} C_5 C_3 \Delta s_2(0) (\frac{m-n}{2})^{-\gamma} \\ &\geq K_0^{-1} K_1^{-1} C_5 C_3 2^{\gamma} (m-n)^{-\gamma} \frac{1}{\sqrt{1+\mu^2}} ||\Delta s(0)|| \\ &\geq K_0^{-2} K_1^{-2} C_5 C_3 2^{\gamma} (m-n)^{-\gamma} \frac{1}{\sqrt{1+\mu^2}} L(f^n(\sigma)). \end{aligned}$$

Therefore, for some $C_6 > 0$,

$$\frac{L(f^m(\sigma))}{L(f^n(\sigma))} \geq C_6 (m-n)^{-\gamma}.$$

To prove the upper bound we use Lemmas 6.3, 6.4 with $0 < t \leq \frac{n+m}{2}$ and $\frac{n+m}{2} < t \leq m$, the upper bound in Lemma 6.7 as well as inequalities (12.15) and (12.16). Then the arguments similar to the preceding yield that, for some $C_7 > 0$,

$$\frac{L(f^m(\sigma))}{L(f^n(\sigma))} \leq C_7 (m-n)^{-\gamma'}.$$

This completes the proof of the lemma. □

12.7 Proof of Theorem 3.1: A Lower Bound for the Tail of the Return Time

In this section we establish a polynomial lower bound on the decay of the tail of the return time that is $m(\{x \in \Lambda \colon \tau(x) > n\})$. Consider the Markov partitions $\tilde{\mathcal{P}}$ and $\mathcal{P}$ for the automorphism A and the map $f_{\mathbb{T}^2}$ respectively and let $\tilde{P} \in \tilde{\mathcal{P}}$ and $P \in \mathcal{P}$ be the elements of the partitions as in Section 12.5.3. Fix the number Q as in Proposition 5.2. We assume that the partition $\mathcal{P}$ and the number r_0 are chosen such that Proposition 5.1 holds and we set again $f := f_{\mathbb{T}^2}$. Finally, we denote by

$$\mathcal{N} = \{n \in \mathbb{N} \colon \text{ there is } x \in P \text{ such that } n = \tau(x)\}.$$

Lemma 7.1. *There exists an integer $Q_1 > 0$ such that for any $N > 0$ one can find $n > N$ with $n \in \mathcal{N}$, an s-subset Λ_ℓ^s with $\tau(\Lambda_\ell^s) = n$ and numbers $0 < m_1 < m_2$ satisfying $m_1 < Q_1$, $n - m_2 < Q_1$ such that $f^k(\Lambda_\ell^s) \cap D_{r_0}^1 = \emptyset$ for all $0 \leq k < m_1$ or $m_2 < k \leq n$ and $f^k(\Lambda_\ell^s) \cap D_{r_0}^1 \neq \emptyset$ for all $m_1 \leq k \leq m_2$.*

Proof. It suffices to show that there is $Q_1 > 0$ such that for any $N > 0$ there is an admissible word of length $n > N$ with $n \in \mathcal{N}$ of the form

$$P\bar{W}_1\bar{P}_i\bar{W}_2P, \tag{12.17}$$

where the words $\bar{W}_1$ and $\bar{W}_2$ are of length $l(\bar{W}_j) < Q_1$ for $j = 1, 2$ and do not contain any of the symbols P or P_k (the element of the Markov partition containing x_k for $k = 1, 2, 3, 4$), and the word $\bar{P}_i$ consists of the symbol P_i which is repeated $n - 2 - l(\bar{W}_1) - l(\bar{W}_2)$ times. Since the map f is topologically conjugate to A, it is enough to find an admissible word of the form (12.17) which consists of the corresponding elements of the partition $\tilde{\mathcal{P}}$.

Note that $A = B^3$ where B is an automorphism of the torus given by the matrix $B = \begin{pmatrix} 1 & 1 \\ 1 & 2 \end{pmatrix}$. Therefore the result would follow if we find an admissible word of the type of (12.17) for the automorphism B. To this end consider the stable and unstable separatrices through the origin and denote the "first" connected component of their intersection with P by γ^s and γ^u respectively. It takes finitely many iterates of G and G^{-1} for each of these curves to completely enter the disk $D^1_{r_0}$. Now for each sufficiently large $n > 0$ with $n \in \mathcal{N}$ there is an s-set Λ^s_ℓ with $\tau(\Lambda^s_\ell) = n$ which completely enters $D^1_{r_0}$ (under iterates of G and G^{-1}) at the same time as γ^s and γ^u respectively. This completes the proof of the lemma. □

Lemma 7.2. *There exists a constant $C_8 > 0$ such that*

$$m(\{x \in \Lambda \colon \tau(x) > n\}) > C_8 n^{-(\gamma-1)},$$

where γ is defined in (12.8).

Proof. Using the conjugacy (12.6), it suffices to prove the lemma for the map G. Write

$$\begin{aligned} m(\{x \in \Lambda \colon \tau(x) > n\}) &= \sum_{N=n+1}^{\infty} m(\{x \in \Lambda \colon \tau(x) = N\}) \\ &= \sum_{N=n+1}^{\infty} \sum_{\Lambda^s_p : \tau(\Lambda^s_p) = N} m(\Lambda^s_p) > \sum_{N=n+1}^{\infty} m(\Lambda^s_\ell), \end{aligned}$$

where Λ^s_ℓ is the set constructed in Lemma 7.1. We wish to obtain a polynomial bound for the measure of the set Λ^s_ℓ.

Given $x \in \Lambda^s_\ell$, denote by $\gamma^s_\ell(x) := \gamma^s(x) \cap \Lambda^s_\ell$ (recall that $\gamma^s(x)$ is the full-length stable curve through x in the element P of the Markov partition). There is $K_1 > 0$ such that

$$m(\Lambda^s_\ell) = m(G^N(\Lambda^s_\ell)) = K_1 L(G^N(\gamma^s_\ell(x))), \tag{12.18}$$

where L stands for the length of the curve.

Let $x_j = G^j(x)$ for $j = 0, \dots, n$. Assume that x enters the region $D^1_{r_0}$ at time k_1 and exits at time k_2, that is,

(1) $G^j(x) \notin D^1_{r_0}$ if $0 \le j < k_1$, or $k_2 < j \le N$;
(2) $G^j(x) \in D^1_{r_0}$ if $k_1 \le j \le k_2$.

Note that for $0 \le j < k_1$ and $k_2 < j \le N$ the curve $G^j(\gamma^s_\ell(x))$ lies in the stable cone for the automorphism A at x_j and indeed, is an admissible manifold for A (i.e., for any $y \in \gamma^s_\ell(x)$ the line $T_y\gamma^s_\ell(x)$ lies in the stable cone at y). So the length of the curve $\gamma^u_\ell(x)$ expands exponentially outside of the region D_{r_0}. Since by Lemma 7.1, $k_1 < Q_1$ and $N - k_2 < Q_1$, we have that

$$L(\gamma^s_\ell(x)) = \lambda^{k_1} L(G^{k_1}(\gamma^s_\ell(x))) \le \lambda^{Q_1} L(G^{k_1}(\gamma^s_\ell(x))) \tag{12.19}$$

and

$$L(G^N(\gamma^s_\ell(x))) = \lambda^{-(N-k_2)} L(G^{k_2}(\gamma^s_\ell(x))) \ge \lambda^{-Q_1} L(G^{k_2}(\gamma^s_\ell(x))), \tag{12.20}$$

where λ is the largest eigenvalue of the matrix A.

By Lemma 5.6 in [21], the time the trajectory spends in $D^1_{r_0} \setminus D^1_{\frac{r_0}{2}}$ is uniformly bounded. Thus, by Lemma 6.8,

$$L(G^{k_2}(\gamma^s_\ell(x))) > C_6(k_2 - k_1)^{-\gamma} L(G^{k_1}(\gamma^s_\ell(x))). \tag{12.21}$$

Since $k_2 - k_1 < N$, combining Equations (12.18)–(12.21) yields

$$\begin{aligned} m(\Lambda^s_\ell) &\ge K_2 L(G^N(\gamma^s_\ell(x))) = K_2\lambda^{-Q_1} L(G^{k_2}(\gamma^s_\ell(x))) \\ &\ge K_2 C_6 \lambda^{-Q_1}(k_2 - k_1)^{-\gamma} L(G^{k_1}(\gamma^s_\ell(x))) \\ &\ge K_2 C_6 \lambda^{-2Q_1}(k_2 - k_1)^{-\gamma} L(\gamma^s_\ell(x)) \ge K_3 N^{-\gamma}, \end{aligned}$$

where $K_2 > 0$ is a constant and $K_3 = K_2 C_6 \lambda^{2Q_1} L(\gamma^u_\ell(x))$.

Note that $\gamma^s_\ell(x)$ is a full-length stable curve in P and hence, has length which is independent of N. It follows that

$$m(\{x \in \Lambda : \tau(x) > n\}) > \sum_{N=n+1}^{\infty} m(\Lambda^s_\ell) > C_8 \frac{1}{n^{\gamma-1}},$$

where $C_8 > 0$ is a constant. The desired lower bound follows. □

12.8 Proof of Theorem 3.1: An Upper Bound for the Tail of the Return Time

In this section we obtain an upper polynomial bound for the decay of the tail of the return time. As before, we assume that the Markov partition and the number r_0 are chosen such that Proposition 5.1 holds. Recall that D_r is the union of the disks D^i_r around the points x_i and P_i is the element of the partition containing x_i,

$i = 1, 2, 3, 4$. We have that $D^i_{r_0} \subset P_i$. Using the conjugacy (12.6), it suffices to establish that upper bound for the map G.

Given an s-set $\Lambda^s_i \subset P$ with $\tau(\Lambda^s_i) = n$, choose any numbers $k = k(\Lambda^s_i)$, $p = p(\Lambda^s_i)$, and two finite collections of numbers $\{k_m \geq 0\}_{m=1,\ldots,p}$ and $\{l_m \geq 0\}_{l=0,\ldots,p}$ such that

(1) $k_1 + k_2 + \cdots + k_p = k$ and $l_1 + l_2 + \cdots + l_{p+1} = n - k$;
(1) the trajectory of the set Λ^s_i under G^j, $0 \leq j \leq n$, consecutively spends l_m-times outside D_{r_0} and k_m-times inside D_{r_0}.

Given $0 < p < k < n$, consider the collections

$$\mathcal{S}_{k,n,p} = \{\Lambda^s_i \subset P\colon \tau(\Lambda^s_i) = n,\ k = k(\Lambda^s_i),\ p = p(\Lambda^s_i)\}.$$

Lemma 8.1. *There are $0 < h < h_{top}(f)$, $\varepsilon_0 > 0$, and $C_9 > 0$ such that $\varepsilon_0 < h_{top}(f) - h$ and*

$$Card\ \mathcal{S}_{k,n,p} \leq C_9 \frac{1}{p^2} e^{(h+\varepsilon_0)(n-k)}.$$

Proof. Note that the cardinality of $\mathcal{S}_{k,n,p}$ does not exceed the number of symbolic words of length n that start and end at P and contain exactly k symbols P_j, $j = 1, 2, 3, 4$. Since A is topologically mixing, given the positions of P_j, the latter is exactly the number of words of length $n - k$ that start and end at P. By Corollary 1.9.12 and Proposition 3.2.5 in [KtH], the number of such words grows exponentially with an exponent that does not exceed $(n - k)h$ where $0 < h < h_{\text{top}}(A)$.

The number of different ways the iterates of Λ^s_i can enter D_{r_0} exactly p times and stay in this set exactly k times does not exceed the number of ways in which the number k can be written as a sum of p positive integers (where order matters) which is equal to $\binom{k-1}{p-1}$. The number of different ways the iterates of Λ^s_i can spend outside D_{r_0} exactly $p+1$ times is equal to the number of ways in which the number $n - k$ can be written as a sum of $p + 1$ positive integers which is $\binom{n-k-1}{p}$. Since iterates of Λ^s_i may enter any of the disks $D^i_{r_0}$, $i = 1, 2, 3, 4$, we obtain

$$\text{Card } \mathcal{S}_{k,n,p} \leq C_9 4^p \binom{k-1}{p-1}\binom{n-k-1}{p} e^{h(n-k)}.$$

Write

$$\text{Card } \mathcal{S}_{k,n,p} \leq \frac{C_9}{p^2} p^2 4^p \binom{k-1}{p-1}\binom{n-k-1}{p} e^{h(n-k)}.$$

To prove the lemma we wish to estimate $p^2 4^p \binom{k-1}{p-1}\binom{n-k-1}{p}$ and we claim that there is $\varepsilon_0 > 0$ such that $\binom{k-1}{p-1} < e^{\varepsilon_0(n-k)}$.

To this end, note that by Propositions 5.1 and 5.2, it takes Λ_i^s at least Q iterates before it enters D_{r_0} again. This implies that $n = k + l_1 + \cdots + l_{p+1} > k + (p+1)Q$, that is, $p + 1 < \frac{n-k}{Q}$.

For a fixed k note that $\binom{k-1}{p-1}$ achieves its maximum when $p - 1 = [\frac{k-1}{2}]$ or $p - 1 = [\frac{k-1}{2}] + 1$. We may assume $p - 1 = \frac{k-1}{2}$. Then using the asymptotic formula $\binom{m}{l} \sim (\frac{me}{l})^l$, we obtain that

$$\binom{k-1}{p-1} < \binom{2p-2}{p-1} < 4^{p-1} = e^{(p-1)\ln 4} < e^{\frac{n-k}{Q}\ln 4}.$$

To estimate $\binom{n-k-1}{p}$ observe that p does not exceed $\frac{n-k}{Q}$. Hence, using the preceding asymptotic formula, we find that

$$\binom{n-k-1}{p} < \binom{n-k}{\frac{n-k}{Q}} < \left(\frac{(n-k)e}{\frac{n-k}{Q}}\right)^{\frac{n-k}{Q}}$$
$$< e^{\frac{n-k}{Q}\ln\frac{(n-k)e}{\frac{n-k}{Q}}} < e^{\frac{n-k}{Q}\ln(Qe)}.$$

Finally, note that

$$p^2 4^p < e^{2\ln p + p\ln 4} < e^{2p + p\ln 4} < e^{\frac{n-k}{Q}(\ln 4 + 2)}.$$

Now, given any sufficiently small $\varepsilon_0 > 0$, one can choose Q large enough so that $\frac{\ln 4 + 2}{Q} + \frac{\ln 4}{Q} + \frac{\ln(Qe)}{Q} < \varepsilon_0$. Combining the preceding estimates we obtain

$$4^p\binom{k-1}{p-1}\binom{n-k-1}{p}e^{h(n-k)} < e^{(n-k)\varepsilon_0}e^{h(n-k)} = e^{(n-k)\varepsilon_0 + h}$$

and hence,

$$\text{Card } S_{k,n,p} \le C_9 \frac{1}{p^2} e^{(h+\varepsilon_0)(n-k)}.$$

This completes the proof of the lemma. □

Lemma 8.2. *There exists $\varepsilon_0 > 0$ such that for any $\Lambda_i^s \in S_{k,n,p}$,*

$$m(\Lambda_i^s) \le C_{10} k^{-\gamma'} e^{(-\log\lambda + \varepsilon_0)(n-k)},$$

where $C_{10} > 0$ is a constant and γ' is given by (12.8).

Proof. Note that by (12.18), $m(\Lambda_i^s) = m(G^n(\Lambda_i^s)) = K_1 L(G^n(\gamma_i^s(x)))$ and that the length of the backward iterates of γ_i^s lying outside the region D_{r_0} are stretched by the largest eigenvalue λ of the matrix A. Note also that every time the iterates of γ_i^s enter the region we have an upper estimate for its length according to Lemma 6.8

(note that we can apply this lemma in the region D_{r_0} since by Lemma 5.6 in [21] the time spent in $D_{r_0} \setminus D_{\frac{r_0}{2}}$ is uniformly bounded). Thus,

$$\begin{aligned} m(\Lambda_i^s) &= m(G^n(\Lambda_i^s)) = K_1 L(G^n(\gamma_i^s(x))) \\ &= K_1 \lambda^{-l_{p+1}} L(G^{n-l_{p+1}}(\gamma_i^s(x))) \\ &\le K_1 C_7 \lambda^{-l_{p+1}} k_p^{-\gamma'} L(G^{n-(l_1+k_1)}(\gamma_i^s(x))) \le \cdots \\ &\le K_1 C_7^p \lambda^{-(l_{p+1}+\cdots+l_1)} k_p^{-\gamma'} k_{p-1}^{-\gamma'} \cdots k_1^{-\gamma'} L(\gamma_i^s(x)). \end{aligned}$$

Note that $\gamma_i^s(x)$ is a full-length stable curve in P and hence, has length independent of n. One can also assume that $k_i \ge 2$ by making r_0 smaller if necessary. This implies

$$k_1 k_2 \cdots k_p \ge k_{\max} 2^{p-1} \ge k_{\max} p \ge \sum_{i=1}^{p} k_i = k,$$

where $k_{\max}$ denotes the largest of k_i's.

In addition, $C_6^p = e^{p \ln C_6} < e^{\frac{n-k}{Q} \ln C_6} < e^{\varepsilon_0 (n-k)}$ for sufficiently small $\varepsilon_0 > 0$ if one chooses Q large. Therefore,

$$m(\Lambda_i^s) < K_1 e^{\varepsilon_0(n-k)} \lambda^{-(n-k)} k^{-\gamma'} < C_{10} k^{-\gamma'} e^{(-\log \lambda + \varepsilon_0)(n-k)}.$$

This completes the proof of the lemma. □

Lemma 8.3. *There exists* $C_{11} > 0$ *such that*

$$m(\{x \in \Lambda : \tau(x) > n\}) < C_{11} n^{-(\gamma'-1)};$$

see (12.8) *for the definition of* γ'.

Note that

$$m(\{x \in \Lambda : \tau(x) = n\}) \le \sum_{k=1}^{n} \sum_{p=1}^{k} \max_{\Lambda_i^s \in \mathcal{S}_{k,n,p}} \{m(\Lambda_i^s)\} \text{Card } \mathcal{S}_{k,n,p}.$$

Therefore, by Lemmas 8.1 and 8.2, we have

$$\begin{aligned} &m(\{x \in \Lambda : \tau(x) = n\}) \\ &\quad \le \sum_{k=1}^{n} \sum_{p=1}^{k} \frac{1}{p^2} C_9 e^{(h+\varepsilon_0)(n-k)} C_{10} e^{(\varepsilon_0 - \log \lambda)(n-k)} k^{-\gamma'} \\ &\quad < C_9 C_{10} \frac{\pi^2}{6} e^{-\delta n} \sum_{k=1}^{n} e^{\delta k} k^{-\gamma'}, \end{aligned} \tag{12.22}$$

where $\delta = 2\varepsilon_0 + \log \lambda - h > 0$ if ε_0 is sufficiently small.

To estimate $\sum_{k=1}^{n} e^{\delta k} k^{-\gamma'}$ set $u_k = e^{\delta k} k^{-\gamma'}$ and note that $u_{k+1} - u_k \sim e^{\delta k} k^{-\gamma'} = u_k$. Since $\sum_{k=1}^{n} u_k$ is positive and diverges, by the Stolz–Cesaro theorem,

$$\sum_{k=1}^{n} u_k \sim \sum_{k=1}^{n} u_{k+1} - u_k = u_{n+1} - u_1 \sim e^{\delta n} n^{-\gamma'}.$$

Therefore,

$$m(\{x \in \Lambda \colon \tau(x) = n\}) \leq C_9 C_{10} e^{-\delta n} \sum_{k=1}^{n} e^{\delta k} k^{-\gamma'} < C_9 C_{10} n^{-\gamma'}.$$

Thus, we have the following estimate of the tail

$$m(\{x \in \Lambda \colon \tau(x) > n\}) = \sum_{k>n} m(\{x \in \Lambda \colon \tau(x) = k\}) < C_{11} n^{-(\gamma'-1)}$$

for some $C_{11} > 0$. This concludes the proof of the Lemma and the upper bound.

12.9 Proof of Theorem 3.1: Carrying the Slowdown Map to a Surface

In this section we show how to carry over the slowdown map of the torus to a measure-preserving diffeomorphism of any surface. Following [12], we will construct the maps $\varphi_1, \varphi_2, \varphi_3$ such that the following diagram is commutative:

$$\begin{array}{ccccccc}
\mathbb{T}^2 & \xrightarrow{\varphi_1} & S^2 & \xrightarrow{\varphi_2} & D^2 & \xrightarrow{\varphi_3} & M \\
\Big\downarrow{\scriptstyle f_{\mathbb{T}^2}} & & \Big\downarrow{\scriptstyle f_{S^2}} & & \Big\downarrow{\scriptstyle f_{D^2}} & & \Big\downarrow{\scriptstyle f_M} \\
\mathbb{T}^2 & \xrightarrow{\varphi_1} & S^2 & \xrightarrow{\varphi_2} & D^2 & \xrightarrow{\varphi_3} & M
\end{array}$$

We stress that, while our construction of the maps φ_1 and φ_2 follows [12], our construction of the map φ_3 is quite different, since we have to deal with finite regularity of the slowdown map.

First, using the slowdown map, we construct a diffeomorphism of the sphere S^2.

Proposition 9.1 (see [12]). *There exists a map $\varphi_1 \colon \mathbb{T}^2 \to S^2$ satisfying:*

(1) φ_1 *is a double-branched covering, is one-to-one on each branch, and C^∞ everywhere except at the points x_i, $i = 1, 2, 3, 4$ where it branches;*
(2) $\varphi_1 \circ I = \varphi_1$ *where $I \colon \mathbb{T}^2 \to \mathbb{T}^2$ is the involution map given by $I(t_1, t_2) = (1 - t_1, 1 - t_2)$;*
(3) φ_1 *preserves area, namely $(\varphi_1)_* m = m_{S^2}$ where m_{S^2} is the area in S^2;*

(4) *there exists a coordinate system in each disk* $D^i_{r_0}$ *such that*

$$\varphi_1(s_1, s_2) = \left(\frac{{s_1}^2 - {s_2}^2}{\sqrt{{s_1}^2 + {s_2}^2}}, \frac{2s_1 s_2}{\sqrt{{s_1}^2 + {s_2}^2}} \right);$$

(5) *The map* $f_{S^2} := \varphi_1 \circ f_{\mathbb{T}^2} \circ \varphi_1^{-1}$ *preserves the area.*

The sphere can be unfolded onto the unit disk D^2 and the map f_{S^2} can be carried over to an area-preserving map f_{D^2} of the disk which is identity on the boundary of the disk. To see this set $p_i = \varphi_1(x_i)$, $i = 1, 2, 3, 4$. In a small neighborhood of the point p_4 we define a map φ_2 by

$$\varphi_2(\tau_1, \tau_2) = \left(\frac{\tau_1 \sqrt{1 - \tau_1^2 - \tau_2^2}}{\sqrt{\tau_1^2 + \tau_2^2}}, \frac{\tau_2 \sqrt{1 - \tau_1^2 - \tau_2^2}}{\sqrt{\tau_1^2 + \tau_2^2}} \right).$$

One can extend φ_2 to an area-preserving C^∞ diffeomorphism (still denoted by φ_2) between $S^2 \setminus \{p_4\}$ and the interior of the unit disk D^2. The map

$$f_{D^2} := \begin{cases} \varphi_2 \circ f_{S^2} \circ \varphi_2^{-1} & \text{on int} D^2 \\ Id & \text{on } \partial D^2 \end{cases} \tag{12.23}$$

is a diffeomorphism of D^2 that preserves area m_{D^2}.

Proposition 9.2. *The maps* f_{S^2} *and* f_{D^2} *are of class of smoothness* $C^{2+2\kappa}$ *where* $\kappa = \frac{\alpha}{1-\alpha}$.

Proof. Using the explicit local expressions for f_{S^2} and f_{D^2} and following the arguments in Proposition 4.2, we find that the maps f_{S^2} and f_{D^2} are Hamiltonian with respect to the area and the Hamiltonian functions are given as

$$H_3(\tau_1, \tau_2) = \frac{\tau_2 h(\sqrt{\tau_1^2 + \tau_2^2})}{\sqrt{\tau_1^2 + \tau_2^2}} \log \lambda$$

and

$$H_4(x_1, x_2) = \frac{x_2 h(\sqrt{1 - x_1^2 - x_2^2})}{\sqrt{x_1^2 + x_2^2}} \log \lambda$$

respectively. Here, as before, $h(u) = u^{\frac{2}{1-\alpha}}$.

To show that the maps f_{S^2} and f_{D^2} are of the desired class of smoothness, we will show that the Hamiltonian functions H_3 and H_4 have Hölder continuous second-order partial derivatives with Hölder exponent 2κ. Since H_3 and H_4 are of the same regularity, we consider only one of them and we set $g(x,y) = y(x^2+y^2)^\delta$ where $\delta = \frac{1}{1-\alpha} - \frac{1}{2}$.

Obviously, $\frac{\partial g}{\partial x}(0,0) = 0$ and

$$\frac{\partial g}{\partial x}(x,y) = \begin{cases} 2\delta xy(x^2+y^2)^{\delta-1}, & (x,y) \neq (0,0); \\ 0, & (x,y) = (0,0). \end{cases}$$

The function $\frac{\partial g}{\partial x}$ is symmetric, so we will only study Hölder continuity of $\frac{\partial^2 g}{\partial x^2}$ as Hölder continuity of $\frac{\partial^2 g}{\partial x \partial y}$ is immediate. Note that

$$\frac{\partial^2 g}{\partial x^2}(x,y) = \begin{cases} 2\delta \frac{y(x^2+y^2)^{1-\delta} - 2(1-\delta)x^2 y(x^2+y^2)^{-\delta}}{(x^2+y^2)^{2-2\delta}} & (x,y) \neq (0,0); \\ 0 & (x,y) = (0,0). \end{cases}$$

Since the function $\frac{\partial^2 g}{\partial x^2}(x,y)$ is differentiable for all $(x,y) \neq (0,0)$, it is Hölder continuous for all pairs of nonzero points (x,y). It remains to show Hölder continuity for pairs of points one of which is zero. We can write

$$\frac{\partial^2 g}{\partial x^2} = K_1 y(x^2+y^2)^{\delta-1} - K_2 x^2 y(x^2+y^2)^{\delta-2},$$

where $K_1 > 0$ and $K_2 > 0$ are some constants. Choose $(x,y) \neq (0,0)$ and note that

$$|y(x^2+y^2)^{\delta-1}| \leq (x^2+y^2)^{\frac{1}{2}}(x^2+y^2)^{\delta-1} = (x^2+y^2)^{\delta-\frac{1}{2}} = d((x,y),(0,0))^{2\delta-1}.$$

Similarly,

$$|x^2 y(x^2+y^2)^{\delta-2}| \leq d((x,y),(0,0))^{2\delta-1}.$$

Hence, $\frac{\partial^2 g}{\partial x^2}$ is Hölder continuous with Hölder exponent

$$2\delta - 1 = 2\Big(\frac{1}{1-\alpha} - \frac{1}{2}\Big) - 1 = 2\kappa.$$

Further, each of the functions $\frac{\partial^2 g}{\partial y^2}$ and $\frac{\partial^2 g}{\partial y \partial x}$ can be written as a linear combinations of functions

$$x(x^2+y^2)^{\delta-1}, \quad y(x^2+y^2)^{\delta-1}, \quad x^2 y(x^2+y^2)^{\delta-2}, \quad y^2 x(x^2+y^2)^{\delta-2} \tag{12.24}$$

and are 0 at the point $(x,y) = (0,0)$. Arguing in this manner, one can show that each function in (12.24) is Hölder continuous with Hölder exponent 2κ. This completes the proof of the proposition. □

Consider a smooth compact connected oriented surface M. It can be cut along closed geodesics in such a way that the resulting surface with boundary is homeomorphic to a regular polygon via a homeomorphism which we denote by T; this is a well-known topological construction.

Let now f be a C^∞ diffeomorphism of the disk, which is identity on the boundary ∂D^2 and is *infinitely flat*, that is, given a sequence $\rho_n \to 0$ and a sequence of open domains $V_n \subset D^2$ satisfying

$$V_n \subset V_{n+1} \text{ and } \bigcup_{n\geq 1} V_n = D^2, \tag{12.25}$$

we have that for every $n \geq 1$,

$$\|f - \mathrm{Id}\|_{C^n(V_n)} \leq \rho_n.$$

For such an f it is shown in [12] (see also [2]) that there is a homeomorphism $\varphi\colon \overline{D^2} \to M$ such that

(1) φ is of class C^∞ in the interior of the disk;
(2) φ is area preserving, namely $h_* m_{D^2} = m_M$;
(3) the map $\varphi \circ f \circ \varphi^{-1}$ is a C^∞ area-preserving diffeomorphism of the surface.

In our case, however, the map $f = f_{D^2}$ is only of class $C^{2+2\kappa}$ and hence is only *finitely flat* at the boundary, that is, there is a sequence of open domains $V_n \subset D^2$, satisfying (12.25), such that for every $0 < \beta < 2 + 2\kappa$,

$$\|f_{D^2} - \mathrm{Id}\|_{C^{1+\beta}(V_n)} \leq (r_{n-1})^{2+2\kappa-\beta}, \tag{12.26}$$

where $r_n = \mathrm{dist}(V_n, \partial D^2)$. This requires us to develop a specific construction of the homeomorphism φ which guarantees that the map f_M is an area-preserving diffeomorphism of class $C^{1+\beta}$ for some $\beta > 0$. More precisely, the following statement holds.

Theorem 9.3. *Given a smooth compact connected oriented surface M and numbers $\frac{1}{9} < \alpha < \frac{1}{4}$ and $0 < \mu < \frac{1}{2}$, there exist $\beta = \beta(\alpha, \mu) > 0$ and a continuous map $\varphi_3\colon \overline{D^2} \to M$ such that*

(1) *the restriction $\varphi_3|\mathrm{int}\, D^2$ is a diffeomorphic embedding;*
(2) *$\varphi_3(\overline{D^2}) = M$;*
(3) *φ_3 preserves area; more precisely, $(\varphi_3)_* m_{D^2} = m_M$ where m_M is the area in M; moreover, $m_M(M \setminus \varphi_3(\mathrm{int}\, D^2)) = 0$;*
(4) *the map $f_M := \varphi_3 \circ f_{D^2} \circ \varphi_3^{-1}$ is a $C^{1+\beta}$ area-preserving diffeomorphism of the surface.*

Proof. One can represent a compact smooth oriented surface M as a regular p-polygon P (the number p is even) whose angles are $\alpha = \frac{\pi(p-2)}{p}$. Let $A_1, A_2, \ldots, A_p$ be vertices of the polygon and O its center. For each $i = 1, \ldots, p$ denote by B_i the

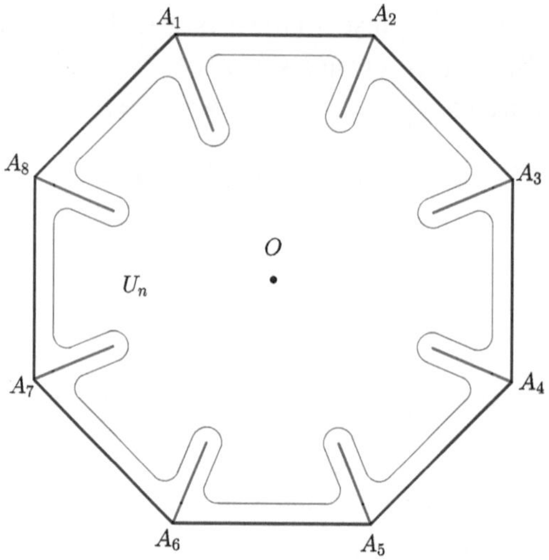

Figure 12.1 The shape of the set U_n

points on the segment A_iO for which $\frac{|A_iB_i|}{|A_iO|} = \frac{1}{3}$. In what follows we assume that $A_{p+1} = A_1$ and $B_{p+1} = B_1$. Denote by

$$P^* := \Big(\bigcup_{i=1}^{p} A_iA_{i+1}\Big) \cup \Big(\bigcup_{i=1}^{p} A_iB_i\Big). \tag{12.27}$$

Note that the complement to P^* is an open simply connected set.

We now construct a homeomorphism from the unit disk D^2 onto P.

Proposition 9.4. *There exist*

(1) *a nested sequence of open simply connected sets* $U_0 \subset U_1 \subset \cdots \subset U_n \subset \cdots$ *satisfying* $\bigcup_n U_n = P \setminus P^*$;
(2) *a sequence of* C^∞ *diffeomorphisms* $h_n \colon U_n \to U_{n+1}$ *for* $n \geq 0$;
(3) *a number* $\beta > 0$

such that setting $h(x) = \lim_{n\to\infty} h_{n-1} \circ \cdots \circ h_1 \circ h_0$, *we have that the map* $f_P \colon P \to P$ *given by*

$$f_P = \begin{cases} (h \circ f_{D^2} \circ h^{-1})(x), & x \in P \setminus P^*, \\ Id & \text{otherwise} \end{cases}$$

is a $C^{1+\beta}$ *diffeomorphism.*

Proof of the Proposition. We split the proof into three steps.

Step 1. We first construct a sequence of open sets U_n (see Figure 12.1).

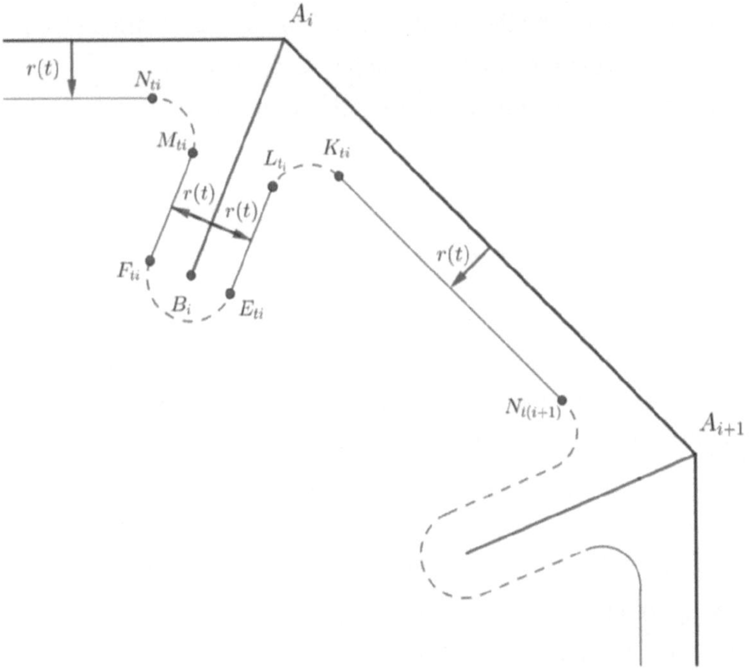

Figure 12.2 The collection of marked points

Fix $n > 0$, $t \in [n, n+1]$ and let $r(t) > 0$ be a strictly monotonically decreasing continuous function on $[1, \infty)$ which will be determined later in Step 4. For $i = 1, \ldots, p$ consider the following collection of points associated with the point A_i (see Figure 12.2):

- K_{ti}, the point that is determined uniquely by the requirements that the angle $\angle(K_{ti}A_iA_{i+1}) = \frac{1}{8}\alpha$ and $\text{dist}(K_{ti}, A_iA_{i+1}) = r(t)$;
- L_{ti}, the point that is determined uniquely by the requirements that the angle $\angle(OA_iL_{ti}) = \frac{1}{8}\alpha$ and $\text{dist}(L_{ti}, A_iO) = r(t)$;
- M_{ti}, the image of L_{ti} under the reflection about the line OA_i;
- N_{ti}, the image of K_{ti} under the reflection about the line OA_i;
- E_{ti}, the point on the line through B_i which is perpendicular to the line A_iB_i and such that $\text{dist}(E_{ti}, B_i) = r(t)$;
- F_{ti}, the image of E_{ti} under the reflection about the line A_iB_i.

We introduce the following curves: for $i = 1, \ldots, p$ let

- $\gamma_{ti}^{(1)}$ be a line segment connecting the points K_{ti} and $N_{t(i+1)}$ where we assume that $N_{t(p+1)} = N_{t1}$;
- $\gamma_{ti}^{(2)}$ be a curve connecting the points K_{ti} and L_{ti} to be determined later in Step 2;

- $\gamma_{ti}^{(3)}$ be a curve connecting the points M_{ti} and N_{ti} to be determined later in Step 2;
- $\gamma_{ti}^{(4)}$ be the line segment connecting the points L_{ti} and E_{ti};
- $\gamma_{ti}^{(5)}$ be the line segment connecting the points M_{ti} and F_{ti};
- $\gamma_{ti}^{(6)}$ be a curve connecting the points E_{ti} and F_{ti} to be determined later in Step 2.

Let τ_t be the curve given by

$$\tau_t = \bigcup_{i=1}^{p} \bigcup_{j=1}^{6} \gamma_{ti}^{(j)}.$$

By construction, τ_t, $t \in [n, n+1]$, is a closed connected continuous curve which bounds an open simply connected domain in the polygon P. We denote this domain by U_t. In particular, U_n is the desired open set.

Step 2. We show how to choose the curves $\gamma_{ti}^{(2)}$ and $\gamma_{ti}^{(6)}$. The curve $\gamma_{ti}^{(3)}$ can be chosen in a similar way.

Let $\varphi, \psi\colon [-1, 1] \to \mathbb{R}$ be two continuous functions satisfying:

(1) $\varphi \in C^\infty$ on $[0, 1]$ and $\psi \in C^\infty$ on $(0, 1)$;
(2) $\varphi(-x) = \varphi(x)$ and $\psi(-x) = \psi(x)$;
(3) $\varphi(0) = a$ where $1 - \tan\frac{\alpha}{4} < a < \cot\frac{\alpha}{4}$ and $\varphi(-1) = \varphi(1) = \cot\frac{\alpha}{8}$;
(4) $\psi(0) = 1$ and $\psi(-1) = \psi(1) = 0$;
(5) $0 < \varphi'(x) \le \cot\frac{\alpha}{4}$ for $0 < x \le 1$ and $-\cot\frac{\alpha}{4} \le \varphi'(x) < 0$ for $-1 \le x < 0$;
(6) $\varphi'(-1) = -\cot\frac{\alpha}{4}$ and $\varphi'(1) = \cot\frac{\alpha}{4}$;
(7) ψ is infinitely vertically flat at -1 and 1.

For $i = 1, \dots, p$ consider the orthogonal coordinate system with origin at A_i whose vertical axis is the bisector of the angle $\angle(B_iA_iA_{i+1})$ and for every $t \in [n, n+1]$ we let $\gamma_{ti}^{(2)}$ be the graph of the function $\varphi_t(x) = r(t)\varphi(\frac{x}{r(t)})$ where $-r(t) \le x \le r(t)$. It is easy to see that $\gamma_{ti}^{(2)}$ is a C^∞ curve that connects the points K_{ti} and L_{ti} and is infinitely tangent to the lines $K_{ti}N_{ti}$ and $L_{ti}E_{ti}$.

Now consider the orthogonal coordinate system with origin at B_i whose vertical axis is the line A_iB_i. We let $\gamma_{ti}^{(6)}$ be the graph of the function $\psi_t(x) = r(t)\psi(\frac{x}{r(t)})$ where $-r(t) \le x \le r(t)$. It is easy to see that $\gamma_{ti}^{(6)}$ is a C^∞ curve that connects the points E_{ti} and F_{ti} and is infinitely tangent to the lines $L_{ti}E_{ti}$ and $M_{ti}F_{ti}$. Hence, with the preceding choice of curves γ_{ti}^{j}, $j = 1, \dots, 6$, the curve τ_t is of class C^∞.

We show that the curves τ_t corresponding to different values of t are disjoint. To this end fix $n \le t_1 < t_2 \le n+1$. It suffices to show that the curves $\gamma_{ti}^{(2)}$, $\gamma_{ti}^{(6)}$, and $\gamma_{ti}^{(3)}$ with $t = t_1$ and $t = t_2$ are disjoint. We will prove this for the curve $\gamma_{ti}^{(2)}$ only as the proof for other curves is similar. By Property (3),

$$\varphi_{t_1}(0) = ar(t_1), \quad \varphi_{t_1}(r(t_1)) = r(t_1)\cot\frac{\alpha}{8}, \quad \varphi'_{t_1}(r(t_1)) = \cot\frac{\alpha}{4}.$$

Similarly,

$$\varphi_{t_2}(0) = ar(t_2), \quad \varphi_{t_2}(r(t_2)) = r(t_2)\cot\frac{\alpha}{8}, \quad \varphi'_{t_2}(r(t_2)) = \cot\frac{\alpha}{4}.$$

In view of Property (5), the desired result would follow if we show that

$$\varphi_{t_1}(r(t_2)) = r(t_1)\varphi\Big(\frac{r(t_2)}{r(t_1)}\Big) \geq \varphi_{t_2}(r(t_2)) = r(t_2)\cot\frac{\alpha}{8}.$$

Setting $x = \frac{r(t_2)}{r(t_1)}$, the above inequality amounts to $\varphi(x) \geq x\cot\frac{\alpha}{8}$ and immediately follows from Properties (3) and (4) of the function φ.

Step 3. We now construct maps h_n. By the Riemann Mapping theorem, there is a C^∞ diffeomorphism $h_0 \colon D^2 \to U_1$. For each $n = 1, 2, 3, \dots$ we will construct maps $h_n \colon U_n \to U_{n+1}$ such that $h_n \big| U_{n-1} = \mathrm{Id}$.

Given two numbers $n - 1 \leq s \leq n$ and $n - 1 \leq t \leq n + 1$ such that $s < t$, we construct a C^∞ diffeomorphism $\hat{h}_{st} \colon \tau_s \to \tau_t$ in the following way.

- $\hat{h}_{st} \colon \gamma_{si}^{(1)} \to \gamma_{ti}^{(1)}$ is a linear map, given by $\hat{h}_{st}(z) = \frac{r(t)}{r(s)}(z)$, $z \in \gamma_{si}^{(1)}$ and $i = 1, \dots, p$;
- $\hat{h}_{st} \colon \gamma_{si}^{(j)} \to \gamma_{ti}^{(j)}$ is a map, given by $\hat{h}_{st}(z) = v$, where $z = (y, \varphi_s(y))$ and $v = (\frac{r(t)}{r(s)}y, \varphi_t(y))$ for $-r(s) \leq y \leq r(s)$, $i = 2, \dots, p, j = 2, \dots, 6$.

Now given $n - 1 \leq s \leq n$, define the map $\hat{h}_s := \hat{h}_{st}$ with $t = 2(s - n + 1) + n - 1$. The desired map $h_n \colon U_n \to U_{n+1}$ is now given as follows: for $A \in U_n$ choose a unique s such that $A \in \tau_s$ with $n - 1 \leq s \leq n$ and then set $h_n(A) = B$ where $B = \hat{h}_s(A) \in \tau_{2(s-n+1)+n-1} \subset U_{n+1}$. It is easy to see that h_n is a C^∞ diffeomorphism.

It follows that the map $h = \lim\limits_{n\to\infty} h_{n-1} \circ \cdots \circ h_1 \circ h_0$ is a well-defined C^∞ diffeomorphism from $\mathrm{int}D^2$ onto $P \setminus P^*$ where P^* is given by (12.27). It also follows from the construction of the map h that there is $C > 0$ such that

$$\|h\|_{C^1} \leq C, \quad \|h^{-1}\|_{C^1} \leq C. \tag{12.28}$$

Step 4. It remains to show that the map $f_P = h \circ f_{D^2} \circ h^{-1}$ is a $C^{1+\beta}$ diffeomorphism for some $\beta > 0$.

Observe that $f_P(P \setminus P^*) = P \setminus P^*, f_P(P^*) = P^*$, and $f_P|P^* = \mathrm{Id}$. In particular, $f_P|P \setminus P^*$ is a C^∞ diffeomorphism. It remains to show that f_P is of class $C^{1+\beta}$ on P^*. To do so, we will show the following:

$$\|f_P - \mathrm{Id}\|_{C^{1+\beta}(U_{n+1}\setminus U_n)} \to 0 \tag{12.29}$$

as $n \to \infty$.

First, we will prove the following lemma.

Lemma 9.5. *Let $r(n)$ be a decreasing sequence such that $0 < r(1) < 1$ and*

$$r(n+1) = r^2(n), \quad 0 < r(1) < 1. \tag{12.30}$$

Then the sequence of open sets $V_n = h^{-1}(U_n)$ satisfies (12.25) *and*

$$V_{n-1} \subset f_{D^2}(V_n) \subset V_{n+1}.$$

Proof of the Lemma. Note that there are $C_2 \geq C_1 > 0$ such that

$$C_1 r(n) \leq \operatorname{dist}(U_n, P^*) \leq C_2 r(n).$$

Furthermore, in view of (12.28) there are $C_4 \geq C_3 > 0$ such that

$$C_3 r(n) \leq \operatorname{dist}(V_n, \partial D^2) \leq C_4 r(n).$$

Since the map f_{D^2} is identity on ∂D^2 and is of class of smoothness $2 + 2\kappa$, we obtain that, for all sufficiently small r_n, any x in the neighborhood $U_{r_n}(\partial D^2)$ and any $\beta > 0$,

$$\operatorname{dist}(x, f_{D^2}(x)) < r(n)^{2+2\kappa-\beta}.$$

Therefore, for some $0 < a < 1$,

$$C_3 r(n) - r(n)^{2+a} \leq \operatorname{dist}(f_{D^2}(x), \partial D^2) \leq C_4 r(n) + r(n)^{2+a}.$$

To prove the desired inclusion we will show that

$$C_4 r(n+2) < C_3 r(n) - r(n)^{2+a} \leq C_4 r(n) + r(n)^{2+a} < C_3 r(n-2).$$

We prove the leftmost inequality. Since $r(n+2) = r^4(n)$, we have

$$\begin{aligned} C_4 r(n+2) < C_3 r(n) - r(n)^{2+a} &\Leftrightarrow \\ C_4 \frac{r(n+2)}{r(n)} < C_3 - r(n)^{1+a} &\Leftrightarrow \\ C_4 r(n)^3 < C_3 - r(n)^{1+a}&. \end{aligned}$$

For large values of n both $C_4 r(n)^3$ and $r(n)^{1+a}$ are small. Hence, the last inequality holds.

We now prove the rightmost inequality (the inequality in the middle is obvious).

$$\begin{aligned} C_4 r(n) + r(n)^{2+a} < C_3 r(n-2) &\Leftrightarrow \\ C_4 + r(n)^{1+a} < C_3 \frac{r(n-2)}{r(n)} &\Leftrightarrow \\ C_4 + r(n)^{1+a} < C_3 r(n)^{-\frac{3}{4}}&. \end{aligned}$$

For large values of n, the left-hand side of the last inequality is close to C_4 while the right-hand side gets large. This completes the proof of the lemma.

The preceding lemma allows us to write

$$\|h \circ (f_{D^2} - \mathrm{Id}) \circ h^{-1}\|_{C^{1+\beta}(U_{n+1}\setminus U_n)} \leq \|h\|_{C^{1+\beta}(V_{n+2}\setminus V_{n-1})} \|f_{D^2} - \mathrm{Id}\|_{C^{1+\beta}(V_{n+1}\setminus V_n)} \|h^{-1}\|_{C^{1+\beta}(U_{n+1}\setminus U_n)}. \tag{12.31}$$

Observe that $\|f_{D^2} - \mathrm{Id}\|_{C^{1+\beta}}$ admits Estimate (12.26) and it remains to estimate the norms $\|h\|_{C^{1+\beta}(V_{n+2}\setminus V_{n-1})}$ and $\|h^{-1}\|_{C^{1+\beta}(U_{n+1}\setminus U_n)}$.

We write $h|_{V_{n+2}} = h_{n+1}|_{U_{n+1}}$, so in order to estimate the norm of h we will estimate the norm of h_{n+1}. In order to estimate the norm of h^{-1} we will need to estimate the norms of h_n^{-1} for each $n = 0, 1, 2, \cdots$.

Further, it suffices to estimate the norm of h restricted to the boundary of the sets U_n. Recall that τ_n, the boundary of U_n, is a union of the curves $\gamma_{ni}^{(j)}$, $i = 1, \dots, p$, $j = 1, \dots, 6$. Note that the map h acts linearly on the curves $\gamma_{ni}^{(1)}$ and hence, the norm of h restricted to these parts of the curve τ_n is bounded.

The curves $\gamma_{ni}^{(2)}$ and $\gamma_{ni}^{(3)}$ are the graphs of the function φ_n and the curves $\gamma_{ni}^{(6)}$ are the graphs of the function ψ_n. We shall only give an estimate of the norm of h restricted to the curves $\gamma_{ni}^{(2)}$, since the estimates of the norm of h restricted to other curves are similar. Also, we can assume that i and $j > 1$ are fixed.

Now for a fixed curve $\gamma_{ni}^{(2)}$ we define the orthogonal coordinate system centered at the vertex A_i with the vertical axis A_iO (recall that O are the graphs of the polygon P). In this coordinate system the map h_n is given by

$$h_n\colon (x, \varphi_n(x)) \to \left(\frac{r(n+1)}{r(n)}x, \varphi_{n+1}(x)\right), \quad -r(n) \leq x \leq r(n).$$

Since φ_n is symmetric, we can further assume that $x > 0$. We have that

$$\|h_n\|_{C^{1+\beta}} = \max\Big(\|h_n\|_{C^0}, \|dh_n\|_{C^0}, \|dh_n\|_{C^\beta}\Big),$$

where

$$\|dh_n\|_{C^\beta} = \max\left(\sup \frac{\|\partial_x h_n(x) - \partial_y h_n(y)\|}{\|x-y\|^\beta}, \sup \frac{\|\partial_x h_n(x) - \partial_y h_n(y)\|}{\|x-y\|^\beta}\right)$$

and $\partial_x h_n$ and $\partial_y h_n$ are the partial derivatives of h with respect to the first and the second variables respectively.

Let us write $h_n = (h_n^{(1)}, h_n^{(2)})$, $y = \varphi_n(x)$ where

$$h_n^{(1)}(x,y) = \frac{r(n+1)}{r(n)}x \text{ and } h_n^{(2)}(x,y) = \varphi_{n+1}(x).$$

Since

$$h_n^{(1)}(x,y) = \frac{r(n+1)}{r(n)}x \leq r(n+1) < 1 \text{ and } h_n^{(2)}(x,y) = \varphi_{n+1}(x) < K_1,$$

we obtain that $\sup \|h_n\| < K_1$.

Further, we have that $\partial_x h_n^{(1)}(x,y) = \frac{r(n+1)}{r(n)} < 1$ and

$$\partial_y h_n^{(2)}(x,y) = \frac{\partial}{\partial y}\Big(\frac{r(n+1)}{r(n)}x\Big) = \frac{\partial}{\partial y}\Big(\frac{r(n+1)}{r(n)}\varphi_n^{-1}(y)\Big)$$
$$= r(n+1)(\varphi^{-1})'\Big(\frac{x}{r(n)}\Big)\frac{1}{r(n)} < 1.$$

Also, $\partial h_n^{(2)}(x,y) = \varphi_{n+1}'(x) < \cot\frac{\alpha}{4}$ and

$$\partial h_n^{(2)}(x,y) = \frac{\partial}{\partial y}\Big(r(n+1)\varphi\Big(\frac{r(n)\varphi^{-1}(\frac{y}{r(n)})}{r(n+1)}\Big)\Big)$$
$$= r(n+1)\varphi'\frac{r(n)}{r(n+1)}\varphi^{-1}\Big(\frac{y}{r(n)}\Big)(\varphi^{-1})'\frac{1}{r(n)} < K_2.$$

Thus, the partial derivatives of the functions h_1 and h_2 are bounded. However, a similar calculation shows that $\|dh_n\|_{C^\beta}$ tends to infinity as $r(n)^{-\beta}$. Therefore, we conclude that

$$\|h\|_{C^{1+\beta}(V_{n+2}\setminus V_{n-1})} \le r(n+1)^{-\beta} = r(n)^{-2\beta}. \tag{12.32}$$

Similar computations hold for $h_n^{-1}\colon U_{n+1} \to U_n$, with the only difference being that the partial derivatives are estimated by $\frac{r(n)}{r(n+1)}$ which is unbounded. Therefore, the Hölder norm of the first derivatives of h_n^{-1} is bounded by

$$\frac{r(n)}{r(n+1)}\frac{1}{r(n)^{-\beta}} = \frac{r(n)^{1-\beta}}{r(n+1)}.$$

It follows that

$$\|h^{-1}\|_{C^{1+\beta}(U_{n+1}\setminus U_n)} = \|h_n^{-1}\circ h_{n-1}^{-1}\circ\cdots\circ h_0^{-1}\|_{C^{1+\beta}(U_{n+1}\setminus U_n)}$$
$$\le \prod_{i=0}^{n}\frac{r(n)^{1-\beta}}{r(n+1)} = \prod_{i=0}^{n} r(n)^{-1-\beta}. \tag{12.33}$$

Since $r(n-1) = r(n)^{\frac{1}{2}}$, we obtain that

$$\prod_{i=0}^{n} r(n)^{-1-\beta} = r(n)^{-(1+\beta)(1+\frac{1}{2}+\frac{1}{4}+\dots+\frac{1}{2^{n+1}})} = r(n)^{-2(1+\beta)(1-\frac{1}{2^{n+2}})}. \tag{12.34}$$

Finally, using (12.31), we find that

$$\|h\circ(f_{D^2}-\mathrm{Id})\circ h^{-1}\|_{C^{1+\beta}(U_n\setminus U_{n-1})}$$
$$\le r(n)^{-2\beta}r(n)^{2+2\kappa-\beta}r(n)^{-2(1+\beta)(1-\frac{1}{2^{n+2}})}$$
$$= r(n)^{2\kappa-5\beta+\frac{1}{2^{n+1}}-\frac{\beta}{2^{n+1}}}.$$

One can choose β such that $2\kappa - 5\beta + \frac{1}{2^{n+1}} - \frac{\beta}{2^{n+1}} > 0$ and conclude that

$$\|h \circ (f_{D^2} - \mathrm{Id}) \circ h^{-1}\|_{C^{1+\beta}(U_n \setminus U_{n-1})} \to 0$$

as $n \to \infty$. Therefore, f_P is tangent to Id near ∂P and hence, the map f_P is of class $C^{1+\beta}$.

Proof of Theorem 9.3. By construction, the map f_P generates via a homeomorphism T a $C^{1+\beta}$ diffeomorphism f_M of the surface M. We construct a C^∞ diffeomorphism $\psi\colon D^2 \to D^2$ such that $\varphi_3 := T \circ h \circ \psi$ is the desired area-preserving diffeomorphism (that is, $(\varphi_3)_* m_{D^2} = m_M$) which can be continuously extended to the closure of D^2.

Denote $\mu = (h^{-1} \circ T^{-1})_* m_M$. Since both m_{D^2} and m_M are normalized Lebesgue measures, we have

$$\int_{D^2} d\,m_{D^2} = 1 = \int_M d\,m_M = \int_{D^2} d\mu.$$

To obtain the desired result it suffices to show that there is a C^∞ diffeomorphism $\psi\colon D^2 \to D^2$ that can be continuously extended to ∂D^2 such that $\psi_*\mu = m_{D^2}$.

Set $\mu_1 = m_{D^2}$ and for $n > 1$ define a sequence of measures μ_n such that

(i) $\mu_n \in C^\infty(D^2)$, that is, the measure μ_n is absolutely continuous with respect to m_{D^2} with density function of class C^∞;

(ii) $\mu_n = \mu$ on $h^{-1}(U_{n-1})$;

(iii) $\int_{h^{-1}(U_n)} d\mu_n = \int_{h^{-1}(U_n)} d\mu$.

It is clear that for any $n \geq 1$, $\int_{D^2} d\mu_n = \int_{D^2} d\mu = 1$.

We need the following version of Moser's theorem (see [9], Lemma 1).

Lemma 9.6. *Let ω and μ be two volume forms on an oriented manifold M and let K be a connected compact set such that the support of $\omega - \mu$ is contained in the interior of K and $\int_K d\omega = \int_K d\mu$. Then there is a C^∞ diffeomorphism $\hat{\psi}\colon M \to M$ such that $\hat{\psi}|(M \setminus K) = Id|(M \setminus K)$ and $\hat{\psi}_*\omega = \mu$.*

Applying Lemma 9.6 to each compact set $K_n = h^{-1}(\overline{U}_n \setminus U_n)$ and volume forms $\mu_{n+1}|(\overline{U}_n \setminus U_n)$ and $\mu_n|(\overline{U}_n \setminus U_n)$, we obtain a C^∞ diffeomorphism $\hat{\psi}_n\colon D^2 \to D^2$ such that $(\hat{\psi}_n)_*\mu_{n+1} = \mu_n$ and $\hat{\psi}_n|h^{-1}(U_{n-1}) = \mathrm{Id}$. Then we let

$$\psi_n = \hat{\psi}_n \circ \cdots \circ \hat{\psi}_1 \quad \text{and} \quad \psi = \lim_{n\to\infty} \psi_n.$$

The construction gives $\hat{\psi}_n(h^{-1}(U_n \setminus U_n)) = h^{-1}(U_n \setminus U_n)$. Recalling that $r(n)$ satisfies (12.30) and using (12.32), (12.33), and (12.34), we find that diam $\hat{\psi}_n^{-1}(U_n) \leq Cd_n$ where $C > 0$ is a constant and d_n is a decreasing sequence of numbers such that $\sum_{n=1}^\infty d_n < \infty$. This implies that $d(x, \hat{\psi}_n(x)) \leq Cd_n$ for any $x \in D^2$. It follows that for any $x \in D^2$ and $n > j > 0$,

$$d(\psi_j(x), \psi_n(x)) \leq \sum_{i=j}^{n-1} d(\psi_i(x), \psi_{i+1}(x))$$
$$\leq \sum_{i=j}^{n-1} d\big(\psi_i(x), \hat{\psi}_i(\psi_i(x))\big) \leq C \sum_{i=j}^{n-1} d_i.$$

This implies that the sequence ψ_n is uniformly Cauchy and hence, ψ is well defined and continuous on D^2. We can also get that $\psi : D^2 \to D^2$ is a C^∞ diffeomorphism.

By construction, we know that $(\psi_n)_* \mu_{n+1} = \mu_1 = m_{D^2}$. Note that $D^2 = \cup_{n\geq 1} h^{-1}(U_n)$. Hence, for any $x \in D^2$ there is $n > 0$ and a neighborhood of x on which $\mu_{n+i} = \mu_n$ for any $i > 0$. It follows that $\psi_* \mu = (\psi_n)_* \mu_n = m_{D^2}$ on the neighborhood and hence, $\psi_* \mu = m_{D^2}$ on D^2. □

12.10 Completion of the Proof of Theorem 3.1

12.10.1 Representing the Map f_M as a Young Diffeomorphism

Consider a smooth compact connected oriented surface M of genius $g \geq 0$ and the diffeomorphism $f_M : M \to M$ given by Statement 4 of Theorem 9.3. In this section we represent the map f_M as a Young diffeomorphism.

Proposition 10.1. *The map f_M is a Young diffeomorphism. More precisely, one can choose the number r_2 in* (12.34) *so small that the collection of s-subsets satisfies Conditions (Y1)–(Y6).*

Proof. First note that we already know that the map f_{T^2} is a Young diffeomorphism, so we can assume that the genius $g \geq 1$. Consider the collection of s-subsets Λ_i^s and the return time $\tau : \Lambda \to \mathbb{N}$ for the map $f_{\mathbb{T}^2}$ defined in Section 12.5.3. Define $\Delta_i^s := \varphi_3(\varphi_2(\varphi_1(\Lambda_i^s)))$ with the return time on M (again denoted by τ) given by $\tau(\varphi_3(\varphi_2(\varphi_1(x))) = \tau(x)$, $x \in \mathbb{T}^2$. Let $\Delta = \bigcup_i \Delta_i^s$. We claim that f_M is a Young diffeomorphism with respect to the collection of s-subsets Δ_i^s.

To prove this we need to check Conditions (Y1)–(Y6). Since the maps φ_i, $i = 1, 2, 3$ are homeomorphisms and (Y1) and (Y2) are satisfied for the map $f_{\mathbb{T}^2}$, then these conditions are also satisfied for f_M. In addition, (Y5) and (Y6) hold true for f_M since the maps φ_i, $i = 1, 2, 3$ preserve the area.

To show (Y3) and (Y4) observe that the element of the Markov partition P in the Young tower representation for the map $f_{\mathbb{T}^2}$ is away from the critical points x_i, $i = 1, 2, 3, 4$. This implies that the map $\varphi_3 \circ \varphi_2 \circ \varphi_1$ is a smooth diffeomorphism from P onto its image. Since the return time function $\tilde{f}_M = f_M^\tau$ is defined on $\Delta \subset P$, (Y4) follows as it holds true for the map $f_{\mathbb{T}^2}$ by Proposition 5.2. Note that (Y3) holds for the map $f_{\mathbb{T}^2}$ for some constant $0 < a < 1$. If the number r_2 in (12.1) is chosen sufficiently small, then the fact that (Y3) holds for the map f_M can be shown by applying the argument in the proof of Proposition 6.2 in [21].

Finally, arguing similarly it is easy to show that the diffeomorphism f_{S^2} of the sphere S^2 is a Young diffeomorphism. This completes the proof of the proposition. □

12.10.2 Lower and Upper Polynomial Bounds on the Decay of Correlations

Let f be a Young diffeomorphism admitting a Young tower with base Λ, s-sets Λ_i^s, and inducing time $\tau = \{\tau_i\}$. Consider the associated Young tower given by (see Section 12.5.4)

$$\hat{Y} = \{(x, k) \in \Lambda \times \mathbb{N} : 0 \leq k < \tau(x)\}.$$

For $k > 0$ let

$$\hat{M}_k = \{(x, \ell) \in \hat{Y} : 0 \leq \ell \leq \min\{k, \tau(x)\}\}.$$

Consider the projection $\pi : \hat{Y} \to M$ given by $\pi(x, \ell) = f^\ell(x)$ and let

$$Y_M = \pi(\hat{Y}), \quad M_k = \pi(\hat{M}_k).$$

The sets M_k are clearly nested and exhaust Y.

To establish upper and lower bounds on the decay of correlations we need the following result, which is a corollary of results in [23] (see also [8] and [22]).

Proposition 10.2. *Assume that*

- *the greatest common divisor of numbers* $\{\tau_i\}$, $gcd\{\tau_i\} = 1$, *where* τ_i *are the values of the function* τ;
- *there is* $C > 0$ *such that for all* $x, y \in \Delta_i^s$ *and* $0 \leq j \leq \tau_i$,

$$d(f^j(x), f^j(y)) \leq C \max\{d(x, y), d(f^{\tau_i}(x), f^{\tau_i}(y))\}; \qquad (12.35)$$

- *there are* $\nu > 0$ *and* $C_1 > 0$ *such that*

$$m(\tau > n) \leq \frac{C_1}{n^\nu}. \qquad (12.36)$$

Then the following statements hold:

(1) *There is* $C_2 > 0$ *such that* $Cor_n(h_1, h_2) \leq \frac{C_2}{n^{\nu-1}}$ *for any* $h_1, h_2 \in C^\rho(M)$.
(2) *For any* $h_1, h_2 \in C^\rho(M)$ *supported in* M_k *for some* $k > 0$, *we have*

$$Cor_n(h_1, h_2) = \sum_{k=n+1}^{\infty} m(\{x : \tau(x) > k\}) \int_M h_1 dm \int_M h_2 dm + r_\nu(n), \qquad (12.37)$$

where $r_\nu(n) = O(R_\nu(n))$ *and*

$$R_\nu(n) = \begin{cases} \frac{1}{n^\nu} & \text{if } \nu > 2, \\ \frac{\log n}{n^2} & \text{if } \nu = 2, \\ \frac{1}{n^{2\nu-2}} & \text{if } 1 < \nu < 2. \end{cases}$$

Moreover, if $\int_M h_1 \int_M h_2 = 0$*, then* $Cor_n(h_1, h_2) = O(1/n^\nu)$.

To apply Proposition 10.2 to the Young diffeomorphism f_M we need to verify the assumptions of this proposition.

To prove the first assumption that $\gcd\{\tau_i\} = 1$, observe that the maps φ_i, $i = 1, 2, 3$ and the map H are homeomorphisms. Hence, it suffices to prove this for the linear map A. This is well known (see, for example, [24]).

To prove the second assumption that the map f_M satisfies (12.35) observe that this is true for the map $f_{\mathbb{T}^2}^{\tau_i}$ of the torus which is smoothly conjugate to the map $f_M^{\tau_i}$.

Finally, to prove the third assumption, observe that by Lemmas 7.2 and 8.3, we have that

$$\frac{C_8}{n^{\gamma-1}} < m(\{x \in \Delta \colon \tau(x) > n\}) < \frac{C_{11}}{n^{\gamma'-1}}, \tag{12.38}$$

where γ and γ' are defined by (12.8). It is easy to see that $\gamma > \gamma' > 2$ for all $0 < \alpha < \frac{1}{4}$ and $0 < \mu < \frac{1}{2}$.

Since the homeomorphisms φ_1, φ_2, and φ_3 are measure-preserving we also have the same estimates for the map f_M^τ. In particular, (12.36) holds with $\nu = \gamma' - 1$ and $C_1 = C_{11}$.

The upper bound on correlations is now an immediate corollary of Statement 1 of Proposition 10.2 where $\nu = \gamma' - 1$. In particular, we have $\gamma_1 = \gamma' - 2 > 0$.

To obtain the lower bound on correlations, we apply Statement 2 of Proposition 10.2 with $\nu = \gamma' - 1$ and obtain for all $h_1, h_2 \in C^\rho(M)$ supported in M_k for some $k > 0$ that

$$\mathrm{Cor}_n(h_1, h_2) = \sum_{k=n+1}^{\infty} m(\{x \colon \tau(x) > k\}) \int_M h_1 dm \int_M h_2 dm + r_{\gamma'}(n), \tag{12.39}$$

where $r_{\gamma'}(n) = O(R_{\gamma'}(n))$ and

$$R_{\gamma'}(n) = \begin{cases} \frac{1}{n^{\gamma'-1}} & \text{if } \gamma' > 3, \\ \frac{\log n}{n^2} & \text{if } \gamma' = 3, \\ \frac{1}{n^{2\gamma'-4}} & \text{if } 2 < \gamma' < 3. \end{cases}$$

Consider the two cases $\gamma' \geq 3$ and $2 < \gamma' < 3$.

Assume first that $\gamma' > 3$, which is true if $\alpha < \frac{1}{6}$. By assumption, $\int_M h_1\, dm \int_M h_2\, dm > 0$ and hence, applying (12.38) and (12.39), we obtain

$$\mathrm{Cor}_n(h_1, h_2) > \frac{K_1}{n^{\gamma-2}} - \frac{K_2}{n^{\gamma'-1}},$$

where $K_1 > 0$ and $K_2 > 0$ are constants. Using definitions of γ and γ' (see (12.8)) and choosing any $0 < \mu < \frac{1}{2}$, one can show that $\gamma - 2 < \gamma' - 1$ for all $0 < \alpha < \frac{1}{6}$.[4] We conclude that for some $C > 0$,

$$\mathrm{Cor}_n(h_1, h_2) > \frac{C}{n^{\gamma-2}}.$$

Now, we consider the case when $\frac{1}{6} < \alpha < \frac{1}{4}$. This implies that $\gamma' > 2$. Depending on the value of μ, we may either have $\gamma' > 3$ or $\gamma' < 3$ and we assume the latter (otherwise we are back to the previous case). With this assumption we have

$$\mathrm{Cor}_n(h_1, h_2) > \frac{K_1}{n^{\gamma-2}} - \frac{K_3}{n^{2\gamma'-4}},$$

where $K_3 > 0$ is a constant. Choosing again $0 < \mu < \frac{1}{2}$, one can show that $\gamma - 2 < 2\gamma' - 4$ holds for all $0 < \alpha < \frac{1}{4}$.[5] Thus we have the desired estimate

$$\mathrm{Cor}_n(h_1, h_2) > \frac{C}{n^{\gamma-2}}$$

for some $C > 0$ and all $0 < \alpha < \frac{1}{4}$. In particular, we have $\gamma_2 = \gamma - 2 > 0$.

12.10.3 The Central Limit Theorem

By Statement 1 of Theorem 3.1, for any Hölder continuous function h satisfying $\int h dm = 0$ we have $\mathrm{Cor}_n(h, h) = O(\frac{1}{n^{\gamma'-1}})$. This implies that the correlation function is summable, when $\gamma' > 2$, that is, when $0 < \alpha < \frac{1}{4}$. The desired result now follows from [14], Theorem 4.1 (see also [23], Theorem 3.1).

12.10.4 The Large Deviation Property

We consider the Young tower Y that represents the map f_M. The upper bound in (12.38) allows us to use Theorem 4.2 in [17] to obtain for $0 < \alpha < \frac{1}{4}$ that, for all sufficiently small $a > 0$,

$$m_M\left(\left|\frac{1}{n}\sum_{i=0}^{n-1} h(f_M^i(x)) - \int h\right| > \varepsilon\right) < C_{h,a}\varepsilon^{-2(\gamma'-2-a)} n^{-(\gamma'-2-a)}.$$

4 One can use a computer-assisted calculation to show that $\gamma - 2 < \gamma' - 1$ for all $0 < \alpha < 0.42\ldots$.
5 Again, a computer-assisted calculation to show that $\gamma - 2 < 2\gamma' - 4$ holds for all $0 < \alpha < 0.36\ldots$.

Moreover, for each such $a > 0$ the constant $C_{h,a}$ depends on the Hölder norm of h continuously.

To get a lower bound we need to check the conditions of Theorem 4.3 in [17]. More precisely, for the set $\hat{Y}_k = \{(x, l) \in \hat{Y} : \tau(x) > k\}$ it must be true that for some k, $m_M(\pi(\hat{Y}_k)) < 1$ where $\pi : \hat{Y} \to M$ is given by $\pi(x, k) = f^k(x)$ as before.

Given k, let us choose a partition element Δ_i in the base Δ of the tower for f_M with $\tau(\Delta_i) \leq k$. Then $\Delta_i \subset \hat{Y} \setminus \hat{Y}_k$ and obviously $\hat{m}_M(\Delta_i) > 0$. Thus $\hat{m}_M(\hat{Y}_k) < 1$ and since π is measure preserving, we obtain $m_M(\pi(\hat{Y}_k)) < 1$. Thus, by Theorem 4.3 in [17], we obtain the lower bound

$$\frac{1}{n^{\gamma'-2+a}} < m_M\left(\left|\frac{1}{n}\sum_{i=0}^{n-1} h(f_M^i(x)) - \int h\right| > \varepsilon\right)$$

for small ε, open and dense subset of Hölder continuous observables h, and infinitely many n.

12.10.5 The Measure of Maximal Entropy (MME)

Recall that the diffeomorphism f_M of the surface M is a Young diffeomorphism and consider the corresponding collection $\{\Delta_i^s\}$ of s-sets. Denote by $\mathcal{S}_n = \{\Delta_i^s : \tau(\Delta_i^s) = n\}$. Since the map f_M is topologically conjugate to the toral automorphism A, the number $\mathcal{S}_n$ for f_M is equal to the number $\mathcal{S}_n$ for A. The latter is known to satisfy $\mathcal{S}_n \leq e^{hn}$ with $h < h_{\text{top}}(A)$ (see [21]). It now follows from [20] (see Theorem 7.1) and [23] that the map f_M possesses a unique MME which has all the desired properties.

References

[1] D. Anosov and A. Katok, *New examples in smooth ergodic theory: Ergodic diffeomorphisms*, Trans. Moscow. Math. Soc., **23** (1970), 1–35.

[2] L. Barreira and Ya. Pesin, *Introduction to smooth ergodic theory*, **148**, Graduate Studies in Mathematics, American Mathematical Society, Providence, RI, 2013.

[3] M. Brin, *Bernoulli diffeomorphisms with n – 1 nonzero nonzero exponents*, Ergod. Theory Dyn. Syst., **1** (1981), 453–465.

[4] M. Brin, J. Feldman, and A. Katok, *Bernoulli diffeomorphisms and group extensions of dynamical systems with non-zero characteristic exponents*, Ann. Math., **113** (1981), 159–179.

[5] H. Bruin and D. Terhesiu, *Regular variation and rates of mixing for infinite measure preserving almost Anosov diffeomorphisms*, Ergod. Theory Dyn. Syst., **40** (2020), 663–698.

[6] D. Dolgopyat and Ya. Pesin, *Every compact manifold carries a completely hyperbolic diffeomorphism*, Ergod. Theory Dyn. Syst., **22** (2002), 409–435.

[7] P. Eslami and C. Liverani, *Mixing rates for symplectic almost Anosov maps*, Nonlinearity, **34** (2021), 3709–3731.

[8] S. Gouëzel, *Sharp polynomial estimates for the decay of correlations*, Israel J. Math, **139** (2004), 29–65.
[9] R. E. Greene and K. Shiohama, *Diffeomorphisms and volume-preserving embeddings of noncompact manifolds*, Trans. Amer. Math. Soc., **255** (1979), 403–414.
[10] H. Hu, *Decay of correlations for piecewise smooth maps with indifferent fixed point*, Ergod. Theory Dyn. Syst., **24** (2004), 495–524.
[11] H. Hu and X. Zhang, *Polynomial decay of correlation for almost Anosov maps*, Ergod. Theory Dyn. Syst., 2017.
[12] A. Katok, *Bernoulli diffeomorphisms on surfaces*, Ann. Math., **110** (1979), 529–547.
[13] A. Katok and B. Hasselblatt. *Introduction to the theory modern theory of dynamical systems*, **54**, Cambridge University Press, 1995.
[14] C. Liverani, *Central limit theorem for deterministic systems*, International Conference on Dynamical Systems (Montevideo, 1995), **362**, Pitman Res. Notes Math. Ser., Longman-Harlow (1996), 56–75.
[15] C. Liverani, B. Saussol, and S. Vaienti. *A probabilistic approach to intermittency*, Ergod. Theory Dyn. Syst., **19** (1999), 671–685.
[16] C. Liverani and M. Martens, *Convergence to equilibrium for intermittent symplectic maps*, Comm. Math. Phys., **260** (2005), 527–556.
[17] I. Melbourne and M. Nicol, *Large deviations for nonuniformly hyperbolic systems*, Trans. Amer. Math. Soc., **360** (2008), 6661–6676.
[18] I. Melbourne and D. Terhesiu, *Decay of correlations for non-uniformly expanding systems with general return times*, Ergod. Theory Dyn. Syst., **34** (2014), 893–918.
[19] Yves Pomeau and Paul Manneville, *Intermittent transition to turbulence in dissipative dynamical systems*, Comm. Math. Phys., **74** (1980), 189–197.
[20] Y. Pesin, S. Senti, and K. Zhang, *Thermodynamics of towers of hyperbolic type*, Trans. Amer. Math. Soc., **368** (2016), 8519–8552.
[21] Y. Pesin, S. Senti, and K. Zhang, *Thermodynamics of the Katok map*, Ergod. Theory Dyn. Syst., **39** (2019), 764–794.
[22] O. Sarig, *Subexponential decay of correlations*, Invent. Math., **150** (2002) 629–653.
[23] F. Shahidi and A. Zelerowicz, *Thermodynamics via inducing*, J. Stat. Phys. **175** (2019) 351–383.
[24] L.-S. Young, *Statistical properties of dynamical systems with some hyperbolicity*, Ann. Math., **147** (1998), 585–650.
[25] L.-S. Young, *Recurrence times and rates of mixing*, Israel J. Math., **110** (1999), 153–188.

13

Linear Cocycles over Hyperbolic Systems

VICTORIA SADOVSKAYA*

Dedicated to the memory of Anatole Katok

13.1 Introduction

Cocycles are a fundamental tool in the study of dynamical systems and group actions. In smooth dynamics, the differential and related objects provide important examples of cocycles. For a hyperbolic or partially hyperbolic diffeomorphism f of a manifold X, the restrictions of the differential to invariant sub-bundles of TX, such as stable, unstable, and center, give examples of so-called linear cocycles. Linear cocycles on trivial bundles can be viewed as $GL(d, \mathbb{R})$-valued cocycles. Another class of examples is given by matrix or operator-valued cocycles over non-smooth hyperbolic systems such as shifts and subshifts of finite type. Locally constant cocycles over these systems correspond to random and Markovian sequences of matrices and operators.

Cohomology, or conjugacy, is a notion of equivalence for cocycles, which can be considered in various regularity classes. The key questions in the study of cocycles include:

- When two cocycles are cohomologous;
- When a measurable conjugacy between two cocycles is continuous;
- When a cocycle is cohomologous to a simpler one; for example, to the identity cocycle or to one with values in a smaller group such as orthogonal or conformal.

Reduction to a smaller group can often be interpreted as existence and regularity of an invariant geometric structure, such as a Riemannian metric or a conformal structure. As abundance of periodic orbits is one of the key features of hyperbolic systems, one may try to deduce properties of cocycles from their periodic data, that is, their products along the periodic orbits of f. Questions include whether conformality at the periodic points implies conformality of the cocycle, and whether conjugacy of the periodic data for two cocycles implies cohomology.

* Supported in part by NSF grant DMS-1764216.

Another important problem in the study of cocycles is estimating their growth and, in particular, the asymptotic growth rates given by the Lyapunov exponents. For hyperbolic systems, approximations of the Lyapunov exponents and growth estimates can sometimes be obtained from the periodic data of the cocycle. Cocycles with one Lyapunov exponent give a generalization of conformal ones.

Many of the results and techniques for hyperbolic systems extend to the partially hyperbolic setting, and periodic approximation of Lyapunov exponents also holds in the non-uniformly hyperbolic case.

The motivation for studying linear cocycles comes in part from the area of local and global rigidity of hyperbolic systems. Global rigidity refers to existence of a smooth conjugacy to an algebraic model, and local rigidity is smoothness of the conjugacy to a C^1-small perturbation when natural obstructions vanish. The results on linear cocycles produced many applications to this area.

The study of linear cocycles over hyperbolic and partially hyperbolic systems is an active area, and a variety of questions have been explored. In this survey we focus on the ones just mentioned. Other directions include the study of hyperbolicity, simplicity of Lyapunov spectrum, and (dis)continuous dependence of Lyapunov exponents on the cocycle. Many results in these areas are related to the invariance principle for cocycles developed in [V08, AV10, ASV13].

The text is organized as follows. In Section 13.2 we give definitions of the main classes of hyperbolic dynamical systems, and of linear and Banach cocycles over these systems. In Section 13.3 we discuss fiber bunching and holonomies of a cocycle, which are used in subsequent sections. Section 13.4 is devoted to the three questions on cohomology of group-valued cocycles over hyperbolic systems mentioned earlier. We state several results, give examples, and outline some of the proofs and constructions. In Section 13.5 we discuss the Lyapunov exponents of a cocycle and their periodic approximation. We also deduce growth estimates for cocycles. In Section 13.6 we consider conformal structures and their relation to uniform quasiconformality of a linear cocycle. We also obtain conformality and isometry for a linear cocycle from its periodic data, and we state an analogous result for the infinite-dimensional setting. In Section 13.7 we discuss cocycles with one Lyapunov exponent, and we show that they have a specific structure. We also give a classification of $GL(2, \mathbb{R})$-valued cocycles with one exponent. Section 13.8 is devoted to cocycles over partially hyperbolic diffeomorphisms and extensions of some of the results for hyperbolic systems to this setting. In Section 13.9 we give some applications of results on linear cocycles to the questions of rigidity for hyperbolic diffeomorphisms and flows.

13.2 Cocycles over Hyperbolic Dynamical Systems

We will consider cocycles over *hyperbolic* dynamical systems. Here we describe the three main classes of such systems; see [KtH] for more details.

Transitive Anosov Diffeomorphisms A diffeomorphism f of a compact connected manifold X is called *Anosov* if there exist a splitting of the tangent bundle TX into a direct sum of two Df-invariant continuous sub-bundles E^s and E^u, a Riemannian metric on X, and continuous functions ν and $\hat{\nu}$ such that

$$\|Df_x(v^s)\| < \nu(x) < 1 < \hat{\nu}(x) < \|Df_x(v^u)\| \tag{13.1}$$

for any $x \in X$ and unit vectors $v^s \in E^s(x)$ and $v^u \in E^u(x)$. The sub-bundles E^s and E^u are called stable and unstable. We assume that f is at least $C^{1+\text{Hölder}}$, and then the sub-bundles are Hölder continuous.

The stable and unstable sub-bundles are tangent to the stable and unstable foliations W^s and W^u respectively. We define the *local stable manifold* of x, $W^s_{\text{loc}}(x)$, as a ball centered at x of radius ρ in the intrinsic metric of $W^s(x)$. We choose ρ sufficiently small so that $W^s_{\text{loc}}(x) \cap W^u_{\text{loc}}(z)$ consists of a single point for any sufficiently close x and z in X, and for every $x \in X$ we have $\|Df|_{E^s(y)}\| \le \nu(x)$ for all y in $W^s_{\text{loc}}(x)$. The first property is called the local product structure of the foliations, and the second one implies that

$$\text{dist}(f(x), f(y)) \le \nu(x) \cdot \text{dist}(x, y) \quad \text{for all } x \in X \text{ and } y \in W^s_{\text{loc}}(x). \tag{13.2}$$

Local unstable manifolds are defined similarly.

A diffeomorphism f is *(topologically) transitive* if there is a point x in X with dense orbit. All known examples of Anosov diffeomorphisms have this property. We note that any transitive Anosov diffeomorphism is *topological mixing*; that is, for any open sets U_1 and U_2 in X, there exists $N \in \mathbb{N}$ such that $f^n(U_1) \cap U_2 \neq \emptyset$ for all $n \ge N$.

Topologically Mixing Diffeomorphisms of Locally Maximal Hyperbolic Sets More generally, let f be a diffeomorphism of a manifold $\mathcal{M}$. A compact f-invariant set $X \subseteq \mathcal{M}$ is called *hyperbolic* if there exist a continuous Df-invariant splitting $T_X\mathcal{M} = E^s \oplus E^u$, and a Riemannian metric and continuous functions ν, $\hat{\nu}$ on an open set $U \supseteq X$ such that (13.1) holds for all $x \in X$. The set X is called *locally maximal* if $X = \bigcap_{n \in \mathbb{Z}} f^{-n}(U)$ for some open set U containing X.

Mixing Subshifts of Finite Type Let M be $k \times k$ matrix with entries from $\{0, 1\}$ such that all entries of M^N are positive for some N. Let

$$X = \{x = (x_n)_{n \in \mathbb{Z}} \; : \; 1 \le x_n \le k \text{ and } M_{x_n, x_{n+1}} = 1 \text{ for every } n \in \mathbb{Z}\}.$$

The shift map $f : X \to X$ is defined by $(fx)_n = x_{n+1}$. The system (X,f) is called a *mixing subshift of finite type*. We fix $\nu \in (0, 1)$ and consider the metric

$$\mathrm{dist}(x,y) = d_\nu(x,y) = \nu^{n(x,y)}, \quad \text{where } n(x,y) = \min\{\,|i| \;:\; x_i \neq y_i\}.$$

The following sets play the role of the local stable and unstable manifolds of x:

$$W^s_{\mathrm{loc}}(x) = \{y \mid x_i = y_i,\ i \geq 0\}, \quad W^u_{\mathrm{loc}}(x) = \{y \mid x_i = y_i,\ i \leq 0\},$$

and we can take $\nu(x) = \nu$ and $\hat{\nu}(x) = \nu^{-1}$.

Lemma 2.1 (Anosov Closing Lemma). *[KtH, 6.4.15–17] Let (X,f) be a hyperbolic system. Then there exist constants $c,\ \delta_0 > 0$ such that for any $x \in X$ and $k \in \mathbb{N}$ with $dist(x,f^k x) < \delta_0$ there exists a periodic point $p \in X$ with $f^k p = p$ such that the orbit segments $x, fx, \ldots, f^k x$ and $p, fp, \ldots, f^k p$ remain close:*

$$dist(f^i x, f^i p) \leq c\, dist(x, f^k x) \quad \text{for every } i = 0, \ldots, k.$$

Moreover,

$$dist(f^i x, f^i p) \leq c\ dist(x, f^k x)\, e^{-\gamma \min\{i, k-i\}} \quad \text{for every } i = 0, \ldots, k, \tag{13.3}$$

where $e^{-\gamma} = \max_{x \in X} \max\{\nu(x), (\hat{\nu}(x))^{-1}\} < 1.$

Linear and Banach Cocycles over Dynamical Systems Let G be a topological group with a complete metric. We primarily consider the group $G = GL(d,\mathbb{R})$ and more generally $GL(V)$, the group of invertible bounded linear operators on a Banach space V. We fix the metric d on the group:

$$d(A,B) = \|A - B\| + \|A^{-1} - B^{-1}\|, \quad \text{where } \|\,.\,\| \text{ is the operator norm.}$$

Definition 2.2. *Let A be a function from X to a group G. The* G-valued cocycle over (X,f) generated by *A is the map $\mathcal{A} : X \times \mathbb{Z} \to G$ defined by $\mathcal{A}(x,0) = Id$ and $\mathcal{A}(x,n) = \mathcal{A}^n_x = A(f^{n-1}x) \circ \cdots \circ A(x)$ and $\mathcal{A}(x,-n) = \mathcal{A}^{-n}_x = (\mathcal{A}^n_{f^{-n}x})^{-1}$ for $n \in \mathbb{N}$.*
Clearly, $\mathcal{A}$ satisfies the cocycle equation $\mathcal{A}^{n+k}_x = \mathcal{A}^n_{f^k x} \circ \mathcal{A}^k_x$.

We refer to cocycles with values in $G = GL(d,\mathbb{R})$ and in $GL(V)$ as *linear cocycles* and *Banach cocycles*, respectively.

A prime example of a linear cocycle is the *derivative cocycle*, namely the differential of a diffeomorphism $f : X \to X$, where the tangent bundle of X is trivial, that is, $TX = X \times \mathbb{R}^d$. Then Df can be viewed as a $GL(d,\mathbb{R})$-valued cocycle:

$$A(x) = \mathcal{A}_x = D_x f \in GL(d,\mathbb{R}) \quad \text{and} \quad \mathcal{A}^n_x = D_x f^n.$$

One can also consider restrictions of Df to invariant sub-bundles of TX, for example the stable and unstable sub-bundles.

Another important class of examples is given by random and Markovian sequences of matrices and operators. They correspond to locally constant cocycles over full shifts or subshifts of finite type.

Standing Assumption *We will always assume that the cocycle $\mathcal{A}$ is* β-Hölder, *that is, its generator A is a Hölder continuous function with exponent β from X to G.*

Hölder continuity of the cocycle is natural in view of the examples just presented. The derivative cocycle of a $C^{1+\text{Hölder}}$ diffeomorphism f is Hölder continuous, and so are its restrictions to the stable and unstable sub-bundles. These sub-bundles are usually only Hölder continuous even for a more regular diffeomorphism [HaW99], and hence $Df|_{E^s}$ and $Df|_{E^s}$ are also only Hölder continuous. Also, symbolic dynamical systems have a Hölder structure, but no smooth one. In addition, Hölder continuity of the cocycle is necessary to develop a meaningful theory, even for $\mathbb{R}$-valued cocycles over hyperbolic systems.

Bundle Setting The derivative cocycle example suggests viewing a linear cocycle $\mathcal{A}$ as an automorphism of a vector bundle $\mathcal{E} = X \times \mathbb{R}^d$ with the fiber map $\mathcal{A}_x : \mathcal{E}_x \to \mathcal{E}_{fx}$.

More generally, one can allow non-trivial bundles, and most of our results extend to this setting. Let (X,f) be a hyperbolic system and $P : \mathcal{E} \to X$ be a finite-dimensional β-Hölder vector bundle over X. A continuous linear cocycle over f is a homeomorphism $\mathcal{A} : \mathcal{E} \to \mathcal{E}$ such that

$$P \circ \mathcal{A} = f \circ P \quad \text{and } \mathcal{A}_x : \mathcal{E}_x \to \mathcal{E}_{fx} \text{ is a linear isomorphism.}$$

Such an $\mathcal{A}$ is called β-Hölder if $\mathcal{A}_x$ depends β-Hölder on x, with proper identification of fibers at nearby points. We refer to Section 2.2 of [KS13] for a detailed description of this setting.

13.3 Fiber Bunching and Holonomies

In the study of cocycles, comparing their iterates along exponentially converging orbits plays an important role. If the group G is Abelian, the products of the form $(\mathcal{A}_y^n)^{-1} \circ \mathcal{A}_x^n$, where $y \in W^s(x)$, can be rearranged as $(\mathcal{A}_y^n)^{-1} \circ \mathcal{A}_x^n = \prod_{i=0}^{n-1} \left(A(f^i(y))^{-1} \circ A(f^i(x)\right)$ and then estimated easily using exponential closeness of the iterates of x and y and Hölder property of the cocycle. In the non-commutative case, estimating such products is more difficult and various assumptions on the growth of the cocycle have been used to ensure their convergence. For linear and Banach cocycles, we use a condition called *fiber bunching*, which means that non-conformality of the cocycle in the fiber is dominated by the expansion and contraction in the base.

Definition 3.1. *The* quasiconformal distortion *of a cocycle* $\mathcal{A}$ *is the function*

$$Q_{\mathcal{A}}(x,n) = \|\mathcal{A}_x^n\| \cdot \|(\mathcal{A}_x^n)^{-1}\| = \frac{\sup\{\, \|\mathcal{A}_x^n(v)\| : \|v\| = 1\,\}}{\inf\{\, \|\mathcal{A}_x^n(v)\| : \|v\| = 1\,\}}, \quad x \in X \text{ and } n \in \mathbb{Z}.$$

The cocycle is uniformly quasiconformal *if* $Q_{\mathcal{A}}(x,n) \le K$ *for all* x *and* n, *and it is* conformal *if* $Q_{\mathcal{A}}(x,n) = 1$ *for all* x *and* n.

In the finite-dimensional case, $\mathcal{A}_x^n$ maps the unit sphere to an ellipsoid, and $Q_{\mathcal{A}}(x,n)$ is the ratio of its largest and smallest semi-axes.

Definition 3.2. *A* β*-Hölder cocycle* $\mathcal{A}$ *over a hyperbolic system* (X,f) *is* fiber bunched *if there exist numbers* $\theta < 1$ *and* L *such that for all* $x \in X$ *and* $n \in \mathbb{N}$,

$$Q_{\mathcal{A}}(x,n) \cdot (\nu_x^n)^\beta < L\theta^n \quad \text{and} \quad Q_{\mathcal{A}}(x,-n) \cdot (\hat{\nu}_x^{-n})^\beta < L\theta^n,$$

where ν *and* $\hat{\nu}$ *are as in* (13.1), $\nu_x^n = \nu(f^{n-1}x)\cdots\nu(x)$ *and* $\hat{\nu}_x^{-n} = (\hat{\nu}_{f^{-n}x}^n)^{-1}$.

Clearly, conformal and uniformly quasiconformal cocycles are fiber bunched, and so are cocycles that are sufficiently close to conformal.

The limits of products $(\mathcal{A}_y^n)^{-1} \circ \mathcal{A}_x^n$ are useful in the study of non-commutative cocycles. These limits are called *holonomies*, and we will define them in Proposition 3.3. In the context of linear cocycles, this notion was introduced in [BV04, V08] and holonomies were further studied in [ASV13, KS13, S15, KS16]. Existence of holonomies was proved in [V08, ASV13] under the stronger "one-step" fiber bunching condition that there is $\theta < 1$ such that

$$\|\mathcal{A}_x\| \cdot \|(\mathcal{A}_x)^{-1}\| \cdot \nu(x)^\beta < \theta \quad \text{for all } x \in X. \tag{13.4}$$

The following result, which also holds in the bundle setting [KS13], gives existence of holonomies under the weakest fiber bunching assumption given in Definition 3.2.

Proposition 3.3. *Let* $\mathcal{A}$ *be a* β*-Hölder linear or Banach cocycle over* (X,f). *If* $\mathcal{A}$ *is fiber bunched, then for every* $x \in X$ *and* $y \in W^s(x)$ *the limit*

$$H_{x,y}^{\mathcal{A},s} = \lim_{n\to\infty} (\mathcal{A}_y^n)^{-1} \circ \mathcal{A}_x^n, \tag{13.5}$$

exists and satisfies

(H1) $H_{x,y}^{\mathcal{A},s} \in GL(V)$ is an invertible linear map from $\mathcal{E}_x$ to $\mathcal{E}_y$;

(H2) $H_{x,x}^{\mathcal{A},s} = \text{Id}$ and $H_{y,z}^{\mathcal{A},s} \circ H_{x,y}^{\mathcal{A},s} = H_{x,z}^{\mathcal{A},s}$, which implies $(H_{x,y}^{\mathcal{A},s})^{-1} = H_{y,x}^{\mathcal{A},s}$;

(H3) $H_{x,y}^{\mathcal{A},s} = (\mathcal{A}_y^n)^{-1} \circ H_{f^nx,f^ny}^{\mathcal{A},s} \circ \mathcal{A}_x^n$ for all $n \in \mathbb{N}$;

(H4) $\|H_{x,y}^{\mathcal{A},s} - \text{Id}\,\| \le c\,\text{dist}(x,y)^\beta$, where c is independent of x and $y \in W_{\text{loc}}^s(x)$, and also $\|(\mathcal{A}_y^n)^{-1} \circ \mathcal{A}_x^n - \text{Id}\,\| \le c\,\text{dist}(x,y)^\beta$ for every $n \in \mathbb{N}$.

The continuous map $H^{\mathcal{A},s} : (x,y) \mapsto H^{\mathcal{A},s}_{x,y}$, where $x \in X$, $y \in W^s(x)$, is called the *stable holonomy* for $\mathcal{A}$. The unstable holonomy $H^{\mathcal{A},u}$ for $\mathcal{A}$ is defined similarly:

$$H^{\mathcal{A},u}_{x,y} = \lim_{n\to\infty} \left((\mathcal{A}^{-n}_y)^{-1} \circ (\mathcal{A}^{-n}_x)\right) = \lim_{n\to\infty} \left(\mathcal{A}^n_{f^{-n}y} \circ (\mathcal{A}^n_{f^{-n}x})^{-1}\right), \text{ where } y \in W^u(x).$$

Sometimes, any continuous map $H^{\mathcal{A},s} : (x,y) \mapsto H^{\mathcal{A},s}_{x,y}$ satisfying (H1–H3) is called a stable holonomy [ASV13], and the one in (13.5) is referred to as the *standard holonomy* to distinguish it [KS16]. It was proved in [KS13] that properties (H1–H4) uniquely define the standard holonomy. However, other maps may exist satisfying (H1–H3) with Hölder exponent lower than β in (H4), as discussed in [KS16] after Corollary 4.9.

Proof. We give a proof under the "one-step" fiber bunching assumption (13.4).

Let $x \in X$ and $y \in W^s_{\text{loc}}(x)$. The key step is to show that the sequence $((\mathcal{A}^n_y)^{-1} \circ \mathcal{A}^n_x)$ is Cauchy. Denoting $x_i = f^i(x)$ and $y_i = f^i(y)$, we obtain

$$\begin{aligned}
(\mathcal{A}^n_y)^{-1} \circ \mathcal{A}^n_x &= (\mathcal{A}^{n-1}_y)^{-1} \circ \left((\mathcal{A}_{y_{n-1}})^{-1} \circ \mathcal{A}_{x_{n-1}}\right) \circ \mathcal{A}^{n-1}_x = \\
&= (\mathcal{A}^{n-1}_y)^{-1} \circ (\text{Id} + r_{n-1}) \circ \mathcal{A}^{n-1}_x = (\mathcal{A}^{n-1}_y)^{-1} \circ \mathcal{A}^{n-1}_x + (\mathcal{A}^{n-1}_y)^{-1} \circ r_{n-1} \circ \mathcal{A}^{n-1}_x = \\
&= \cdots = \text{Id} + \sum_{i=0}^{n-1} (\mathcal{A}^i_y)^{-1} \circ r_i \circ \mathcal{A}^i_x, \quad \text{where } r_i = (\mathcal{A}_{y_i})^{-1} \circ \mathcal{A}_{x_i} - \text{Id}.
\end{aligned}$$

Since $y \in W^s_{\text{loc}}(x)$, for each $i \in \mathbb{N}$ we can estimate

$$\text{dist}(x_i, y_i) \le \text{dist}(x,y) \cdot \nu^i_y = \text{dist}(x,y) \cdot \nu(y_0)\nu(y_1)\cdots \nu(y_{i-1}).$$

Since $\mathcal{A}$ is Hölder continuous with exponent β, we have

$$\|r_i\| \le \|(\mathcal{A}_{y_i})^{-1}\| \cdot \|\mathcal{A}_{x_i} - \mathcal{A}_{y_i}\| \le c_1 \,\text{dist}(x_i, y_i)^\beta \le c_1 (\text{dist}(x,y)\, \nu^i_y)^\beta.$$

Also,

$$\frac{\|\mathcal{A}_{x_k}\|}{\|\mathcal{A}_{y_k}\|} \le 1 + \frac{\|\mathcal{A}_{x_k} - \mathcal{A}_{y_k}\|}{\|\mathcal{A}_{y_k}\|} \le 1 + c_2 (\text{dist}(x_k, y_k))^\beta,$$

and we estimate

$$\|(\mathcal{A}^i_y)^{-1}\| \cdot \|\mathcal{A}^i_x\| \le \|(\mathcal{A}_y)^{-1}\| \cdot \|(\mathcal{A}_{y_1})^{-1}\| \cdots \|(\mathcal{A}_{y_{i-1}})^{-1}\| \cdot \|\mathcal{A}_x\| \cdot \|\mathcal{A}_{x_1}\| \cdots \|\mathcal{A}_{x_{i-1}}\|$$

$$\le \prod_{k=0}^{i-1} \|\mathcal{A}_{y_k}\|\, \|(\mathcal{A}_{y_k})^{-1}\| \cdot \prod_{k=0}^{i-1} \frac{\|\mathcal{A}_{x_k}\|}{\|\mathcal{A}_{y_k}\|} < \prod_{k=0}^{i-1} \theta\, \nu(y_k)^{-\beta} \cdot \prod_{k=0}^{i-1} \left(1 + c_2\, (\text{dist}(x_k, y_k))^\beta\right).$$

Since the distance between x_k and y_k decreases exponentially, the second product is uniformly bounded and we obtain $\|(\mathcal{A}^i_y)^{-1}\| \cdot \|\mathcal{A}^i_x\| \le c_3\, \theta^i\, (\nu^i_y)^{-\beta}$.

It follows that for every $i \geq 0$,

$$\|(\mathcal{A}_y^i)^{-1} \circ r_i \circ \mathcal{A}_x^i\| \leq \|(\mathcal{A}_y^i)^{-1}\| \cdot \|\mathcal{A}_x^i\| \cdot \|r_i\| \leq$$
$$\leq c_3\, \theta^i\, (\nu_y^i)^{-\beta} \cdot c_1(\text{dist}(x,y)\, \nu_y^i)^\beta = c_4\, \text{dist}(x,y)^\beta\, \theta^i.$$

Therefore, for every $n \in \mathbb{N}$,

$$\|\text{Id} - (\mathcal{A}_y^n)^{-1} \circ \mathcal{A}_x^n\| \leq \sum_{i=0}^{n-1} \|(\mathcal{A}_y^i)^{-1} \circ r_i \circ \mathcal{A}_x^i\| \leq c_4\, \text{dist}(x,y)^\beta \sum_{i=0}^{n-1} \theta^i \leq c\, \text{dist}(x,y)^\beta.$$

Also, since

$$\|(\mathcal{A}_y^{n+1})^{-1} \circ \mathcal{A}_x^{n+1} - (\mathcal{A}_y^n)^{-1} \circ \mathcal{A}_x^n\| = \|(\mathcal{A}_y^n)^{-1} \circ r_n \circ \mathcal{A}_x^n\| \leq c_4\, \text{dist}(x,y)^\beta\, \theta^n,$$

the sequence $\{(\mathcal{A}_y^n)^{-1} \circ \mathcal{A}_x^n\}$ is Cauchy and hence it has a limit $H_{xy}^s \in GL(V)$. Since the convergence is uniform on the set of pairs (x,y) where $y \in W_{\text{loc}}^s(x)$, the map H^s is continuous. It is easy to see that the maps H_{xy}^s satisfy (H2, H3, H4). The map H_{xy}^s can be extended to any $y \in W^s(x)$ using (iii).

Existence of holonomies under the weaker fiber bunching of Definition 3.2 can be deduced by considering iterates of the cocycle and using uniqueness of holonomies; see [S15, Proposition 4.4] for details. □

13.4 Cohomology of Group-Valued Cocycles over Hyperbolic Systems

13.4.1 Introduction and Main Questions

Cohomology of Hölder continuous cocycles over hyperbolic systems has been extensively studied, starting with the seminal work of A. Livšic [Li71, Li72], where he established the following result. Parts (i) and (ii) are referred to as the Periodic Point and Measurable Livšic Theorems, respectively.

Theorem 4.1 (Livšic Theorem). *Let $f : X \to X$ be a transitive Anosov diffeomorphism, and let $\alpha : X \to \mathbb{R}$ be a Hölder continuous function. Then*

(i) *If $\sum_{i=0}^{n-1} \alpha(f^i(p)) = 0$ whenever $f^n(p) = p$, then there exists a Hölder continuous function $\varphi : X \to \mathbb{R}$ such that $\alpha(x) = \varphi(f(x)) - \varphi(x)$ for all $x \in X$.*

(ii) *Let μ be an ergodic invariant measure with full support and local product structure. If there exists a μ-measurable function $\tilde{\varphi}$ such that $\alpha(x) = \tilde{\varphi}(f(x)) - \tilde{\varphi}(x)$ for μ-a.e. $x \in X$, then there exists a Hölder continuous function $\varphi : X \to \mathbb{R}$ such that $\alpha(x) = \varphi(f(x)) - \varphi(x)$ for all $x \in X$ and $\tilde{\varphi}(x) = \varphi(x)$ μ-a.e.*

The conclusions are that α is a *coboundary*; more specifically, it is Hölder continuously cohomologous to the zero cocycle. The notion of *cohomology* can be considered in various categories. For cocycles with values in a group G it is defined as follows.

Definition 4.2. *Cocycles $\mathcal{A}$ and $\mathcal{B}$ are (measurably, continuously)* cohomologous *if there exists a (measurable, continuous) function $C : X \to G$ such that*

$$\mathcal{A}_x^n = C(f^n x) \circ \mathcal{B}_x^n \circ C(x)^{-1} \quad \textit{for all } x \in X \textit{ and } n \in \mathbb{Z}.$$

The function C is called a conjugacy *between $\mathcal{A}$ and $\mathcal{B}$, or a* transfer map.

In addition to the case of $G = \mathbb{R}$, Livšic also established a result similar to Theorem 4.1 for Abelian G. Following his work, the questions of cohomology to the identity cocycle have been considered for non-Abelian groups, where they proved to be much more difficult.

For a G-valued cocycle $\mathcal{A}$, we call its *periodic data* the set of its products along the periodic orbits in the base, that is,

$$\mathcal{A}_P = \{\mathcal{A}_p^n : p = f^n p, \ n \in \mathbb{N}\}.$$

A number of results extending Theorem 4.1 (i) were obtained [Li72, NT95, PWa01, LWi10] under additional assumptions on the group G or on the growth of the cocycle, such as compactness of the group or fiber-bunching–type conditions for the cocycle; see [KtN] for more details and references. Without such assumptions, the results were established in the following cases.

Theorem 4.3. *Let $\mathcal{A}$ be a Hölder continuous cocycle over (X, f) with values in one of the following groups:*

$GL(d, \mathbb{R})$, and more generally Lie groups and their closed subgroups (Kalinin [K11]*);*

The group of invertible elements of a Banach algebra (Grabarnik, Gyusinsky [GrGu17]*);*

A group of diffeomorphisms of a compact manifold (Avila, Kocsard, Liu [AKL18]*).*

If $\mathcal{A}$ has identity periodic data, that is, $\mathcal{A}_P = \{e_G\}$, then $\mathcal{A}$ is Hölder cohomologous to the identity cocycle.

Extending Theorem 4.1 (ii), Livšic established Hölder continuity of a measurable conjugacy to the identity cocycle for compact groups and groups with a bi-invariant metric [Li72]. Further results for Lie groups were obtained under additional assumptions such as boundedness of conjugacy or bunching of the cocycle [PaP97, NiP99, PWa01], and for $GL(d, \mathbb{R})$ without extra assumptions by Butler [Bt18].

Now we consider cohomology of two arbitrary Hölder continuous G-valued cocycles $\mathcal{A}$ and $\mathcal{B}$ over a hyperbolic system. This general problem is the most relevant for applications in the study of hyperbolic systems, as the derivative cocycle or its restrictions to the stable and unstable sub-bundles are not cohomologous to the identity cocycles. If G is an Abelian group, one can consider the cocycle $\mathcal{A} \circ \mathcal{B}^{-1}$ and thus reduce the problem to the case when $\mathcal{B}$ is the identity cocycle, namely $\mathcal{B}_x = e_G$, which was resolved by Livšic. For non-Abelian G, however, the general problem does not reduce to the special case $\mathcal{B} = e_G$ and is much more difficult. Moreover, the answers to some natural questions are negative in general. The main focus of the research has been on the following generalizations of the Livšic theorem.

Question 4.4. *Suppose that $\mathcal{A}$ and $\mathcal{B}$ have conjugate periodic data, that is, whenever $p = f^n p$, $\mathcal{A}_p^n = C(p) \circ \mathcal{B}_p^n \circ C^{-1}(p)$ for some $C(p) \in G$. Does it follow that $\mathcal{A}$ and $\mathcal{B}$ are continuously cohomologous?*

Clearly, continuous cohomology implies conjugacy of the periodic data. As the answer to Question 4.4 is negative in general, one may ask:

Question 4.5. *Suppose that $\mathcal{A}$ and $\mathcal{B}$ have equal periodic data, that is, $\mathcal{A}_p^n = \mathcal{B}_p^n$ whenever $f^n p = p$. Does it follow that $\mathcal{A}$ and $\mathcal{B}$ are continuously cohomologous?*

Finally, the following question extends the second part of the Livšic theorem.

Question 4.6. *Let C be a measurable conjugacy between $\mathcal{A}$ and $\mathcal{B}$. Is C continuous? More precisely, does it coincide with a continuous conjugacy almost everywhere?*

As in the Livšic theorem, measurability should be understood with respect to an ergodic invariant measure with full support and local product structure, for example the measure of maximal entropy or the invariant volume.

Positive answers for Questions 4.6 and 4.5, as well as some results for cocycles with conjugate data, were obtained by Parry and Pollicott [PaP97, Pa99] for compact G and, somewhat more generally, by Schmidt [Schm99] for cocycles with "bounded distortion." First results outside this setting, including the following proposition, were obtained in [S13] for certain types of $GL(2,\mathbb{R})$-valued cocycles.

Proposition 4.7. [S13] *Let $f : X \to X$ be a transitive Anosov diffeomorphism, and let $\mathcal{A}$ and $\mathcal{B}$ be cocycles over (X,f) given by*

$$\mathcal{A}_x = k(x)\begin{bmatrix} 1 & \alpha(x) \\ 0 & 1 \end{bmatrix} \quad \text{and} \quad \mathcal{B}_x = l(x)\begin{bmatrix} 1 & \beta(x) \\ 0 & 1 \end{bmatrix},$$

where $k(x)$, $l(x) \neq 0$ for all x, and α, β are not cohomologous to 0.

(i) *If $\mathcal{A}$ and $\mathcal{B}$ have conjugate periodic data with $C(p)$ continuous at a periodic point p_0, then $\mathcal{A}$ and $\mathcal{B}$ are Hölder continuously cohomologous.*
(ii) *Any measurable conjugacy between $\mathcal{A}$ and $\mathcal{B}$ is Hölder continuous.*

(iii) *$\mathcal{A}$ and $\mathcal{B}$ are (measurably or Hölder) cohomologous if and only if there exist Hölder functions $\varphi(x)$ and $s(x)$ and a constant $c \neq 0$ such that $k(x)/l(x) = \varphi(fx)/\varphi(x)$ and $\alpha(x) - c\beta(x) = s(fx) - s(x)$ for all $x \in X$.*

We note that the cocycles in the proposition are fiber bunched, and that statements (i) and (ii) hold more generally for $GL(2,\mathbb{R})$-valued cocycles with one exponent at each periodic point.

General results on cohomology of fiber-bunched cocycles were obtained in [S15] and [S17] for linear and Banach cocycles, respectively.

13.4.2 Continuous Conjugacy from Periodic Data

As the following example demonstrates, the answer to Question 4.4 is negative in general, even if the cocycles are fiber bunched and $C(p)$ is bounded.

Example 4.8. *[S13] There exist cocycles $\mathcal{A}$ and $\mathcal{B}$ with generators $\mathcal{A}_x = \begin{bmatrix} 1 & \alpha(x) \\ 0 & 1 \end{bmatrix}$ and $\mathcal{B}_x = \begin{bmatrix} 1 & \beta(x) \\ 0 & 1 \end{bmatrix}$ arbitrarily close to the identity, such that $\mathcal{A}$ and $\mathcal{B}$ have conjugate periodic data with $C(p)$ uniformly bounded, but are not measurably cohomologous.*

The Construction. By Proposition 4.7, $\mathcal{A}$ and $\mathcal{B}$ are measurably cohomologous only if there exist $c \neq 0$ and a Hölder function $s(x)$ such that $\alpha(x) - c\beta(x) = s(fx) - s(x)$. By the Livšic theorem, the latter is equivalent to the fact that $\alpha(p,n) - c\beta(p,n) = 0$ for every periodic point $p = f^n p$, where $\alpha(p,n) = \alpha(p) + \alpha(fp) + \cdots + \alpha(f^{n-1}p)$.

Let $\varepsilon > 0$ and let α and β be Hölder continuous functions such that, for two periodic points $p_1 = f^{n_1}(p_1)$ and $p_2 = f^{n_2}(p_2)$,

$$\alpha(f^i p_1) = \beta(f^i p_1), \quad 0 \leq i \leq n_1 - 1, \quad \alpha(f^i p_2) = 2\beta(f^i p_2), \quad 0 \leq i \leq n_2 - 1,$$

and $0 < \beta(x) \leq \alpha(x) \leq 2\beta(x) < \varepsilon$ for all x. The function β can be chosen constant.

As $\alpha(p_1, n_1) = \beta(p_1, n_1)$ and $\alpha(p_2, n_2) = 2\beta(p_2, n_2)$, there is no constant c such that $\alpha(p,n) - c\beta(p,n) = 0$ for every periodic p.

Since $\alpha(p,n) > 0$ and $\beta(p,n) > 0$ at every periodic point $p = f^n p$, the functions α and β are not cohomologous to 0, and the matrices $\mathcal{A}_p^n$ and $\mathcal{B}_p^n$ are conjugate by the matrix $C(p) = \operatorname{diag}(\alpha(p,n)/\beta(p,n), 1)$. Since $1 \leq \alpha(p,n)/\beta(p,n) \leq 2$ for every p, $C(p)$ is uniformly bounded. □

Thus one has to make an assumption on $C(p)$ stronger than boundedness. If $C(p)$ is Hölder continuous, extending it to a function C on X and conjugating $\mathcal{B}$ by this extension reduces Question 4.4 to Question 4.5. We obtain a positive answer to Question 4.4 under a weaker assumption that $C(p)$ is Hölder continuous at one periodic point.

Assumptions. In Theorem 4.9, Corollary 4.10, Proposition 4.11, and Corollary 4.12, $\mathcal{A}$ and $\mathcal{B}$ are β-Hölder $GL(d,\mathbb{R})$ or $GL(V)$-valued cocycles over a hyperbolic system.

We note that in the following theorem, fiber bunching is assumed only for $\mathcal{A}$, since fiber bunching for $\mathcal{B}$ can be obtained from the periodic data. This is discussed in Section 13.5.5.

Theorem 4.9. *[S15, S17] Suppose that $\mathcal{A}$ is fiber bunched, and $\mathcal{B}$ has conjugate periodic data such that for some periodic point p_0, $d(C(p),C(p_0)) \leq c\, dist(p,p_0)^\beta$ for every periodic point p. Then $\mathcal{A}$ and $\mathcal{B}$ are β-Hölder cohomologous. More precisely, there exists a unique β-Hölder continuous conjugacy $\bar{C}$ between $\mathcal{A}$ and $\mathcal{B}$ such that $\bar{C}(p_0) = C(p_0)$. Moreover, if $\mathcal{A}$ and $\mathcal{B}$ take values in a closed subgroup G_0 of $GL(d,\mathbb{R})$ or $GL(V)$ and $C(p_0) \in G_0$, then $\bar{C}(x) \in G_0$ for all x.*

The conjugacy $\bar{C}(p)$ does not necessarily coincide with $C(p)$ for $p \neq p_0$. For example, let $\mathcal{B}$ be the identity cocycle and let $\mathcal{A}_x = \bar{C}(fx) \circ \mathcal{B}_x \circ \bar{C}(x)^{-1} = \bar{C}(fx) \circ \bar{C}(x)^{-1}$, where $\bar{C} : X \to GL(d,\mathbb{R})$ is a Hölder continuous function with $\bar{C}(p_0) = \text{Id}$. Then $\mathcal{A}_p^n = \mathcal{B}_p^n = \text{Id}$ whenever $p = f^n p$, and thus we can take $C(p) = \text{Id}$ for each p.

As a corollary, we obtain a complete positive answer to Question 4.5 when one of the cocycles is fiber bunched. For $G = GL(d,\mathbb{R})$, this result was independently obtained by Backes [Bac15] under a stronger, uniform version of bunching for both cocycles.

Corollary 4.10. [S15, S17] *If a cocycle $\mathcal{A}$ is fiber bunched and $\mathcal{B}_p^n = \mathcal{A}_p^n$ whenever $f^n p = p$, then $\mathcal{A}$ and $\mathcal{B}$ are β-Hölder cohomologous.*

Our approach to the proof is inspired by [Pa99, Schm99, PWa01]. An important role is played by the following observation.

Proposition 4.11. [S15] *Let $\mathcal{A}$ and $\mathcal{B}$ be two fiber-bunched cocycles and let C be a β-Hölder continuous conjugacy between $\mathcal{A}$ and $\mathcal{B}$. Then C intertwines the holonomies for $\mathcal{A}$ and $\mathcal{B}$, that is,*

$$H_{x,y}^{\mathcal{A},s/u} = C(y) \circ H_{x,y}^{\mathcal{B},s/u} \circ C(x)^{-1} \quad \textit{for every } x \in X \textit{ and } y \in W^{s/u}(x).$$

We note that this result holds for C with the same Hölder exponent β as the one used in the fiber-bunching condition. However, there are examples of fiber-bunched cocycles and conjugacy C that has lower Hölder exponent and does not intertwine their holonomies [KS16].

Proof. Let $x \in X$ and $y \in W^s(x)$. By iterating x and y forward the problem reduces to the case of $y \in W^s_{\text{loc}}(x)$. Since $\mathcal{A}(x) = C(fx) \circ \mathcal{B}_x \circ C(x)^{-1}$, we have

$$\begin{aligned}(\mathcal{A}^n_y)^{-1} \circ \mathcal{A}^n_x &= C(y) \circ (\mathcal{B}^n_y)^{-1} \circ C(f^n y)^{-1} \circ C(f^n x) \circ \mathcal{B}^n_x \circ C(x)^{-1} = \\ &= C(y) \circ (\mathcal{B}^n_y)^{-1} \circ (\text{Id} + r_n) \circ \mathcal{B}^n_x \circ C(x)^{-1} = \\ &= C(y) \circ (\mathcal{B}^n_y)^{-1} \circ \mathcal{B}^n_x \circ C(x)^{-1} + C(y) \circ (\mathcal{B}^n_y)^{-1} \circ r_n \circ \mathcal{B}^n_x \circ C(x)^{-1}.\end{aligned} \tag{13.6}$$

Since C is Hölder continuous and $y \in W^s_{\text{loc}}(x)$, we can estimate

$$\begin{aligned}\|r_n\| &= \|C(f^n y)^{-1} \circ C(f^n x) - \text{Id}\| \le \|C(f^n y)^{-1}\| \cdot \|C(f^n x) - C(f^n y)\| \le \\ &\le c_1 \, \text{dist}(f^n x, f^n y)^\beta \le c_1 \, (\nu^n_y)^\beta.\end{aligned}$$

Properties (H3) and (H4) of holonomies imply that there is a constant c_2 such that

$$\|\mathcal{B}^n_x\| \le c_2 \|\mathcal{B}^n_y\| \quad \text{for all } x \in X,\ y \in W^s_{\text{loc}}(x), \text{ and } n \in \mathbb{N}.$$

Using the preceding estimates and fiber bunching of the cocycle $\mathcal{B}$, we obtain

$$\begin{aligned}\|(\mathcal{B}^n_y)^{-1} \circ r_n \circ \mathcal{B}^n_x\| &\le \|(\mathcal{B}^n_y)^{-1}\| \cdot \|r_n\| \cdot c_3 \, \|\mathcal{B}^n_y\| \le \\ &\le c_4 \, \|(\mathcal{B}^n_y)^{-1}\| \cdot \|\mathcal{B}^n_y\| \cdot (\nu^n_y)^\beta \le c_5 \theta^n \to 0 \quad \text{as } n \to \infty.\end{aligned}$$

Therefore the last term in (13.6) tends to 0.

Since $\lim_{n\to\infty} (\mathcal{A}^n_y)^{-1} \circ \mathcal{A}^n_x = H^{\mathcal{A},s}_{x,y}$ and $\lim_{n\to\infty} (\mathcal{B}^n_y)^{-1} \circ \mathcal{B}^n_x = H^{\mathcal{B},s}_{x,y}$, passing to the limit in (13.6) we obtain intertwining of the holonomies. □

Let C be as in the proposition and let $C(x_0)$ be given for some $x_0 \in X$. It follows that for any $y \in W^s(x_0)$,

$$C(y) = H^{\mathcal{B},s}_{y,x} \circ C(x_0) \circ H^{\mathcal{A},s}_{x,y}. \tag{13.7}$$

Since the stable manifold $W^s(x_0)$ is dense in X and C is Hölder continuous, such C is uniquely determined by its value at x_0.

Outline of the Proof of Theorem 4.9. Since $\mathcal{A}$ is fiber bunched and $\mathcal{B}$ has conjugate periodic data with bounded $C(p)$, the cocycle $\mathcal{B}$ is also fiber bunched by Proposition 5.11. Therefore both cocycles have holonomies.

First, let us consider the case when the point p_0 is fixed. We conjugate the cocycle $\mathcal{B}$ by $C(p_0)$ and thus reduce the question to that of two cocycles with the same value at p_0 and $C(p_0) = \text{Id}$. We still denote them $\mathcal{A}$ and $\mathcal{B}$. Following (13.7) we construct conjugacies between $\mathcal{A}$ and $\mathcal{B}$ along the stable and unstable manifolds of p_0 as

$$\begin{aligned}C^s(x) &= H^{\mathcal{A},s}_{p_0,x} \circ C(p_0) \circ H^{\mathcal{B},s}_{x,p_0} = H^{\mathcal{A},s}_{p_0,x} \circ H^{\mathcal{B},s}_{x,p_0} \quad \text{for } x \in W^s(p_0), \\ C^u(x) &= H^{\mathcal{A},u}_{p_0,x} \circ C(p_0) \circ H^{\mathcal{B},u}_{x,p_0} = H^{\mathcal{A},u}_{p_0,x} \circ H^{\mathcal{B},u}_{x,p_0} \quad \text{for } x \in W^u(p_0).\end{aligned}$$

It is easy to verify that they are indeed conjugacies on the corresponding manifolds. The main part of the proof is to show that if x is a homoclinic point for p_0, that is, $x \in W^s(p_0) \cap W^u(p_0)$, then

$$H^{\mathcal{A},s}_{x,p_0} \circ H^{\mathcal{A},u}_{p_0,x} = H^{\mathcal{B},s}_{x,p_0} \circ H^{\mathcal{B},u}_{p_0,x}, \quad \text{that is } C^s(x) = C^u(x) \stackrel{def}{=} \bar{C}(x). \tag{13.8}$$

To do this, we consider the orbit segment $\{f^{-m}x, \dots, x, \dots f^n(x)\}$ with sufficiently large m and n. Since both $f^{-m}x$ and $f^n(x)$ are close to p_0, by Anosov Closing Lemma 2.1 there exists a periodic point q close to p_0 whose iterates follow this orbit segment. Since q is close to p_0, $C(q)$ is Hölder close to $C(p_0) = \text{Id}$. Using the assumption that $C(q)$ conjugates the periodic values of $\mathcal{A}$ and $\mathcal{B}$ at q and carefully estimating the orbit products, one can establish (13.8) in the limit as $m, n \to \infty$.

After (13.8) is proven, it is not difficult to show that the function $\bar{C}$ is β-Hölder continuous on the set of homoclinic points, and hence it can be extended to a Hölder continuous function on X. It is clear from the construction that if $\mathcal{A}(x), \mathcal{B}(x) \in G_0$ for all x, and $C(p_0) \in G_0$, then the holonomies and $\bar{C}$ also take values in G_0.

In the case of a periodic point $p_0 = f^m(p_0)$, the preceding argument gives a conjugacy $\bar{C}$ between the cocycles $\mathcal{A}^m$ and $\mathcal{B}^m$ over the map f^m. Then an additional argument shows that $\bar{C}$ is also a conjugacy between $\mathcal{A}$ and $\mathcal{B}$. □

We also obtained a result for a constant linear cocycle and its perturbation *without the fiber bunching assumption.* It is useful in the study of rigidity of Anosov automorphisms.

Corollary 4.12. [S15] *Let $\mathcal{A}$ be a constant $GL(d,\mathbb{R})$-valued cocycle, and let $\mathcal{B}$ be a Hölder continuous cocycle sufficiently close to $\mathcal{A}$. If $\mathcal{A}$ and $\mathcal{B}$ have conjugate periodic data and $C(p)$ is Hölder continuous at a fixed or periodic point p_0, then they are Hölder continuously cohomologous.*

Outline of the Proof. Let $A(x) = A$ be the generator of $\mathcal{A}$. Let $\rho_1 < \cdots < \rho_\ell$ be the distinct moduli of the eigenvalues of A and let $\mathbb{R}^d = E^A_1 \oplus \cdots \oplus E^A_\ell$ be the corresponding invariant splitting into the direct sum of the generalized eigenspaces. We denote $A_i = A|E^A_i$. It follows that, for any $\varepsilon > 0$, there exists C_ε such that

$$C_\varepsilon^{-1}(\rho_i - \varepsilon)^n \le \|A_i^n u\| \le C_\varepsilon(\rho_i + \varepsilon)^n \quad \text{for any unit vector } u \in E^A_i,$$

and hence the cocycle $\mathcal{A}_i$ generated by A_i is fiber bunched for any $\beta > 0$. Moreover, any cocycle $\mathcal{B}$ with generator B sufficiently C^0 close to A has the corresponding invariant splitting $\mathbb{R}^d = E^B_1(x) \oplus \cdots \oplus E^B_\ell(x)$, which is close to that of $\mathcal{A}$ and is β-Hölder for some $\beta > 0$. The corresponding restrictions $\mathcal{B}_i$ satisfy similar estimates and hence are also fiber bunched. Since the conjugacy $C(p)$ maps $E^A_i(p)$ to $E^B_i(p)$, the cocycles $\mathcal{A}_i$ and $\mathcal{B}_i$ have conjugate periodic data. Hence by Theorem 4.9

they are conjugate via a Hölder continuous function C_i, and we obtain a conjugacy between $\mathcal{A}$ and $\mathcal{B}$ as the direct sum of C_i. □

13.4.3 Hölder Continuity of a Measurable Conjugacy

Fiber bunching of two cocycles does not guarantee continuity of a measurable conjugacy between them. An example of $GL(2,\mathbb{R})$-valued cocycles that are measurably but not continuously cohomologous was constructed in [PWa01]. In the example, both generators can be made arbitrarily close to the identity, and hence the cocycles are fiber bunched.

Example 4.13. [PWa01] *Arbitrarily close to the identity, there exist smooth functions* $A(x) = \begin{bmatrix} \alpha(x) & \beta \\ 0 & 1 \end{bmatrix}$ *and* $B(x) = \begin{bmatrix} \alpha(x) & 0 \\ 0 & 1 \end{bmatrix}$ *such that the corresponding cocycles* $\mathcal{A}$ *and* $\mathcal{B}$ *are measurably, but not continuously cohomologous.*

Outline of the Construction. Let f be an Anosov automorphism, and let x_0 be a fixed point. Let $\alpha(x)$ be a smooth function such that $\alpha(x_0) = 1$ and $1 - \varepsilon < \alpha(x) < 1$ for all $x \neq x_0$, and let $\beta = \varepsilon$. Since the matrices $A(x_0)$ and $B(x_0)$ are not conjugate, the cocycles $\mathcal{A}$ and $\mathcal{B}$ are not continuously cohomologous.

A measurable conjugacy is constructed in the form $C(x) = \begin{bmatrix} 1 & c(x) \\ 0 & 1 \end{bmatrix}$. Then

$$A(x) = C(fx)B(x)C(x)^{-1} \text{ is equivalent to } \begin{bmatrix} \alpha(x) & \beta \\ 0 & 1 \end{bmatrix}$$

$$= \begin{bmatrix} \alpha(x) & c(fx) - \alpha(x)c(x) \\ 0 & 1 \end{bmatrix}.$$

A function c such that $c(fx) = \beta + \alpha(x)c(x)$ is obtained as a series. Let

$$c_m(x) = \beta \cdot \left(1 + \alpha(f^{-1}x) + \alpha(f^{-1}x)\alpha(f^{-2}x) + \cdots + \alpha(f^{-1}x)\cdots\alpha(f^{-m}x)\right).$$

By the Birkhoff Ergodic Theorem, $(\alpha(f^{-1}x)\cdots\alpha(f^{-m}(x))^{1/m} \to \alpha < 1$ a.e. It follows that the sequence $\{c_m(x)\}$ converges to a limit $c(x)$ a.e., and the function c is measurable as a limit of continuous functions. The functions c_m satisfy the equation $c_m(fx) = \beta + \alpha(x)c_{m-1}(x)$, and passing to the limit it follows that $c(fx) = \beta + \alpha(x)c(x)$. □

Since the answer to Question 4.6 for fiber-bunched cocycles is negative in general, we make a stronger assumption that one of the cocycles is uniformly quasiconformal.

Theorem 4.14. [S15] *Let* $\mathcal{A}$ *and* $\mathcal{B}$ *be linear cocycles, and let* μ *be an ergodic* f*-invariant measure on* X *with full support and local product structure. Suppose that* $\mathcal{A}$ *is fiber bunched* *and $\mathcal{B}$ is uniformly quasiconformal. Then any μ-measurable conjugacy between $\mathcal{A}$ and $\mathcal{B}$ is β-Hölder continuous, that is, it coincides with a β-Hölder continuous conjugacy on a set of full measure.*

A measure has local product structure if it is locally equivalent to the product of its conditional measures on the local stable and unstable manifolds. Examples of ergodic measures with full support and local product structure include the measure of maximal entropy, more generally Gibbs (equilibrium) measures of Hölder continuous potentials, and the invariant volume if it exists [PWa01].

Outline of the Proof of Theorem 4.14. Let C be a μ-measurable conjugacy between $\mathcal{A}$ and $\mathcal{B}$. The key step in the proof is to show that C intertwines holonomies of $\mathcal{A}$ and $\mathcal{B}$ on a set of full measure; more precisely, there exists a set $Y \subset X$, $\mu(Y) = 1$, such that

$$H^{\mathcal{A},s}_{x,y} = C(y) \circ H^{\mathcal{B},s}_{x,y} \circ C(x)^{-1} \quad \text{for all } x, y \in Y \text{ such that } y \in W^s(x). \tag{13.9}$$

This yields $C(y) = H^{\mathcal{A},s}_{x,y} \circ C(x) \circ H^{\mathcal{B},s}_{x,y}$, and then Hölder continuity of the holonomies easily implies essential Hölder continuity of C along W^s_{loc}; that is,

$$d(C(x), C(y)) \leq c\, \text{dist}(x,y)^{\beta} \quad \text{for all } x, y \in Y \text{ such that } y \in W^s_{\text{loc}}(x),$$

where c does not depend on x and y. Similarly one obtains essential Hölder continuity of C along W^u_{loc}. Then global Hölder continuity of C is obtained using the local product structure of stable and unstable manifolds and the local product structure of the measure.

To establish (13.9) we consider $x \in X$ and $y \in W^s_{\text{loc}}(x)$ and, as in the proof of Proposition 4.11, we obtain

$$(\mathcal{A}^n_y)^{-1} \circ \mathcal{A}^n_x = C(y) \circ (\mathcal{B}^n_y)^{-1} \circ \mathcal{B}^n_x \circ C(x)^{-1} + C(y) \circ (\mathcal{B}^n_y)^{-1} \circ r_n \circ \mathcal{B}^n_x \circ C(x)^{-1}, \tag{13.10}$$

where

$$\|r_n\| \leq \|C(f^n y)^{-1}\| \cdot \|C(f^n x) - C(f^n y)\|.$$

Since C is μ-measurable, by Lusin's theorem there exists a compact set $S \subset X$ with $\mu(S) > 1/2$ such that C is uniformly continuous on S and hence $\|C\|$ and $\|C^{-1}\|$ are bounded on S. Let Y be the set of points in X for which the frequency of visiting S equals $\mu(S) > 1/2$. By the Birkhoff Ergodic Theorem $\mu(y) = 1$. If x and y are in Y, there exists a sequence $\{n_i\}$ such that $f^{n_i}x$ and $f^{n_i}y$ are in S for all i. It follows that $\|r_{n_i}\| \to 0$ as $i \to \infty$. The product

$$\|(\mathcal{B}^n_y)^{-1}\| \cdot \|\mathcal{B}^n_x\| \leq \|H^{\mathcal{B},s}_{f^n x, f^n y}\| \cdot \|(\mathcal{B}^n_x)^{-1}\| \cdot \|H^{\mathcal{B},s}_{x,y}\| \cdot \|\mathcal{B}^n_x\|$$

is uniformly bounded since the cocycle $\mathcal{B}$ is uniformly quasiconformal. Thus for every x and y in Y such that $y \in W^s_{\text{loc}}(x)$, the second term in (13.10) tends to 0 along a subsequence, and (13.9) follows. □

The next result extends Theorem 4.14 to the infinite-dimensional setting. One of the difficulties here is that the space $(GL(V), d)$ is not separable even if V is. To use the tools of the theory of measurable functions, such as Lusin's theorem, we work with the strong operator topology, namely the topology of pointwise convergence. We assume that $\mathcal{B}$ takes values in a precompact set, which for a finite-dimensional V is equivalent to uniform boundedness of $\mathcal{B}$ in $(GL(V), d)$.

Theorem 4.15. [S17] *Let $\mathcal{A}$ and $\mathcal{B}$ be $GL(V)$-valued cocycles. Suppose that the Banach space V is separable, $\mathcal{A}$ is fiber bunched and $\mathcal{B}$ takes values in a subset of $GL(V)$ that is precompact in the strong operator topology. Let μ be an ergodic invariant measure with full support and local product structure. Then any μ-measurable conjugacy between $\mathcal{A}$ and $\mathcal{B}$ coincides with a β-Hölder continuous conjugacy on a set of full measure.*

The μ-measurability of the conjugacy means that the preimage of each Borel set in $GL(V)$ is μ-measurable and the conjugacy equation holds μ-almost everywhere. We note that Borel σ-algebra in $GL(V)$ is the same for the metric d and for the strong operator topology.

13.5 Periodic Approximation of Lyapunov Exponents and Growth Estimates for Cocycles

13.5.1 Definitions and Results

First we consider a β-Hölder continuous linear cocycle $\mathcal{A}$ on a finite-dimensional vector bundle $\mathcal{E}$ over a hyperbolic system (X,f). We discuss Lyapunov exponents of $\mathcal{A}$ with respect to an f-invariant ergodic Borel probability measure μ on X. The vector bundle $\mathcal{E}$ is trivial on a set of full measure and hence $\mathcal{A}$ can be viewed as a $GL(d,\mathbb{R})$-valued cocycle on a set of full measure. Continuity of the cocycle implies that the integrability assumption in the following theorem is satisfied, and so we obtain the Lyapunov exponents of $\mathcal{A}$ and the corresponding Lyapunov decomposition of $\mathcal{E}$ on a set of full measure.

Theorem 5.1 (Oseledets' Multiplicative Ergodic Theorem [MET]). [O68]
Let f be an invertible ergodic measure-preserving transformation of a Lebesgue probability space (X, μ). Let $\mathcal{A}$ be a measurable $GL(d,\mathbb{R})$-valued cocycle over f satisfying $\log \|\mathcal{A}_x\| \in L^1(X,\mu)$ and $\log \|\mathcal{A}_x^{-1}\| \in L^1(X,\mu)$.

Then there exist numbers $\lambda_1 < \cdots < \lambda_\ell$, an f-invariant set Λ with $\mu(\Lambda) = 1$, and an $\mathcal{A}$-invariant Lyapunov decomposition $\mathcal{E}_x = \mathcal{E}_x^1 \oplus \cdots \oplus \mathcal{E}_x^\ell$ for $x \in \Lambda$ such that

(i) $\lim\limits_{n\to\pm\infty} \frac{1}{n} \log \|\mathcal{A}_x^n v\| = \lambda_i$ for any $i = 1, \ldots, \ell$ and any $0 \neq v \in \mathcal{E}_x^i$, and

(ii) $\lim\limits_{n\to\pm\infty} \frac{1}{n} \log |\det \mathcal{A}_x^n| = \sum_{i=1}^{\ell} m_i\lambda_i,$ where $m_i = \dim \mathcal{E}_x^i$.

The numbers $\lambda_1, \dots, \lambda_\ell$ are called the *Lyapunov exponents* of $\mathcal{A}$ with respect to μ. The exponents and the decomposition depend on the choice of μ.

By Lyapunov exponents of $\mathcal{A}$ at a periodic point p we mean the Lyapunov exponents of $\mathcal{A}$ with respect to the invariant measure μ_p on the orbit of p. They equal $(1/k)$ of the logarithms of the absolute values of the eigenvalues of $\mathcal{A}_p^k$, where k is a period of p. The following result on periodic approximation of Lyapunov exponents for finite-dimensional $\mathcal{E}$ was established by Kalinin.

Theorem 5.2. [K11] *Let (X,f) be a hyperbolic dynamical system, let $\mathcal{A}$ be a Hölder continuous linear cocycle over f, and let μ be an ergodic invariant measure for f.*

Then the Lyapunov exponents $\lambda_1 \le \dots \le \lambda_d$ of $\mathcal{A}$ with respect to μ, listed with multiplicities, can be approximated by the Lyapunov exponents of $\mathcal{A}$ at periodic points. More precisely, for any $\varepsilon > 0$ there exists a periodic point $p \in X$ for which the Lyapunov exponents $\lambda_1^{(p)} \le \dots \le \lambda_d^{(p)}$ of $\mathcal{A}$ satisfy $|\lambda_i - \lambda_i^{(p)}| < \varepsilon$ for $i = 1, \dots, d$.

For $GL(V)$-valued cocycles with infinite-dimensional V, there is no Oseledets' MET in general, but **upper and lower Lyapunov exponents** $\lambda(\mathcal{A}, \mu)$ and $\chi(\mathcal{A}, \mu)$ of $\mathcal{A}$ with respect to μ can be defined as follows.

$$\begin{aligned} \lambda(\mathcal{A}, \mu) &= \lim_{n\to\infty} \frac{1}{n} \log \|\mathcal{A}_x^n\| \quad \text{for } \mu\text{-a.e. } x \in X, \\ \chi(\mathcal{A}, \mu) &= \lim_{n\to\infty} \frac{1}{n} \log \|(\mathcal{A}_x^n)^{-1}\|^{-1} \quad \text{for } \mu\text{-a.e. } x \in X. \end{aligned} \tag{13.11}$$

These limits exist and are constant μ almost everywhere. Indeed, the sequence of functions $a_n(x) = \frac{1}{n} \log \|\mathcal{A}_x^n\|$ satisfies

$$a_{n+m}(x) \le a_m(x) + a_n(f^m x) \quad \text{for all } x \in X \text{ and } n, m \in \mathbb{N}, \tag{13.12}$$

that is, $\{a_n(x)\}$ is a *subadditive cocycle*. By the Subadditive Ergodic Theorem for μ-a.e. $x \in X$

$$\lim_{n\to\infty} \frac{a_n(x)}{n} = \lim_{n\to\infty} \frac{a_n(\mu)}{n} = \inf\left\{ \frac{a_n(\mu)}{n} : n \in \mathbb{N}\right\}, \quad \text{where } a_n(\mu) = \int_X a_n(x)\, d\mu.$$

This limit is called the *exponent* of the subadditive cocycle. A similar argument applies to the second limit in (13.11)

If V is finite-dimensional, then $\lambda(\mathcal{A}, \mu)$ and $\chi(\mathcal{A}, \mu)$ are precisely the largest and smallest of the Lyapunov exponents given by Oseledets' MET, that is, $\lambda(\mathcal{A}, \mu) = \lambda_\ell$ and $\chi(\mathcal{A}, \mu) = \lambda_1$.

In the infinite-dimensional setting, it is not always possible to approximate $\lambda(\mathcal{A}, \mu)$ and $\chi(\mathcal{A}, \mu)$ by Lyapunov exponents at periodic points. The following proposition is based on an example by Gurvits of a pair of operators whose joint spectral radius is greater than the generalized spectral radius [Gu95].

Proposition 5.3. [KS17] *There exists a locally constant cocycle $\mathcal{A}$ over a full shift on two symbols and an ergodic invariant measure μ such that $\lambda(\mathcal{A},\mu) > \sup_{\mu_p} \lambda(\mathcal{A},\mu_p)$, where the supremum is taken over all uniform measures μ_p on periodic orbits.*

Nonetheless, the upper and lower exponents can be approximated in terms of norms at periodic points as follows.

Theorem 5.4. [KS17] *Let (X,f) be a hyperbolic system, let μ be an ergodic f-invariant measure on X, and let $\mathcal{A}$ be a Hölder continuous $GL(V)$-valued cocycle over f. Then for any $\varepsilon > 0$ there exists a periodic point $p = f^k p$ such that*

$$\left| \lambda(\mathcal{A},\mu) - \frac{1}{k}\log \|\mathcal{A}_p^k\| \right| < \varepsilon \quad \text{and} \quad \left| \chi(\mathcal{A},\mu) - \frac{1}{k}\log \|(\mathcal{A}_p^k)^{-1}\|^{-1} \right| < \varepsilon.$$

Moreover, for any $N \in \mathbb{N}$ there exists such $p = f^k p$ with $k > N$.

We note that for the measures μ_p on the orbit of $p = f^k p$,

$$\begin{aligned}\lambda(\mathcal{A},\mu_p) &= \lim_{m\to\infty} \frac{1}{mk}\log \|(\mathcal{A}_p^k)^m\| = \frac{1}{k}\log \lim_{m\to\infty} \|(\mathcal{A}_p^k)^m\|^{1/m} = \\ &= \frac{1}{k}\log(\text{spectral radius of } \mathcal{A}_p^k) \le \frac{1}{k}\log\|\mathcal{A}_p^k\|,\end{aligned} \tag{13.13}$$

and the inequality can be strict, even in the finite-dimensional case. Also, $\chi(\mathcal{A},\mu_p) \ge \frac{1}{k}\log\|(\mathcal{A}_p^k)^{-1}\|^{-1}$. Thus, for a point p as in the theorem we have one-sided estimates

$$\lambda(\mathcal{A},\mu_p) < \lambda(\mathcal{A},\mu) + \varepsilon \quad \text{and} \quad \chi(\mathcal{A},\mu_p) > \chi(\mathcal{A},\mu) - \varepsilon.$$

Remark 5.5. *Theorems 5.2 and 5.4 can be strengthened to conclude the existence of a periodic point $p = f^k p$ which gives simultaneous approximation for finitely many cocycles $\mathcal{A}_{(i)}$, $i = 1,\ldots,m$, over f with values in $GL(d_i,\mathbb{R})$ and $GL(V_i)$, respectively.*

In [KS18] we extended the periodic approximation results in [WaSu10, K11, KS17] to non-uniformly hyperbolic setting. We recall that a measure is *hyperbolic* if all its Lyapunov exponents for the derivative cocycle Df are nonzero, and a periodic point p is *hyperbolic* if the corresponding measure μ_p is hyperbolic.

Theorem 5.6. [KS18] *Let f be a $C^{1+\text{Hölder}}$ diffeomorphism of a compact manifold X, let μ be a hyperbolic ergodic f-invariant measure on X, and let $\mathcal{A}$ be a $GL(V)$-valued Hölder continuous cocycle over f.*

(i) *If $V = \mathbb{R}^d$, then the Lyapunov exponents $\lambda_1 \le \cdots \le \lambda_d$ of $\mathcal{A}$ with respect to μ, listed with multiplicities, can be approximated by the Lyapunov exponents of $\mathcal{A}$ at a hyperbolic periodic point p as in Theorem 5.2.*

(ii) *For any Banach space V, the upper and lower Lyapunov exponents λ and χ of $\mathcal{A}$ with respect to μ can be approximated as in Theorem 5.4, where $p = f^k p$ is a hyperbolic periodic point.*

13.5.2 Outline of the Proof of Theorem 5.4

We consider a full measure set $\Lambda = \Lambda_\mu$ of points $x \in X$ such that

$$\lim_{n\to\infty} \frac{1}{n} \log \|\mathcal{A}^n_{f^{-n}x}\| = \lambda = \lim_{n\to\infty} \frac{1}{n} \log \|\mathcal{A}^n_x\| \text{ and}$$
$$\lim_{n\to\infty} \frac{1}{n} \log \|\mathcal{A}^{-n}_x\| = -\chi = \lim_{n\to\infty} \frac{1}{n} \log \|(\mathcal{A}^n_x)^{-1}\|.$$

We construct a certain version of Lyapunov, or adapted, norm which allows us to control the norms of $\mathcal{A}_x$ and $(\mathcal{A}_x)^{-1}$.

Let γ be as in the Anosov Closing Lemma 2.1 and β be the Hölder exponent of $\mathcal{A}$. We fix ε such that $0 < \varepsilon < \varepsilon_0 = \beta\gamma/4$. For a point $x \in \Lambda$ we define the ε-*Lyapunov norm* $\|.\|_x = \|.\|_{x,\varepsilon}$ on V_x as follows. For $u \in V_x$,

$$\|u\|_x = \|u\|_{x,\varepsilon} = \sum_{n=0}^{\infty} \|\mathcal{A}^n_x(u)\|\, e^{-(\lambda+\varepsilon)n} + \sum_{n=1}^{\infty} \|\mathcal{A}^{-n}_x(u)\|\, e^{(\chi-\varepsilon)n}. \tag{13.14}$$

By the definition of Λ, both series converge exponentially.

For any points $x, y \in \Lambda$ and any linear map $A : V_x \to V_y$, we denote its operator norm with respect to the Lyapunov norms by

$$\|A\|_{y\leftarrow x} = \sup\{\|A(u)\|_{y,\varepsilon} : u \in V_x,\ \|u\|_{x,\varepsilon} = 1\}$$

and estimate it using (13.14)

$$\|\mathcal{A}_x\|_{fx\leftarrow x} \le e^{\lambda+\varepsilon} \quad\text{and}\quad \|\mathcal{A}^{-1}_x\|_{f^{-1}x\leftarrow x} \le e^{-\chi+\varepsilon}. \tag{13.15}$$

Since the Lyapunov norm depends only measurably on x, we provide a comparison with the standard norm. In [KS17, Proposition 3.1] we showed that there exists an f-invariant set $\mathcal{R} \subset \Lambda$ with $\mu(\mathcal{R}) = 1$ and a measurable function $K = K_\varepsilon$ such that, for all $x \in \mathcal{R}$,

$$\|u\| \le \|u\|_{x,\varepsilon} \le K(x)\|u\| \quad \text{for all } u \in \mathcal{E}_x, \quad \text{and} \tag{13.16}$$

$$K(x)e^{-\varepsilon|n|} \le K(f^n x) \le K(x)e^{\varepsilon|n|} \quad \text{for all } n \in \mathbb{Z}. \tag{13.17}$$

Also, for any points $x, y \in \mathcal{R}$ the inequality (13.16) yields

$$\|A\|_{y\leftarrow x} \le K(y)\|A\| \quad\text{and}\quad \|A\| \le K(x)\|A\|_{y\leftarrow x}. \tag{13.18}$$

We let $\mathcal{R}_t = \{x \in \mathcal{R} : \; K(x) \le t\}$. Then $\mu(\mathcal{R}_t) \to 1$ as $t \to \infty$.

An important role in the arguments is played by the following estimate of $\|\mathcal{A}_y^n\|$ along an orbit close to a regular one. We recall that $\varepsilon < \beta\gamma$, where γ is as in Lemma 2.1.

Lemma 5.7. *There exists a constant* $c = c(\mathcal{A}, \beta\gamma - \varepsilon)$ *such that for any point* x *in* $\mathcal{R}_t$ *with* $f^m x$ *in* $\mathcal{R}_t$ *and any point* $y \in X$ *such that the orbit segments* $x, fx, \dots, f^m x$ *and* $y, fy, \dots, f^m y$ *satisfy with some* $\delta > 0$

$$dist(f^i x, f^i y) \le \delta e^{-\gamma \min\{i, m-i\}} \quad \textit{for every } i = 0, \dots, m \tag{13.19}$$

we have for all $0 \le n \le m$

$$\|\mathcal{A}_y^n\| \le t\, \|\mathcal{A}_y^n\|_{f^n x \leftarrow x} \le t\, e^{c\, t\delta^\beta} e^{n(\lambda+\varepsilon)}. \tag{13.20}$$

A similar estimate can be obtained for $\|(\mathcal{A}_y^n)^{-1}\|$. We note that y may not be a regular point, and so we consider the norm of the linear map $\mathcal{A}_y^n$ with respect to the Lyapunov norms at x and $f^n x$.

Proof. We denote $x_i = f^i x$ and $y_i = f^i y$, $i = 0, \dots, m$, and use (13.15) to estimate the Lyapunov norm for $0 < n \le m$:

$$\begin{aligned} \|\mathcal{A}_y^n\|_{x_n \leftarrow x_0} &\le \prod_{i=0}^{n-1} \|\mathcal{A}_{y_i}\|_{x_{i+1} \leftarrow x_i} \le \prod_{i=0}^{n-1} \|\mathcal{A}_{x_i}\|_{x_{i+1} \leftarrow x_i} \cdot \|(\mathcal{A}_{x_i})^{-1} \circ \mathcal{A}_{y_i}\|_{x_i \leftarrow x_i} \\ &\le e^{n(\lambda+\varepsilon)} \prod_{i=0}^{n-1} \| \,(\mathcal{A}_{x_i})^{-1} \circ \mathcal{A}_{y_i}\, \|_{x_i \leftarrow x_i} = e^{n(\lambda+\varepsilon)} \prod_{i=0}^{n-1} \| \,\mathrm{Id} + \Delta_i\, \|_{x_i \leftarrow x_i}. \end{aligned} \tag{13.21}$$

Since $\mathcal{A}_x$ is β-Hölder and $\|(\mathcal{A}_x)^{-1}\|$ is uniformly bounded,

$$\|\Delta_i\| \le \|(\mathcal{A}_{x_i})^{-1}\| \cdot \|\mathcal{A}_{y_i} - \mathcal{A}_{x_i}\| \le M' \mathrm{dist}(x_i, y_i)^\beta \le M' \left(\delta e^{-\gamma \min\{i, m-i\}}\right)^\beta,$$

where the constant M' depends only on the cocycle $\mathcal{A}$. Since both x and $f^m x$ are in $\mathcal{R}_t$ we have $K(x_i) \le t e^{\varepsilon \min\{i, m-i\}}$ by (13.17), so using the first part of (13.18) we conclude that

$$\|\Delta_i\|_{x_i \leftarrow x_i} \le K(x_i)\|\Delta_i\| \le t e^{\varepsilon \min\{i,m-i\}} M' \delta^\beta e^{-\gamma\beta \min\{i,m-i\}} = M' t\, \delta^\beta\, e^{(\varepsilon - \beta\gamma) \min\{i,m-i\}}.$$

Taking the logarithm in (13.21) and using the preceding inequality, we obtain

$$\begin{aligned} \log\, \|\mathcal{A}_y^n\|_{x_n \leftarrow x_0} - n(\lambda + \varepsilon) &\le \sum_{i=0}^{n-1} \log\left(\| \,\mathrm{Id} + \Delta_i\, \|_{x_i \leftarrow x_i}\right) \le \sum_{i=0}^{n-1} \log\left(1 + \| \,\Delta_i\, \|_{x_i \leftarrow x_i}\right) \\ &\le \sum_{i=0}^{n-1} \| \,\Delta_i\, \|_{x_i \leftarrow x_i} \le M' t\delta^\beta \sum_{i=0}^{n-1} e^{(\varepsilon - \alpha\gamma) \min\{i,m-i\}} \le M' t\delta^\beta \cdot 2 \sum_{i=0}^{\infty} e^{(\varepsilon - \beta\gamma) i} = c\, t\delta^\beta, \end{aligned}$$

since $\varepsilon < \beta\gamma$. The constant c depends only on $\mathcal{A}$ and $(\beta\gamma - \varepsilon)$. We conclude that

$$\|\mathcal{A}_y^n\|_{x_n \leftarrow x_0} \leq e^{c\,t\delta^\beta} e^{n(\lambda+\varepsilon)}. \tag{13.22}$$

Since $K(x_0) \leq t$, using the second part of (13.18) we can also estimate the standard norm

$$\|\mathcal{A}_y^n\| \leq K(x_0)\|\mathcal{A}_y^n\|_{x_n \leftarrow x_0} \leq te^{c\,t\delta^\beta} e^{n(\lambda+\varepsilon)}.$$

□

Now we can obtain a point $p = f^k p$ such that $\frac{1}{k} \log \|\mathcal{A}_p^k\| \leq \lambda(\mathcal{A}, \mu) + 2\varepsilon$ as follows. We take $\mathcal{R}_t$ with a sufficiently large t and $x \in \mathcal{R}_t$ so that $f^k x \in \mathcal{R}_t$ and is sufficiently close to x to apply the Anosov Closing Lemma. Then there exists $p = f^k p$ satisfying (13.19). It follows from (13.20) that for a sufficiently large k we have

$$\frac{1}{k} \log \|\mathcal{A}_p^k\| \leq \frac{1}{k} \log(t\, e^{c\,t\delta^\alpha}) + (\lambda + \varepsilon) < \lambda + 2\varepsilon. \tag{13.23}$$

Now we outline how to find a point $p = f^k p$ for which $\frac{1}{k} \log \|\mathcal{A}_p^k\|$ also satisfies a corresponding lower estimate. In addition to Lemma 5.7, this part of the argument uses an approach developed in [GrGu17] and the Karlsson–Margulis Lemma [KaM99, Proposition 4.2]. The latter provides indices n for which a subadditive cocycle behaves almost additively in a certain sense. For future reference, we formulate its generalization for the case of several cocycles [KS18].

Proposition 5.8. [KaM99, KS18] *Let $a_n^{(1)}(x), \ldots, a_n^{(m)}(x)$ be integrable subadditive cocycles with exponents $\lambda^{(1)} > -\infty, \ldots, \lambda^{(m)} > -\infty$, respectively, over an ergodic measure-preserving system (X, f, μ). Then there exists a set $E \subset X$ with $\mu(E) = 1$ such that for each $x \in E$ and each $\varepsilon > 0$ there exists an integer $L = L(x, \varepsilon)$ so that the set $S = S(x, \varepsilon, L)$ of integers n satisfying the following condition is infinite:*

$$a_n^{(j)}(x) - a_{n-i}^{(j)}(f^i x) \geq (\lambda^{(j)} - \varepsilon)i \quad \textit{for all } L \leq i \leq n \textit{ and } 1 \leq j \leq m. \tag{13.24}$$

We apply this proposition to the subadditive cocycle $a_n(x) = \log \|\mathcal{A}_x^n\|$, consider a typical point $x \in \mathcal{R}_t$, and take a sufficiently large $n \in S(x, \varepsilon, L)$. Then we have

$$\|\mathcal{A}_x^n\| \cdot \|\mathcal{A}_{f^i x}^{n-i}\| \geq e^{(\lambda-\varepsilon)i} \quad \text{for all } i \text{ such that } L \leq i \leq n. \tag{13.25}$$

Now we would like to find k so that $f^k x \in \mathcal{R}_t$ and is sufficiently close to x to apply Anosov Closing Lemma, and also so that the ratio k/n is bounded, specifically,

$$n(1 + \sigma) \leq k \leq n(1 + 2\sigma), \quad \text{where } \sigma = 4\varepsilon/(\beta\gamma).$$

Existence of such k relies on [GrGu17, Lemma 8] or its refinement [KS18, Lemma 5.1].

The upper bound on k ensures that $\|\mathcal{A}_p^k\|$ and $\|\mathcal{A}_p^n\|$ are roughly comparable, so that it suffices to estimate $\|\mathcal{A}_p^n\|$ and show that it must grow as $\|\mathcal{A}_x^n\|$, with exponential rate at least $\lambda - \varepsilon$. For this it suffices to prove that $\|\mathcal{A}_x^n - \mathcal{A}_p^n\| \le \frac{1}{2}\|\mathcal{A}_x^n\|$. Thus the main part of the argument is to estimate $\|\mathcal{A}_x^n - \mathcal{A}_p^n\|$. For this we write

$$\mathcal{A}_x^n - \mathcal{A}_p^n = \mathcal{A}_{x_1}^{n-1} \circ (\mathcal{A}_x - \mathcal{A}_p) + (\mathcal{A}_{x_1}^{n-1} - \mathcal{A}_{p_1}^{n-1}) \circ \mathcal{A}_p =$$

$$= \cdots = \sum_{i=0}^{n-1} \mathcal{A}_{x_{i+1}}^{n-(i+1)} \circ (\mathcal{A}_{x_i} - \mathcal{A}_{p_i}) \circ \mathcal{A}_p^i, \quad \text{and hence}$$

$$\|\mathcal{A}_x^n - \mathcal{A}_p^n\| \le \sum_{i=0}^{n-1} \|\mathcal{A}_{x_{i+1}}^{n-(i+1)}\| \cdot \|\mathcal{A}_{x_i} - \mathcal{A}_{p_i}\| \cdot \|\mathcal{A}_p^i\|.$$

The third term in the product can be estimated by Lemma 5.7, $\|\mathcal{A}_p^i\| \le t\, e^{c\, t\delta^\alpha} e^{i(\lambda+\varepsilon)}$. The second term is estimated using Hölder continuity of $\mathcal{A}$ and exponential closeness of the trajectories

$$\|\mathcal{A}_{x_i} - \mathcal{A}_{p_i}\| \le c\, \text{dist}(f^i x, f^i p)^\beta \le c\, e^{-\gamma\beta \min\{i,k-i\}} \delta^\beta.$$

To estimate the first term, we use our choice of n and apply (13.25) to get

$$\|\mathcal{A}_{x_{i+1}}^{n-(i+1)}\| \le \|\mathcal{A}_x^n\| \cdot e^{-(i+1)(\lambda-\varepsilon)} \quad \text{for all } L \le i \le n.$$

Combining these inequalities and estimating the first L terms separately, we obtain

$$\|\mathcal{A}_x^n - \mathcal{A}_p^n\| \le c(\delta)\, \|\mathcal{A}_x^n\| \sum_{i=0}^{n-1} e^{2i\varepsilon} e^{-\gamma\beta \min\{i,k-i\}}, \quad \text{where } c(\delta) \to 0 \text{ as } \delta \to 0.$$

To complete the argument we need an estimate for the sum independent of k and n. This follows from the lower bound $k \ge (1+\sigma)n$, which ensures that

$$\beta\gamma \min\{i, k-i\} \ge 4\varepsilon i \quad \text{for } i = 0, \dots, n.$$

This completes the outline of the estimate from below, $\frac{1}{k} \log \|\mathcal{A}_p^k\| \ge \lambda - c'\varepsilon$. The upper estimate (13.23) also holds for the same x and k since both $x, f^k x$ are in $\mathcal{R}_t$. This completes the approximation of the upper Lyapunov exponent λ in Theorem 5.4.

The preceding arguments allow us to obtain simultaneous approximation of the upper exponents for several cocycles $\mathcal{A}_{(1)}, \dots, \mathcal{A}_{(m)}$ as in Remark 5.5. The only modification needed is to consider the intersection of the corresponding regular sets $\mathcal{R}_{t,(j)}$ for all $\mathcal{A}_{(j)}$ and use Proposition 5.8 for m subadditive cocycles $a^{(j)}(x) = \log \|\mathcal{A}_{(j)}\|$, $j = 1, \dots, m$. The same argument with minor modifications can also be applied to the cocycle $\mathcal{A}^{-1}$ to obtain a simultaneous approximation of the lower exponent χ. □

13.5.3 Outline of the Proof of Theorem 5.2

In place of the Lyapunov norm given by (13.14), we use the norm generated by the usual Lyapunov scalar product $\langle \cdot, \cdot \rangle_{x,\varepsilon}$ defined as follows. For a fixed $\varepsilon > 0$ and for any x in the set Λ of the Lyapunov regular points from Theorem 5.1,

$$\langle u, v\rangle_{x,\varepsilon} = 0 \quad \text{if } u \in \mathcal{E}_x^i,\ v \in \mathcal{E}_x^j,\ i \neq j, \text{ and}$$

$$\langle u, v\rangle_{x,\varepsilon} = d \sum_{n\in\mathbb{Z}} \langle \mathcal{A}_x^n u, \mathcal{A}_x^n v\rangle\, e^{-2\lambda_i n - \varepsilon|n|} \quad \text{if } u, v \in \mathcal{E}_x^i,\ i = 1, \ldots, m.$$

The properties of this norm are given in [BaPe, sections 3.5.1–3.5.3].

We take $\mathcal{R}_t' = \{x \in \Lambda : K'(x) \leq t\}$, where $K'(x)$ is a function similar to $K(x)$ which gives the estimate (13.16) for this Lyapunov norm.

First we approximate the largest exponent λ_ℓ. We again take $x \in \mathcal{R}_t'$ and k so that $f^k x \in \mathcal{R}_t'$, k and t are large enough, and $f^k x$ is sufficiently close to x. Then we obtain a periodic point $p = f^k p$ by the Anosov Closing Lemma. As in the approximation of λ from above in Theorem 5.4, we obtain estimate (13.23), which, using (13.13), also implies the desired upper estimate

$$\lambda_\ell(\mathcal{A}, \mu_p) \leq \frac{1}{k} \log \|\mathcal{A}_p^k\| < \lambda_\ell + 2\varepsilon.$$

The lower estimate $\lambda_\ell(\mathcal{A}, \mu_p) > \lambda_\ell - 3\varepsilon$ is obtained by constructing a $\mathcal{A}_p^k$-invariant cone $K \subseteq \mathcal{E}_p$, where vectors are expanded by a factor of at least $e^{k(\lambda_\ell - 3\varepsilon)}$; see [K11]. The cone K is defined as a cone of vectors whose direction is close, in the Lyapunov metric at x, to the Lyapunov subspace $\mathcal{E}_x^\ell$ corresponding to λ_ℓ. Then for any vector in K its forward Lyapunov exponent is at least $\lambda_\ell - 3\varepsilon$, and hence the same is true for $\lambda_\ell(\mathcal{A}, \mu_p)$.

To approximate all Lyapunov exponents of $\mathcal{A}$, we consider cocycles $\wedge^i \mathcal{A}$ induced by $\mathcal{A}$ on the i-fold exterior powers $\wedge^i \mathbb{R}^d$, for $i = 1, \ldots, d$. The largest Lyapunov exponent of $\wedge^i \mathcal{A}$ is $(\lambda_d + \cdots + \lambda_{d-i+1})$, where $\lambda_1 \leq \cdots \leq \lambda_d$ are the Lyapunov exponents of $\mathcal{A}$ listed with multiplicities. If a periodic point $p = f^k p$ satisfies

$$|(\lambda_d + \cdots + \lambda_{d-i+1}) - (\lambda_d^{(p)} + \cdots + \lambda_{d-i+1}^{(p)})| \leq 3\varepsilon \quad \text{for } i = 1, \ldots, d,$$

then we obtain the approximation $|\lambda_i - \lambda_i^{(p)}| \leq 3d\varepsilon$ for all $i = 1, \ldots, d$, completing the proof of Theorem 5.2.

A similar argument shows that one can obtain a simultaneous approximation of all Lyapunov exponents for several cocycles. □

13.5.4 Growth Estimates for Cocycles

As a corollary of Theorem 5.4, we obtain growth estimates for the norm and the quasiconformal distortion of a cocycle $\mathcal{A}$ in terms of the growth at periodic points.

Corollary 5.9. [S17] *Let $\mathcal{A}$ be a Hölder continuous Banach cocycle over f. Then*

(i) $\lim_{n\to\infty} \sup\left\{\|\mathcal{A}_x^n\|^{1/n} : x \in X\right\} = \limsup_{k\to\infty} \sup\left\{\|\mathcal{A}_p^k\|^{1/k} : p = f^k p \in X\right\}.$
In particular, if for some numbers C and s we have $\|\mathcal{A}_p^k\| \le Ce^{sk}$ whenever $p = f^k p$, then for each $\varepsilon > 0$ there exists a number C_ε such that

$$\|\mathcal{A}_x^n\| \le C_\varepsilon\, e^{(s+\varepsilon)n} \quad \textit{for all } x \in X \textit{ and } n \in \mathbb{N}.$$

(ii) $\lim_{n\to\pm\infty} \sup\left\{Q_{\mathcal{A}}(x,n)^{1/|n|} : x \in X\right\} = \limsup_{k\to\infty} \sup\left\{Q_{\mathcal{A}}(p,k)^{1/k} : p = f^k p \in X\right\}.$
In particular, if for some numbers C and s we have $Q_{\mathcal{A}}(p,k) \le Ce^{sk}$ whenever $p = f^k p$, then for each $\varepsilon > 0$ there exists a number C'_ε such that

$$Q_{\mathcal{A}}(x,n) \le C'_\varepsilon\, e^{(s+\varepsilon)|n|} \quad \textit{for all } x \in X \textit{ and } n \in \mathbb{Z}.$$

The number

$$\hat{\lambda}(\mathcal{A}) = \lim_{n\to\infty} \sup\left\{\frac{1}{n}\log\|\mathcal{A}_x^n\| : x \in X\right\}$$

gives the maximal exponential growth rate for a cocycle $\mathcal{A}$. Part (i) of Corollary 5.9 can also be deduced from [GrGu17], where the authors obtained a generalization of Livšic periodic point theorem for Banach cocycles and in particular showed that for each $\varepsilon > 0$ there exists a periodic point $p = f^k p$ such that $\frac{1}{k}\log\|\mathcal{A}_p^k\| > \hat{\lambda}(\mathcal{A}) - \varepsilon$.
Outline of the Proof of (ii). Let $\hat{\sigma}(\mathcal{A})$ and $\hat{\sigma}_P(\mathcal{A})$ be the logarithms of the left-hand side and the right-hand side of (ii), and let

$$q_n(x) = \log Q_{\mathcal{A}}(x,n) = \log\|\mathcal{A}_x^n\| + \log\|(\mathcal{A}_x^n)^{-1}\|.$$

The sequence $\{q_n(x)\}$ is a subadditive cocycle over f, and its exponent with respect to an ergodic measure μ is $\gamma(q,\mu) = \lambda(\mathcal{A},\mu) - \chi(\mathcal{A},\mu)$. The sequence of numbers $\hat{q}_n = \sup_{x\in X} q_n(x)$ is subadditive and hence $\hat{\gamma}(q) = \lim_{n\to\infty}(\hat{q}_n/n)$ exists. It follows from [Schr98] that $\hat{\gamma}(q) = \sup_\mu \gamma(q,\mu)$, where the supremum is taken over all ergodic f-invariant measures; see also [KS13, Proposition 4.9] for a direct proof of a similar result. Then, using Theorem 5.4, we obtain

$$\lim_{n\to\infty} \sup\left\{\log Q(x,n)^{1/n} : x \in X\right\} = \hat{\gamma}(q) = \sup_\mu \gamma(q,\mu) = \sup_\mu(\lambda(\mathcal{A},\mu) - \chi(\mathcal{A},\mu))$$

$$\le \sup\left\{\frac{1}{k}\log\|\mathcal{A}_p^k\| + \frac{1}{k}\log\|(\mathcal{A}_p^k)^{-1}\| : p = f^k p,\ k \ge N\right\}$$

$$= \sup\left\{\frac{1}{k}\log Q_{\mathcal{A}}(p,k) : p = f^k p,\ k \ge N\right\}.$$

It follows that $\hat{\sigma}(\mathcal{A}) \le \hat{\sigma}_P(\mathcal{A})$, and the opposite inequality clearly holds.

Now, suppose that $Q(p,k) \le Ce^{sk}$ whenever $p = f^k p$. Then we have

$$s \ge \hat{\sigma}_P(\mathcal{A}) = \hat{\sigma}(\mathcal{A}) = \hat{\gamma}(q) = \lim_{n\to\infty} \hat{q}_n/n.$$

It follows that for each $\varepsilon > 0$, there exists $N \in \mathbb{N}$ such that $\hat{q}_n \le (s+\varepsilon)n$ for all $n > N$ and hence $Q(x,n) \le e^{(s+\varepsilon)n}$ for all $x \in X$ and $n > N$. Taking

$$C'_\varepsilon = \max\{\, Q(x,n) : \; x \in X \text{ and } 1 \le n \le N\,\},$$

we obtain $Q(x,n) \le C'_\varepsilon e^{(s+\varepsilon)n}$ for all $x \in X$ and $n \in \mathbb{N}$. The statements for negative n follows since $Q(x,-n) = Q(f^{-n}x,n)$. □

In the finite-dimensional case, using Theorem 5.2, we can similarly obtain growth estimates using the Lyapunov exponents rather than norms.

Corollary 5.10. [K11, S15] *Let $\mathcal{A}$ be a Hölder continuous linear cocycle over (X,f).*

(i) *If for some number s we have $\lambda_\ell(\mathcal{A},\mu_p) \le s$ for every invariant measure μ_p on a periodic orbit, then for each $\varepsilon > 0$ there exists a number C_ε such that*

$$\|\mathcal{A}_x^n\| \le C_\varepsilon\, e^{(s+\varepsilon)n} \quad \textit{for all } x \in X \textit{ and } n \in \mathbb{N}.$$

(ii) *If for some s we have $\lambda_\ell(\mathcal{A},\mu_p) - \lambda_1(\mathcal{A},\mu_p) \le s$ for every invariant measure μ_p on a periodic orbit, then for each $\varepsilon > 0$ there exists a number C'_ε such that*

$$Q_{\mathcal{A}}(x,n) \le C'_\varepsilon\, e^{(s+\varepsilon)|n|} \quad \textit{for all } x \in X \textit{ and } n \in \mathbb{Z}.$$

13.5.5 Obtaining Fiber Bunching from Periodic Data

We recall that in Theorem 4.9 we assumed that fiber bunching for only one of the cocycles as fiber bunching for the other one can be deduced from the conjugacy of their periodic data. Now we explain how this is done using the results on approximation of Lyapunov exponents.

Proposition 5.11. [S15, S17] *Let $\mathcal{A}$ and $\mathcal{B}$ be Hölder continuous cocycles over (X,f).*

(i) *If $\mathcal{A}$ and $\mathcal{B}$ take values in $GL(d,\mathbb{R})$, $\mathcal{A}$ is fiber bunched, and $\mathcal{B}$ has conjugate periodic data, then $\mathcal{B}$ is also fiber bunched.*
(ii) *If $\mathcal{A}$ and $\mathcal{B}$ take values in $GL(V)$, $\mathcal{A}$ is fiber bunched, and $\mathcal{B}$ has conjugate periodic data with bounded $C(p)$, then $\mathcal{B}$ is also fiber bunched.*

Part (ii) follows from the next result that establishes fiber bunching of a cocycle from fiber bunching of its periodic data. Part (i) follows from a similar characterization of fiber bunching in the finite-dimensional case using Lyapunov exponents at periodic orbits [S15, Corollary 4.2].

Proposition 5.12. [S17] *Let $\mathcal{B}$ be a β-Hölder cocycle over (X,f). Suppose that there exist numbers $\tilde{\theta} < 1$ and $\tilde{L}$ such that whenever $f^n p = p$,*

$$Q_{\mathcal{B}}(p,n) \cdot (\nu_p^n)^\beta < \tilde{L}\tilde{\theta}^n \quad \textit{and} \quad Q_{\mathcal{B}}(p,-n) \cdot (\hat{\nu}_p^{-n})^\beta < \tilde{L}\tilde{\theta}^n. \tag{13.26}$$

Then $\mathcal{B}$ is fiber bunched.

Outline of the Proof. If functions $\nu(x) = \nu$ and $\hat{\nu}(x) = \hat{\nu}$ are constant, then the result follows immediately from Corollary 5.9 (ii). In general, one can apply an argument as in the proof of Corollary 5.9 to the subadditive cocycle $\tilde{q}_n(x) = \log Q_{\mathcal{B}}(x,n) \cdot \log \nu_n(x)^\beta$ and use simultaneous approximation of exponents as in Remark 5.5. □

13.6 Boundedness, Conformality, and Reductions

It is often useful to determine whether a cocycle is cohomologous to one with values in a smaller group, such as orthogonal or conformal, since cocycles with values in these groups are relatively well understood. Such reduction problems are highly non-trivial: Schmidt asked in [Schm99] whether a $GL(d,\mathbb{R})$-valued cocycle $\mathcal{A}$ with $\|\mathcal{A}_x^n\|$ uniformly bounded is Hölder cohomologous to an orthogonal one. In [KS10] we showed this assuming boundedness of the periodic data only. Moreover, we established a general criterion for a linear cocycle to be isometric or conformal in terms of its periodic data.

In this section we view a linear cocycle $\mathcal{A}$ as an automorphism of a vector bundle $\mathcal{E}$ over X with the fiber map $\mathcal{A}_x : \mathcal{E}_x \to \mathcal{E}_{fx}$, and we do not assume triviality of the bundle.

13.6.1 Conformal Structures

A conformal structure on $\mathbb{R}^d$, $d \geq 2$, is a class of proportional inner products. Since an inner product identifies with a symmetric positive definite $d \times d$ matrix, we can uniquely represent a conformal structure by such a matrix with determinant 1. The group $GL(d,\mathbb{R})$ acts on $\mathcal{C}^d$ via

$$D(C) = (\det D^T D)^{-1/d} \cdot D^T C D, \quad \text{where } C \in \mathcal{C}^d \text{ and } D \in GL(d,\mathbb{R}).$$

Its subgroup $SL(d,\mathbb{R})$ acts transitively on $\mathcal{C}^d$ since every $C \in \mathcal{C}^d$ can be expressed as $C = D(\mathrm{Id})$, where D is the positive square root of C. The stabilizer of the identity matrix in $SL(d,\mathbb{R})$ is $SO(d,\mathbb{R})$, and thus $\mathcal{C}^d$ can be identified with $SL(d,\mathbb{R})/SO(d,\mathbb{R})$. It is well known that $\mathcal{C}^d = SL(d,\mathbb{R})/SO(d,\mathbb{R})$ is a Riemannian symmetric space with a certain $GL(d,\mathbb{R})$-invariant metric of nonpositive curvature; see [T86, p. 327].

For a vector bundle $\mathcal{E} \to X$ we can consider a bundle $\mathcal{C}$ over X whose fiber $\mathcal{C}_x$ is the space of conformal structures on $\mathcal{E}_x$. Using a background Riemannian metric on $\mathcal{E}$, the space $\mathcal{C}_x$ can be identified with the space of symmetric positive linear operators on $\mathcal{E}_x$ with determinant 1. We equip the fibers of $\mathcal{C}$ with the Riemannian metric as in the preceding. A continuous (measurable) section σ of $\mathcal{C}$ is called a continuous (measurable) *conformal structure* on $\mathcal{E}$. A measurable conformal structure σ is called *bounded* if the distance between $\sigma(x)$ and $\tau(x)$ is bounded on X for a continuous conformal structure τ on X.

An invertible linear map $A : \mathcal{E}_x \to \mathcal{E}_y$ induces an isometry from $\mathcal{C}_x$ to $\mathcal{C}_y$ via

$$A(C) = (\det(A^*A))^{1/d} \cdot (A^{-1})^* \, C A^{-1},$$

where $C \in \mathcal{C}_x$ is a conformal structure viewed as an operator, and A^* is the adjoint of A. If $\mathcal{A} : \mathcal{E} \to \mathcal{E}$ is a linear cocycle over f, we say that a conformal structure σ on $\mathcal{E}$ is $\mathcal{A}$-*invariant* if

$$\mathcal{A}_x(\sigma(x)) = \sigma(fx) \quad \text{for all } x \in X.$$

13.6.2 Conformality and Uniform Quasiconformality

Let $\mathcal{A}$ be a f linear cocycle over (X,f). We recall that its quasiconformal distortion is

$$Q_{\mathcal{A}}(x,n) = \|\mathcal{A}_x^n\| \cdot \|(\mathcal{A}_x^n)^{-1}\| = \frac{\max\{\, \|\mathcal{A}_x^n(v)\| : \ v \in \mathcal{E}_x, \ \|v\| = 1\,\}}{\min\{\, \|\mathcal{A}_x^n(v)\| : \ v \in \mathcal{E}_x, \ \|v\| = 1\,\}}, \quad x \in X, \, n \in \mathbb{Z}.$$

The cocycle is called *uniformly quasiconformal* if $Q_{\mathcal{A}}(x,n) \le K$ for all x and n, and it is called *conformal* if $Q_{\mathcal{A}}(x,n) = 1$ for all x and n.

Clearly, $\mathcal{A}$ is conformal with respect to a Riemannian metric on $\mathcal{E}$ if and only if it preserves the conformal structure associated with this metric. We note that the notion of uniform quasiconformality does not depend on the choice of a continuous metric on $\mathcal{E}$. So if $\mathcal{A}$ preserves a continuous or bounded conformal structure on $\mathcal{E}$, then $\mathcal{A}$ is uniformly quasiconformal on $\mathcal{E}$ with respect to any continuous metric on $\mathcal{E}$. The theorem that follows shows that the converse is also true.

Theorem 6.1. [S02, KS10] *Let $\mathcal{A} : \mathcal{E} \to \mathcal{E}$ be a Hölder continuous linear cocycle over a hyperbolic dynamical system (X,f). If $\mathcal{A}$ is uniformly quasiconformal, then it preserves a Hölder continuous conformal structure on $\mathcal{E}$, equivalently, $\mathcal{A}$ is conformal with respect to a Hölder continuous Riemannian metric on $\mathcal{E}$.*

The first step of the proof is obtaining a bounded measurable conformal structure. We apply observations made by Sullivan and Tukia for quasiconformal group actions.

Proposition 6.2. *Let f be a homeomorphism of a compact metric space X and let $\mathcal{A} : \mathcal{E} \to \mathcal{E}$ be a continuous linear cocycle over f. If $\mathcal{A}$ is uniformly quasiconformal, then it preserves a bounded Borel measurable conformal structure σ on $\mathcal{E}$.*

Proof. Let τ be a continuous conformal structure on $\mathcal{E}$. We denote by $\tau(x)$ the conformal structure on $\mathcal{E}_x$, $x \in X$. We consider the set

$$S(x) = \{\, (\mathcal{A}^{-n}_{f^n x})(\tau(f^n x)) : \; n \in \mathbb{Z} \}$$

in $\mathfrak{C}_x$, the space of conformal structures on $\mathcal{E}_x$. Since $\mathcal{A}$ is uniformly quasiconformal, the sets $S(x)$ have uniformly bounded diameters. Since the space $\mathfrak{C}_x$ is simply connected and has nonpositive curvature, every non-empty bounded set in $\mathfrak{C}_x$ is contained in a unique closed ball of the smallest radius; see for example [T86, Lemma E]. The center of this ball is called a circumcenter of the set. For each $x \in X$ we denote by $\sigma(x)$ the circumcenter of $S(x)$.

It follows from the construction that the conformal structure σ is $\mathcal{A}$-invariant and its distance from τ is bounded. For any $k \geq 0$, the set

$$S_k(x) = \{\, (\mathcal{A}^{-n}_{f^n x})(\tau(f^n x)) : \; |n| \leq k\}$$

depends continuously on x in Hausdorff distance, and so does the center $\sigma_k(x)$ of the smallest ball containing $S_k(x)$. Since $S_k(x) \to S(x)$ as $k \to \infty$ for any x, the conformal structure σ is the pointwise limit of continuous conformal structures $\sigma_k(x)$. Hence σ is Borel measurable. □

To complete the proof of Theorem 6.1, we use Theorem 6.3 to obtain Hölder continuity of σ. The theorem applies since uniform quasiconformality of $\mathcal{A}$ implies its fiber bunching.

Theorem 6.3. [KS13] *Let (X,f) be a hyperbolic dynamical system, let $\mathcal{A} : \mathcal{E} \to \mathcal{E}$ be a β-Hölder fiber-bunched linear cocycle over f, and let μ be an ergodic f-invariant probability measure with full support and local product structure. Then any $\mathcal{A}$-invariant μ-measurable conformal structure on $\mathcal{E}$ is β-Hölder.*

Outline of the Proof. We recall that a fiber-bunched cocycle has stable and unstable holonomies; see Proposition 3.3. The main part of the proof is establishing essential invariance of the conformal structure under holonomies.

Lemma 6.4. *Let H^s be the stable holonomy for a linear cocycle $\mathcal{A}$. If τ is a μ-measurable $\mathcal{A}$-invariant conformal structure, then τ is essentially H^s-invariant; namely there is a set $G \subset X$ of full measure such that $\tau(y) = H^s_{xy}(\tau(x))$ for all $x, y \in G$ such that $y \in W^s_{loc}(x)$.*

Proof. Let $x_i = f^i x$. Since $\mathcal{A}^n_y$ induces an isometry and τ is $\mathcal{A}$-invariant, we obtain

$$\begin{aligned}\text{dist}\,(\tau(y), H^s_{xy}(\tau(x))) &= \text{dist}(\mathcal{A}^n_y(\tau(y)), \mathcal{A}^n_y H^s_{xy}(\tau(x))) = \\ &= \text{dist}(\tau(y_n), H^s_{x_n y_n} \mathcal{A}^n_x(\tau(x))) = \text{dist}(\tau(y_n), H^s_{x_n y_n}(\tau(x_n))) \leq \\ &\leq \text{dist}(\tau(y_n), \tau(x_n)) + \text{dist}(\tau(x_n), H^s_{x_n y_n}(\tau(x_n))).\end{aligned}$$

Since τ is μ-measurable, by Lusin's theorem there exists a compact set $S \subset X$ with $\mu(S) > 1/2$ on which τ is uniformly continuous and hence bounded. Let G be the set of points in X for which the frequency of visiting S equals $\mu(S) > 1/2$. By the Birkhoff Ergodic Theorem, $\mu(G) = 1$.

Suppose that both x and y are in G. Then there exists a sequence $\{n_i\}$ such that both x_{n_i} and y_{n_i} are in S. Since $y \in W^s_{\text{loc}}(x)$, $\text{dist}(x_{n_i}, y_{n_i}) \to 0$ and hence $\text{dist}(\tau(x_{n_i}), \tau(y_{n_i})) \to 0$ by uniform continuity of τ on S. Since $\|H^s_{x_{n_i} y_{n_i}} - \text{Id}\| \to 0$ and τ is bounded on S, we obtain

$$\text{dist}(\tau(x_{n_i}), H^s_{x_{n_i} y_{n_i}}(\tau(x_{n_i}))) \to 0.$$

Here we used the fact that for any linear map X sufficiently close to the identity we have $\text{dist}\,(\tau, X(\tau))) \leq k(\tau) \cdot \|X - \text{Id}\,\|$, where $k(\tau)$ is bounded on bounded sets in $\mathcal{C}^d$ [KS10]. We conclude that $\text{dist}\,(\tau(y), H^s_{xy}(\tau(x))) = 0$ and thus τ is essentially H^s-invariant. □

Similarly, τ is essentially H^u-invariant. By Hölder continuity of holonomies (H4) this implies essential Hölder continuity of τ along the stable and unstable leaves:

$$\text{dist}\,(\tau(x), \tau(y)) \leq C\,\text{dist}(x, y)^\beta \quad \text{if } y \in W^{s/u}_{\text{loc}}(x) \text{ and both } x, y \in G.$$

To complete the proof of Theorem 6.3 we use the local product structure of μ and the local product structure of the stable and unstable foliations to show that τ coincides μ-a.e. with a Hölder continuous conformal structure on $\text{supp}\,\mu = X$. □

13.6.3 Conformality and Isometry from Periodic Data

If a cocycle $\mathcal{A}$ over an Anosov diffeomorphism f is conformal or uniformly quasi-conformal, then clearly there exists a constant C_{per} such that $Q_{\mathcal{A}}(p, n) \leq C_{per}$ for every periodic point p and n such that $p = f^n p$. The next theorem shows that the converse is also true. A similar result was also obtained in [LWi10] under an extra bunching-type assumption.

Theorem 6.5. [KS10] *Let $\mathcal{A} : \mathcal{E} \to \mathcal{E}$ be a Hölder continuous linear cocycle over a hyperbolic system (X, f).*

(i) *If there exists a constant C_{per} such that $Q_{\mathcal{A}}(p,n) \le C_{per}$ whenever $f^n p = p$, then $\mathcal{A}$ is conformal with respect to a Hölder continuous Riemannian metric on $\mathcal{E}$.*

(ii) *If $\mathcal{A}$ has bounded periodic data, that is, there exists a constant C'_{per} such that $\max\{\|\mathcal{A}_p^n\|, \|(\mathcal{A}_p^n)^{-1}\|\} \le C'_{per}$ whenever $f^n p = p$, then $\mathcal{A}$ is an isometry with respect to a Hölder continuous Riemannian metric on $\mathcal{E}$.*

Remark 6.6. *For a cocycle on a trivial bundle $X \times \mathbb{R}^d$ given by $A : X \to GL(d,\mathbb{R})$ the theorem implies Hölder cohomology to a cocycle with values in the conformal or orthogonal subgroup. Indeed, let $C(x)$ be the unique positive square root of the symmetric positive definite matrix that defines the Riemannian metric at x. Then C is a desired Hölder continuous conjugacy.*

Proof of Theorem 6.5.

(i) The assumption and Proposition 5.12 yield fiber bunching of $\mathcal{A}$. By Theorem 6.1 it suffices to show that $\mathcal{A}$ is uniformly quasiconformal on $\mathcal{E}$. Let $z \in X$ be a point with dense orbit $\mathcal{O} = \{f^k z : k \in \mathbb{Z}\}$. We show that the quasiconformal distortion $Q_{\mathcal{A}}(z,k)$ is uniformly bounded in $k \in \mathbb{Z}$. Since $Q_{\mathcal{A}}(x,k)$ is continuous on X for each k, this implies uniform quasiconformality.

Let δ_0 be as in the Anosov Closing Lemma, and let $0 < \delta < \delta_0$. We consider two points of $\mathcal{O}$ with $k_1 < k_2$ and $\text{dist}(f^{k_1}z, f^{k_2}z) < \delta$, and we denote $x = f^{k_1}z$ and $n = k_2 - k_1$. Then there exists $p \in X$ with $f^n p = p$ such that $\text{dist}(f^i x, f^i p) \le c\delta$ for $i = 0, \dots, n$. Let $y \in W^s_{\text{loc}}(p) \cap W^u_{\text{loc}}(x)$. By Proposition 3.3 (H4),

$$\|(\mathcal{A}_p^n)^{-1} \circ \mathcal{A}_y^n - \text{Id}\,\| \le c'\delta^\beta \quad \text{and} \quad \|(\mathcal{A}_y^{-n})^{-1} \circ \mathcal{A}_x^{-n} - \text{Id}\,\| \le c'\delta^\beta.$$

It follows that if δ is sufficiently small, then

$$Q_{\mathcal{A}}(y,n)/Q_{\mathcal{A}}(p,n) \le (1 + c'\delta^\beta)/(1 - c'\delta^\beta) \le 2 \quad \text{and} \quad Q_{\mathcal{A}}(x,n)/Q_{\mathcal{A}}(y,n) \le 2.$$

Thus $Q_{\mathcal{A}}(x,n) \le 4\, Q_{\mathcal{A}}(p,n) \le 4C_{per}$.

We take $m > 0$ such that the set $\{f^j z;\ |j| \le m\}$ is δ-dense in X, and set

$$Q_m = \max\{Q_{\mathcal{A}}(z,j) :\ |j| \le m\}.$$

Then for any $k > m$ there exists j, $|j| \le m$, such that $\text{dist}(f^k z, f^j z) \le \delta$ and hence

$$Q_{\mathcal{A}}(z,k) \le Q_{\mathcal{A}}(z,j) \cdot Q_{\mathcal{A}}(f^j z, k-j) \le Q_m \cdot 4C_{per}.$$

The case of $k < -m$ is considered similarly. Thus $Q_{\mathcal{A}}(z,k)$ is uniformly bounded.

(ii) By (i), the cocycle is conformal with respect to a Hölder continuous Riemannian metric g on $\mathcal{E}$. Hence there exists a positive Hölder continuous function $a(x)$ such that

$$\|\mathcal{A}_x(v)\|_{g(fx)} = a(x) \cdot \|v\|_{g(x)} \quad \text{for each } x \in X \text{ and each } v \in \mathcal{E}_x.$$

We seek a positive Hölder continuous function $\varphi(x)$ such that the renormalized metric $\tilde{g}(x) = g(x)/\varphi(x)$ is invariant, that is,

$$\varphi(fx) \cdot \|\mathcal{A}_x(v)\|_{\tilde{g}(fx)} = a(x)\varphi(x) \cdot \|v\|_{\tilde{g}(x)}, \text{ equivalently, } a(x) = \varphi(fx)/\varphi(x).$$

Let p be a periodic point of a period n. The boundedness assumption for *all* periods does not allow $a(p)a(fp)\cdots a(f^{n-1}p) = \|\mathcal{A}_p^n\|_{g(p)}$ to be greater or less than one. Hence $a(p)a(fp)\cdots a(f^{n-1}p) = 1$ whenever $p = f^n$, and by the Livšic theorem the equation $a(x) = \varphi(fx)/\varphi(x)$ has a Hölder continuous solution. □

Part (ii) can also be obtained without using (i) as follows. First, similarly to (i) we can show that boundedness of the periodic data implies boundedness of the cocycle, that is, of the set $\{\mathcal{A}_x^n : x \in X,\ n \in \mathbb{Z}\}$. Then an analog of Proposition 6.2 can be proven for inner products instead of conformal structures. This would give a bounded measurable invariant family of inner products. Its Hölder continuity can be established as in Theorem 6.3.

One may try to make weaker assumptions on the periodic data; for example, that for each p, there is a uniform bound $C(p)$ on $Q_{\mathcal{A}}(p,n)$ for all periods n. This is equivalent to each of the following three statements:

$\mathcal{A}_p^n$ is diagonalizable over $\mathbb{C}$ with its eigenvalues equal in modulus; $\mathcal{A}_p^n$ is conjugate to a conformal linear map.

There exists a $\mathcal{A}_p^n$-invariant conformal structure on $\mathcal{E}_p$. In fact, the periodic assumption in the first part of Theorem 6.5 is equivalent to having such conformal structures for all periodic points uniformly bounded. Without such a boundedness assumption, the theorem does not hold in dimension higher than two, as the following example demonstrates.

Proposition 6.7. [KS10] *Let (X,f) be a hyperbolic system and let $\mathcal{E} = X \times \mathbb{R}^d$, $d \geq 3$. For any $\varepsilon > 0$ there exists a Lipschitz continuous linear cocycle $\mathcal{A} : \mathcal{E} \to \mathcal{E}$, with its generator ε-close to the identity, such that for all periodic points $p \in X$ the return maps $\mathcal{A}_p^n : \mathcal{E}_p \to \mathcal{E}_p$ are conjugate to orthogonal maps, but $\mathcal{A}$ is not conformal with respect to any continuous Riemannian metric on $\mathcal{E}$.*

Outline of the Construction. Let $\mathcal{E} = X \times \mathbb{R}^3$ and $\mathcal{A}_x = \begin{bmatrix} \cos\alpha(x) & -\sin\alpha(x) & \varepsilon \\ \sin\alpha(x) & \cos\alpha(x) & 0 \\ 0 & 0 & 1 \end{bmatrix}$.

Let S be a closed f-invariant set in X without periodic points and let $\alpha : X \to \mathbb{R}$ be a Lipschitz continuous function satisfying

$$\alpha(x) = 0 \text{ for } x \in S \text{ and } 0 < \alpha(x) \leq \varepsilon \text{ for } x \notin S\,.$$

Then for $x \in S$, $\mathcal{A}_x^n = \begin{bmatrix} 1 & 0 & n\varepsilon \\ 0 & 1 & 0 \\ 0 & 0 & 1 \end{bmatrix}$ and so $Q_{\mathcal{A}}(x,n) \to \infty$ as $n \to \infty$.

Thus $\mathcal{A}$ cannot be conformal with respect to a continuous Riemannian metric on $\mathcal{E}$.

At $p = f^n p$, the map $\mathcal{A}_p^n$ has eigenvalues of modulus 1 and is diagonalizable if the 2×2 rotation block has complex eigenvalues; that is, the rotation angle $\alpha(p) + \cdots + \alpha(f^{n-1}p)$ does not equal πk. This can be ensured for all periodic points by slightly modifying the function α, if necessary. □

The next result shows that the weaker assumption of existence of invariant conformal structures at periodic points suffices for two-dimensional bundles.

Theorem 6.8. [KS10] *Let $\mathcal{A} : \mathcal{E} \to \mathcal{E}$ be a Hölder continuous linear cocycle over a hyperbolic system (X,f). Suppose that the fibers of $\mathcal{E}$ are* two-dimensional.

(i) *If, for each periodic point $p \in X$, the return map $\mathcal{A}_p^n : \mathcal{E}_p \to \mathcal{E}_p$ is diagonalizable over $\mathbb{C}$ and its eigenvalues are equal in modulus, then $\mathcal{A}$ is conformal with respect to a Hölder continuous Riemannian metric on $\mathcal{E}$.*

(ii) *If, for each periodic point $p \in X$, the return map $\mathcal{A}_p^n : \mathcal{E}_p \to \mathcal{E}_p$ is diagonalizable over $\mathbb{C}$ and its eigenvalues are of modulus 1, then $\mathcal{A}$ is isometric with respect to a Hölder continuous Riemannian metric on $\mathcal{E}$.*

Outline of the Proof. The proof relies on the fact that $\mathcal{A}$ as in the theorem is either conformal or has a one-dimensional continuous invariant sub-bundle; see Corollary 7.5 in the next section. In the first case we immediately obtain (i).

In the latter case, for any point $p = f^n p$ there is an invariant line for $\mathcal{A}_p^n : \mathcal{E}_p \to \mathcal{E}_p$. This implies that the eigenvalues of $\mathcal{A}_p^n$ are real and by the assumption of the theorem, they are either λ, λ or $\lambda, -\lambda$. If $\mathcal{A}$ is orientation preserving, the former is always the case. It follows that $\mathcal{A}_p^n = \lambda \cdot \mathrm{Id}$ since it is diagonalizable. Hence $Q_{\mathcal{A}}(p,n) = 1$ and conformality of $\mathcal{A}$ follows from Theorem 6.5. If $\mathcal{E}$ is not orientable, we can pass to a double cover. If $\mathcal{A}$ is orientation reversing, we can consider cocycle $\mathcal{A}^2$ over f^2. Thus we can always obtain an orientation-preserving cocycle $\mathcal{A}'$ that is conformal. This implies uniform quasiconformality of the original cocycle $\mathcal{A}$, and conformality of $\mathcal{A}$ follows from Theorem 6.1.

The second part can be established as in the proof of Theorem 6.5. Indeed, the assumption implies that for any periodic point p, the map $\mathcal{A}_p^n$ is conjugate to an orthogonal matrix and hence there exists a constant $C(p)$ such that $\max\{\|\mathcal{A}_p^n\|, \|(\mathcal{A}_p^n)^{-1}\|\} \le C(p)$ for any period n of p. □

13.6.4 Infinite-Dimensional Setting

In [KS18-2] we obtained an analog of Theorem 6.5 for infinite-dimensional setting. The finite-dimensional boundedness result was extended in two directions: boundedness and pre-compactness, as the latter does not follow automatically. For a Banach cocycle $\mathcal{A}$, we consider the periodic data set $\mathcal{A}_P$ and the set of all values $\mathcal{A}_X$,

$$\mathcal{A}_P = \{\mathcal{A}_p^k : p = f^k p,\ p \in X,\ k \in \mathbb{N}\} \quad \text{and} \quad \mathcal{A}_X = \{\mathcal{A}_x^n : x \in X,\ n \in \mathbb{Z}\}.$$

Theorem 6.9. [KS18-2] *Let $\mathcal{A}$ be a Hölder continuous Banach cocycle over (X,f).*

(i) *If there exists a constant C_{per} such that $Q_{\mathcal{A}}(p,k) \le C_{per}$ whenever $f^k p = p$, then $\mathcal{A}$ is uniformly quasiconformal.*
(ii) *If the set $\mathcal{A}_P$ is bounded in $(GL(V), d)$, then so is the set $\mathcal{A}_X$.*
(iii) *If the set $\mathcal{A}_P$ has compact closure in $(GL(V), d)$, then so does the set $\mathcal{A}_X$, moreover, there exists a Hölder continuous family of norms $\|.\|_x$ on V such that $\mathcal{A}_x : (V, \|.\|_x) \to (V, \|.\|_{fx})$ is an isometry for each $x \in X$.*

We note that the closures in (iii) are not the same in general. For example, if $\mathcal{A}$ is a coboundary, that is, it is generated by $A(x) = C(fx) \circ C(x)^{-1}$ for a function $C : X \to GL(V)$, then $\mathcal{A}_P = \{\mathrm{Id}\}$ while $\mathcal{A}_X$ is usually not.

The proofs of uniform quasiconformality and boundedness of the cocycle are similar to the proof of uniform quasiconformality in the first part of Theorem 6.5. Part (iii) in the infinite-dimensional case requires substantially different arguments. Indeed, Theorem 6.1 and Proposition 6.2 rely on the fact that the space of conformal structures and the space of Euclidean norms have a structure of a symmetric space of nonpositive curvature, but in the infinite-dimensional case there is no analogous metric structure. Instead, we consider a natural distance on the set of norms, but the resulting space is not separable so we work with a small subset. The following general result yields a measurable invariant family of norms and then we show its continuity.

Proposition 6.10. *Let f be a homeomorphism of a metric space X and let $\mathcal{A}$ be a continuous Banach cocycle over (X,f). If the set $\mathcal{A}_X$ has compact closure in $GL(V)$, then there exists a bounded Borel measurable family of norms $\|.\|_x$ on V such that $\mathcal{A}_x : (V, \|.\|_x) \to (V, \|.\|_{fx})$ is an isometry for each $x \in X$.*

13.7 Cocycles with One Lyapunov Exponent

Now we return to the finite-dimensional case and consider linear cocycles with one Lyapunov exponent, that is, satisfying $\lambda(\mathcal{A}, \mu) = \chi(\mathcal{A}, \mu)$, where $\lambda = \lambda_\ell$ and $\chi = \lambda_1$ are the largest and the smallest Lyapunov exponents given by Oseledets' Multiplicative Ergodic Theorem. Clearly, this is a broader class than conformal or uniformly quasiconformal cocycles. We considered cocycles with one exponent in [KS09, KS10], and the results in these papers indicated that such cocycles should exhibit some rigidity. In [KS13] we showed that it is true in a strong sense by obtaining a continuous version of Zimmer's Amenable Reduction Theorem.

When $\mathcal{A}$ has more than one Lyapunov exponent, the invariant sub-bundles $\mathcal{E}^i$ given Oseledets' MET are measurable but not necessarily continuous. The next theorem establishes continuity of measurable invariant sub-bundles for fiber-bunched cocycles with one exponent. It is a corollary of results by Avila and Viana in [AV10].

Theorem 7.1. *Let (X,f) be a hyperbolic system, let $\mathcal{A} : \mathcal{E} \to \mathcal{E}$ be a β-Hölder fiber-bunched linear cocycle over (X,f), and let μ be an ergodic f-invariant probability measure with full support and local product structure. If $\lambda(\mathcal{A},\mu) = \chi(\mathcal{A},\mu)$, then any μ-measurable $\mathcal{A}$-invariant sub-bundle of $\mathcal{E}$ is β-Hölder.*

Now we obtain a structure theorem for cocycles with one exponent.

Theorem 7.2 (Continuous Amenable Reduction). *[KS13] Let (X,f) be a hyperbolic system and let $\mathcal{A} : \mathcal{E} \to \mathcal{E}$ be a β-Hölder linear cocycle over f. Suppose that for every periodic point $p = f^n p$ the invariant measure μ_p on its orbit satisfies $\lambda(\mathcal{A},\mu_p) = \chi(\mathcal{A},\mu_p)$, that is, all eigenvalues of $\mathcal{A}_p^n$ are equal in modulus. Then there exists a flag of β-Hölder $\mathcal{A}$-invariant sub-bundles*

$$\{0\} = \mathcal{E}^0 \subset \mathcal{E}^1 \subset \cdots \subset \mathcal{E}^{j-1} \subset \mathcal{E}^j = \mathcal{E} \tag{13.27}$$

and β-Hölder Riemannian metrics on the factor bundles $\mathcal{E}^i/\mathcal{E}^{i-1}$, $i = 1,\dots,j$, so that the factor-cocycles induced by $\mathcal{A}$ on $\mathcal{E}^i/\mathcal{E}^{i-1}$ are conformal. Moreover, there exists a positive β-Hölder function $\phi : X \to \mathbb{R}$ such that the factor-cocycles of $\phi\mathcal{A}$ on $\mathcal{E}^i/\mathcal{E}^{i-1}$ are isometries.

In the case when $\mathcal{E}_1 = \mathcal{E}$, the cocycle $\mathcal{A}$ itself is conformal on $\mathcal{E}$ with respect to some β-Hölder continuous Riemannian metric. If there are $d = \dim \mathcal{E}_x$ continuous vector fields that give bases for all $\mathcal{E}^i$, then the theorem implies that $\mathcal{A}$ is continuously cohomologous to a cocycle with values in a "standard" maximal amenable subgroup of $GL(d,\mathbb{R})$. However, triviality of $\mathcal{E}$ alone is insufficient for such reduction even if $\mathcal{E} = \mathbb{T}^2 \times \mathbb{R}^2$ since invariant sub-bundles may be non-orientable [S13].

Outline of the Proof of Theorem 7.2. The periodic assumptions imply that the cocycle is fiber bunched by Corollary 5.10(ii) and has one exponent for each ergodic measure by Theorem 5.2. We take μ to be the measure of maximal entropy, or the invariant volume if it exists, and trivialize the bundle $\mathcal{E}$ on a set of full measure, that is, we measurably identify $\mathcal{E}$ with $X \times \mathbb{R}^d$ and view $\mathcal{A}$ as a $GL(d,\mathbb{R})$-valued cocycle. Thus we can use Zimmer's Amenable Reduction [BaPe, Theorem 3.5.9].

Theorem 7.3 (Zimmer's Amenable Reduction). *Let f be an ergodic transformation of a measure space (X,μ) and let $A : X \to GL(d,\mathbb{R})$ be a measurable function.*

Then there exists a measurable function $C : X \to GL(d,\mathbb{R})$ such that the function $B(x) = C^{-1}(fx)\,A(x)\,C(x)$ takes values in an amenable subgroup G of $GL(d,\mathbb{R})$.

There are 2^{d-1} *standard* maximal amenable subgroups of $GL(d,\mathbb{R})$ [Mo79]. They correspond to the distinct compositions of d, $d_1 + \cdots + d_k = d$, and each group consists of all block-triangular matrices of the form

$$\begin{bmatrix} B_1 & * & \dots & * \\ 0 & B_2 & \ddots & \vdots \\ \vdots & \ddots & \ddots & * \\ 0 & \dots & 0 & B_k \end{bmatrix}$$

where each diagonal block B_i is a $d_i \times d_i$ conformal matrix, namely a scalar multiple of an orthogonal matrix. Any amenable subgroup of $GL(d,\mathbb{R})$ has a finite index subgroup which is contained in a conjugate of one of these standard subgroups. Thus we may assume that G has a finite index subgroup G_0 which is contained in one of the standard subgroups.

We concentrate on the simplest case when $G_0 = G$. Then the sub-bundle V^i spanned by the first $d_1 + \cdots + d_i$ coordinate vectors in $\mathbb{R}^d$ is B-invariant for $i = 1,\dots,k$. Denoting $\mathcal{E}^i_x = C(x)V^i$ we obtain the corresponding flag of measurable $\mathcal{A}$-invariant sub-bundles

$$\mathcal{E}^1 \subset \mathcal{E}^2 \subset \cdots \subset \mathcal{E}^k = \mathcal{E} \quad \text{with} \quad \dim \mathcal{E}^i = d_1 + \cdots + d_i.$$

By Theorem 7.1 we may assume that the sub-bundles $\mathcal{E}^i$ are Hölder continuous. Since $B_1(x)$ is a conformal matrix for μ-a.e. x, the push forward by C of the standard conformal structure on V^1 is invariant under the restriction of $\mathcal{A}$ to $\mathcal{E}^1$ and hence β-Hölder continuous by Theorem 6.3.

Similarly, we consider the factor-bundles $\mathcal{E}^i/\mathcal{E}^{i-1}$ over X with the natural induced cocycle $\mathcal{A}^{(i)}$. Since the matrix of the map induced by B on $V^i/V^{i-1} = \mathbb{R}^{d_i}$ is B_i, it preserves the standard conformal structure on $\mathbb{R}^{d_i}$. Pushing it forward by C we obtain a measurable conformal structure τ_i on $\mathcal{E}^i/\mathcal{E}^{i-1}$ invariant under $\mathcal{A}^{(i)}$. The holonomies $H^{\mathcal{A},s}$ and $H^{\mathcal{A},u}$ induce holonomies for $\mathcal{A}^{(i)}$ on $\mathcal{E}^i/\mathcal{E}^{i-1}$. By Lemma 6.4, we conclude that τ_i is essentially invariant under these holonomies and hence is also β-Hölder continuous on X.

Now we outline the argument for the case when $G_0 \neq G$. An example illustrating this case is when G_0 is the full diagonal subgroup and G is its normalizer, which is the finite extension of G_0 that contains all permutations of the coordinate axes.

In the general case, G_0 still has the invariant flag $V^1 \subset V^2 \subset \cdots \subset V^k = \mathbb{R}^d$ with conformal structures on the factors. The elements of G may not preserve this flag; instead there may be several images of it under G:

$$V^1_{(1)} \subset V^2_{(1)} \subset \cdots \subset V^k_{(1)}$$

$$\dots$$

$$V^1_{(\ell)} \subset V^2_{(\ell)} \subset \cdots \subset V^k_{(\ell)}.$$

These flags are mapped by C to the corresponding measurable flags on $\mathcal{E}$ as before. The sub-bundles in the flags are not $\mathcal{A}$-invariant individually, but $\mathcal{A}$ preserves the union of subspaces at each level of the flag $C(V^i_{(1)}) \cup \cdots \cup C(V^i_{(\ell)})$, $i = 1,\dots,k-1$. For each level of the flag we show that the union is essentially holonomy invariant and hence Hölder continuous. The union may not split into individual invariant sub-bundles since it may "twist" along non-trivial loops in X. Thus, to obtain actual continuous sub-bundles one may need to pass to a special finite cover of X. These sub-bundles are not necessarily invariant, as $\mathcal{A}$ may permute them, but they are invariant under an iterate of $\mathcal{A}$. In this way we can obtain the general result Theorem 7.4.

Theorem 7.4. [KS13] *Let* (X,f) *be a hyperbolic system, let* μ *be an ergodic* f*-invariant probability measure with full support and local product structure, and let* $\mathcal{A} : \mathcal{E} \to \mathcal{E}$ *be a* β*-Hölder continuous linear cocycle over* f. *Suppose that* $\mathcal{A}$ *is fiber bunched and* $\lambda(\mathcal{A}, \mu) = \chi(\mathcal{A}, \mu)$. *Then there exists a finite cover* $\tilde{\mathcal{A}} : \tilde{\mathcal{E}} \to \tilde{\mathcal{E}}$ *of* $\mathcal{A}$ *and* $N \in \mathbb{N}$ *such that* $\tilde{\mathcal{A}}^N$ *satisfies the following property. There exist a flag of* β*-Hölder continuous* $\tilde{\mathcal{A}}^N$*-invariant sub-bundles*

$$\{0\} = \tilde{\mathcal{E}}^0 \subset \tilde{\mathcal{E}}^1 \subset \cdots \subset \tilde{\mathcal{E}}^{k-1} \subset \tilde{\mathcal{E}}^k = \tilde{\mathcal{E}}$$

and β*-Hölder continuous conformal structures on the factor bundles* $\tilde{\mathcal{E}}^i/\tilde{\mathcal{E}}^{i-1}$, $i = 1, \dots, k$, *invariant under the factor-cocycles induced by* $\tilde{\mathcal{A}}^N$.

To complete the proof of Theorem 7.2, we claim that the assumption on the periodic data implies that $\mathcal{A}$ is conformal on the sub-bundle spanned by the union $C(V^1_{(1)}) \cup \cdots \cup C(V^1_{(\ell)})$, and then similarly for the induced cocycles on the factors. Indeed, the scalar cocycles that give expansion/contraction of $\mathcal{A}$, more precisely, of an iterate of the lift of $\mathcal{A}$ as earlier, are all continuously cohomologous by the Livšic theorem, since the exponents at each periodic point are the same. This yields uniform quasiconformality and hence conformality of $\mathcal{A}$ on the span. The same reasoning shows that the scalar cocycles of expansion/contraction for the different conformal factors of $\mathcal{A}$ are all cohomologous to one scalar function, whose inverse gives the function ϕ in the theorem. □

We note that the assumption of Theorems 7.2 implies that $\mathcal{A}$ has one exponent with respect to *any* measure, while in Theorem 7.4 the cocycle has one exponent with respect to just *one* measure μ.

Now we apply the theorems to cocycles on bundles with two-dimensional fibers. Then Theorem 7.2 yields the following.

Corollary 7.5. *Let* $\mathcal{A}$ *and* (X,f) *be as in Theorem 7.2, and suppose that the fibers of* $\mathcal{E}$ *are two-dimensional. Then either* $\mathcal{A}$ *is conformal with respect to a* β*-Hölder Riemannian metric on* $\mathcal{E}$ *or* $\mathcal{A}$ *has a one-dimensional Hölder continuous invariant sub-bundle.*

The proof of Theorem 7.4 can be used to obtain the following two-dimensional result.

Corollary 7.6. [GKS20] *Let* $\mathcal{A}$ *and* (X,f) *be as in Theorem 7.4, and suppose that the fibers of* $\mathcal{E}$ *are two-dimensional. Then at least one of the following holds:*

(1) *$\mathcal{A}$ is conformal with respect to a Hölder continuous Riemannian metric on $\mathcal{E}$;*
(2) *$\mathcal{A}$ preserves a Hölder continuous one-dimensional sub-bundle;*
(3) *$\mathcal{A}$ preserves a Hölder continuous field of two transverse lines.*

We note that in the previous corollary every measure has one exponent, which in case (3) would yield conformality.

Using Theorem 13.27 we also obtain the following estimates.

Corollary 7.7 (Polynomial growth of the quasiconformal distortion and norm). *Let $\mathcal{A} : \mathcal{E} \to \mathcal{E}$ be a Hölder continuous linear cocycle over a hyperbolic system (X,f). Suppose that for every f-periodic point p the invariant measure μ_p on its orbit satisfies $\lambda(\mathcal{A}, \mu_p) = \chi(\mathcal{A}, \mu_p)$. Then there exists $m < \dim \mathcal{E}_x$ and C such that*

$$Q_{\mathcal{A}}(x,n) \leq Cn^{2m} \quad \textit{for all } x \in X \textit{ and } n \in \mathbb{Z}.$$

Moreover, if $\lambda(\mathcal{A}, \mu_p) = \chi(\mathcal{A}, \mu_p) = 0$ for every μ_p, then there exists $m < \dim \mathcal{E}_x$ and C such that

$$\|\mathcal{A}_x^n\| \leq C|n|^m \quad \textit{for all } x \in X \textit{ and } n \in \mathbb{Z}.$$

One can take $m = j - 1$, which is the number of non-trivial sub-bundles in (13.27).

13.7.1 Classification of $GL(2,\mathbb{R})$-Valued Cocycles with One Exponent

The structure Theorems 7.2 and 7.4 and Corollary 7.5 can be used to study cohomology of cocycles. First such results were obtained in the two-dimensional case before the theorems in higher dimensions were established. In [KS10] we proved criteria for conformality and isometry, and hence cohomology to a conformal or orthogonal cocycle; see Theorem 6.8. Then a complete classification up to Hölder cohomology of $GL(2,\mathbb{R})$-valued cocycles with one exponent at each periodic point was obtained in [S13]. It showed that such cocycles can be viewed as either elliptic or parabolic.

Theorem 7.8. [S13] *Let $f : X \to X$ is a transitive Anosov diffeomorphism, and let $\mathcal{A}$ be an orientation-preserving $GL(2,\mathbb{R})$-valued cocycle over (X,f). Suppose that for each periodic point $p = f^n p$ in X, the eigenvalues of the matrix $\mathcal{A}_p^n$ are equal in modulus. Then $\mathcal{A}$ belongs to exactly one of the following five types.*

I. *If $\mathcal{A}$ preserves exactly one Hölder continuous sub-bundle, which is orientable, then $\mathcal{A}$ is Hölder cohomologous to a cocycle $\mathcal{A}'$ with the generator $A'(x) = k(x)\begin{bmatrix} 1 & \alpha(x) \\ 0 & 1 \end{bmatrix}$, where $k(x) \neq 0$ and α is not cohomologous to 0.*

I$'$. *If $\mathcal{A}$ preserves exactly one Hölder continuous sub-bundle, which is not orientable, then there exists a cocycle $\mathcal{A}'$ as in I such that the lifts of $\mathcal{A}$ and $\mathcal{A}'$ to a double cover are Hölder cohomologous.*

II. *If $\mathcal{A}$ preserves more than one orientable Hölder continuous sub-bundle, then $\mathcal{A}$ is Hölder cohomologous to $\mathcal{A}'$ with $A'(x) = k(x) \cdot Id$, where $k(x) \neq 0$.*

II$'$. *If $\mathcal{A}$ preserves more than one non-orientable Hölder continuous sub-bundle, then there exists a cocycle $\mathcal{A}'$ as in II such that the lifts of $\mathcal{A}$ and $\mathcal{A}'$ to a double cover are Hölder cohomologous.*

III. *If $\mathcal{A}$ has no invariant Hölder continuous sub-bundles, then $\mathcal{A}$ is Hölder cohomologous to $\mathcal{A}'$ with $A'(x) = k(x)\, R(\alpha(x))$, where $k(x) > 0$, $R(\alpha(x))$ is the rotation by $\alpha(x)$, and $\alpha : X \to \mathbb{R}/2\pi\mathbb{Z}$ is such that α is not cohomologous to 0 in $\mathbb{R}/\pi\mathbb{Z}$.*

Examples of cocycles as in I$'$ and I$''$ were also constructed in [S13]. We note that for cocycles as in the theorem, any measurable invariant sub-bundle is Hölder continuous; thus measurable sub-bundles can be considered in the statement.

Explicit necessary and sufficient conditions for cohomology of "model" cocycles within each class were also obtained in [S13]; see Proposition 4.7. In particular these results yielded the following.

Corollary 7.9. [S13] *Let μ be an f-invariant ergodic measure with full support and local product structure, and let $\mathcal{A}$ and $\mathcal{B}$ be two cocycles as in Theorem 7.8. Then any μ-measurable conjugacy between $\mathcal{A}$ and $\mathcal{B}$ is Hölder continuous. In particular, the Hölder classification in the theorem coincides with the measurable one.*

We note that cocycles of types I and I$'$ are not uniformly quasiconformal. Hence Theorem 4.14 on continuity of measurable conjugacy does not apply to them.

13.8 Cocycles over Partially Hyperbolic Diffeomorphisms

Many of the ideas and techniques that were used for hyperbolic systems extend to a partially hyperbolic setting. A diffeomorphism f of X is called *partially hyperbolic* if there exists a non-trivial Df-invariant splitting of the tangent bundle $TX = E^s \oplus E^c \oplus E^u$, a Riemannian metric on X, and continuous functions $\nu < 1 < \hat{\nu}$, γ, $\hat{\gamma}$ such that for any $x \in X$ and any unit vectors $v^{s/c/u}$ in $E^{s/c/u}(x)$,

$$\|Df(v^s)\| < \nu(x) < \gamma(x) < \|Df(v^c)\| < \hat{\gamma}(x) < \hat{\nu}(x) < \|Df(v^u)\|. \qquad (13.28)$$

The sub-bundles E^s, E^u, and E^c are called stable, unstable, and center, and E^s and E^u are tangent to the foliations W^s and W^u. A partially hyperbolic diffeomorphism is called *accessible* if any two points in X can be connected by an su-path, namely a concatenation of finitely many subpaths lying in a single leaf of W^s or W^u, and f is called *center bunched* if $\nu < \gamma\hat{\gamma}^{-1}$ and $\hat{\nu}^{-1} < \gamma\hat{\gamma}^{-1}$.

Assumptions. In this section, f is an accessible center bunched partially hyperbolic C^2 diffeomorphism preserving a volume μ, and $\mathcal{A} : \mathcal{E} \to \mathcal{E}$ is a β-Hölder linear cocycle over (X,f). The volume μ is ergodic for such f by [BuW10].

Problems of continuity for invariant objects and measurable conjugacies are natural in this setting. The following theorem extends the corresponding results for hyperbolic systems.

Theorem 8.1. [KS13, ASV13] *Let f, μ, and $\mathcal{A}$ be as in the Assumptions.*

(i) *If $\mathcal{A}$ is fiber bunched, then any $\mathcal{A}$-invariant μ-measurable conformal structure on $\mathcal{E}$ is continuous, that is, coincides μ-a.e. with a continuous one.*

(ii) *If $\mathcal{A}$ is uniformly quasiconformal, then it preserves a continuous conformal structure on $\mathcal{E}$; equivalently, $\mathcal{A}$ is conformal with respect to a continuous Riemannian metric on $\mathcal{E}$.*

(iii) *If $\mathcal{A}$ is fiber bunched and has one Lyapunov exponent with respect to μ, then any μ-measurable $\mathcal{A}$-invariant sub-bundle of $\mathcal{E}$ is continuous.*

Note that only continuity of measurable objects is obtained in the partially hyperbolic case. Hölder continuity can be proved under a stronger accessibility assumption.

Outline of the Proof. Fibers bunching for cocycles over partially hyperbolic systems is defined in the same way as for hyperbolic ones; see Definition 3.2. It implies existence of stable and unstable holonomies for $\mathcal{A}$ as in Proposition 3.3. Lemma 6.4 also extends directly and yields essential invariance of a measurable conformal structure τ under the stable and unstable holonomies of $\mathcal{A}$. In the proof of Theorem 8.1, the local product structure of stable and unstable manifolds is replaced by accessibility. In this case the global continuity of τ on X follows from [ASV13, Theorem E] or [W13, Theorem 4.2]. Part (ii) follows from part (i) and Proposition 6.2. Essential invariance of the sub-bundle in part (iii) is obtained using results in [AV10, ASV13]. □

Using Theorem 8.1 we extended Theorem 7.4 on structure of cocycles with one exponent to partially hyperbolic setting [KS13].

Theorem 8.2. [KS13] *Let f, μ, and $\mathcal{A}$ be as in the Assumptions. Suppose that $\mathcal{A}$ is fiber bunched and $\lambda(\mathcal{A},\mu) = \chi(\mathcal{A},\mu)$ for the invariant volume μ.*

Then there exists a finite cover $\tilde{\mathcal{A}} : \tilde{\mathcal{E}} \to \tilde{\mathcal{E}}$ of $\mathcal{A}$ and $N \in \mathbb{N}$ such that $\tilde{\mathcal{A}}^N$ satisfies the following property. There exist a flag of continuous $\tilde{\mathcal{A}}^N$-invariant sub-bundles

$$\{0\} = \tilde{\mathcal{E}}^0 \subset \tilde{\mathcal{E}}^1 \subset \cdots \subset \tilde{\mathcal{E}}^{k-1} \subset \tilde{\mathcal{E}}^k = \tilde{\mathcal{E}} \tag{13.29}$$

and continuous conformal structures on the factor bundles $\tilde{\mathcal{E}}^i/\tilde{\mathcal{E}}^{i-1}$, $i = 1, \dots, k$, *invariant under the factor-cocycles induced by* $\tilde{\mathcal{A}}^N$.

Now we consider cohomology of linear cocycles over partially hyperbolic systems. Continuity of a measurable conjugacy for real-valued cocycles was obtained by Wilkinson in [W13]. The next theorem extends this result to linear cocycles.

Theorem 8.3. *[KS16] Let f and μ be as in the Assumptions, and let $\mathcal{A}$ and $\mathcal{B}$ be Hölder continuous linear cocycles over (X, f). If $\mathcal{A}$ is fiber bunched and $\mathcal{B}$ is uniformly quasiconformal, then any μ-measurable conjugacy C between $\mathcal{A}$ and $\mathcal{B}$ coincides on a set of full volume with a continuous conjugacy.*

The proof uses the techniques of the hyperbolic setting as in the proof of Theorem 4.14 to obtain essential Hölder continuity along stable and unstable foliations and then concludes global continuity using results from [ASV13, W13]. Global Hölder continuity for scalar cocycles was obtained in [W13]. For linear cocycles this is only known under a stronger accessibility assumption [KS16].

Partially hyperbolic systems may have no periodic points. In analogs of Questions 4.4 and 4.5 for these systems, the role of periodic data is played by *cycle weights (or functionals)*. A cycle weight is a composition of holonomies of a cocycle along a closed *su*-path. More precisely, let $H^{\mathcal{A},s}$ and $H^{\mathcal{A},u}$ be the stable and unstable holonomies for a cocycle $\mathcal{A}$. Let $P = P_{x_0} = \{x_0, x_1, \dots, x_{k-1}, x_k = x_0\}$ be an *su*-cycle, that is a closed *su*-path in X. The *cycle weight* of P is

$$\mathcal{H}_{x_0}^{\mathcal{A},P} = H_{x_{k-1} x_0} \circ \cdots \circ H_{x_1 x_2} \circ H_{x_0 x_1},$$

where $H_{x_i x_{i+1}} = H^{s/u}_{x_i x_{i+1}}$ if $x_{i+1} \in W^{s/u}(x_i)$.

In case of real-valued cocycles, $\mathcal{H}_{x_0}^{\mathcal{A},P}$ is also called a cycle functional. This notion was introduced in [KtKo96] to show that for systems with strong accessibility, a real-valued cocycle has zero cycle weights if and only if it is cohomologous to a constant cocycle. This was extended in some cases to non-commutative cocycles [KtNT00], but as we showed in [KS16], having trivial cycle weights is not necessary in general for continuous cohomology to a constant cocycle. In [KS16] we obtained necessary and sufficient conditions for cohomology of a cocycle to a constant one, as well as the following result for cohomology of two arbitrary cocycles in terms of cycle weights.

Theorem 8.4. *Let f be an accessible center bunched partially hyperbolic diffeomorphism. Let $\mathcal{A}$ and $\mathcal{B}$ be β-Hölder linear or Banach cocycles over (X, f) with holonomies $H^{\mathcal{A}}$ and $H^{\mathcal{B}}$. Suppose that there exist a fixed point x_0 and $C_{x_0} \in GL(V)$ such that*

$$\mathcal{A}_{x_0} = C_{x_0} \circ \mathcal{B}_{x_0} \circ C_{x_0}^{-1} \quad \text{and} \quad \mathcal{H}_{x_0}^{\mathcal{A},P} = C_{x_0} \circ \mathcal{H}_{x_0}^{\mathcal{B},P} \circ C_{x_0}^{-1} \text{ for every su-cycle } P_{x_0}.$$

Then there exists a continuous conjugacy C between $\mathcal{A}$ and $\mathcal{B}$ with $C(x_0) = C_{x_0}$.

For the case when x_0 is not fixed, we established similar conditions that guarantee existence of a continuous conjugacy C with $C(x_0) = C_{x_0}$.

13.9 Applications to Rigidity Problems for Hyperbolic Systems

13.9.1 Local Rigidity

It is well known that if g is an Anosov diffeomorphism and f is sufficiently C^1 close to g, then f is also Anosov and f is topologically conjugate to g, that is, there is a homeomorphism h of X such that $g = h^{-1} \circ f \circ h$. In general, h is only Hölder continuous. A necessary condition for it to be C^1 is that the derivatives of the return maps of f and g at the corresponding periodic points are conjugate, specifically,

$$D_p g^n = (D_p h)^{-1} \circ D_{h(p)} f^n \circ D_p h \quad \text{whenever } p = f^n p.$$

An Anosov diffeomorphism g is said to be *locally rigid* if, for any C^1-small perturbation f, this condition is also sufficient for the conjugacy to be a C^1 diffeomorphism. The problem of local rigidity has been extensively studied and Anosov diffeomorphisms with one-dimensional stable and unstable sub-bundles were shown to be locally rigid [L87, LM88, L92]. In higher dimensions, examples where the periodic condition is not sufficient were constructed by de la Llave [L92, L02]. We recall that L is *irreducible* if it has no rational invariant subspaces, equivalently, if its characteristic polynomial is irreducible over $\mathbb{Q}$. Examples in [G08] showed that irreducibility of L is a necessary condition for local rigidity, except when L is conformal on the stable and unstable sub-bundles. In the latter case, local rigidity was obtained for some systems in [L02] and for broader classes [KS03, KS09] as a corollary of global rigidity results; see Theorem 9.7 and Remark 9.8.

Local rigidity was proved in [G08] for an irreducible Anosov toral automorphism $L : \mathbb{T}^d \to \mathbb{T}^d$ with real eigenvalues of distinct moduli, as well as for some nonlinear systems with similar structure. In [GKS11, S15] we used results on conformality and cohomology of linear cocycles to establish the following local rigidity results.

Theorem 9.1. [S15, GKS11] *Let $L : \mathbb{T}^d \to \mathbb{T}^d$ be an irreducible Anosov automorphism and let f be a C^1-small perturbation of L such that for each periodic point $p = f^n p$ there is a matrix $C(p)$ such that $D_p f^n = C(p) \circ L^n \circ C(p)^{-1}$. Suppose that either*

(i) *No three of the eigenvalues of L have the same modulus [GKS11], or*

(ii) *$C(p)$ is Hölder continuous at a periodic point p_0, that is,*
$d(C(p), C(p_0)) \leq c\, dist(p, p_0)^\beta$ for every periodic point p [S15].

Then f is $C^{1+\text{Hölder}}$ conjugate to L.

We note that f in this theorem has an invariant volume. Indeed, the conjugacy assumption implies $\det D_p f^n = \det L^n = 1$ whenever $p = f^n p$ and existence of density which gives an invariant volume follows from the Livšic Theorem 4.1(i).

Outline of the Proof of Theorem 9.1. The proofs of (i) and (ii) differ only in the way we obtain conformality of Df on certain invariant sub-bundles. Conformality of these restrictions plays an important role in establishing smoothness of the conjugacy.

We denote by $E^{u,L}$ the unstable sub-bundle of L. Let $1 < \rho_1 < \cdots < \rho_\ell$ be the distinct moduli of the unstable eigenvalues of L, and let $E^{u,L} = E_1^L \oplus \cdots \oplus E_\ell^L$ be the corresponding splitting of the unstable sub-bundle. Since f is C^1 close to L, f is also Anosov, and its unstable sub-bundle $E^{u,f}$ splits into a direct sum of ℓ invariant Hölder continuous sub-bundles close to the corresponding sub-bundles for L: $E^{u,f} = E_1^f \oplus \cdots \oplus E_\ell^f$. Let $\mathcal{A}_i = L|E_i^L$ and $\mathcal{B}_i = Df|E_i^f$.

Since L is irreducible, all its eigenvalues are simple and thus L and its restrictions $\mathcal{A}_i$ are diagonalizable over $\mathbb{C}$. Since the eigenvalues of $\mathcal{A}_i$ are of the same modulus, $\mathcal{A}_i$ is conformal in some norm. Since $C(p)$ maps $E_i^L(p)$ to $E_i^f(p)$, the cocycle $\mathcal{B}_i$ has conjugate periodic data and hence is conformal at the periodic points.

In [GKS11], conformality of $\mathcal{B}_i$ at the periodic points together with the assumption that the sub-bundles E_i^L and E_i^f are either one- or two-dimensional allows us to conclude that, by Theorem 6.8, the cocycle $\mathcal{B}_i$ is conformal. In higher dimensions, conformality at the periodic points does not imply conformality. In (ii) we can use Corollary 4.12 to conclude that $\mathcal{A}_i$ and $\mathcal{B}_i$ are Hölder continuously cohomologous, and so in particular $\mathcal{B}_i$ is also conformal.

After conformality of Df on each sub-bundle E_i^f is obtained, we consider the topological conjugacy h close to the identity between L and f. If h maps foliation W_i^f tangent to E_i^f to the corresponding linear foliation W_i^L, the conformality allows us to obtain that h is Lipschitz along the leaves of W_i^f. This also uses the fact that $\mathcal{A}_i$ and $\mathcal{B}_i$, or at least their norms, are continuously cohomologous. Then we can differentiate the conjugacy equation $h \circ f = L \circ h$ almost everywhere with respect to volume. The derivative $D(h|W_i^f)$ is then a measurable conjugacy between $\mathcal{A}_i$ and $\mathcal{B}_i$ and hence is Hölder by Theorem 4.14. This implies that h is $C^{1+\text{Hölder}}$ along W_i^f, and having this for each foliation W_i^f yields that h is $C^{1+\text{Hölder}}$ along $W^{u,f}$. Then regularity of h along $W^{s,f}$ is obtained similarly and regularity on $\mathbb{T}^d$ follows.

Establishing that $h(W_i^f) = W_i^L$ for each i is done using an inductive process within stable and unstable foliations. The base is given by the slowest foliation

W_1. The inductive step also uses Hölder conjugacy between $\mathcal{A}_i$ and $\mathcal{B}_i$ established at the previous steps, as well as their conformality, to obtain conformality of the holonomies of certain foliations between the leaves of W_i^f. □

Since L is linear, its Lyapunov exponents are the same for all invariant measures. The conjugacy of periodic data of f and L implies that Lyapunov exponents of f with respect to any f-invariant measure μ equal to the Lyapunov exponents of L by Theorem 5.2 on their periodic approximation.

Saghin and Yang obtained smoothness of the conjugacy h for a volume-preserving perturbation f of an irreducible L if f and L have the same Lyapunov exponents with respect to the volume and all Lyapunov exponents are *simple* [SaY19]. The following theorem allows for double Lyapunov exponents and extends the Saghin–Yang result to a much broader class of irreducible hyperbolic automorphisms. We note, however, that these results rely on the fact that volume is the measure of maximal entropy for L, and they do not hold in general when L is replaced by a nonlinear system.

Theorem 9.2. [GKS20] *Let $L : \mathbb{T}^d \to \mathbb{T}^d$ be an Anosov automorphism such that no three of its eigenvalues have the same modulus and L^4 is irreducible. Let f be a volume-preserving C^2 diffeomorphism of $\mathbb{T}^d$ sufficiently C^1-close to L. If the Lyapunov exponents of f with respect to the volume are the same as the Lyapunov exponents of L, then f is $C^{1+\text{Hölder}}$ conjugate to L.*

In fact, it suffices to assume that L is irreducible and has no pairs of eigenvalues of the form $\lambda, -\lambda$ or $i\lambda, -i\lambda$, where λ is real, which follows if L^4 is irreducible. This assumption implies that L has no invariant lines or pairs of lines. The same property holds for the derivative at a fixed point for any small perturbation f. This implies that for the cocycle $\mathcal{B}_i = D(h|W_i^f)$, as in the proof of Theorem 9.1, options (2) and (3) in Corollary 7.6 are impossible and hence $\mathcal{B}_i$ is conformal. The other significant difference from the proof of Theorem 9.1 is that cohomology of the norms of $\mathcal{A}_i$ and $\mathcal{B}_i$ is obtained using the Jacobian of the conjugacy h restricted to W_i^f. The latter relies on establishing that h maps the conditional measures of the volume for f on W_i^f to those for L. Otherwise, the structure of the proof is similar.

13.9.2 Smoothness of the Stable/Unstable Sub-bundles and Global Rigidity

The results on conformality of linear cocycles in Section 13.6 were motivated in part by the study of rigidity for conformal Anosov systems. The initial results were established in [S02] for the restrictions of Df to the stable and unstable sub-bundles. They allowed us to obtain regularity of the foliations and *global rigidity*, that is existence of a conjugacy to an algebraic model, for some classes of Anosov systems.

Theorem 9.3. [S02] *Let f be a transitive C^∞ Anosov diffeomorphism of X. Suppose that Df is uniformly quasiconformal on E^s with* $\dim E^s \geq 2$. *Then*

(i) *Df is conformal with respect to a Riemannian metric on E^s which is Hölder continuous on X and C^∞ along the leaves of the stable foliation.*
(ii) *The holonomy maps of the unstable foliation are conformal and E^u is C^∞.*

A similar result holds for a topologically mixing C^∞ Anosov flow that is uniformly quasiconformal on the strong stable sub-bundle E^s. It yields conformality on E^s, conformality of the holonomies of the weak stable foliation, and C^∞ regularity of the weak stable sub-bundle [S02]. The proof is also similar.
Outline of the Proof of Theorem 9.3.

(i) Since Df is uniformly quasiconformal on E^s, it preserves a Hölder continuous conformal structure τ on E^s. We show its smoothness along the leaves using a non-stationary linearization. Taking a smooth normalization of τ, we obtain a desired Riemannian metric on E^s.

The following statement is a generalization of a one-dimensional result established by Katok and Lewis in [KtL91].

Proposition 9.4 (Non-stationary linearization). [S02] *Let f be a diffeomorphism of a compact Riemannian manifold X, and let W be a continuous invariant foliation with C^∞ leaves. Suppose that $\|Df|_{TW}\| < 1$, and there exist $C > 0$ and $0 < \theta < 1$ such that*

$$\| (Df^n|_{T_xW})^{-1} \| \cdot \| Df^n|_{T_xW} \|^2 \leq C\theta^n \quad \text{for all } x \in X \text{ and } n \in \mathbb{N}.$$

Then for every $x \in X$ there exists a C^∞ diffeomorphism $h_x : W(x) \to T_xW$ such that $h_{fx} \circ f = Df_x \circ h_x$, $h_x(x) = 0$, $D_xh_x = Id$, and h_x depends continuously on x in C^∞ topology.

Applying the proposition to f yields a non-stationary linearization $\{h_x\}$ along W^s.

For each $x \in X$ we extend the conformal structure $\tau(x)$ to a constant conformal structure σ on TE^s_x. That is, for each $t \in E^s_x$ we denote by $\sigma(t)$ the push forward of $\tau(x)$ by the translation from 0 to t.

Lemma 9.5. *Each $h_x : W^s(x) \to E^s_x$ is conformal with respect to τ and σ; that is, $h_x(\tau(y)) = \sigma(h_x(y))$ for any $y \in W^s(x)$, and hence τ is C^∞ along the leaves of W^s.*

Proof. By continuity of τ and the properties of h_x, for each $\varepsilon > 0$ there exists $n > 0$ such that

$$\operatorname{dist}(h_{f^nx}(\tau(f^ny)), \sigma(h_{f^nx}(f^n(y))) < \varepsilon.$$

To obtain the following equalities, we note that Df^n induces an isometry between the spaces of conformal structures, τ is f-invariant, σ is Df-invariant, and $h_x(y) = Df^{-n}(h_{f^nx}(f^ny))$. Thus,

$$\begin{aligned}\varepsilon &> \operatorname{dist}(\, h_{f^nx}(\tau(f^ny)),\ \sigma(h_{f^nx}(f^ny))\,)\\ &= \operatorname{dist}(\, Df^{-n}(h_{f^nx}(\tau(f^ny))),\ Df^{-n}(\sigma(h_{f^nx}(f^ny)))\,)\\ &= \operatorname{dist}(\, Df^{-n}(h_{f^nx}(f^n(\tau(y))),\ \sigma(Df^{-n}(h_{f^nx}(f^ny)))\,)\\ &= \operatorname{dist}(\, h_x(\tau(y)),\ \sigma(h_x(y))\,).\end{aligned}$$

It follows that $h_x(\tau(y)) = \sigma(h_x(y))$. Since h is a C^∞ diffeomorphism and σ is constant, τ is C^∞ along the leaves of W^s. □

(ii) Let x and y be two nearby points in X. We consider the holonomy map of the unstable foliation $\mathcal{H}_{x,y} : W^s_{\text{loc}}(x) \to W^s_{\text{loc}}(y)$:

$$z \in W^s_{\text{loc}}(x) \mapsto \mathcal{H}_{x,y}(z) = W^s_{\text{loc}}(y) \cap W^u_{\text{loc}}(z).$$

First we show that the holonomy maps $\mathcal{H}_{x,y}$ are conformal, that is, $\mathcal{H}_{x,y}(\tau(z)) = \tau(\mathcal{H}_{x,y}(z))$. By the C^r-section theorem [HPuSh], conformality of Df on E^s implies that the unstable sub-bundle is C^1, and hence the holonomy maps are uniformly C^1 [PSW97]. It suffices to consider the case when $z = x$ and $y \in W^u(x)$ so that $y = \mathcal{H}_{x,y}(z)$. We iterate by f^{-1} and note that $\mathcal{H}_{x,y} = f^n \circ \mathcal{H}_{f^{-n}x,f^{-n}y} \circ f^{-n}$ and that $D_{f^{-n}x}\mathcal{H}_{f^{-n}x,f^{-n}y}$ is close to identity for large n since $f^{-n}y$ is close to $f^{-n}x$. Since τ is continuous, $\tau(f^{-n}x)$ is close to $\tau(f^{-n}y)$. Thus, $\mathcal{H}_{f^{-n}x,f^{-n}y}(\tau(f^{-n}x))$ is close to $\tau(f^{-n}y)$. Since Df^n induces an isometry between the spaces of conformal structures on $E^s(f^{-n}y)$ and on $E^s(y)$, we conclude that $\mathcal{H}_{x,y}(\tau(x))$ is close to $\tau(y)$. By letting $n \to \infty$, we obtain $\mathcal{H}_{x,y}(\tau(z)) = \tau(\mathcal{H}_{x,y}(z))$.

Now we show that $\mathcal{H}_{x,y}$ are C^∞. Using the non-stationary linearization coordinates, we view it as the map

$$G_{x,y} = h_y \circ \mathcal{H}_{x,y} \circ h_x^{-1} : \ E^s_x \to E^s_y,$$

Then $G_{x,y}$ is a conformal C^1 diffeomorphism of $\mathbb{R}^d$ defined on a neighborhood of 0, and hence it is C^∞. Indeed, if $d > 2$, it is Möbius, and if $d = 2$, it is complex analytic. Therefore $\mathcal{H}_{x,y}$ is C^∞ and it follows that E^u is C^∞ [PSW97]. □

We say that an Anosov diffeomorphism is uniformly quasiconformal if it is uniformly quasiconformal on both E^u and E^s. Theorem 9.3 implies that for such diffeomorphisms both E^u and E^s are C^∞. In general, smoothness of both E^u and E^s is not known to imply smooth conjugacy to algebraic model, although it is conjectured. However, the implication is known to hold under additional assumptions; for example, that f is symplectic [BFL92] or that f preserves a smooth affine connection [BL93]. Thus we obtain the following result for a symplectic diffeomorphism

f. For such f, conformality on the stable sub-bundle implies conformality on the unstable one, and vice versa.

Theorem 9.6. [S02] *Let f be a C^∞ symplectic Anosov diffeomorphism of a compact manifold X, $\dim X \geq 4$. Suppose that f is uniformly quasiconformal on the unstable sub-bundle. Then a lift of f to a finite cover of X is C^∞ conjugate to an Anosov automorphism of a torus.*

Without symplecticity, we obtained the following general result in [KS03].

Theorem 9.7 (Global rigidity of uniformly quasiconformal Anosov diffeomorphisms). [KS03] *Let f be a transitive C^∞ Anosov diffeomorphism of a compact manifold X which is uniformly quasiconformal on the stable and unstable sub-bundles. Suppose either that both sub-bundles have dimension at least three, or that they have dimension at least two and X is an infranilmanifold. Then a lift of f to a finite cover of X is C^∞ conjugate to an Anosov automorphism of a torus.*

Outline of the Proof. Theorem 9.3 implies that both E^u and E^s are C^∞ and there is a C^∞ Riemannian metric on X for which f is conformal on both E^u and E^s. Under the assumptions of the theorem, we showed that the holonomy maps $\mathcal{H}_{x,y}$ are globally defined affine conformal maps between the leaves. Then we proved that the non-stationary linearization coordinates h^s_x and h^u_x depend C^∞ on x. This allowed us to construct a smooth invariant affine connection. Existence of such a connection together with the smoothness of E^u and E^s implies the conjugacy by the result in [BL93]. □

Remark 9.8. *Uniform quasiconformality or conformality of f can be established from its periodic data. Results of this type were obtained in [KS03, KS09] and later extended to general cocycles in Theorems 6.5 and 6.8.*

13.9.3 Global Rigidity of Anosov Flows

Continuous-time conformal Anosov systems were studied by Kanai in the special case of geodesic flows. He proved in [Kn93] that the geodesic flow of a compact Riemannian manifold of negative curvature of dimension at least three is C^2 conjugate to the geodesic flow of a manifold of constant negative curvature under the assumption that either the flow preserves a continuous conformal structure on the strong stable sub-bundle, or the flow satisfies 1/2 pinching and preserves a bounded measurable conformal structure on the strong stable sub-bundle. The next theorem generalizes this result to the case of contact flows.

Theorem 9.9. [S02] *Let φ^t be a C^∞ contact Anosov flow on a compact manifold $\mathcal{M}$, $\dim \mathcal{M} \geq 5$. If the flow is uniformly quasiconformal on the strong stable sub-bundle, then* *it is essentially, that is up to a time change of a specific form, C^∞ conjugate to the geodesic flow of a manifold of constant negative curvature.*

This theorem follows from the analog of Theorem 9.3 for flows and the result of Benoist, Foulon, and Labourie on rigidity of contact Anosov flows with smooth stable and unstable sub-bundles in [BFL92]. In a particular case when φ^t is a geodesic flow of a compact Riemannian manifold $\mathcal{N}$ of negative curvature of dimension at least three, the theorem yields that φ^t is conjugate to the geodesic flow of a manifold of constant negative curvature. Then it follows from the main theorem of Besson, Courtois, and Gallot in [BCG95] that $\mathcal{N}$ is a manifold of constant negative curvature.

We showed in [S02] that *any* smooth time change of a uniformly quasiconformal Anosov flow is also uniformly quasiconformal. Therefore, if the flow is not assumed contact, then it could be any smooth time change of the geodesic flow of a hyperbolic manifold or of the suspension flow of a conformal hyperbolic automorphism of a torus. Rigidity of uniformly quasiconformal Anosov diffeomorphisms and flows was further studied by Fang in [Fa04, Fa07]. In particular, the following classification was obtained in higher dimensions.

Theorem 9.10. [Fa07] *Let φ^t be a C^∞ topologically transitive uniformly quasiconformal Anosov flow such that the strong stable and unstable sub-bundles E^u and E^s are at least three-dimensional. Then up to finite covers, φ^t is C^∞ conjugate to a time change of either the geodesic flow of a hyperbolic manifold or the suspension of a hyperbolic automorphism of a torus.*

In the context of geodesic flows, stronger results on rigidity of Lyapunov spectrum were recently obtained by Butler [Bt17]. They rely on techniques from [S02] and analogs of structure results for cocycles with one Lyapunov exponent, Theorems 7.2 and 7.4. We note that $1/4$-pinching of the sectional curvature for geodesic flows ensures fiber bunching of stable and unstable differential for the geodesic flow.

Theorem 9.11. [Bt17] *Let $\mathcal{M}$ be a closed negatively curved Riemannian manifold with* $\dim \mathcal{M} \geq 3$.

- **(i)** *If at each periodic orbit the geodesic flow φ^t has one unstable Lyapunov exponent, then $\mathcal{M}$ has constant curvature.*
- **(ii)** *Suppose that $\mathcal{M}$ has relatively $1/4$-pinched sectional curvatures. Let μ be an ergodic measure with full support and local product structure invariant under the geodesic flow φ^t. If μ has one unstable Lyapunov exponent, then $\mathcal{M}$ has constant curvature.*

Further results on Lyapunov spectrum rigidity for geodesic flows on negatively curved symmetric spaces were obtained in [Bt19].

References

[AKL18] A. Avila, A. Kocsard, X. Liu. *Livšic theorem for diffeomorphism cocycles.* Geom. Funct. Anal. 28 (2018), 943–964.

[ASV13] A. Avila, J. Santamaria, M. Viana. *Holonomy invariance: rough regularity and applications to Lyapunov exponents.* Asterisque 358 (2013), 13–74.

[AV10] A. Avila, M. Viana. *Extremal Lyapunov exponents: an invariance principle and applications.* Invent. Math. 181 (2010), 115–189.

[Bac15] L. Backes. *Rigidity of fiber bunched cocycles.* Bul. Brazilian Math. Soc. 46 (2015), 163–179.

[BaPe] L. Barreira, Ya. Pesin. *Nonuniformly hyperbolicity: dynamics of systems with nonzero Lyapunov exponents.* Encyclopedia of Mathematics and Its Applications 115, Cambridge University Press, 2007.

[BFL92] Y. Benoist, P. Foulon, F. Labourie. *Flots d'Anosov à distributions stable et instable différentiables.* J. Amer. Math. Soc, 5 (1992), 33–74.

[BL93] Y. Benoist, F. Labourie. *Sur les difféomophismes d'Anosov affines à feuilletages stable et instable diffŕentiables.* Invent. Math. 11 (1993), 285–308.

[BCG95] G. Besson, G. Courtois, S. Gallot. *Entropies et rigidités des espaces localement symétriques de courbure strictement négative.* Geom. Funct. Anal. 5 (1995), 731–800.

[BV04] C. Bonatti, M. Viana. *Lyapunov exponents with multiplicity 1 for deterministic products of matrices.* Ergod. Theory Dyn. Syst. 24 (2004) 1295–1330.

[BuW10] K. Burns, A. Wilkinson. *On the ergodicity of partially hyperbolic systems.* Ann. Math. 171 (2010) 451–489.

[Bt17] C. Butler. *Rigidity of equality of Lyapunov exponents for geodesic flows.* J. Differ. Geom. 109 (2018), 39–79.

[Bt18] C. Butler. *Measurable rigidity of the cohomological equation for linear cocycles over hyperbolic systems.* Israel J. Math. 227 (2018), 27–61.

[Bt19] C. Butler. *Characterizing symmetric spaces by their Lyapunov spectra.* Preprint. arXiv:1709.08066

[Fa04] Y. Fang. *Smooth rigidity of uniformly quasiconformal Anosov flows.* Ergod. Theory Dyn. Syst. 24 (2004), 1937–1959.

[Fa07] Y. Fang. *On the rigidity of quasiconformal Anosov flows.* Ergod. Theory Dyn. Syst. 27 (2007), 1773–1802.

[G08] A. Gogolev. *Smooth conjugacy of Anosov diffeomorphisms on higher dimensional tori.* J. Mod. Dyn. 2 (2008), 645–700.

[GKS11] A. Gogolev, B. Kalinin, V. Sadovskaya. *Local rigidity for Anosov automorphisms.* Math. Res. Lett. 18 (2011), 843–858.

[GKS20] A. Gogolev, B. Kalinin, V. Sadovskaya. *Local rigidity of Lyapunov spectrum for toral automorphisms.* Isr. J. Math. 238 (2020), 389–403.

[GrGu17] G. Grabarnik, M. Guysinsky. *Livšic theorem for Banach rings.* Discrete Contin. Dyn. Syst. 37 (2017), 4379–4390.

[Gu95] L. Gurvits. *Stability of discrete linear inclusion.* Linear Algebra Appl. 231 (1995), 47–85.

[HaW99] B. Hasselblatt, A. Wilkinson. *Prevalence of non-Lipschitz Anosov foliations.* Ergod. Theory Dyn. Syst. 19 (1999), 643–656.

[HPuSh] M. Hirsch, C. Pugh, M. Shub. *Invariant manifolds.* Lecture Notes in Mathematics 583, Springer-Verlag, 1977.

[K11] B. Kalinin. *Livšic theorem for matrix cocycles.* Ann. Math. 173 (2011), 1025–1042.

[KS03] B. Kalinin, V. Sadovskaya. *On local and global rigidity of quasiconformal Anosov diffeomorphisms.* J. Inst. Math. Jussieu 2 (2003), 567–582.

[KS09] B. Kalinin, V. Sadovskaya. *On Anosov diffeomorphisms with asymptotically conformal periodic data.* Ergod. Theory Dyn. Syst. 29 (2009), 117–136.

[KS10] B. Kalinin, V. Sadovskaya. *Linear cocycles over hyperbolic systems and criteria of conformality.* J. Mod. Dyn. 4 (2010), 419–441.

[KS13] B. Kalinin, V. Sadovskaya. *Cocycles with one exponent over partially hyperbolic systems.* Geom. Dedicata 167 (2013), 167–188.

[KS16] B. Kalinin, V. Sadovskaya. *Holonomies and cohomology for cocycles over partially hyperbolic diffeomorphisms.* Discrete Contin. Dyn. Syst. 36 (2016), 245–259.

[KS17] B. Kalinin, V. Sadovskaya. *Periodic approximation of Lyapunov exponents for Banach cocycles.* Ergod. Theory Dyn. Syst. 39 (2019), 689–706. Published online in 2017, doi: https://doi.org/10.1017/etds.2017.43.

[KS18-2] B. Kalinin, V. Sadovskaya. *Boundedness, compactness, and invariant norms for Banach cocycles over hyperbolic systems.* Proc. Am. Math. Soc. 146 (2018), 3801–3812.

[KS18] B. Kalinin, V. Sadovskaya. *Lyapunov exponents of cocycles over non-uniformly hyperbolic systems.* Discrete Contin. Dyn. Syst. 38 (2018), 5105–5118.

[Kn93] M. Kanai. *Differential-geometric studies on dynamics of geodesic and frame flows.* Japan. J. Math., 19 (1993), 1–30.

[KaM99] A. Karlsson, G. Margulis. *A multiplicative ergodic theorem and nonpositively curved spaces.* Comm. Math. Phys. 208 (1999), 107–123.

[KtH] A. Katok, B. Hasselblatt. *Introduction to the modern theory of dynamical systems.* Encyclopedia of Mathematics and Its Applications 54, Cambridge University Press, 1995.

[KtKo96] A. Katok, A. Kononenko. *Cocycle stability for partially hyperbolic systems.* Math. Res. Lett. 3 (1996), 191–210.

[KtL91] A. Katok, J. Lewis. *Local rigidity for certain groups of toral automorphisms.* Israel J. Math. 75 (1991), 203–241.

[KtN] A. Katok, V. Nitica. *Rigidity in Higher Rank Abelian Group Actions: Volume 1, Introduction and Cocycle Problem,* Cambridge University Press, 2011.

[KtNT00] A. Katok, V. Nitica, A. Török. *Non-abelian cohomology of abelian Anosov actions.* Ergod. Theory Dyn. Syst. 20 (2000), 259–288.

[Li71] A. N. Livšic. *Homology properties of Y-systems.* Math. Zametki 10 (1971), 758–763.

[Li72] A. N. Livšic. *Cohomology of dynamical systems.* Math. USSR Izvestija 6 (1972), 1278–1301.

[L87] R. de la Llave. *Invariants for smooth conjugacy of hyperbolic dynamical systems II.* Commun. Math. Phys. 109 (1987), 368–378.

[L92] R. de la Llave. *Smooth conjugacy and SRB measures for uniformly and nonuniformly hyperbolic systems.* Commun. Math. Phys. 150 (1992), 289–320.

[L02] R. de la Llave. *Rigidity of higher dimensional conformal Anosov systems.* Ergod. Theory Dyn. Syst. 22 (2002), 1845–1870.

[LM88] R. de la Llave, R. Moriyón. *Invariants for smooth conjugacy of hyperbolic dynamical systems IV.* Commun. Math. Phys. 116 (1988), 185–192.

[LWi10] R. de la Llave, A. Windsor. *Livšic theorem for non-commutative groups including groups of diffeomorphisms, and invariant geometric structures.* Ergod. Theory Dyn. Syst. 30 (2010), 1055–1100.

[Mo79] C. Moore. *Amenable subgroups of semi-simple groups and proximal flows.* Isr. J. Math. 34 (1979), 121–138.

[NiP99] M. Nicol, M. Pollicott. *Measurable cocycle rigidity for some non-compact groups.* Bull. London Math. Soc. 31 (1999), 592–600.

[NT95] V. Nitica, A. Török. *Cohomology of dynamical systems and rigidity of partially hyperbolic actions of higher-rank lattices.* Duke Math. J. 79 (1995), 751–810.

[O68] V. Oseledets. *A multiplicative ergodic theorem. Liapunov characteristic numbers for dynamical systems.* Trans. Mosc. Math. Soc. 19 (1968), 197–221.

[Pa99] W. Parry. *The Livšic periodic point theorem for non-Abelian cocycles.* Ergod. Theory Dyn. Syst. 19 (1999), 687–701.

[PaP97] W. Parry, M. Pollicott. *The Livšic cocycle equation for compact Lie group extensions of hyperbolic systems.* J. London Math. Soc. (2) 56 (1997), 405–416.

[PWa01] M. Pollicott, C. P. Walkden. *Livšic theorems for connected Lie groups.* Trans. Amer. Math. Soc. 353 (2001), 2879–2895.

[PSW97] C. Pugh, M. Shub, A. Wilkinson. *Hölder foliations.* Duke Math. J. 86 (1997), 517–546.

[S02] V. Sadovskaya. *On uniformly quasiconformal Anosov systems.* Math. Res. Lett. 12 (2005), 425–441.

[S13] V. Sadovskaya. *Cohomology of* $GL(2,\mathbb{R})$*-valued cocycles over hyperbolic systems.* Discrete Contin. Dyn. Syst. 33 (2013), 2085–2104.

[S15] V. Sadovskaya. *Cohomology of fiber bunched cocycles over hyperbolic systems.* Ergod. Theory Dyn. Syst. 35 (2015), 2669–2688.

[S17] V. Sadovskaya. *Fiber bunching and cohomology for Banach cocycles over hyperbolic systems.* Discrete Contin. Dyn. Syst. 37 (2017), 4959–4972.

[SaY19] R. Saghin, J. Yang. *Lyapunov exponents and rigidity of Anosov automorphisms and skew products.* Adv. Math. 355 (2019), 106764.

[Schm99] K. Schmidt. *Remarks on Livšic theory for non-Abelian cocycles.* Ergod. Theory Dyn. Syst. 19 (1999), 703–721.

[Schr98] S. J. Schreiber. *On growth rates of subadditive functions for semi-flows.* J. Differ. Equ. 148 (1998), 334–350.

[T86] P. Tukia. *On quasiconformal groups.* J. Anal. Math. 46 (1986), 318–346.

[V08] M. Viana. *Almost all cocycles over any hyperbolic system have nonvanishing Lyapunov exponents.* Ann. Math. (2) 167 (2008), 643–680.

[WaSu10] Z. Wang, W. Sun. *Lyapunov exponents of hyperbolic measures and hyperbolic periodic orbits*. Trans. Amer. Math. Soc. 362 (2010), 4267–4282.

[W13] A. Wilkinson. *The cohomological equation for partially hyperbolic diffeomorphisms.* Asterisque 358 (2013) 75–165.